Computational Paradigm Techniques for Enhancing Electric Power Quality

Computational Paradigm Techniques for Enhancing Electric Power Quality

L. Ashok Kumar
S. Albert Alexander

CRC Press is an imprint of the
Taylor & Francis Group, an **informa** business

CRC Press
Taylor & Francis Group
6000 Broken Sound Parkway NW, Suite 300
Boca Raton, FL 33487-2742

First issued in paperback 2020

ISBN 13: 978-0-367-57067-5 (pbk)
ISBN 13: 978-1-138-33699-5 (hbk)

Library of Congress Cataloging-in-Publication Data

Names: Kumar, L. Ashok, author. | Albert Alexander, S. author.
Title: Computational paradigm techniques for enhancing electric power quality / L. Ashok Kumar and S Albert Alexander.
Description: First edition. | New York, NY : CRC Press/Taylor & Francis Group, 2019. | Includes bibliographical references and index.
Identifiers: LCCN 2018033182 | ISBN 9781138336995 (hardback : acid-free paper) | ISBN 9780429442711 (ebook)
Subjects: LCSH: Electric power production--Quality control--Data processing.
Classification: LCC TK1010 .K86 2019 | DDC 621.31/042--dc23
LC record available at https://lccn.loc.gov/2018033182

Visit the Taylor & Francis Web site at
http://www.taylorandfrancis.com

and the CRC Press Web site at
http://www.crcpress.com

Contents

Contents

Preface

Power Quality has been a problem bristling with snags ever since electrical power was invented. It has become a well-researched area of interest in recent years because of the electrical appliances (load) it affects. The electric current that the customers' appliances draw from the supply network flows through the impedances of the supply system and causes a voltage drop, which affects the voltage delivered to the customer. Hence, both the voltage quality and the current quality are important. The power distribution supplier is responsible for the voltage quality and the customer is accountable for the quality of electric current that they draw from the utility. The power system electromagnetic phenomena that affect the power quality are categorized as transient, short-duration variations, long-duration variations, and waveform distortions. A waveform distortion is defined as a steady state deviation from an ideal sine wave of a power frequency that is principally characterized by the spectral content of the deviation.

Power Quality is intended as a useful text for undergraduate, postgraduate, and research students. It is also useful for practitioner engineers and industrial personnel. Power quality improvement techniques are vital to ensure the equipment safety and cost reduction sought in the electrical system. Hence, in lieu of a conventional system, intelligent and more advanced techniques are required for the improvement of power quality. The main objective of the book is to prove to the readers the need for power quality improvement in real-time systems. Keeping this fact in mind, a detailed review is presented for the power quality indices along with detailed description, algorithm formulation, simulation results, and corresponding analysis and experimental study. The techniques covered extensively in this book provide a platform for students and researchers to understand the problems and the path to be travelled in alleviating those problems.

L. Ashok Kumar
S. Albert Alexander

Acknowledgments

The authors are always thankful to the Almighty for their perseverance and achievements. The authors owe their gratitude to Shri L. Gopalakrishnan, Managing Trustee, PSG Institutions, and all the trustees of Kongu Vellalar Institute of Technology Trust, Perundurai. The authors also owe their gratitude to Dr. R. Rudramoorthy, Principal, PSG College of Technology, Coimbatore, India, and Prof. S. Kuppuswami, Principal, Kongu Engineering College, Perundurai, India, for their wholehearted cooperation and great encouragement in this successful endeavor.

I, Dr. L. Ashok Kumar would like to take this opportunity to acknowledge those people who helped me in completing this book. I am thankful to all my research scholars and students who are doing their project and research work with me. But the writing of this book is possible mainly because of the support of my family members, parents, and sisters. Most importantly, I am very grateful to my wife, Y. Uma Maheswari, for her constant support during writing. Without her, all these things would not be possible. I would like to express my special gratitude to my daughter, A. K. Sangamithra, for her smiling face and support; it helped a lot in completing this work.

I, Dr. S. Albert Alexander would like to take this opportunity to acknowledge those people who helped me in completing this book. I am thankful to all my research scholars and students who are doing their project and research work with me. But the writing of this book is possible mainly because of the support of my family members, parents, and brothers. Most importantly, I am very grateful to my wife, A. Lincy Annet, for her constant support during writing. Without her, all these things would not be possible. I would like to express my special gratitude to my son, A. Albin Emmanuel, for his smiling face and support; it helped a lot in completing this work.

Authors

L. Ashok Kumar is a Postdoctoral Research Fellow from San Diego State University, California. He is a recipient of the BHAVAN fellowship from the Indo-US Science and Technology Forum and SYST Fellowship from DST, Government of India. His current research focuses on integration of renewable energy systems in the smart grid and wearable electronics. He has 3 years of industrial experience and 19 years of academic and research experience. He has published 167 technical papers in international and national journals and presented 157 papers at national and international conferences. He has completed 23 Government of India-funded projects, and currently 5 projects are in progress. His PhD work on wearable electronics earned him a National Award from ISTE, and he has received 24 awards on the national level. Ashok Kumar has seven patents to his credit. He has guided 92 graduate and postgraduate projects. He is a member and in prestigious positions in various national forums. He has visited many countries for institute/industry collaboration and as a keynote speaker. He has been an invited speaker in 178 programs. Also, he has organized 72 events, including conferences, workshops, and seminars. He completed his graduate program in Electrical and Electronics Engineering from the University of Madras; his post-graduate program from PSG College of Technology, India; and his Master's in Business Administration from IGNOU, New Delhi. After completion of his graduate degree, he joined as project engineer for Serval Paper Boards Ltd., Coimbatore (now ITC Unit, Kovai). Presently, he is working as a Professor and Associate HoD in the Department of EEE, PSG College of Technology and also doing research work in wearable electronics, smart grids, solar PV, and wind energy systems. He is also a Certified Chartered Engineer and BSI-Certified ISO 500001 2008 Lead Auditor. He has authored the following books in his areas of interest: (1) *Computational Intelligence Paradigms for Optimization Problems Using MATLAB®/SIMULINK®*, CRC Press; (2) *Solar PV and Wind Energy Conversion Systems—An Introduction to Theory, Modeling with MATLAB/SIMULINK, and the Role of Soft Computing Techniques*—Green Energy and Technology, Springer, USA; (3) *Electronics in Textiles and Clothing: Design, Products and Applications,* CRC Press; (4) *Power Electronics with MATLAB,* Cambridge University Press, London; and (5) *Automation in Textile Machinery: Instrumentation and Control System Design Principles*—CRC Press, Taylor & Francis Group, USA, ISBN 9781498781930, April 2018. He has also published the following monographs: (1) *Smart Textiles,* (2) *Information Technology for Textiles,* and (3) *Instrumentation & Textile Control Engineering.*

S. Albert Alexander is a Postdoctoral Research Fellow from Northeastern University, Boston, Massachusetts. He is a recipient of Raman Research Fellowship from the University Grants Commission (Government of India). His current research focuses on fault diagnostic systems for solar energy conversion systems and smart grids. He has 12 years of academic and research experience. He has published 15 technical papers in international and national journals and presented 19 papers at national and international conferences. He has completed 4 Government of India-funded projects. His PhD work on power quality earned him a National Award from ISTE, and he has received 20 awards on the national level. He has guided 33 graduate and postgraduate projects. He is a member and in prestigious positions in various national forums. He has been an invited speaker in 150 programs. Also, he has organized 15 events, including faculty development programs, workshops, and seminars. He completed his graduate program in Electrical and Electronics Engineering from Bharathiar University and his postgraduate program from Anna University, India. Presently he is working as an Associate Professor in the Department of EEE, Kongu Engineering College and also doing research work in smart grids, solar PV, and power quality improvement techniques. He has authored the following books in his areas of interest: (1) *Basic Electrical, Electronics and Measurement Engineering* and (2) *Special Electrical Machines.*

Abbreviations

IEEE	The Institute of Electrical and Electronics Engineers
ANSI	American National Standards Institute
CW	continuous wave
ESD	Electrostatic discharge
NEMP	Nuclear electromagnetic pulse
RMS	Root mean square
UPS	Uninterruptible power supply
EFT	Electrical Fast Transient
TVSS	Transient voltage surge suppression
SVC	static VAR controller
PCC	point of common coupling
SMPS	switched-mode power supplies
THD	total harmonic distortion
TDD	Total demand distortion
EMI	electromagnetic interference
pu	per unit
MOV	metal-oxide varistor
PIV	peak inverse voltage
SCR	silicon-controlled rectifier
TVSS	transient voltage surge suppressors
MCOV	maximum continuous operating voltage
LIPC	Low-impedance power conditioner
LPS	Utility System Lightning Protection
BIL	basic impulse insulation level
DSTATCOM	Distribution static compensator
VSI	Voltage source inverter
CSI	Current source inverter
PWM	Pulse width modulation
DVR	Dynamic voltage restorer
FACTS	Flexible AC transmission system
MATLAB	Matrix laboratory
UPF	Unity power factor
RES	Renewable Energy Systems
MLI	Multilevel inverter
PV	photovoltaic
NPC	neutral point-clamped
DCMLI	diode-clamped multilevel inverter
FCMLI	flying capacitor multilevel inverter
CMLI	Cascaded multilevel inverter
FC	flying capacitor
SDCs	separate DC sources
MPPT	Maximum power point tracking
ANN	Artificial neural networks

FLC	fuzzy logic controller
GA	Genetic algorithm
PSO	particle swarm optimization
BO	bees optimization
MMC	modified multilevel converter
SHE	Selective harmonic elimination
NR	Newton–Raphson
OPWM	Optimal pulse-width modulation
STC	standard test conditions
LMBPN	Levenberg Marquart back-propagation
BPN	back propagation
PPG	programmable pulse generator
FFT	Fast Fourier transform
DDR	double data rate
PQA	power quality analyzer
LMS	least mean square
PLL	phase-locked loop
ShAPF	shunt active power filter
APQC	Active power quality conditioners
APLC	active power line conditioners
IRPC	instantaneous reactive power compensator
ACO	ant colony optimization
SRF	synchronous reference frame
CCM	Continuous conduction mode
DCM	Discontinuous conduction mode
MMF	Magneto motive force
IGBT	insulated-gate bipolar transistors
NEC	National Electrical Code
DG	Distributed generation
CSC	Current source converter
VSC	Voltage source converter
NTSHAPF	novel type series hybrid active power filter
VFD	Variable frequency drive
SRF	synchronous reference frame theory
SCD	Source current detection
LVD	Load voltage detection
CVD	Current and voltage detection
ADALINE	Adaptive linear neuron
PMSG	permanent magnet synchronous generator
BESS	battery energy storage system
DCC	direct current control
ICC	indirect current control
LQR	Linear Quadratic Regulator
SVM	Space-Vector Modulation
EAF	electric arc furnaces
TCR	thyristor-controlled reactors
LCI	load-commutated inverter
SCR	silicon-controlled rectifier
TCSC	thyristor-controlled series capacitor or compensator

TCPST	thyristor-controlled phase-shifting transformer
IPFC	Interline power flow controller
UPFC	Unified power flow controller
STATCOM	Static synchronous compensator
SSSC	Static synchronous series compensator
EMP	Electromagnetic pulses
UNIPEDE	Union Internationale des Producteurs et Distributeurs d'Energie Electrique

1

Introduction

1.1 General Classes of Power Quality Problems

Many different types of power quality measurement devices exist and it is important for employees in different areas of power distribution, transmission, and processing to use the same language and measurement techniques.

The IEEE Standards Coordinating Committee 22 (IEEE SCC22) has framed power quality standards in the United States. The Industry Applications Society and the Power Engineering Society along with IEEE played a major role in framing standards. The International Electro technical Commission (IEC) classifies electromagnetic phenomena into the sets presented in Table 1.1.

The power quality standard for IEC was developed by monitoring electric power quality for U.S. industries. *Sag* is a synonym to the IEC term *dip*. The category *short-duration variations* includes *voltage dips*, *swell*, and *short interruptions*. The word *swell* is an exact opposite to sag (dip). The category *long-duration variation* deals with American National Standards Institute (ANSI) C84.1 limits. The broadband conducted phenomena are under the category of noise. The category *waveform distortion* contains *harmonics, interharmonics, DC in AC networks*, and *notching* phenomena. The IEEE Standard 519–1992 explains the concept related to harmonics.

Table 1.2 shows the electromagnetic phenomena categorization related to power quality community. The listed phenomena in the table can be further listed in detail by appropriate attributes.

The following attributes can be used for steady-state phenomena:

- Amplitude
- Frequency
- Spectrum
- Modulation
- Source impedance
- Notch depth
- Notch area

For non-steady-state phenomena, other attributes may be required:

- Rate of rise
- Amplitude
- Duration
- Spectrum
- Frequency
- Rate of occurrence
- Energy potential
- Source impedance

TABLE 1.1

Principal Phenomena Causing Electromagnetic Disturbances as Classified by IEC

- Conducted low-frequency phenomena
  1. Harmonics, interharmonics
  2. Signal system (power line carrier)
  3. Voltage fluctuations (flicker)
  4. Voltage dips and interruptions
  5. Voltage imbalance (unbalance)
  6. Power frequency variations
  7. Induced low-frequency voltages
  8. DC in AC networks
- Radiated low-frequency phenomena
  1. Magnetic fields
  2. Electric fields
- Conducted high-frequency phenomena
  1. Induced Continuous Wave (CW) voltages or currents
  2. Unidirectional transients
  3. Oscillatory transients
- Radiated high-frequency phenomena
  1. Magnetic fields
  2. Electric fields
  3. Electromagnetic fields
  4. Continuous waves
  5. Transients
- Electrostatic Discharge Phenomena (EDP)
- Nuclear Electro Magnetic Pulse (NEMP)

TABLE 1.2

Categories and Characteristics of Power System Electromagnetic Phenomena

Categories	Typical Spectral Content	Typical Duration	Typical Voltage
Transient Impulsive			
Nanosecond	5-ns rise	<50 ns	
Microsecond	1-μs rise	50 ns–1 ms	
Millisecond	0.1-ms rise	> 1 ms	
Oscillatory			
Low frequency	<5 kHz	0.3–50 ms	0–4 pu
Medium frequency	5–500 kHz	20 μs	0–8 pu
High frequency	0.5–5 MHz	5 μs	0–4 pu
Short-duration variations			
Instantaneous			
Interruption		0.5–30 cycles	<0.1 pu
Sag (dip)		0.5–30 cycles	0.1–0.9 pu
Swell		0.5–30 cycles	1.1–1.8 pu
Momentary			
Interruption		30 cycles–3 s	<0.1 pu
Sag (dip)		30 cycles–3 s	0.1–0.9 pu
Swell		30 cycles–3 s	1.1–1.4 pu
Temporary			
Interruption		3 s–1 min	<0.1 pu
Sag (dip)		3 s–1 min	0.1–0.9 pu
Swell		3 s–1 min	1.1–1.2 pu

Table 1.2 shows each category of electromagnetic phenomena regarding typical spectral content, duration, and magnitude. The categories and their descriptions provide the cause of power quality problems.

1.2 Types of Power Quality Problems

Defining and understanding the diverse power quality problems helps to prevent and solve those problems. The type of power quality problem is identified by the signature or characteristics of the disturbance. The variation in behavior of the sine wave, i.e., voltage, current, and frequency, recognizes the type of power quality problem. The most common type of power quality problem is voltage sag. Table 1.3 shows the sources, causes and effects of typical power quality problems.

TABLE 1.3

Summary of Power Quality Problems

Example Wave Shape or RMS Variation	Causes	Sources	Effects	Examples of Power Conditioning Solutions
	Impulsive transients (Transient disturbance)	• Lightning • Electrostatic discharge • Load switching • Capacitor switching	• Destroys computer chips and TV regulators	• Surge arresters • Filters • Isolation transformers
	Oscillatory transients (Transient disturbance)	• Line/cable switching • Capacitor switching • Load switching	• Destroys computer chips and TV regulators	• Surge arresters • Filters • Isolation transformers
	Sags/swells (RMS disturbance)	• Remote system faults	• Motors stalling and overheating • Computer failures • ASDs shutting down	• Ferroresonant transformers • Energy storage technologies • Uninterruptible Power Supply (UPS)
	Interruptions (RMS disturbance)	• System protection • Breakers • Fuses • Maintenance	• Loss production • Shutting down of equipment	• Energy storage technologies • UPS • Backup generators
	Undervoltages/ overvoltages (steady-state variation)	• Motor starting • Load variations • Load dropping	• Reduces life of motors and lightning filaments	• Voltage regulators • Ferroresonant transformers
	Harmonic distortion (steady-state variation)	• Nonlinear loads • System resonance	• Overheating transformers and motors • Fuses blow • Relays trip • Meters malfunction	• Active or passive filters • Transformers with neglecting zero sequence components
	Voltage flicker (steady-state variation)	• Intermittent loads • Motor starting • Arc furnaces	• Lights flicker • Irritation	• Static VAR systems

1.2.1 Voltage Sags (Dips)

IEEE Standard P1564 gives the recommended indices and procedures for characterizing voltage sag performance and comparing performance across different systems. Also, a new IEC Standard 61000-2-8 titled "Environment – Voltage Dips and Short Interruptions" has come recently. This standard warrants considerable discussion within the IEEE to avoid conflicting methods of characterizing system performance in different parts of the world.

Voltage sags are named as *voltage dips* in Europe. Decline in voltage for a short time is defined by IEEE as voltage sag. The voltage sag lasts for 0.5 cycles to 1 minute. The voltage magnitude between 10%–90% of the normal Root Mean Square (RMS) voltage is stated as voltage sag. The RMS or effective value of a sine wave is equal to the square root of the average of the squares of all the instantaneous values of a cycle and is equivalent to $(1/\sqrt{2})$ times the peak value of the sine wave, as shown in Figure 1.1.

Figure 1.2 shows that voltage comes back to normal value after a 0.12 second voltage sag. Voltage sag occurs only for short periods.

Voltage sags occurs on transmission and distribution systems by both utilities and end users. For example, the utility power system is affected by transformer failure that results in voltage sag. The energy is supplied by power system for faults. The fault in the utility power system causes voltage sag. As the fault is cleared, the voltage comes to normal. Sags due to transmission system lasts about 6 cycles (or 0.10 seconds). Distribution faults occur for more time than transmission faults.

The most frequent power quality problem affecting industrial and commercial end users is voltage sags. They decrease the energy being delivered to the end user and cause computers to fail, adjustable-speed drives to shut down, and motors to stall and overheat.

Some of the equipment that provides solutions to voltage sag problems is ferroresonant, i.e., constant voltage transformers; Dynamic Voltage Restorers (DVRs), superconducting energy storage devices, flywheels, written pole motor-generator sets, and Uninterruptible Power Supplies (UPS).

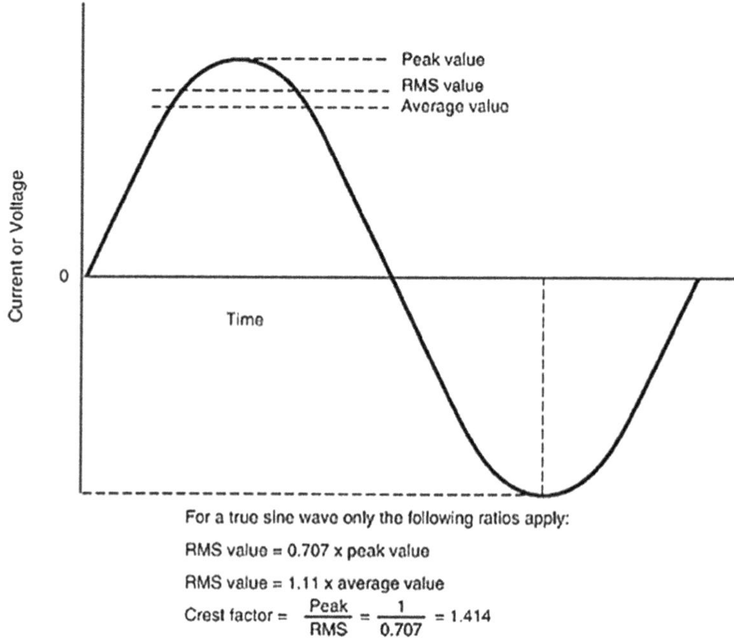

FIGURE 1.1 Sine wave values.

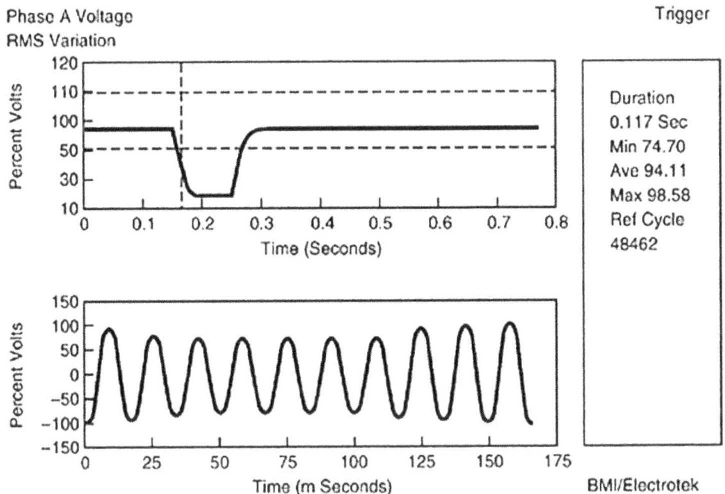

FIGURE 1.2 Voltage sag plot.

1.2.2 Voltage Swells

According to IEC 61000-4-30:2008, a voltage event can have more than one category. IEEE Standard 1159-1995 defines voltage events as dips, swells, or interruptions based on the magnitude and duration of the event.

Swell is defined as 110% of the nominal voltage and lasts for less than one minute. The occurrence of voltage swell is rare. The causes of voltage swell are single-line-to-ground (SLG) faults including lightning or a tree striking a live conductor. The voltage swell overheats the equipment and reduces the life of the equipment. Figure 1.3 demonstrates a typical voltage swell due to SLG fault occurring in an adjacent phase. Figure 1.4 presents an example of a tree growing into a power line, which causes a SLG fault.

1.2.3 Long-Duration Overvoltages

IEEE standard 1159 uses different magnitude and duration thresholds to distinguish between the different types of voltage events. IEC 60038:1983 defines a set of standard voltages for use in low-voltage and high-voltage AC electricity supply systems.

Long-duration overvoltages are like voltage swells except that duration is greater than one minute.

Capacitor switching is one of the major causes of overvoltage. On switching the capacitor, the utility's system voltage is enhanced. The next cause of overvoltage is the reduction of load. Overvoltages mainly

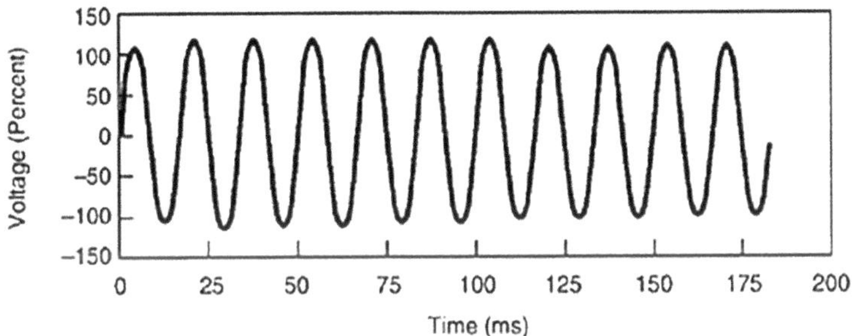

FIGURE 1.3 Voltage swell plot.

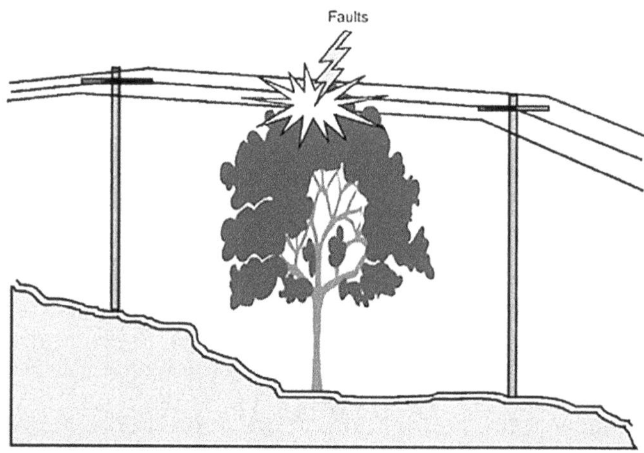

FIGURE 1.4 Single-line-to-ground fault due to a tree.

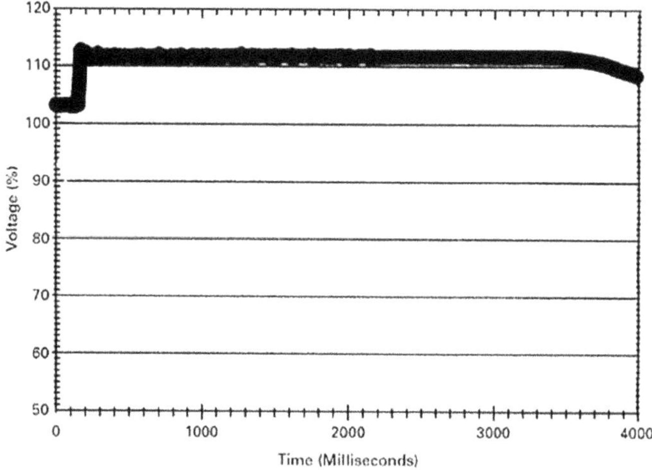

FIGURE 1.5 Overvoltage plot.

occur at lightly loaded evening conditions on high-voltage systems. The error in setting transformer tap is also a cause. Lighting filaments and motors wear out quickly due to extended overvoltage. The use of inductors during light load conditions and correct setting of transformer taps are used to prevent overvoltages. Figure 1.5 shows a plot of overvoltage versus time.

1.2.4 Undervoltages

Undervoltages occur during drop in 90% of the nominal voltage for more than one minute. They are sometimes denoted as "brownouts". Undervoltages are identified by end users when their lights dim and their motors slow down.

The cause of undervoltage is overload on the utility's system, occurring when there is very cold or hot weather or the loss of a major transmission line serving a region. Overloading on distribution system also can cause undervoltages. Sometimes utilities intentionally cause undervoltages to reduce the load during heavy load conditions. As load is voltage times current (kW = $V \times I$), as voltage reduces the overall load reduces. Undervoltages affect the sensitive computer equipment to read data incorrectly and stall motors. Undervoltages can be prevented by constructing more generation and transmission lines on utilities. Figure 1.6 shows a plot of undervoltage versus time.

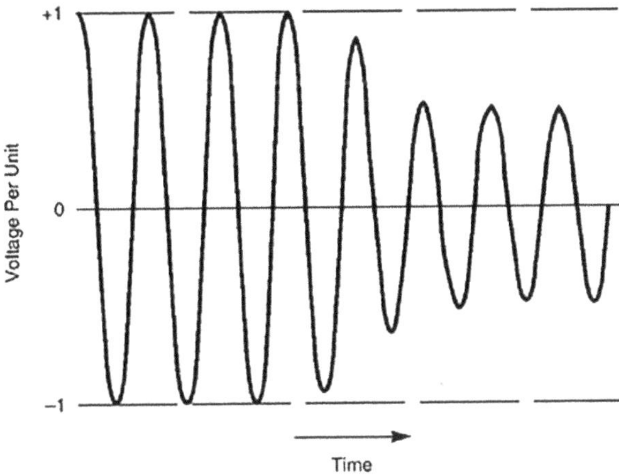

FIGURE 1.6 Undervoltage plot.

1.2.5 Interruptions

Interruptions are a drop in nominal voltage of less than 10% or complete loss of voltage. IEEE Standard 1159-1995 defines three types of interruptions. They are classified by the time duration of the interruption: momentary, temporary, and long-duration.

Momentary interruptions are due to complete loss of voltage on more than one phase conductor for a time period between 8 ms and 3 seconds. A temporary or short-duration interruption is due to a drop in nominal voltage below 10% for a time period between 3 seconds and 1 minute. Long-duration or sustained interruptions last longer than 1 minute. Figure 1.7 shows a momentary interruption.

Interruption can result in loss of production in an office, retail market, or industrial factory. The loss of electrical service and the time required to return back to electrical service causes lost production. Some types of events cannot "ride through" even short interruptions. "Ride through" is the ability of equipment to tolerate the power disturbance for a particular time. For example, in a plastic injection molding plant, a short interruption of 0.5 second takes 6 hours to restore production.

The solution to the interruptions includes on-site and off-site alternative sources of electrical supply. An end user may install on-site sources, such as UPSs with battery or motor-generator sets, whereas a utility may offer an off-site source that comprises two feeders with a high-speed switch that switches to the alternate one when one fails.

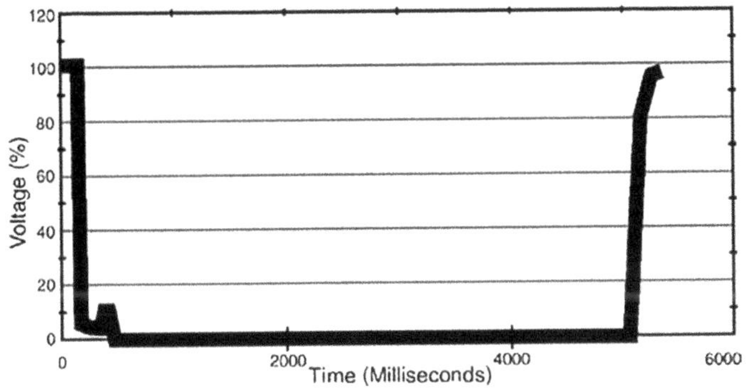

FIGURE 1.7 Momentary, temporary, and long-duration interruption plots.

1.2.6 Transients

IEC 61000-4-2 addresses one of the most common forms of transients in electronic systems: Electro Static Discharge (ESD). ESD results from conditions that allow the buildup of electrical charge from contact and separation of two non-conductive materials.

IEC 61000-4-4 – Electrical Fast Transient (EFT) standard. EFT disturbances are common in industrial environments where electromechanical switches are used to connect and disconnect inductive loads.

IEEE Std C37.09b-2010 (Amendment to IEEE Std C37.09-1999) – IEEE Standard Test Procedure for AC High-Voltage Circuit Breakers Rated on a Symmetrical Current Basis – Amendment 2: To Change the Description of Transient Recovery Voltage for Harmonization with IEC 62271-100.

A rapid increase or decrease in current or voltage is called a transient. Transients destroy computer chips and TV. They often dissipate quickly. There are mostly two types of transients: impulsive and oscillatory.

The time taken by transients to increase to peak value and decrease to normal value determines the type of transient. For example, page 13 of IEEE Standard 1159-1995, Copyright © 1995 describes an impulsive transient caused by a lightning strike. In this case the transient current increases to its peak value of 2000 V in 1.2 μs and declines to half its peak value in 50 μs. Transients are reduced by resistive components of the electrical transmission and distribution system. Lightning strikes are the most common cause of impulsive transients. Figure 1.8 illustrates an impulsive current transient caused by lightning.

Lightning arresters mounted on transmission and distribution systems and in substations are used by utilities. Transient Voltage Surge Suppression (TVSS) or battery-operated UPSs in homes, offices, or factories are used by utility customers. If the impulsive transients are not stopped, they can interrelate with capacitive components of the power system. Capacitors cause the impulsive transients to resonant and converted to oscillatory transients.

Oscillatory transients do not drop quickly like impulsive transients. They fluctuate for 0.5 to 3 cycles and reach 2 times the nominal voltage or current. Switching of equipment and power lines on the utility's power system also causes oscillatory transients. Figure 1.9 explains a typical low-frequency oscillatory transient caused by the energization of a capacitor bank.

1.2.7 Voltage Unbalance

Voltage unbalance, according to IEEE 112-1991, is the maximum deviation from the average phase voltage, referring to the average of the phase voltage.

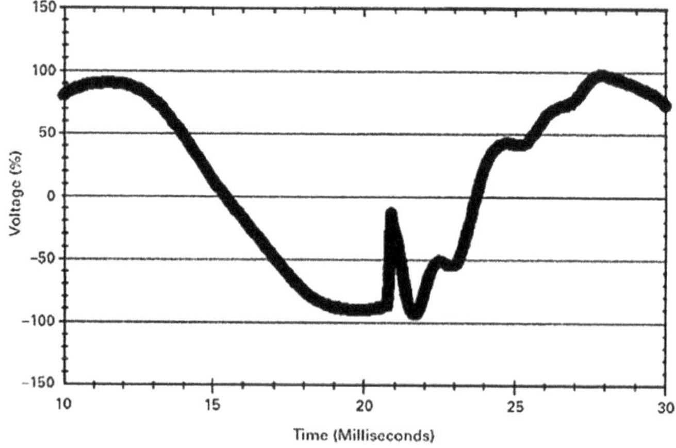

FIGURE 1.8 Impulsive transient plot.

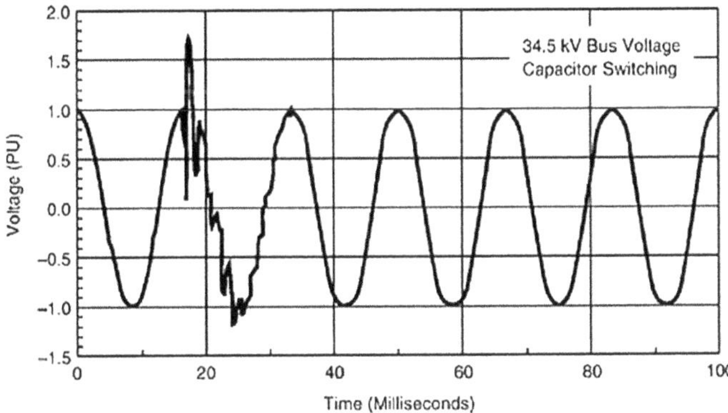

FIGURE 1.9 Oscillatory transient plot.

Voltage unbalance or imbalance is the deviation of average voltage of each phase from all three phases. It can be stated by the formula:

$$\text{Voltage unbalance} = 100 \times \frac{\text{max. deviation from average voltage}}{\text{average voltage}} \tag{1.1}$$

where average voltage = (sum of voltage of each phase)/3.

The tolerance for a voltage unbalance is 2%. The motors and transformers will overheat if the voltage unbalance is greater than 2%. This is because in an induction device, such as a motor or transformer, current unbalance varies as the cube of the voltage unbalance applied to the terminals. Main causes of voltage unbalance are capacitor banks not operating properly, single phasing of equipment, and connecting more single-phase loads on one phase than another. Continuously monitoring the voltage unbalance provides the necessary data to analyze and eliminate the cause of the unbalance.

1.2.8 Voltage Fluctuations

IEEE 1453-2004 Standard Recommended Practice for Measurement and Limits of Voltage Fluctuations and Associated Light Flicker on AC Power Systems. This recommended practice provides specifications for measurement of voltage fluctuations on electric power systems that cause noticeable illumination changes from lighting equipment (flicker, lamp flicker, or voltage flicker) and recommends acceptable levels for 120 V, 60 Hz, and 230 V 50 Hz AC electric power systems. IEC 61000-4-15 is the standard for a functional design specification for flicker measuring apparatus intended to indicate the correct flicker perception level for all practical voltage fluctuation waveforms.

Voltage fluctuations are rapid changes in voltage with a voltage magnitude of 0.95 to 1.05 of nominal voltage. Devices like electric arc furnaces and welders that have continuous, rapid changes in load current cause voltage fluctuations. Incandescent and fluorescent lights blink rapidly due to voltage fluctuations. This blinking of lights is stated as "flicker". Light intensity changes occur at frequencies of 6 to 8 Hz and are visible to the human eye. They can cause people to have headaches and become stressed and irritable. Sensitive equipment starts to malfunction due to this effect.

The solution to voltage fluctuations is to use an effective static VAR controller (SVC) that controls the voltage fluctuation frequency by controlling the amount of reactive power being supplied to the arc furnace. Figure 1.10 shows voltage fluctuations that produce flicker.

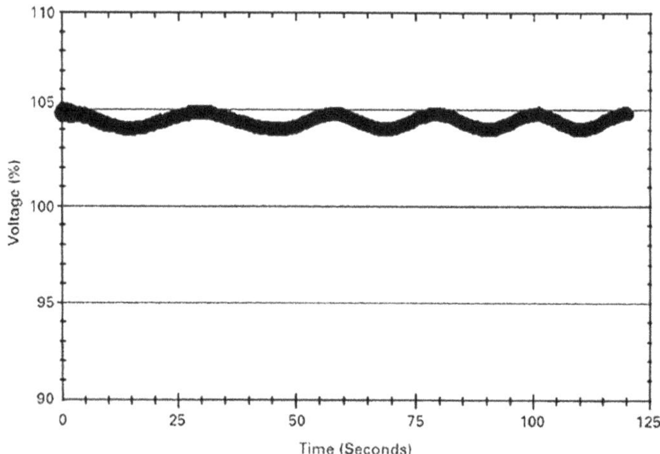

FIGURE 1.10 Voltage fluctuation (flicker) plot.

1.2.9 Harmonics

IEEE 519-1992, Recommended Practices and Requirements for Harmonic Control in Electric Power Systems, established limits on harmonic currents and voltages at the point of common coupling (PCC), or point of metering. Harmonics standard IEC 61000-3-2 Ed. 3 2005 assesses sets the limit for equipment that draws input current ≤16A per phase. Equipment that draws current >16A and ≤75A per phase is covered by IEC/TS 61000-3-12. Harmonics measurement and evaluation methods for both standards are governed by IEC 61000-4-7.

The distortion in sine waveform is the major source of harmonics. The use of nonlinear equipment also causes harmonics. Figure 1.11 shows the architecture of a standard sine wave. An analysis of the sine wave architecture gives an understanding of the basic anatomy of harmonics.

Harmonics are integral multiples of the fundamental frequency of the sine wave. They add to the fundamental waveform and distort it. They can be 2, 3, 4, 5, 6, 7, etc. times the fundamental. For example, the third harmonic is 60 Hz times 3, or 180 Hz, and the sixth harmonic is 60 Hz times 6, or 360 Hz. The waveform in Figure 1.12 shows how harmonics distort the sine wave.

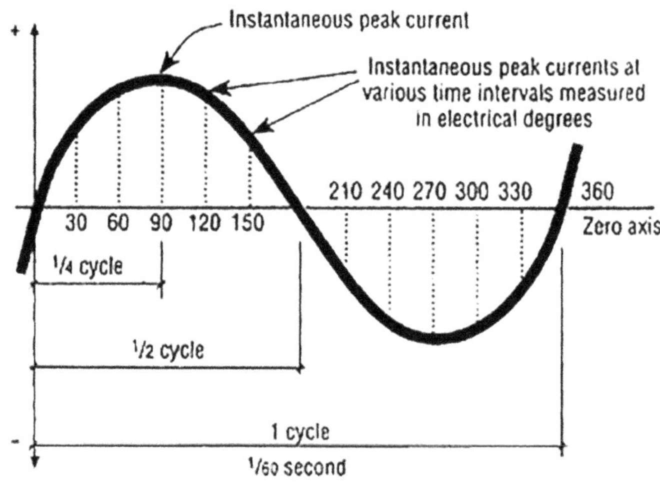

FIGURE 1.11 Sine wave architecture.

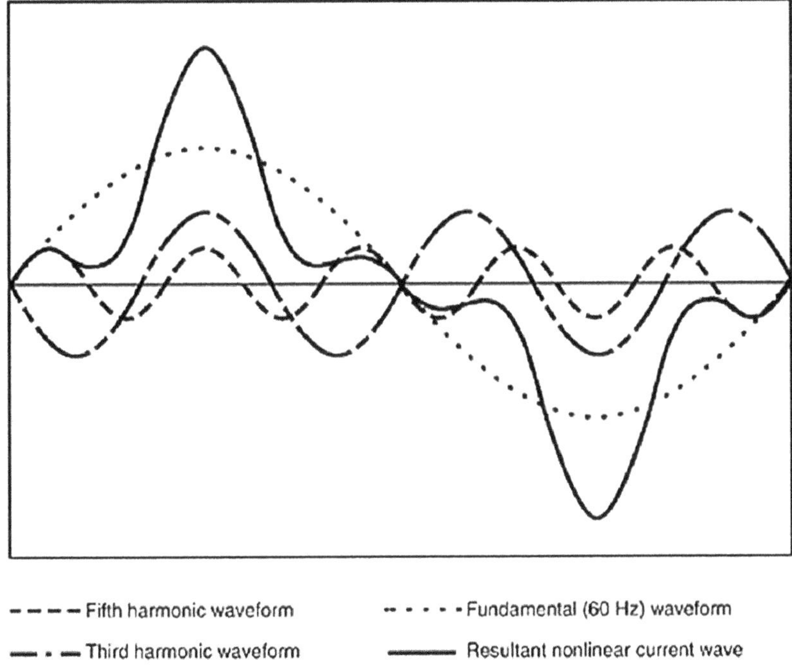

FIGURE 1.12 Composite harmonic waveform.

Harmonics are produced by nonlinear loads, like adjustable speed drives, solid-state heating controls, electronic ballasts for fluorescent lighting, switched-mode power supplies in computers, static UPS systems, electronic and medical test equipment, rectifiers, filters, and electronic office machines. Nonlinear loads cause harmonic currents to change from a sinusoidal current to a nonsinusoidal current. Therefore, the sinusoidal current waveform is distorted.

The shape of the distorted wave is the cumulative addition of fundamental and various harmonics. Table 1.4 illustrates the various nonlinear loads and the corresponding harmonic waveforms they generate.

Harmonic voltages are the outcome of the harmonic currents interacting with the impedance of the power system according to Ohm's law:

$$V = \frac{I}{Z} \tag{1.2}$$

where V = voltage, I = current, and Z = impedance

Harmonic currents and voltages have a harmful effect on utility and end-user equipment. Some of the effects are overheating of transformers, power cables, and motors; inadvertent tripping of relays; and incorrect measurement of voltage and current by meters. Harmonics increase iron losses in transformers and causes rotor heating and reduced torque. Table 1.5 shows the effect of harmonics on various types of equipment.

Harmonics cause power quality problems, not only on the end user or the utility serving the end user but also on other end users. For example, a third harmonic generated by a transformer was injected into a utility's system and transmitted to a city miles away and caused the digital clocks to show the wrong time. Section 1.6 of IEEE 519 discusses the effects of harmonics. This section explains how harmonic currents increase heating in motors, transformers, and power cables. The ratio of harmonic

TABLE 1.4

Nonlinear Loads and Their Current Waveforms

Type of Load	Typical Waveform	Current Distortion
Single Phase Power Supply		80% (high 3rd)
Semiconverter		high 2nd,3rd, 4th at partial loads
6 Pulse Converter, capacitive smoothing, no series inductance		80%
6 Pulse Converter, capacitive smoothing with series inductance > 3%, or dc drive		40%
6 Pulse Converter with large inductor for current smoothing		28%
12 Pulse Converter		15%
ac Voltage Regulator		varies with firing angle
Fluorescent Lighting		20%

Courtesy of EPRI, Palo Alto, CA.

TABLE 1.5
Effects of Harmonics on Equipment

Equipment	Harmonic Effects	Results
Capacitors	• Capacitor impedance decreases with increasing frequency, so capacitors act as sinks where harmonics converge; capacitors do not, however, generate • Supply system inductance can resonate with capacitors at some harmonic frequency, causing large currents and voltages to develop • Dry capacitors cannot dissipate heat very well and are therefore more susceptible to damage from harmonics • Breakdown of dielectric material • Capacitors used in computers are particularly susceptible, since they are often unprotected by fuses or relays • As a general rule of thumb, untuned capacitors and power-switching devices are incompatible	• Heating of capacitors due to increased dielectric losses • Short circuits • Fuse failure • Capacitor explosion
Transformers	• Voltage harmonics cause higher transformer voltage and insulation stress; normally not a significant problem	• Transformer heating • Reduced life • Increased copper and iron losses • Insulation stress • Stress
Motors	• Increased losses • Harmonic voltages produce magnetic fields rotating at a speed corresponding to the harmonic frequency	• Motor heating • Mechanical vibrations and noise • Pulsating torques • Increased copper and iron losses in stator and rotor windings, from 5%–10% • Reduced efficiency • Reduced life • Voltage stress on insulation of motor windings
Electromechanical induction disk relays	• Additional torque components are produced and may alter the time delay characteristics of the relays	• Incorrect tripping of relays • Incorrect readings
Circuit breakers	• Blowout coils may not operate properly in the presence of harmonic currents	• Failure to interrupt currents • Breaker failure
Watt-hour meters, overcurrent relays	• Harmonics generate additional torque on the induction disk, which can cause improper operation since these devices are calibrated for accurate operation on the fundamental frequency only	• Incorrect readings
Electronic and computer-controlled equipment	• Electronic controls are often dependent on the zero crossing or on the voltage peak for proper control; however, harmonics can significantly alter these parameters, thus adversely affecting operation	• Maloperation of control and protection equipment • Premature equipment failure • Erratic operation of static drives and robots

Source: Ontario Hydro Energy Inc. (www.ontariohydroenergy.com).

current or voltage to the fundamental current or voltage shows the extent of harmonics. For example, IEEE 519 suits an upper current distortion limit of 5% to prevent overheating of transformers. The maximum overvoltage for transformers is 5% at rated load and 10% at no load. Tolerance for electronic equipment is 5%.

IEEE 519 sets limits on Total Harmonic Distortion (THD) for the utility side of the meter and Total Demand Distortion (TDD) for the end-user side of the meter. The utility is responsible for the voltage distortion at the PCC between the utility and the end user. By using THD the amount of harmonic current injected into the utility system is found. The THD can be calculated as follows:

$$V_{\text{THD}} = \frac{\sqrt{\sum_{h=2}^{50}}}{V_1} = \sqrt{\left(\frac{V_2}{V_1}\right)^2 + \left(\frac{V_3}{V_1}\right)^2 + \ldots \left(\frac{V_n}{V_1}\right)^2} \quad (1.3)$$

where V_1 is the fundamental voltage value and $V_n = V_2, V_3, V_4$, etc. = harmonic voltage value.

The THD can be used to describe distortion in both current and voltage waves. Mostly THD usually refers to distortions in the voltage wave. For example, the fundamental component for each harmonic is third harmonic distortion = 6/120 × 100% = 50%, fifth harmonic = 9/120 × 100% = 7.5%, and seventh harmonic = 3/120 × 100% = 2.5%. The THD would be calculated as follows:

$$\text{THD} = \sqrt{(0.5)^2 + (0.75)^2 + (0.25)^2} = 0.093 \text{ or } 0.3\%$$

Now the value is greater than 5%; therefore, some mitigating device like a filter is required.

TDD is used to calculate the current distortions caused by harmonic currents in the end-user facilities. TDD of the current I is calculated by the formula

$$\text{TDD} = \frac{\sqrt{\sum_{h=1}^{h=\infty} (I_h)^2}}{I_L} \quad (1.4)$$

where:

I_L is the rms value of maximum demand load current
h is the harmonic order (1, 2, 3, 4, etc.)
I_h is the rms load current at the harmonic order h

Harmonic filters or chokes are used to reduce electrical harmonics just as shock absorbers reduce mechanical harmonics. Filters contain capacitors and inductors in series. There are two types of filters: static and active. Static filters do not change their value. Active filters change their value to fit the harmonic to be filtered. Harmonics can be eliminated by using isolation transformers.

1.2.10 Electrical Noise

The IEEE-469-1988 Recommended Practice for Voice Frequency Electrical-noise Test of Distribution transformer standard provides instruction for the testing of distribution of transformers as sources of voice-frequency noise. These tests measure the degree to which a transformer may contribute to electrical noise in communication circuits that are physically paralleling the power-supply circuits feeding the transformer.

Electrical noise according to our opinion is the audible crackling noise that emanates from high-voltage power lines or the low throbbing hum of an energized transformer. This type of noise can affect our life quality as much as our power quality. According to power quality electrical noise is caused by a low-voltage, high-frequency (but lower than 200-Hz) signal superimposed on the 60-Hz fundamental waveform. This type of electrical noise may be transmitted through the air or wires. This noise is caused by high-voltage lines, arcing from operating disconnect switches, start-up of large motors, radio and TV stations, switched mode power supplies, loads with solid-state rectifiers, fluorescent lights, and electronic devices.

Electrical noise can damage telecommunication equipment, electronic equipment, radio, and TV reception. There are two ways of solving the electrical noise problem. One way is to eliminate the source of the electrical noise. Another way is to stop or reduce the electrical noise from being transmitted. Electrical noise can be reduced by the use of multiple conductors or installation of corona rings in high-voltage lines. Grounding equipment and the service panel to a common point can remove electrical noise from ground loops. This prevents interferes with communication signals.

The Electro Magnetic Interference (EMI) type of noise is reduced by shielding the sensitive equipment from the source of the electrical noise or simply moving the source of EMI far away. For example, the electromagnetic fields from a tabletop fluorescent lamp near a computer screen will cause the lines on the screen to wiggle. The wiggles will stop if the fluorescent light is moved far away. Figure 1.13 shows the electrical noise plot.

1.2.11 Transient Overvoltage

Capacitor switching and lightning are the causes of transient overvoltage. Some power electronic devices generate significant transients when they switch.

1.2.11.1 Capacitor Switching

Capacitor switching occurs more frequently on utility systems. Capacitors are used to deliver reactive power to correct the power factor, which reduces losses and supports the voltage on the system. The use of capacitors is economical compared to the use of rotating machines and electronic var compensators. Thus, the presence of capacitors on power systems is quite common.

That the use of capacitors causes oscillatory transients when switched is one of the major drawbacks. Some capacitors are energized all the time (a fixed bank), while others are switched according to load levels. Various control measures are considered when capacitors are switched: time, temperature, voltage, current, and reactive power.

Mostly capacitor-switching occurs at the same time each day. On distribution feeders with industrial loads, capacitors are frequently switched by time clock in hope of an increase in load with the beginning of the working day.

Figure 1.14 shows the one-line diagram of a typical utility feeder capacitor-switching situation. When the switch contacts are closed, a transient similar to the one in Figure 1.15 can be observed up-line from

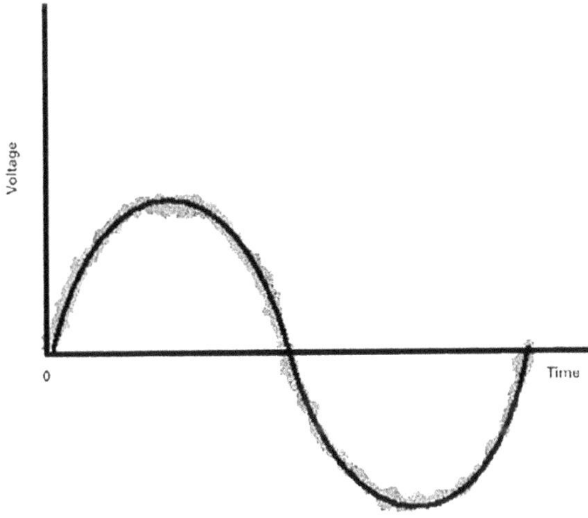

FIGURE 1.13 Electrical noise plot.

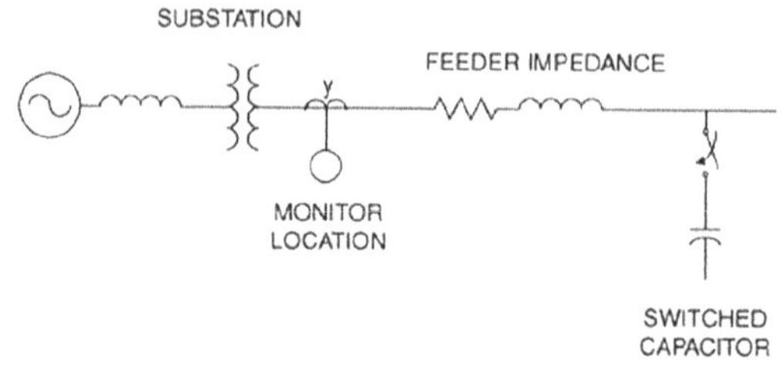

SUBSTATION
FEEDER IMPEDANCE
MONITOR LOCATION
SWITCHED CAPACITOR

FIGURE 1.14 One-line diagram of a capacitor-switching operation corresponding to the waveform in Figure 1.15.

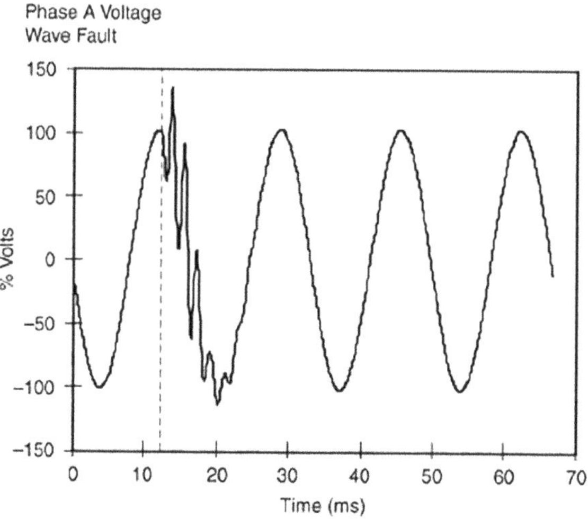

FIGURE 1.15 Typical utility capacitor-switching transient reaching 134% voltage, observed upline from the capacitor.

the capacitor at the monitored location. In this case, at a point near the system, a voltage peak capacitor switch is closed. This is common for many types of switches because the insulation across the switch contacts tends to break down when the voltage across the switch contacts is at a maximum value. At this instant zero voltage is maintained across the capacitor. The system voltage at the capacitor location is greatly pulled down to zero and rises as the capacitor begins to charge towards the system voltage. The capacitor voltage overshoots and rings at the natural frequency of the system due to the inductive nature of the power system source. At the monitoring location the initial variation in voltage will not go completely to zero because of the impedance between the observation point and the switched capacitor.

Depending on system damping the overshoot will produce a transient between 1.0 and 2.0 per unit (pu). In this case the transient recorded at the monitoring location is about 1.34 pu. Utility capacitor-switching transients are in the 1.3- to 1.4-pu range. The transient shown in the oscillogram spreads into the local power system and will usually pass through distribution transformers into consumer load services. The amount is related to the turns ratio of the transformer. The voltage may be magnified on the load side due to the presence of the capacitor. The transients up to 2.0 pu are not generally damaging to the system insulation, but they can often cause malfunction of electronic power conversion devices. The transient also interferes with the gate triggering pulses of thyristors.

Figure 1.16 shows the phase current observed for the capacitor-switching incident described in the preceding text. The transient current flowing in the feeder peaks at nearly 4 times the load current.

1.2.11.2 Magnification of Capacitor-Switching Transients

Addition of power factor correction capacitors at the customer location may increase the impact of utility capacitor-switching transients on end-user equipment. The transient is normally not higher than 2.0 pu

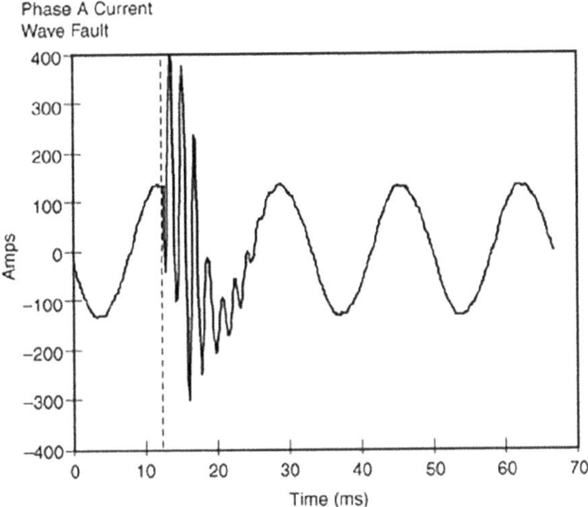

FIGURE 1.16 Feeder current associated with capacitor-switching event.

on the primary distribution system, but ungrounded capacitor banks may yield somewhat higher values. Load-side capacitors can increase this transient overvoltage at the end-user bus for certain low-voltage capacitor and step-down transformer sizes. The circuit of concern for this phenomenon is presented in Figure 1.17. Transient overvoltages on the end-user side may reach as high as 3.0–4.0 pu on the low-voltage bus under these conditions.

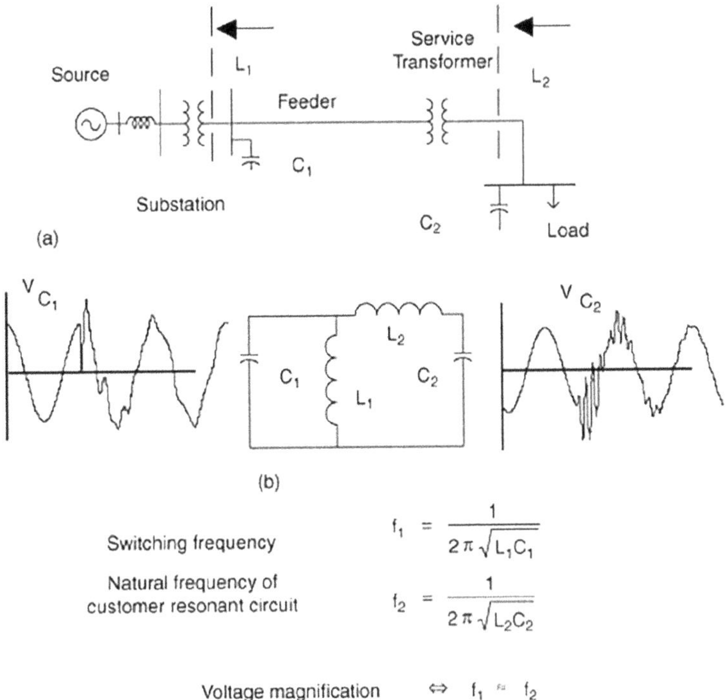

FIGURE 1.17 Voltage magnification of capacitor bank switching. (a) Voltage magnification at customer due to energizing capacitor on utility system and (b) Equivalent circuit.

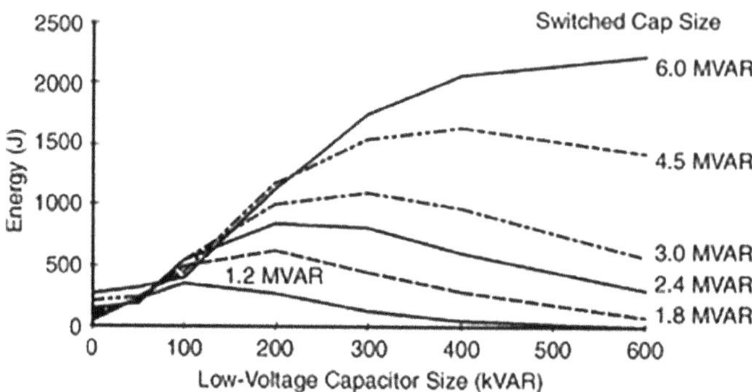

FIGURE 1.18 Arrester energy duty caused by magnified transient.

Utility capacitor-switching transients are magnified due to the presence of a wide range of transformer and capacitor sizes. One method to solve this is to control the transient overvoltage at the utility capacitor. This is possible by using synchronous closing breakers or switches with preinsertion resistors.

High-energy surge arresters can be used to limit the transient voltage magnitude at the consumer bus. Figure 1.18 shows the required arrester energy for a range of low-voltage capacitor sizes. Newer high-energy metal-oxide varistor (MOV) arresters for low-voltage applications can bear the stress from 2 to 4 kJ.

The arresters can only limit the transient to the arrester protective level. This will be approximately 1.8 times the normal peak voltage (1.8 pu). This arrester cannot protect sensitive electronic equipment that might only have a withstand capability of 1.75 pu [1200-V peak inverse voltage (PIV) rating of many silicon-controlled rectifiers (SCRs) used in the industrial environment].

The voltage magnification transient can be limited by the harmonic filters as well. An inductance in series with the power factor correction bank will decline the transient voltage at the customer bus to acceptable levels. There are multiple benefits including providing correction for the displacement power factor, controlling harmonic distortion levels within the facility, and limiting the issues regarding magnified capacitor-switching transients.

Adjustable-speed motor drives are severely affected by the transient. It is cost efficient to place line reactors in series with the drives to block the high-frequency magnification transient. Many types of drives have this protection inherently provided by default, either through an isolation transformer or through a DC bus reactance.

1.2.11.3 Restrikes during Capacitor Deenergizing

When a grounded-wye capacitor bank is energized using a mechanical oil-filled switch, the switching occurs at or near the system voltage peak. At this instant, the capacitor voltage is zero while the system voltage is near maximum. The maximum potential difference between contacts is 1.0 pu. Insulation across the switch tends to break down giving rise to an electric arc due to the slow-moving nature of the mechanical switch and the large potential difference. It permits capacitive current to flow, thereby energizing the capacitor before the actual switch contacts close or touch. This fact is known as pre-strike. Pre-strike occurs in switches without closing control. The resulting peak transient voltage is not greater than that when the capacitor is intentionally energized at the system voltage peak.

When arcs between parting contacts are re-established after initial current extinction causing unintended re-energizing of the capacitor occurs, the phenomenon is called restrike. Due to trapped charges after the initial extinction, the voltage is above 2 pu. Successful capacitor deenergizing does not cause any transient overvoltage.

Restrikes can occur multiple times and in series over a short time period. They can also occur over a longer time span. Restrike causes problems to both the power system and the switching device itself. A potentially damaging capacitor restrike case is illustrated in Figure 1.19 using a time-domain simulation model.

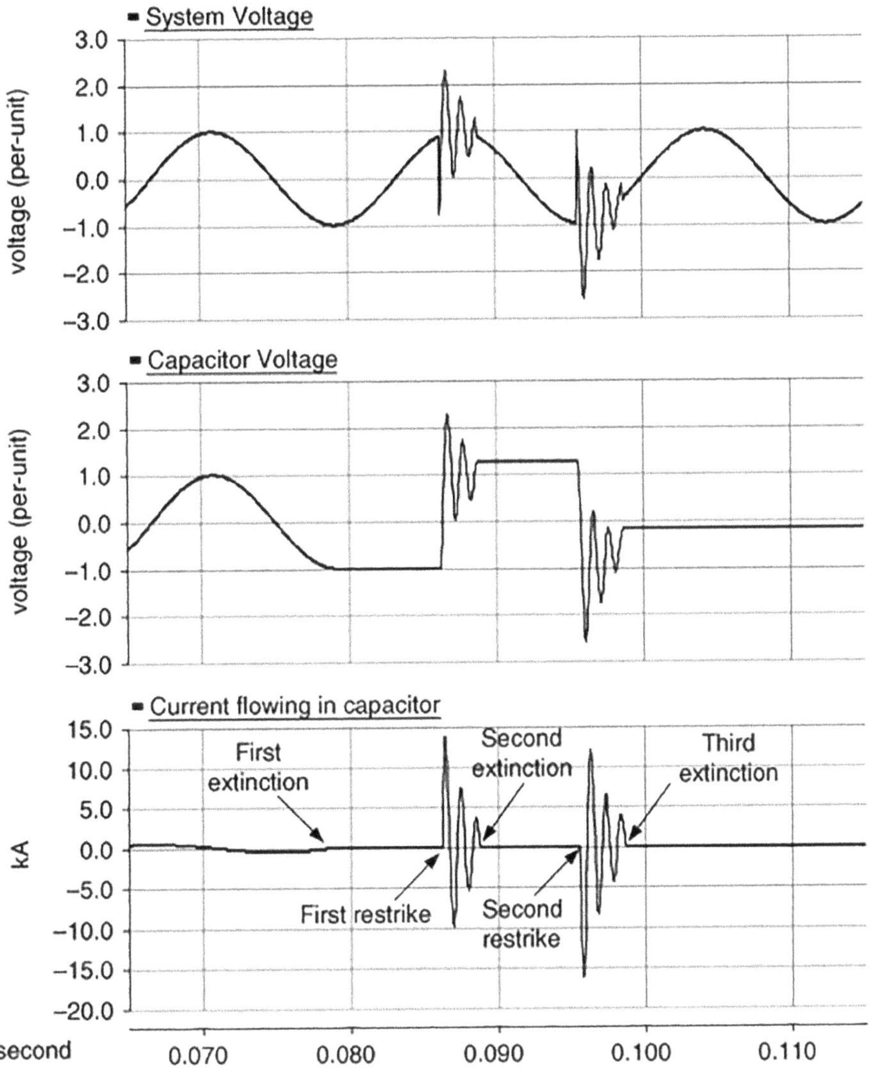

FIGURE 1.19 System and capacitor voltages along with capacitor current during a successive restrike while deenergizing the capacitor.

Consider a grounded-wye capacitor being deenergized. System and capacitor voltages along with current flowing in the capacitor are shown in Figure 1.19. Following the first current extinction, the capacitor maintains a voltage of −1.0 pu because of the trapped charges. The first restrike occurs at the next instant of system peak voltage causing system and capacitor voltages to overshoot above 2.0 pu. Some circuit breakers may be able to interrupt the high-frequency current at its first zero-crossing following the reignition. The capacitor voltage magnitude depends on where in time the second current extinction takes place. The capacitor voltage would be higher if it happens in one of the first few zero-crossings of the inrush current compared to those of the last few zero-crossings. If the potential difference across contacts exceeds the withstand-dielectric-breakdown voltage a second restrike occurs. This second restrike has the potential of causing an even higher overvoltage transient. The capacitor units may blow due to the overvoltage transient caused by the second restrike.

The worst-case restrikes occur at the system voltage peak. A voltage waveform of a restrike during the opening of a 34.5-kV, 9.6-Mvar substation capacitor bank is shown in Figure 1.20. The cause

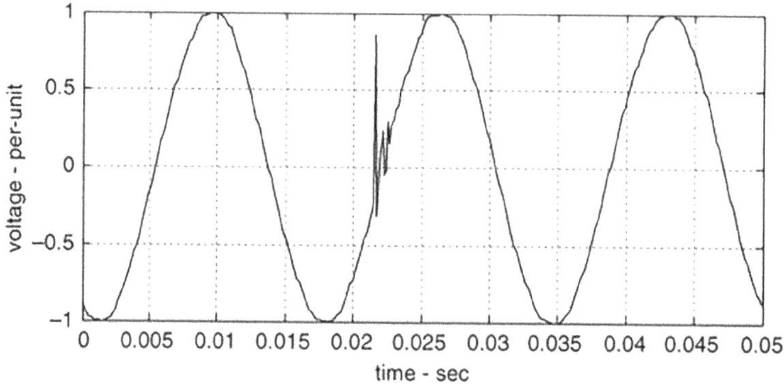

FIGURE 1.20 A voltage waveform during a restrike on opening of a 34.5-kV, 9.6-Mvar capacitor.

of the restrike is due to the build-up of burrs in the contacts of the circuit switcher. Power quality monitors have been used to sense the incidence of restrikes and mitigate the problem so as to avoid switch failures.

1.2.12 Lightning

Lightning is a powerful source of impulsive transients. Figure 1.21 presents the pictorial representation of the places where lightning can strike, resulting in lightning currents being spread from the power system into loads.

The straight conduction path occurs during a direct strike to a live wire, either on the primary or the secondary side of the transformer. Lightning generates very high overvoltages. Very similar transient overvoltages can be produced by lightning currents flowing through ground conductor paths. The lightning currents can enter the grounding system through several paths. Common paths are indicated by the dotted lines in Figure 1.21; these include the primary ground, the secondary ground, and the structure of the load facilities. The strikes to the primary phase are transferred to the ground circuits through the arresters on the service transformer.

Most of the surge current in the end may be dissipated into the ground connection closest to the strike; there will be substantial surge currents flowing in other connected ground conductors in the first few microseconds of the strike.

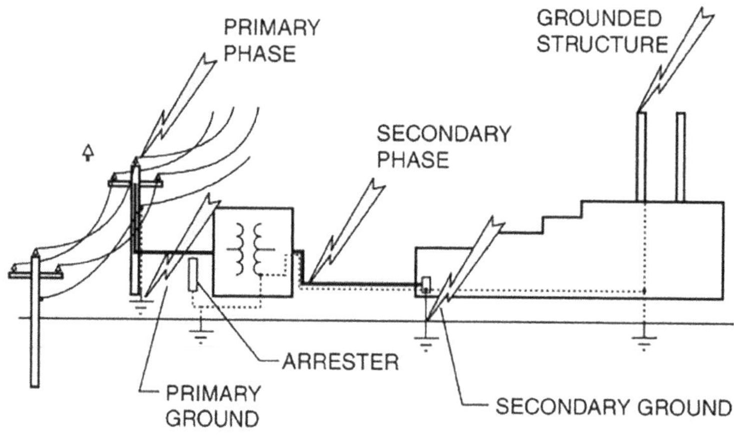

FIGURE 1.21 Lightning strike locations in which lightning impulses will be conducted into load facilities.

Line flashover near the strike point is caused by a direct strike to a phase conductor. It also causes voltage sags and interruptions. The lightning surge can be directed a considerable distance along utility lines and cause multiple flashovers at pole and tower structures as it passes. If the line flashes over at the location of the strike, the tail of the impulse is generally truncated. Arresters near the strike may not survive because of the severity of the strike. Lightning does not have to actually strike a conductor to inject impulses into the power system. Lightning strikes near the line and induce an impulse by the collapse of the electric field. Lightning may also simply strike the ground near a facility causing the local ground reference to rise significantly.

Many investigators in this field postulate that lightning surges enter loads from the utility system through the interwinding capacitance of the service transformer as illustrated in Figure 1.22. The lightning impulse travels faster than the inductance of the transformer windings, blocking the first part of the wave from passing through by the turns ratio. The interwinding capacitance may offer a ready path for the high-frequency surge. The voltage on the secondary terminals is much higher than what the turns ratio of the windings would suggest.

The design of the transformer dictates the degree to which capacitive coupling occurs. Generally, not all transformers have a straightforward high-to-low capacitance. The winding-to-ground capacitance may be greater than the winding-to-winding capacitance, and more of the impulse may actually be coupled to ground than to the secondary winding. In some case, the resulting transient is a very short single impulse, or train of impulses, due to the quick charging of the interwinding capacitance. Arresters on the secondary winding do not face any difficulty dissipating the energy in such a surge, but the rates of rise can be high. This is likely not due to capacitive coupling through the service transformer but for conduction around the transformer through the grounding systems as shown in Figure 1.23. This is

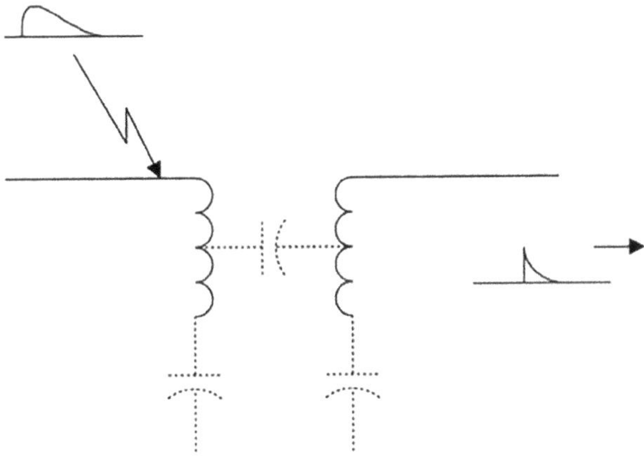

FIGURE 1.22 Coupling of impulses through the interwinding capacitance of transformers.

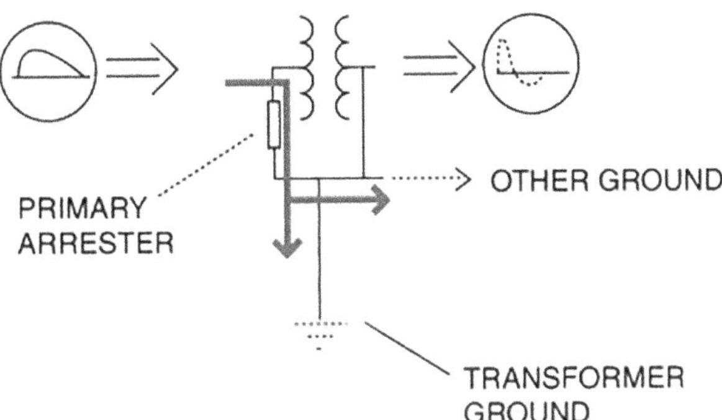

FIGURE 1.23 Lightning impulse bypassing the service transformer through ground connections.

a particular problem when the load system offers a better ground and much of the surge current flows through conductors in the load facility on its way to ground.

The principle of power quality problems with lightning strike currents entering the ground system are:

- The potential of the local ground is increased by several kilovolts. Sensitive electronic equipment such as a computer connected to the base phone system through a modem can fail when subjected to the lightning surge voltages.
- High voltages are induced while passing through the cable on the way towards a better ground.

Lightning causes more flashovers than the researchers expect. Lightning strike current wave fronts are faster than earlier anticipation of the researcher. Durations of some strikes may also be longer than reported by earlier researchers. Therefore, there is a need to design lightning arresters that have required capacity to handle large lightning strikes.

1.2.13 Ferroresonance

The name *ferroresonance* refers to a unique kind of resonance that includes capacitance and iron-core inductance. Disturbances are caused when the magnetizing impedance of a transformer is connected in series with a system capacitor. This occurs when there is an open-phase conductor. Ferroresonance and resonance are different terms. Resonance occurs at resonance frequency with high sinusoidal voltages and current.

Consider a simple series *RLC* circuit presented in Figure 1.24. Ignoring the existence of the resistance *R* for the moment, the current flowing in the circuit can be stated as follows

$$I = \frac{E}{j\left(X_L - |X_C|\right)} \tag{1.5}$$

where E = driving voltage, X_L = reactance of L, and Xc = reactance of C

When $X_L = |X_C|$, a series-resonant circuit is formed, and the equation draws an infinitely large current that in reality would be limited by R.

An alternate solution to the series *RLC* circuit can be found by writing two equations expressing the voltage across the inductor, i.e.,

$$\begin{aligned} \upsilon &= jX_L I \\ \upsilon &= Ej|X_C|I \end{aligned} \tag{1.6}$$

where υ is a voltage variable. Figure 1.25 represents the graphical solution of these two equations for two different reactances, X_L and X_L'. X_L' represents the series-resonant condition. The voltage across inductor *E is given by* the intersection point between the capacitive and inductive lines $_L$. The voltage across

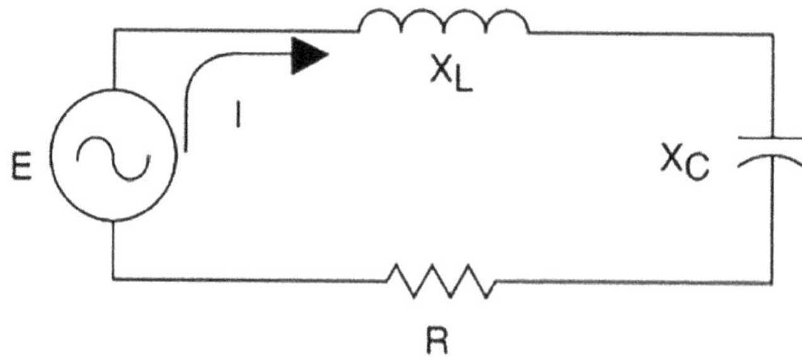

FIGURE 1.24 Simple series RLC circuit.

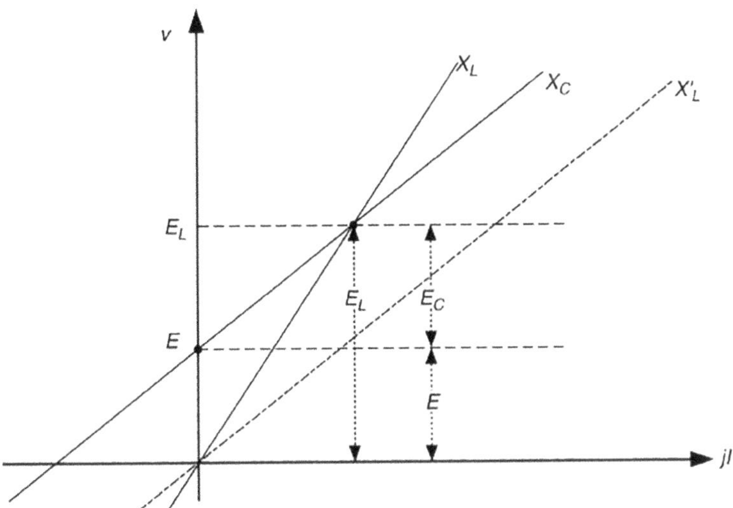

FIGURE 1.25 Graphical solution to the linear LC circuit.

capacitor E_C is determined as represented in Figure 1.25. At resonance condition, the two lines will intersect at infinitely greater voltage and current as the $|X_C|$ line is parallel to the X_L line.

Now, let us assume that the inductive element in the circuit has a nonlinear reactance characteristic similar to that found in transformer magnetizing reactance. Figure 1.26 represents the graphical solution of the equations following the methodology just presented for linear circuits. The above diagram explains well the ferroresonance phenomena.

There are three intersections between the capacitive reactance line and the inductive reactance curve. Intersection 2 will results in an unstable solution, and this operating point gives rise to some of the hectic

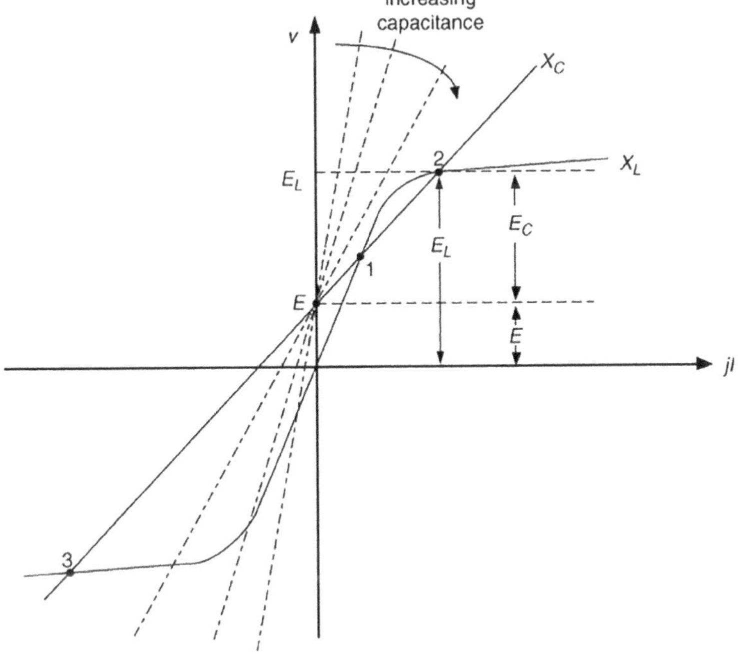

FIGURE 1.26 Graphical solution to the nonlinear LC circuit.

behavior of ferroresonance. Intersections 1 and 3 are stable and will exist in the steady state condition. Intersection 3 results in high voltages and high currents.

Figures 1.27 and 1.28 illustrate the ferroresonant voltages that can result from this simple series circuit. In this example the inductive characteristic was assumed to be the same for each case and the capacitance was varied to achieve a different operating point after an initial transient that drives the system into resonance. During unstable cases, voltage exceeds 4.0 pu, while the stable case results in voltages slightly over 2.0 pu.

For a small amount of capacitance, the $|X_C|$ line is very sharp, causing an intersection point on the third quadrant only. This can yield a range of voltages from less than 1.0 pu to voltages like those shown in Figure 1.28.

When C is very large, the intersection of capacitive reactance line occurs at points 1 and 3. One operating state is of low voltage and lagging current (intersection 1), and the other is of high voltage and leading current (intersection 3). Depending on the applied voltage the operating points during ferroresonance can oscillate between intersection points 1 and 3.

Ferroresonance occurs when unloaded transformers are isolated by underground cables of a certain range of lengths. The minimum length of cable necessary to cause ferroresonance differs with the system voltage level. For all distribution voltage levels, varying from 40 to 100 nF per 1000 feet (ft), the

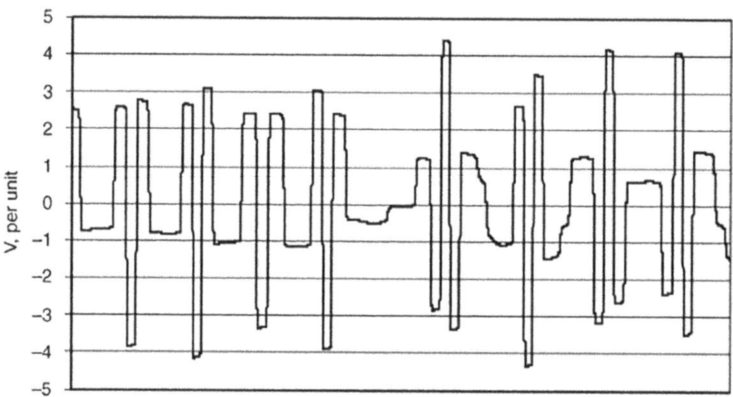

FIGURE 1.27 Example of unstable, chaotic ferroresonance voltages.

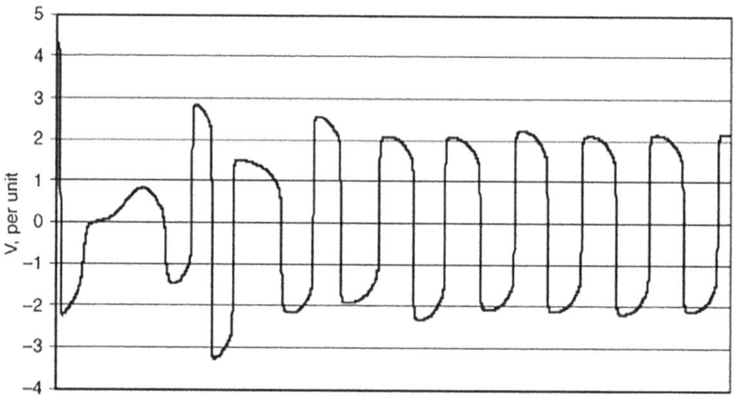

FIGURE 1.28 Example of ferroresonance voltages settling into a stable operating point (intersection 3) after an initial transient.

capacitance of cables is more or less the same. The magnetizing reactance of a 35-kV-class distribution transformer is higher (the curve is steeper) than a 15-kV-class transformer. For delta-connected transformers, ferroresonance can happen for less than 100 ft of cable. For this reason, many utilities avoid this connection on cable-fed transformers. The grounded wye-wye transformer has converted the most commonly used connection in underground systems. It is stronger but affected by ferroresonance because most units use a three-legged or five-legged core design that couples the phases magnetically. The most commonly occurring events leading to ferroresonance are

- Manual switching of an unloaded, cable-fed, three-phase transformer where only one phase is closed (Figure 1.29a).
- Manual switching of an unloaded, cable-fed, three-phase transformer where one of the phases is open (Figure 1.29b). This may happen during energization or deenergization.
- One or two phases open due to the blowing of one or two riser-pole fuses. Single-phase recloser scan also causes this condition. When they sense this condition, many modern commercial loads have transferred the load to backup systems. This specification leaves the transformer without any load to damp out the resonance.

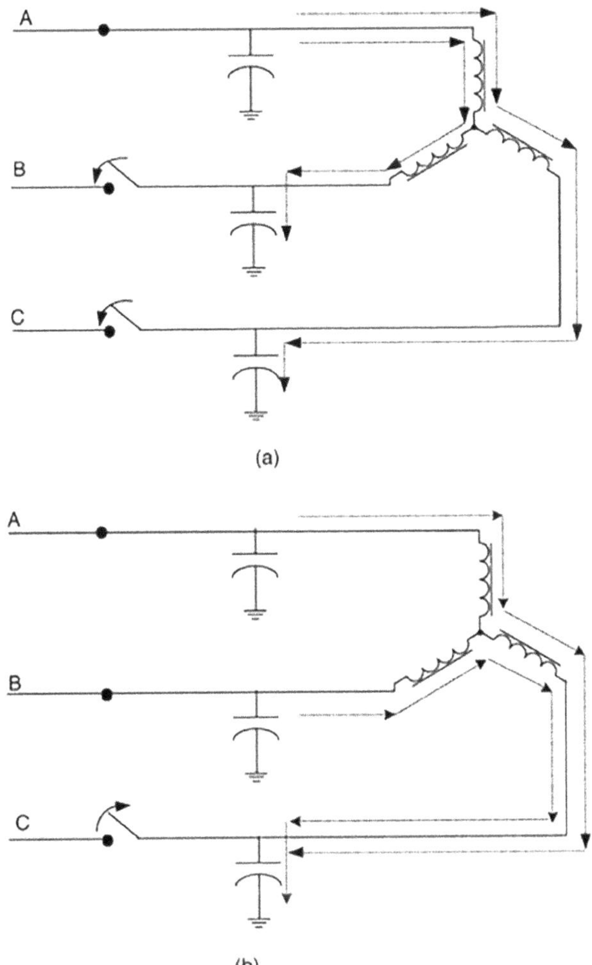

FIGURE 1.29 Common system conditions where ferroresonance may occur: (a) one phase closed, (b) one phase open.

There are certain system conditions that help to increase the likelihood of ferroresonance:

- Higher distribution voltage levels, most notably 25- and 35-kV-class systems
- Transferring of lightly loaded and unloaded transformers
- Ungrounded transformer primary connections
- Very lengthy underground cable circuits
- Cable damage and manual switching during construction of underground cable systems
- Weak systems, i.e., low short-circuit currents
- Low-loss transformers
- Three-phase systems with single-phase switching devices

Ferroresonance occurs at all higher voltage levels in a distribution system. Ferroresonance is limited by proportion of losses, magnetizing reactance, and capacitance but can still occur.

Common indicators of ferroresonance are as follows:

1. Audible Noise: An audible noise is heard during ferroresonance, like that of a large bucket of bolts being shaken, whining, a buzzer, or an anvil chorus pounding on the transformer enclosure from within. The noise is due to the magnetostriction of the steel core being driven into saturation. This noise is louder than the normal hum of a transformer. Most electrical system operating employees are able to recognize it immediately upon first hearing it.
2. Overheating: Due to overheating of transformer it is driven deep into saturation. The charring or bubbling of the paint on the top of the tank is due to stray flux heating. This condition does not indicate that the unit is damaged, but damage can occur in this situation if ferroresonance has continued sufficiently long to cause overheating of some of the larger internal connections. The ferroresonance mode and transformer design is used to determine how the transformer will respond.
3. High overvoltages and surge arrester failure: Surge arresters are affected by both overvoltages and ferroresonance. They are designed to intercept brief overvoltages and clamp them to an acceptable level. Surge arresters continuously withstand several overvoltage events, but there is a definite limit to their energy absorption capabilities. Most of the time the failure of the arrester is due to ferroresonance.
4. Flicker: IEC Standard 61000-4-15 defines the measurement procedure and monitor requirements for characterizing flicker. The IEEE flicker task force working on Standard P1453 is set to adopt the IEC standard as its own.

   On the occurrence of ferroresonance the voltage magnitude may fluctuate violently. This causes the light bulbs to flicker in a secondary circuit. Some electronic devices may be very susceptible to such voltage expeditions. If it is continued for long time it can shorten the expected life of the equipment or may cause immediate failure.
5. Other switching transients: When a switch is closed connecting a line to the power system, line energization transients occur. They commonly involve higher-frequency content than capacitor energizing transients. The transients are produced by a combination of traveling-wave effects and the interaction of the line capacitance and the system equivalent source inductance. The distributed nature of the capacitance and inductance of the transmission or distribution line induce traveling waves. Line-energizing transients frequently die out in about 0.5 cycle.

   The energization transients on distribution feeder circuits are composed of a grouping of line energizing transients, transformer energizing inrush characteristics, and load inrush characteristics. Figure 1.30 shows a typical case in which the monitor was located on the line side of the switch. A small amount of "hash" appears on the front of the waveform and the initial transient frequency is above 1.0 kHz. The distortion caused by the transformer inrush current contains a number of low-order harmonic components, including the second and fourth harmonics.

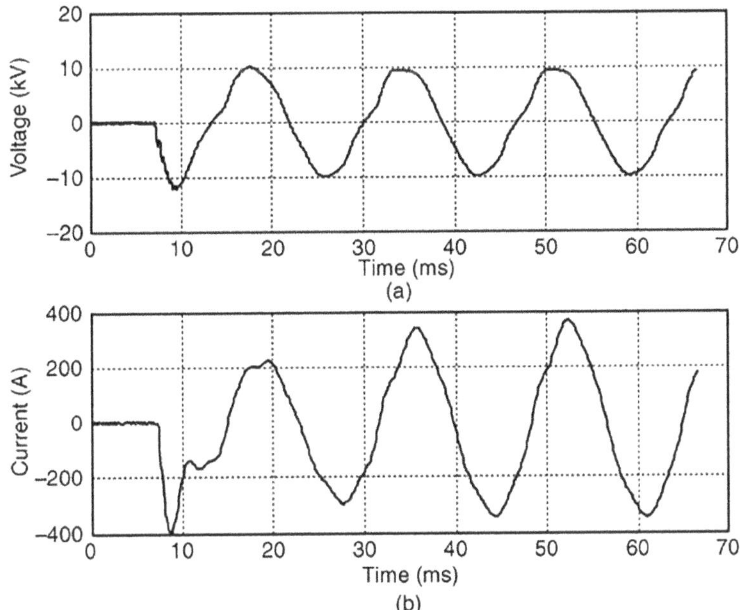

FIGURE 1.30 Energizing a distribution feeder: (a) voltage and (b) current waveforms.

This is identified by the lack of symmetry in the voltage waveform. The first peak of the current waveform shows the simple characteristic of magnetizing inrush, which is consequently swamped by the load inrush current.

End-user equipment is not affected by line energizing transients. Inductive chokes and surge protective devices are used to protect the equipment from high-frequency components. The example shown in Figure 1.30 is relatively benign and should pose few problems. Conditions with fewer loads may exhibit much more oscillatory behavior.

SLG fault is one of the common sources for overvoltages. The voltage rises on effectively grounded four-wire, multigrounded neutral systems are generally not more than 15%–20%. The voltage rise may reach 40% to 50% in the system with neutral reactors. Once the fault is cleared this voltage disappears.

1.3 Principles of Overvoltage Protection

The fundamental principles of overvoltage protection of load equipment are

1. Limit the voltage across sensitive insulation.
2. Distract the surge current away from the load.
3. Prevent the surge current from entering the load.
4. Bond grounds together at the equipment.
5. Reduce, or prevent, surge current from flowing between grounds.
6. Create a low-pass filter using limiting and blocking principles.

Figure 1.31 illustrates these principles, which are applied to protect from a lightning strike. Surge arresters and transient voltage surge suppressors (TVSSs) are used to limit the voltage that can appear between two points in the circuit. Varistors are able to absorb the surge or divert it to ground independently of the rest of the system. Surge currents are like power currents and should obey Kirchhoff's laws. They must

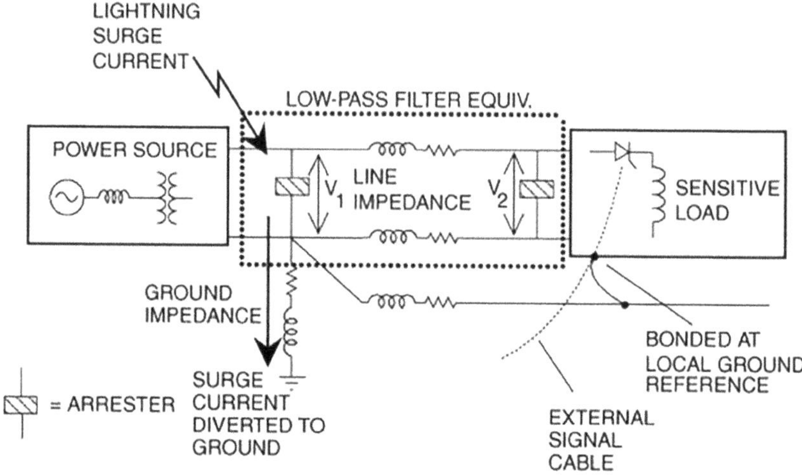

FIGURE 1.31 Demonstrating the principles of overvoltage protection.

flow in a complete circuit, and they cause a voltage drop in every conductor through which they pass. Surge suppressors are frequently connected to the local ground. Local ground may not remain at zero potential during transient impulse events.

Surge suppression devices are located as closely as possible to the critical insulation with a minimum of lead length on all terminals. Arresters are located at the main panels and subpanels, and arresters applied at the point where the power line enters the load equipment are generally the most effective in protecting that particular load. In some cases, it is located inside the load device.

In Figure 1.31 the first arrester is connected from the line to the neutral-ground bond at the service entrance. This connection limits the line voltage V^1 from rising too high relative to the neutral and ground voltage at the panel. During voltage-limiting action, it provides a low-impedance path for the surge current to travel onto the ground lead. Hence, the potential of the whole power system is raised with respect to that of the remote ground by the voltage drop across the ground impedance.

During this situation, most of the surge energy will be discharged through the first arrester directly into ground. In that case, the arrester becomes a surge "diverter." It means a suitable path into which the current can be diverted.

In this Figure 1.31, there is another possible way for the surge current: the signal cable indicated by the dotted line and bonded to the safety ground. If this is connected to another device with ground elsewhere, then there will be some amount of surge current flowing down the safety ground conductor. The first arrester is far away to provide adequate load protection. Therefore, there is a need for second arrester near the load. It is attached "line to neutral" so that it only protects against normal mode transients. Note that the signal cable is bonded to the local ground reference at the load just before the cable enters the cabinet. This creates an unwanted ground loop. However, it is essential to achieve protection of the load and the low-voltage signal circuits. Otherwise, the potential with respect to the signal circuit raises by several kilovolts. Many loads have multiple power and signal cables connected to them. Lightning strike may raise the potential of one ground much higher than the others. This can cause a flashover across the insulation that is between the two ground references or cause physical harm to operators. This phenomenon is a common reason for failure of electronic devices. The situation occurs in TV receivers connected to cables, computers connected to modems, computers with widespread peripherals powered from various sources, and in manufacturing facilities with networked machines.

Necessary control actions are taken to block the surge current. Blocking a high-frequency transient is easier by placing an inductor, or choke, in series with the load. The inductor drops the high surge voltage. One must cautiously consider that high voltage could damage the insulation of both the inductor and the loads. The blocking function is commonly combined with the voltage-limiting function to form a *low-pass filter* in which there is a shunt-connected voltage-limiting device on either side of the series choke. Figure 1.31 illustrates how such a circuit naturally occurs when there are arresters on both ends of the line feeding the load. The line offers the blocking function in proportion to its length. Such a circuit has very useful overvoltage protection characteristics.

Many facilities have multiple ground paths. For example, there may be a driven ground at the service entrance or substation transformer and a second ground at a water well that actually creates a better ground. The amount of current flowing between the grounds may be reduced by improving all the planned grounds at the service entrance and nearby on the utility system.

1.3.1 Devices for Overvoltage Protection

1.3.1.1 Surge Arresters and Transient Voltage Surge Suppressors

By limiting the maximum voltage, arresters and TVSSs protect equipment from transient overvoltages, and the terms are sometimes used interchangeably. The TVSSs, which are generally associated with devices used at the load equipment, will sometimes have more surge-limiting elements than an arrester, which most commonly consists only of MOV blocks. An arrester may have more energy-handling capability; however, the distinction between the two is distorted by common language usage.

The elements that make up these devices can be classified by two different modes of operation, *crowbar* and *clamping*.

Crowbar devices are usually open devices that conduct current during overvoltage transients. Once the device conducts, the line voltage will jump down to nearly zero due to the short circuit imposed across the line. These devices are usually contrived with a gap filled with air or a special gas. The gap arcs over when a sufficiently high overvoltage transient appears. Once the gap arcs over, usually power frequency current, or "follow current," will continue to flow in the gap until the next current zero. The shortcomings of these devices are that the power frequency voltage drops to zero or to a very low value for at least one-half cycle and will cause some loads to drop off-line unnecessarily.

Clamping devices for AC circuits are nonlinear resistors (varistors) that conduct only very low amounts of current until an overvoltage occurs. They conduct heavily, and their impedance drops hastily with increasing voltage. These devices effectively conduct increasing amounts of current (and energy) to limit the voltage rise of a surge. They have an advantage over gap-type devices in that the voltage is not reduced below the conduction level when they begin to conduct the surge current. Zener diodes are also used in this application. Example characteristics of MOV arresters for load systems are shown in Figures 1.32 and 1.33.

MOV arresters have two essential ratings. The first is maximum continuous operating voltage (MCOV), which must be greater than the line voltage and should be at least 125% of the system nominal

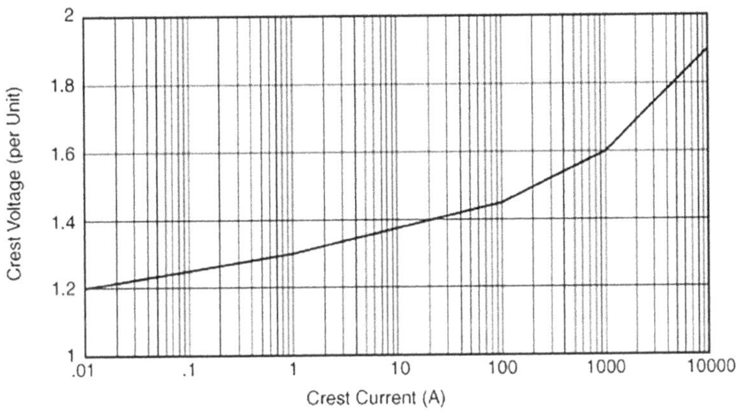

FIGURE 1.32 Crest voltage versus crest amps.

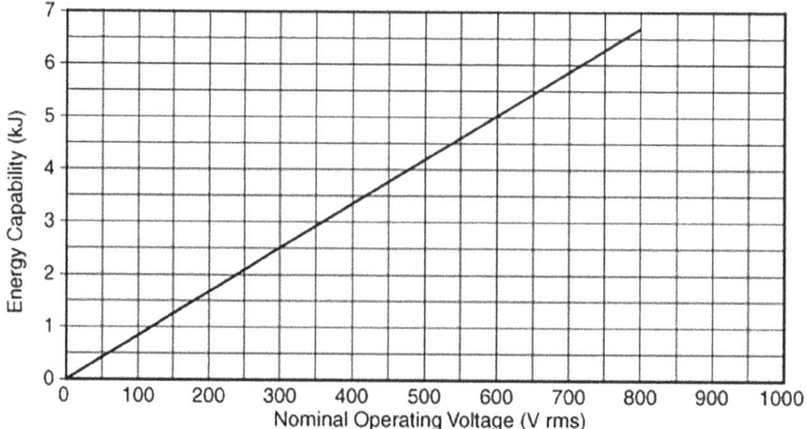

FIGURE 1.33 Energy capability versus operating voltage.

voltage. The second rating is the energy dissipation rating (in joules). MOVs are available in a wide range of energy ratings. Figure 1.33 depicts the typical energy-handling capability versus operating voltages.

1.3.1.2 Isolation Transformers

Figure 1.34 shows a diagram of an isolation transformer which attenuates high-frequency noise and transients. However, some common-mode and normal-mode noise can still arrive at the load. An electrostatic shield, as shown in Figure 1.35, is capable of eliminating common-mode noise. However, some normal-mode noise can still reach the load due to magnetic and capacitive coupling.

The principal feature of isolation transformers for electrically isolating the load from the system for transients is their leakage inductance. Therefore, high-frequency noise and transients should be kept from reaching the load and any load-generated noise and transients are also kept from reaching the rest of the power system. Voltage notching caused by power electronic switching is one example of a problem that can be limited to the load side by an isolation transformer. Capacitor-switching and lightning transients arising from the utility system can be attenuated, thereby avoiding nuisance tripping of adjustable-speed drives and other equipment.

An additional use of isolation transformers is that they permit the user to define a new ground reference, or *separately derived system*. This new neutral-to-ground bond limits the neutral-to-ground voltages at sensitive equipment.

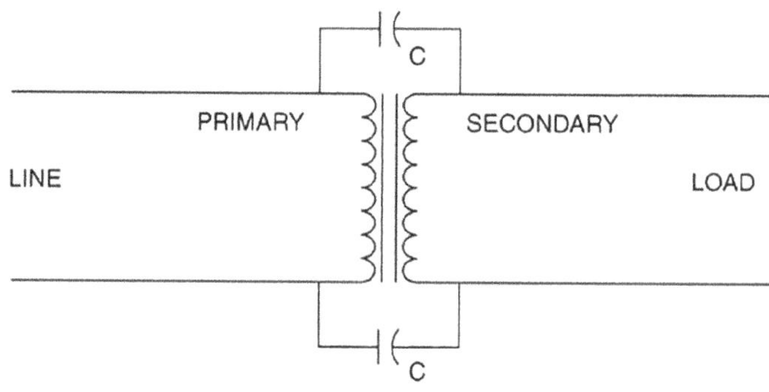

FIGURE 1.34 Isolation transformer.

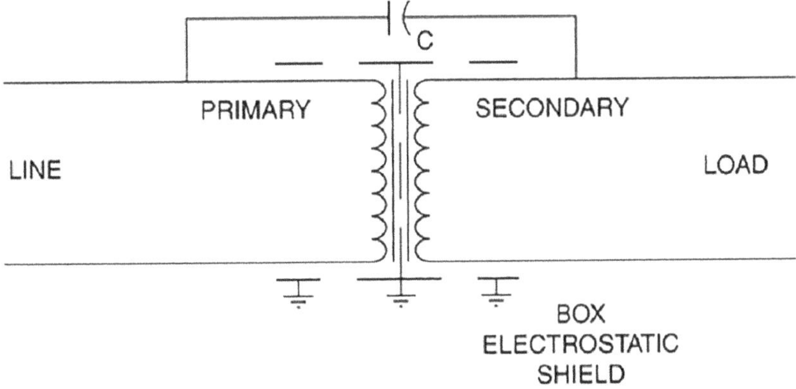

FIGURE 1.35 Isolation transformer with electrostatic shield.

1.3.1.3 Low-Pass Filters

Low-pass filters employ the pi-circuit principle illustrated in Figure 1.36 to accomplish even better protection for high-frequency transients. For general usage in electric circuits, low-pass filters are composed of series inductors and parallel capacitors. This *LC* combination offers a low-impedance path to ground for selected resonant frequencies. In surge protection usage, voltage clamping devices are connected in parallel to the capacitors. In some designs, there are no capacitors.

Figure 1.36 shows a common hybrid protector that combines two surge suppressors and a low-pass filter to offer maximum protection. It uses a gap-type protector on the front end to handle high-energy transients and the low-pass filter limits transfer of high-frequency transients. The inductor blocks high-frequency transients and forces them into the first suppressor. The capacitor limits the rate of rise and the nonlinear resistor (MOV) clamps the voltage magnitude at the protected equipment. Other variations on this design will employ MOVs on both sides of the filters and may have capacitors on the front end as well.

1.3.1.4 Low-Impedance Power Conditioners

Low-impedance power conditioners (LIPCs) are used to interface with the switch-mode power supplies present in electronic equipment. LIPCs differ from isolation transformers in that these conditioners have much lower impedance and have a filter as part of their design (Figure 1.37). The filter is located on the output side and protects against disturbances such as high-frequency, source-side, common-mode, and normal-mode disturbances (i.e., noise and impulses). Note the new neutral-to-ground connection that can be made on the load side because of the persistence of an isolation transformer. But, low- to medium-frequency transients (capacitor switching) can cause problems for LIPCs and the transient can be magnified by the output filter capacitor.

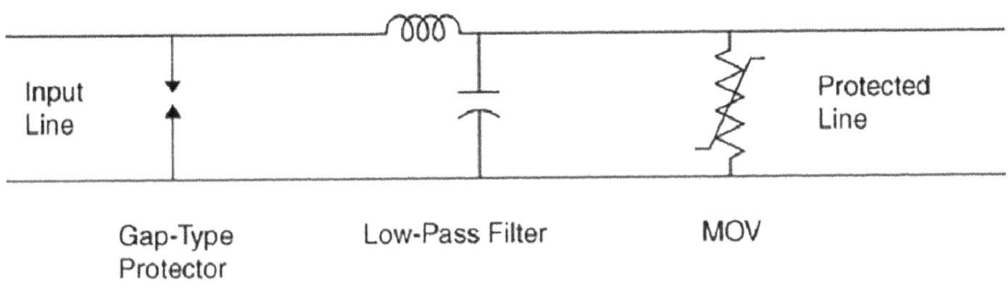

FIGURE 1.36 Hybrid transient protector.

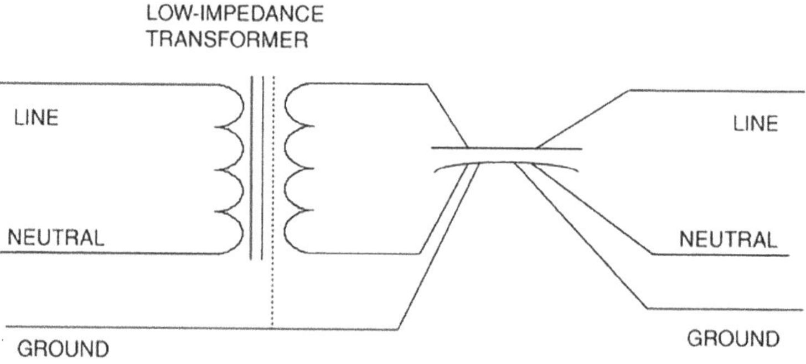

FIGURE 1.37 Low-impedance power conditioner.

1.3.1.5 Utility Surge Arresters

The three most common surge arrester technologies employed by utilities are depicted in Figure 1.38. The recent arresters manufactured today use a MOV as the chief voltage-limiting element. The main ingredient of a MOV is zinc oxide (ZnO), which is combined with several proprietary constituents to achieve the required characteristics and durability. Older-technology arresters use silicon carbide (SiC) as the energy-dissipating non-linear resistive element. The discharge voltages for each of these three technologies are shown in Figure 1.39.

Originally, arresters were little more than spark gaps, which would result in a fault each time the gap sparked over. Also, the spark over transient injects a very sharp fronted voltage wave into the apparatus being protected, which was responsible for many insulation failures. The addition of a SiC nonlinear resistance in series with a spark gap corrected some of these complexities. It permits the spark gap to clear and reseal without causing a fault and reduced the spark over transient to perhaps 50% of the total spark overvoltage (Figure 1.39a). However, insulation failures were still responsible on this front-of-wave transient. Also, there is considerable power-follow current after spark over, which heats the SiC material and erodes the gap structures, ultimately leading to arrester failures or loss of protection.

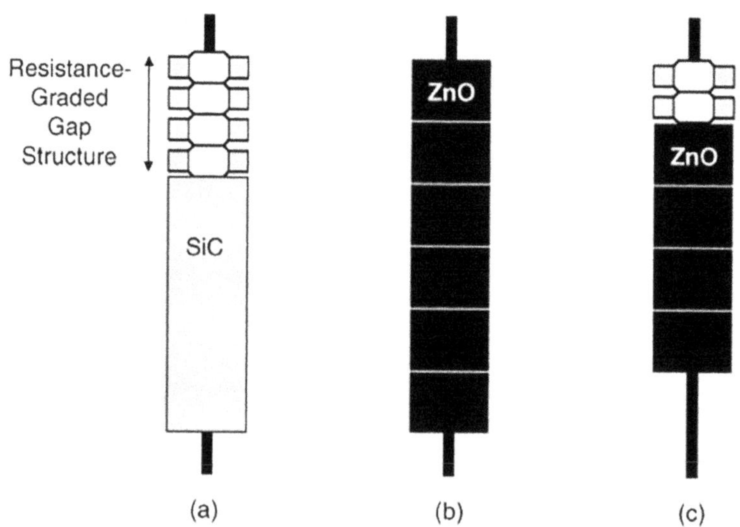

FIGURE 1.38 Three common utility surge arrester technologies. (a) Gapped Silicon Carbide (b) Gapless MOV and (c) Gapped MOV.

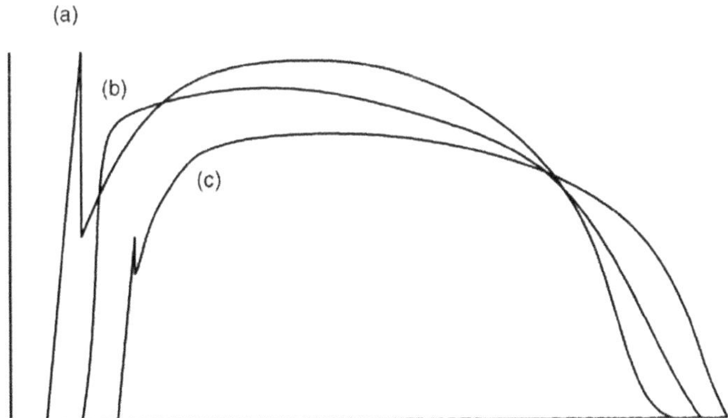

FIGURE 1.39 Comparative lightning wave discharge voltage characteristics for an 8 × 20 μ wave corresponding to the utility surge arrester technologies in Figure 1.38.

Gaps are essential with the SiC because in economical SiC elements the required discharge voltage is unable to hold continuous system operating voltage. The advancement of MOV technology facilitates the exclusion of the gaps. This technology could endure uninterrupted system voltage without gaps and still offer a discharge voltage comparable to the SiC arresters (see Figure 1.39b). By the late 1980s, SiC arrester technology was being phased out in favor of the gapless MOV technology. The gapless MOV employs better discharge characteristic without the objectionable spark over transient. The greater part of utility distribution arresters manufactured today is made of this design.

The gapped MOV technology was introduced commercially about 1990 and has attained acceptance in some applications where there is need for increased protective margins. By combining resistance-graded gaps (with SiC grading rings) and MOV blocks, this arrester technology has some very fascinating, and counterintuitive, characteristics. It has a minor lightning-discharge voltage (Figure 1.39c), but has a higher transient overvoltage withstand characteristic than a gapless MOV arrester. To attain the required protective level for lightning, gapless MOV arresters typically initiate to conduct heavily for low-frequency transients at about 1.7 pu. There are some system conditions where the switching transients will go beyond this value for several cycles and cause failures. Also, applications such as aging underground cable systems demand lower lightning-discharge characteristics.

The gapped MOV technology eliminates about one-third of the MOV blocks and replaces them with a gap structure having a lightning spark over roughly one-half of the old SiC technology. The smaller number of MOV blocks yields a lightning-discharge voltage typically 20%–30% less than a gapless MOV arrester. Because of the capacitive and resistive interaction of the grading rings and MOV blocks, most of the front-of-wave impulse voltage of lightning transients appear across the gaps. They spark over very early into the MOV blocks, yielding a minor spark over transient on the front. For switching transients, the voltage divides by resistance ratios and most of it appears first across the MOV blocks, which hold off conduction until the gaps spark over. This enables this technology to accomplish a transient overvoltage withstand of approximately 2.0 pu in typical designs. In addition, the energy dissipated in the arrester is less than dissipated by gapless designs for the same lightning current because of the lower voltage discharge of the MOV blocks and also there is no power-follow current because there is sufficient MOV capability to block the flow. This reduces the erosion of the gaps. In several ways, this technology holds the promise of yielding a more capable and durable utility surge arrester.

Utility surge arresters are made in various sizes and ratings. The three basic rating classes are designated distribution, intermediate, and station in increasing order of their energy-handling capability. Most of the arresters applied on primary distribution feeders are distribution class. Within this class, there are both small-block and heavy-duty designs. One common omission to this is that sometimes intermediate- or station-class arresters are applied at riser poles to attain a better protective characteristic (lower discharge voltage) for the cable.

1.3.2 Utility Capacitor-Switching Transients

This section portrays how utilities can deal with troubles related to capacitor-switching transients.

1.3.2.1 Switching Times

Capacitor-switching transients are very familiar and usually not destructive. However, the timing of switching may be adverse for some sensitive industrial loads. For example, if the load picks up the same time each day, the utility may choose to switch the capacitors coincident with that load increase. There have been several cases where this coincides with the commencement of a work shift and the resultant transient causes several adjustable-speed drives to shut down shortly after the process starts. One simple and economical solution is to establish if there is a switching time that might be more acceptable. For example, it may be possible to switch on the capacitor a few minutes before the beginning of the shift and before the load actually picks up. It may not be desirable then, but probably will not hurt anything. If this cannot be worked out, other, more expensive, solutions will have to be found.

1.3.2.2 Pre-insertion Resistors

Pre-insertion resistors, known as closing resistors, can diminish the capacitor-switching transient noticeably. The first peak of the transient is usually the most destructive. The suggestion is to insert a resistor into the circuit momentarily so that the first peak is damped significantly. This is an established technology and is quite effective.

Figure 1.40 shows one example of a capacitor switch with pre-insertion resistors to reduce transients. The pre-insertion is accomplished by the movable contacts sliding past the resistor contacts first before mating with the main contacts. As an outcome, the resistor is part of the circuit for about 4–15 μs (or 30%–90% of a 60-Hz cycle). The equivalent circuit relating the action of pre-insertion is illustrated in Figure 1.41.

When a movable contact assembles the resistor contact, the resistor is in the circuit (S1 is closed, while S2 is open). When the movable contact mates with the main contact, the S2 contact is closed,

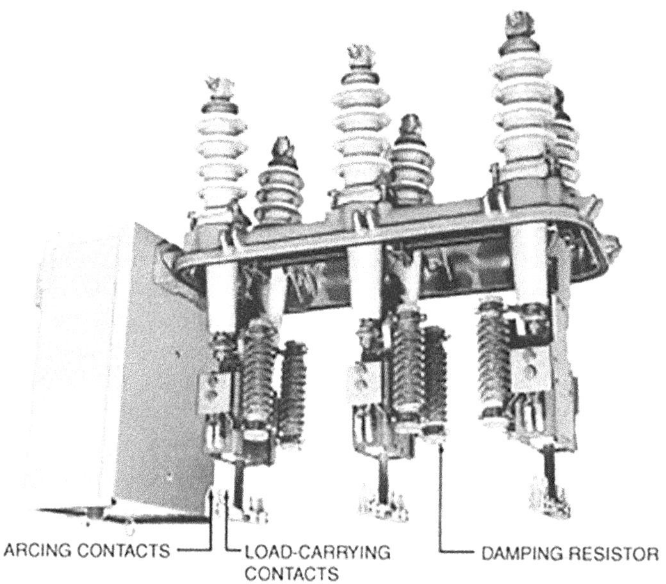

FIGURE 1.40 Capacitor switch with preinsertion resistors.

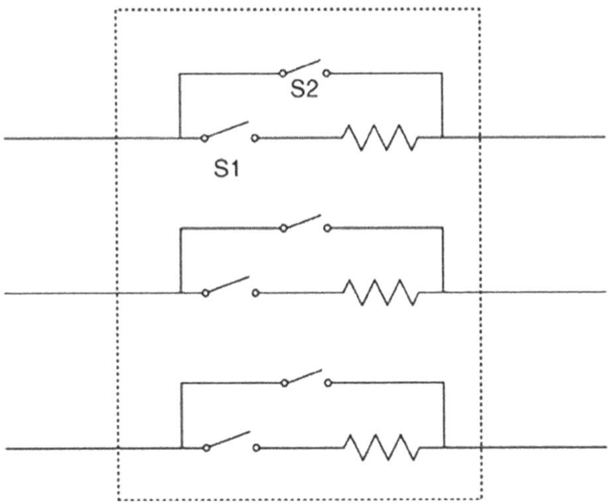

FIGURE 1.41 An equivalent circuit of switches with pre-insertion resistors.

TABLE 1.6

Peak Transient Overvoltages Due to Capacitor Switching with and without Pre-insertion Resistor

Size (kvar)	Available Short Circuit, kA	Without Resistor (pu)	With 6.4Ω Resistor (pu)
900	4	1.95	1.55
900	9	1.97	1.45
900	14	1.98	1.39
1200	4	1.94	1.50
1200	9	1.97	1.40
1200	14	1.98	1.34
1800	4	1.92	1.42
1800	9	1.96	1.33
1800	14	1.97	1.28

Courtesy of Cooper Power Systems, Waukesha, WI.

which fundamentally removes the pre-insertion resistor from the 60-Hz circuit. The effectiveness of the resistors is reliant on capacitor size and available short-circuit current at the capacitor location. Table 1.6 shows expected maximum transient overvoltages upon energization for various conditions, both with and without the pre-insertion resistors. These are the maximum values expected; average values are typically 1.3 to 1.4 pu without resistors and 1.1. to 1.2 p.u. with resistors.

Figure 1.42 illustrates a simulated voltage waveform corresponding to the energizing of a 115-kV, 42-Mvar capacitor using switches without pre-insertion resistors. The peak transient overvoltage is 1.8 pu. A general size of pre-insertion is between 70 and 80Ω. By means of 80-n pre-insertion resistors and with pre-insertion time of one-half of a cycle, the peak transient overvoltage is decreased to 0.13 pu. A pre-insertion resistor also reduces the capacitive inrush current during energizing significantly.

Switches with preinsertion reactors have also been established for this purpose. The inductor is helpful in limiting the higher-frequency components of the transient. This helps the transient to damp out quickly.

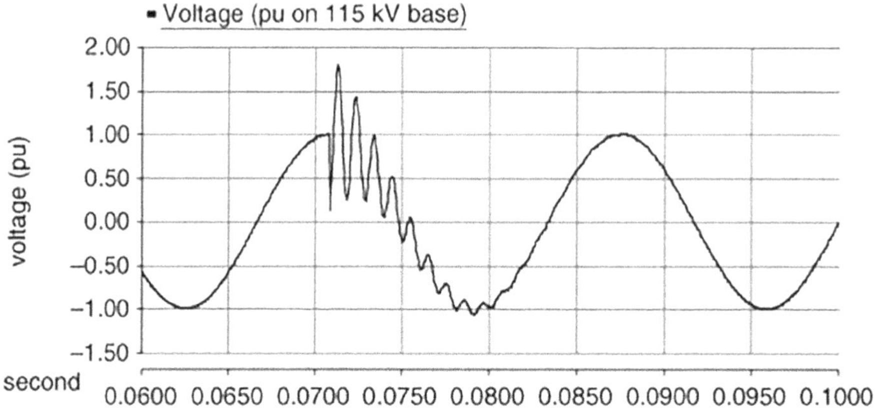

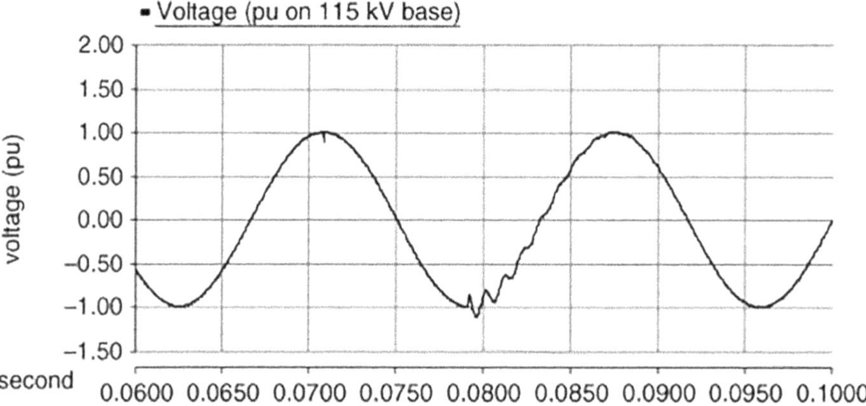

FIGURE 1.42 Simulated voltage waveforms during capacitor energizing using switches without (top) and with (bottom) preinsertion resistors.

1.3.2.3 Synchronous Closing

Synchronous closing stops transients by timing the contact closure such that the system voltage closely matches the capacitor voltage at the instant the contacts mate. This avoids the step change in voltage that normally occurs when capacitors are energized, causing the circuit to oscillate. A successful synchronous closing needs the switch to maintain a sufficient dielectric strength to withstand the system voltage until its contacts touch.

A grounded-wye capacitor is energized at voltage zero. For an ungrounded-wye capacitor, the first phase is energized at voltage zero, the second phase is energized when its voltage is identical to the voltage of the first phase, and the third phase is energized at voltage zero. The peak overvoltage transient is below 1.2 pu. Figure 1.43 shows simulated voltage waveforms due to capacitor energizing. The first phase (top) is energized with switches without closing control, while the second and third phases are with controlled closing. The timing error for the second phase is negligible. The timing error for the third phase is 1 ms, resulting in a clear oscillatory transient voltage with a peak of 1.12 pu.

Actual three-phase voltage and current waveforms of a 150-Mvar, 345-kV grounded-wye capacitor are shown in Figure 1.44. The capacitor was energized by means of switches equipped with synchronous closing control. The timing errors are negligible. The peak inrush currents in all three phases are strangely low (below 1 kA peak) because the instantaneous voltage at the switching instant is near zero. Synchronous closing control reduces both oscillatory voltage transients and inrush currents during energizing significantly.

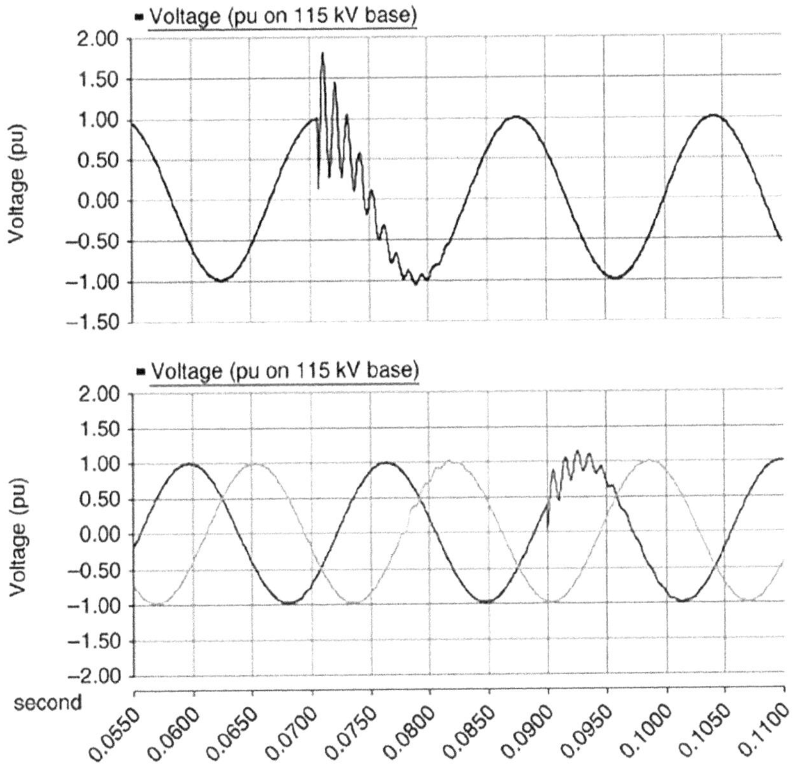

FIGURE 1.43 Simulated voltage waveforms during capacitor energizing using switches equipped without (top) and with (bottom) synchronous closing control.

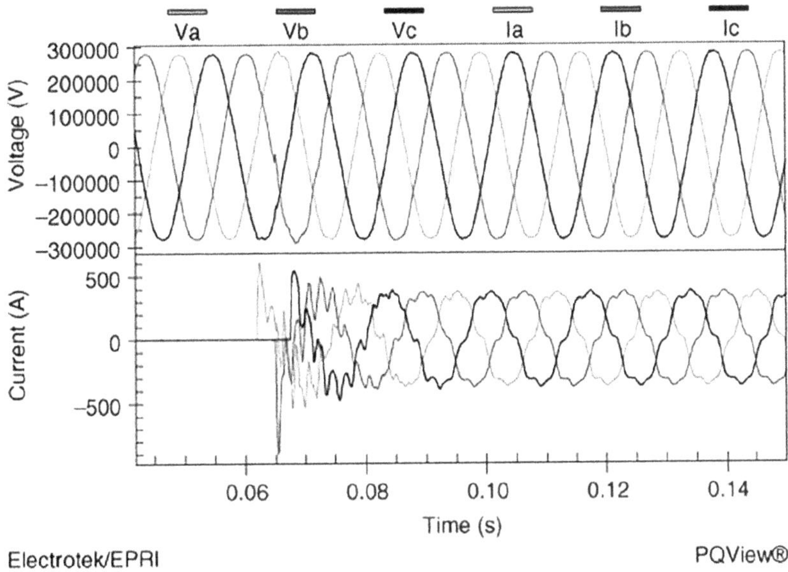

FIGURE 1.44 Actual voltage (V, top) and inrush current (A, bottom) waveforms during capacitor energizing using switches equipped synchronous closing control.

FIGURE 1.45 Synchronous closing breaker.

Figure 1.45 shows one example of a circuit breaker designed for this purpose. This breaker would normally be applied on the utility sub transmission or transmission system (72- and 145-kV classes). This is a three-phase SF_6 breaker that procedures a specially designed operating mechanism with three independently controllable drive rods. It is capable of closing within 1 ms of voltage zero. The real performance of the breaker is sampled to adjust the pole timing for future operations to compensate for wear and changes in mechanical characteristics.

Figure 1.46 shows a vacuum switch made for this purpose. It is applied on 46-kV-class capacitor banks. It contains three independent poles with separate controls. The timing for synchronous

FIGURE 1.46 Synchronous closing capacitor switch.

FIGURE 1.47 One pole of a synchronous closing switch for distribution capacitor banks.

closing is determined by expecting an upcoming voltage zero. Its achievement is dependent on the reliable operation of the vacuum switch. The switch reduces capacitor inrush currents by an order of magnitude and voltage transients to about 1.1 pu. A similar switch may also be used at distribution voltages.

Figure 1.47 shows one phase of a three-phase synchronous switch used for distribution capacitor banks. This particular technology uses a vacuum switch encapsulated in a solid dielectric. Each of the switches designated here requires a sophisticated microprocessor-based control.

1.3.2.4 Capacitor Location

A switched capacitor may be too close to a sensitive load or at a location where the transient overvoltages tend to be much higher. Often, it may be possible to move the capacitor down line or to another branch of the circuit and reduce the problem. The strategy is either to make more damping with more resistance in the circuit or to get more impedance between the capacitor and the sensitive load.

The success of this strategy will depend on a number of factors.

1.3.2.5 Utility System Lightning Protection

The IEC 62561-1 standard summarizes the test procedure and pass requirements for connection components used in a Lightning Protection System (LPS) such as lightning conductor clamps, bonding and earthing clamps, bridging components, pipe clamps, and equipotential bonding bars.

Many power quality problems arise from lightning. Lightning strike to the line causes voltage sags and interruptions. Here are some plans for utilities to use to decrease the impact of lightning.

1.3.2.6 Shielding

Lines are shielded by installing a grounded neutral wire over the phase wires. This will capture most lightning strikes before they strike the phase wires. This can help but will not prevent line flashovers because of the possibility of back flashovers.

Shielding overhead utility lines is common at transmission voltage levels and in substations but is not common on distribution lines. On distribution circuits, the grounded neutral wire is normally installed below the phase conductors to facilitate the connection of line-to-neutral-connected equipment such as transformers and capacitors.

Shielding is not as simple as adding a wire and grounding it every few poles. When lightning strikes the shield wire, the voltages at the top of the pole will still be tremendously high and could cause back flashovers to the line. To minimize this possibility, the path of the ground lead down the pole must be carefully chosen to maintain adequate clearance with the phase conductors. Also, the grounding resistance shows an important role in the magnitude of the voltage and must be maintained as low as possible.

Figure 1.48 illustrates this concept. It is not rare for a few spans near the substation to be shielded. The substation is normally shielded anyway, and this helps prevent high-current faults close to the substation that can damage the substation transformer and breakers. It is also common near substations for distribution lines to be under built on transmission or sub transmission structures. Since the transmission is shielded, this provides shielding for the distribution as well, provided suitable clearance can be maintained for the ground lead.

1.3.2.7 Line Arresters

Generally, lines flash over first at the pole insulators. Therefore, preventing insulator flashover will reduce the interruption and sag rate considerably. Neither shielding nor line arresters will avoid all flashovers from lightning.

As shown in Figure 1.49, the arresters bleed off some of the strike current as it passes along the line. The amount that an individual arrester drains off will depend on the grounding resistance. The idea is

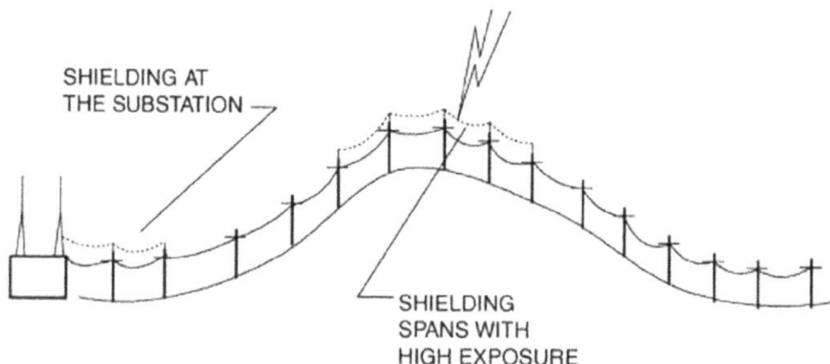

FIGURE 1.48 Shielding a portion of a distribution feeder to reduce the incidence of temporary lightning-induced faults.

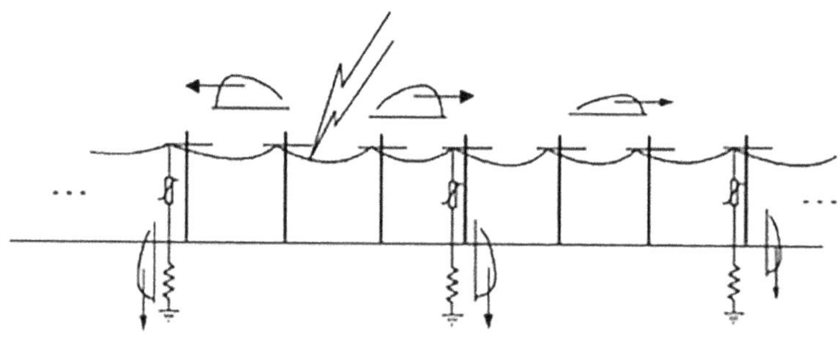

FIGURE 1.49 Periodically spaced line arresters help prevent flashovers.

FIGURE 1.50 Typical polymer-housed utility distribution arrester for overhead line applications.

to place the arresters sufficiently close to prevent the voltage at unprotected poles in the middle from exceeding the basic impulse insulation level (BIL) of the line insulators.

Some services place line arresters only on the top phase when one phase is attached higher than the others.

Figure 1.50 shows a typical utility arrester that is used for overhead line protection applications. This model consists of MOV blocks encapsulated in a polymer housing that is resistant to sunlight and other natural elements.

There are already sufficient arresters on many lines in densely populated areas in North America to achieve sufficient line protection. These arresters are on the distribution transformers, which are installed close together and in sufficient numbers in these areas to help protect the lines from flashover.

1.4 Origin of Short Interruptions

This category encompasses the IEC category of *voltage dips and short interruptions*. Each type of variation can be designated as *instantaneous, momentary*, or *temporary*, depending on its duration.

Short-duration voltage variations are caused by fault conditions, the energization of large loads that require high starting currents, or intermittent loose connections in power wiring. Depending on the fault location and the system conditions, the fault can cause either temporary voltage drops (*sags*), voltage rises (*swells*), or a complete loss of voltage (*interruptions*). The fault condition can be close to or remote from the point of interest. In either case, the impact on the voltage during the actual fault condition is of the short-duration variation until protective devices operate to clear the fault.

1.4.1 Terminology

1.4.1.1 Interruption

An *interruption* occurs when the supply voltage or load current decreases to less than 0.1 pu for a period of time not exceeding 1 minutes. Interruptions can be the result of power system faults, equipment failures, and control malfunctions. The interruptions are measured by their duration since the voltage magnitude is always less than 10% of nominal. The duration of an interruption due to a fault on the utility system is determined by the operating time of utility protective devices. Instantaneous reclosing generally will limit the interruption caused by a nonpermanent fault to less than 30 cycles. Delayed reclosing of the protective device may cause a momentary or temporary interruption. The duration of an interruption due to equipment malfunctions or loose connections can be irregular.

Some interruptions may be preceded by voltage sag when these interruptions are due to faults on the source system. The voltage sag occurs between the time a fault initiates and the protective device

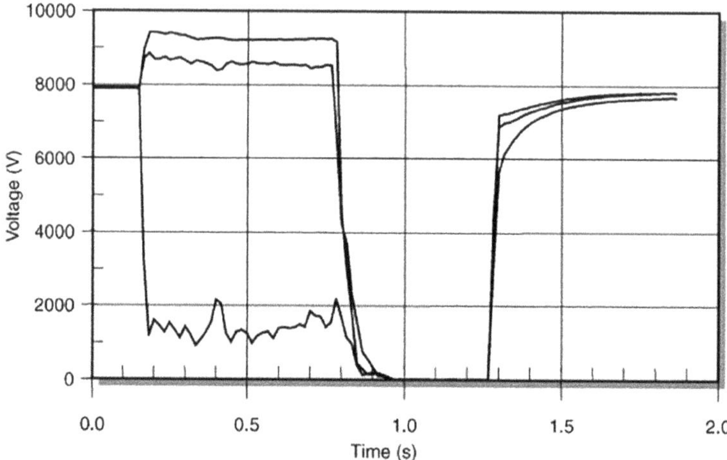

FIGURE 1.51 Three-phase rms voltages for a momentary interruption due to a fault and subsequent recloser operation.

operates. Figure 1.51 shows such a momentary interruption during which voltage on one phase sags to about 20% for about 3 cycles and then drops to zero for about 1.8 seconds until the recloser closes back in.

1.4.1.2 Sags (Dips)

A *sag* is a decrease to between 0.1 and 0.9 pu in rms voltage or current at the power frequency for durations from 0.5 cycle to 1 minutes. The power quality community has used the term *sag* for many years to describe a short-duration voltage decrease. Although the term has not been formally defined, it has been increasingly accepted and used by utilities, manufacturers, and end users. The IEC definition for this phenomenon is *dip*. The two terms are considered interchangeable, with *sag* being the preferred synonym in the U.S. power quality community.

Terminology used to describe the magnitude of voltage sag is often confusing. A "20% sag" can refer to a sag that results in a voltage of 0.8 or 0.2 pu. The preferred terminology would be one that leaves no doubt as to the resulting voltage level: "a sag to 0.8 pu" or "a sag whose magnitude was 20%." When not specified otherwise, a 20% sag will be considered an event during which the rms voltage decreased by 20% to 0.8 pu. The nominal, or base, voltage level should also be specified.

Voltage sags are usually associated with system faults but can also be caused by energization of heavy loads or starting of large motors. Figure 1.52 shows typical voltage sag that can be associated with a SLG fault on another feeder from the same substation. 80% sag exists for about 3 cycles until the substation breaker is able to interrupt the fault current. Typical fault clearing times range from 3 to 30 cycles, depending on the fault current magnitude and the type of over current protection.

Figure 1.53 illustrates the effect of a large motor starting. An induction motor will draw 6 to 10 times its full load current during start-up. If the current magnitude is large relative to the available fault current in the system at that point, the resulting voltage sag can be significant. In this case, the voltage sags immediately to 80% and then gradually returns to normal in about 3 seconds. Note the difference in time frame between this and sags due to utility system faults.

Until recent efforts, the duration of sag events has not been clearly defined. Typical sag duration is defined in some publications as ranging from 2 ms (about one-tenth of a cycle) to a couple of minutes. Undervoltages that last less than one-half cycle cannot be characterized effectively by a change in the rms value of the fundamental frequency value. Therefore, these events are considered *transients*. Undervoltages that last longer than 1 minutes can typically be controlled by voltage regulation

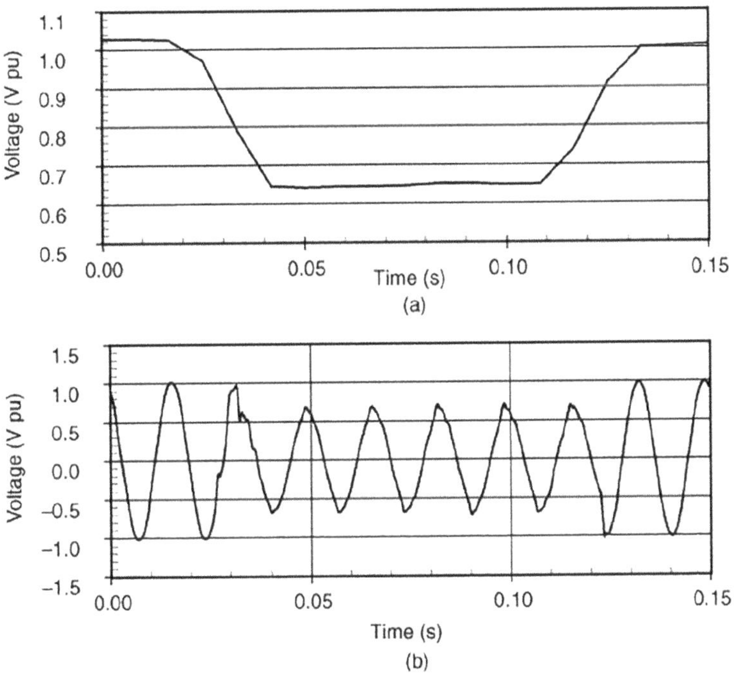

FIGURE 1.52 Voltage sag caused by an SLG fault. (a) RMS waveform for voltage sag event. (b) Voltage sag waveform.

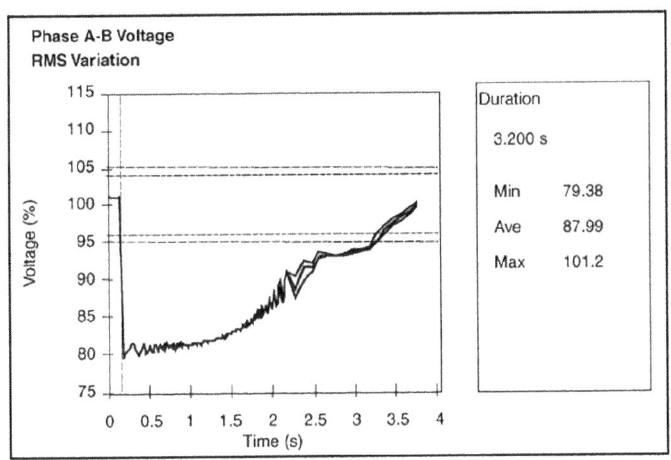

FIGURE 1.53 Temporary voltage sag caused by motor starting.

equipment and may be associated with causes other than system faults. Therefore, these are classified as long-duration variations.

Sag durations are subdivided here into three categories—instantaneous, momentary, and temporary—which coincide with the three categories of interruptions and swells. These durations are intended to correspond to typical utility protective device operation times as well as duration divisions recommended by international technical organizations.

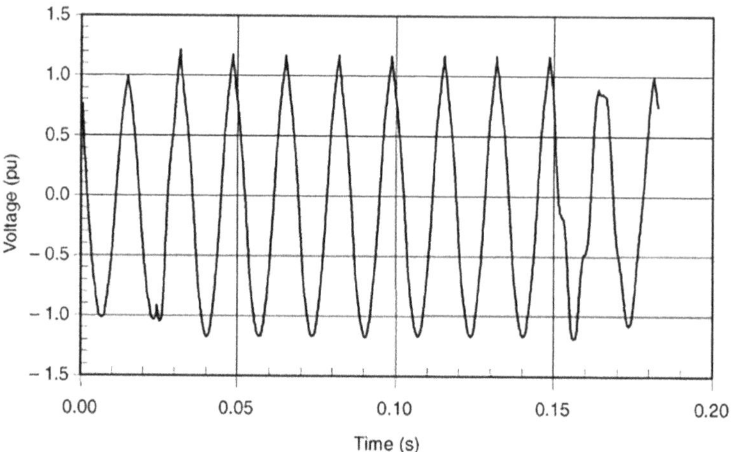

FIGURE 1.54 Instantaneous voltage swell caused by an SLG fault.

1.4.1.3 Swells

A *swell* is defined as an increase to between 1.1 and 1.8 pu in rms voltage or current at the power frequency for durations from 0.5 cycle to 1 minutes. As with sags, swells are usually associated with system fault conditions, but they are not as common as voltage sags. One way that a swell can occur is from the temporary voltage rise on the unfaulted phases during an SLG fault. Figure 1.54 illustrates a voltage swell caused by an SLG fault. Swells can also be caused by switching off a large load or energizing a large capacitor bank.

Swells are characterized by their magnitude (rms value) and duration. The severity of a voltage swell during a fault condition is a function of the fault location, system impedance, and grounding. On an ungrounded system, with infinite zero-sequence impedance, the line-to-ground voltages on the ungrounded phases will be 1.73 pu during an SLG fault condition. Close to the substation on a grounded system, there will be little or no voltage rise on the unfaulted phases because the substation transformer is usually connected delta-wye, providing a low-impedance zero-sequence path for the fault current. Faults at different points along four-wire, multigrounded feeders will have varying degrees of voltage swells on the unfaulted phases. A 15% swell, like that shown in Figure 1.54, is common on U.S. utility feeders. The term *momentary overvoltage* is used by many writers as a synonym for the term *swell.*

1.5 Monitoring of Short Interruptions

1.5.1 Sag

The overview of the main aspects of voltage sags presented above shows that monitoring plays an absolutely crucial role in the assessment of voltage sags. It is through monitoring that the causes of sags can be discerned and their characteristics defined, so that the possible consequences are estimated. Monitoring of voltage sags is also the main source of data for calculating sag indices. As previously stated, sag monitoring is the central topic within the scope of this investigation. A general summary of the monitoring aspects most relevant to this chapter is given next. First, two types of power quality monitoring should be distinguished. One, on-site monitoring, is aimed at assessing the power quality at particular sites. Reasons for this type of monitoring include characterization of specific problems and verification of compliance of power quality contracts. Only one monitoring device installed at the site of interest is required. Another type, system-level monitoring, is intended to estimate the overall power quality of the entire system. This approach requires monitoring usually at a large number of sites and estimating the voltages at sites without monitors, i.e., power quality state estimation (PQSE).

Monitoring of voltage sags at the system level is the factual framework for this research and therefore only this approach will be discussed further. The main aspects that need to be considered when executing a system-wide sag monitoring campaign are monitor placement and monitor connection.

Monitor placement is the first aspect. Ideally, monitors would be placed at all locations throughout the system to completely understand the overall power quality. Such deployment of monitors, however, may be unaffordable and the challenges related to data management, analysis, and interpretation can be significant. Nevertheless, full coverage is usually not required since measurements from several strategic locations can be used to characterize the sag performance of the whole system. Hence, careful selection of monitoring locations based on monitoring objectives is of paramount importance. By recognizing the most important influencing factors and considering site categorization it should be possible to install monitors in a targeted way such that full coverage of the network is achieved. The sag data from monitored sites can then be mapped on to non-monitored sites and verified based on known fault statistics.

It was stated before that the main cause of voltage sags is short-circuit faults. With this in mind, the procedure to estimate sags at sites without monitors involves finding a fault location that produces voltages and currents that most closely match the measurements from the few available monitors.

Monitor connection is the second aspect. Voltage sag statistics in three-phase systems can be collected through monitors connected phase-to-phase or phase-to-ground. Power quality monitoring standards do not prescribe a monitor connection. However, a joint report from CIGRE and CIRED recommends the use of phase-to-phase voltages for system indices derived from monitoring HV and EHV sites because they give a better indication of the voltage experienced by the end-user equipment than phase-to-ground voltages. If monitors are connected phase-to-ground, various methods are used to estimate the phase-to-phase rms voltages from phase-to-neutral voltages.

1.5.2 Swell

A sag or swell is a decrease or increase in the rms value of the voltage ranging from a half cycle to a few seconds. The largest cause of problems from the utility side is voltage sags. Sags or swells can occur within a plant at the point of use and may be unrelated to the quality of power at the service entrance. These types of disturbances can lead to loss of production, etc., and the recorder being used should be able to capture these events. It should be noted that sags/swells can occur and be outside normal operating limits and not cause any problems. It is therefore important to know what levels of abnormal voltage and for how long specific equipment will tolerate it. In the early 1980s, the Computer Business Manufacturers Association (CBEMA, and now the ITI Council) established a susceptibility profile curve to aid manufacturers in the design of power supply protection circuits. This curve has since become a standard reference within the industry. Enetics recorders will plot these voltage events on the CBEMA/ITIC curve highlighting whether this is the cause of problems.

1.5.3 Influence of Equipment

1.5.3.1 Single Phase Tripping

Reclosers in general, and specifically reclosers with single-phase tripping capability, have the ability to improve distribution system reliability significantly not just on the distribution feeder but also in wind generation, cogeneration, various motor loads, oil wells, agricultural entities, small industrial sites, and at substations that feed moderate rural, residential, and commercial loads. The intelligence built into today's microprocessor-controlled reclosers allows the devices to dynamically operate on a single-phase or three-phase basis depending on the applications and conditions. Loop systems can further enhance the reliability of a power system through sectionalizing or removal of faulted line sections.

With an increased focus on service reliability, many utilities are looking at their distribution protection practices to determine if they can improve on the methods they have used historically for standard distribution protection. Although modern power distribution systems use single-phase protective devices such as fuses and reclosers liberally, these devices are typically found on laterals and taps off of the main three-phase feeder.

Utilities have been reluctant to consider single-pole tripping on the main three-phase line for a variety of reasons, including a desire to protect three-phase loads, difficulty coordinating devices along the feeder, and loss of sensitivity of the protective devices for low-magnitude faults. Each of these concerns is valid, but because of the obvious benefits to reliability brought about by single-pole tripping, utilities are looking for solutions to these problems to minimize any detrimental effects. In the past, utilities have achieved single-pole tripping by using three single-phase hydraulic reclosers grouped together. Because these devices were purely per-phase protection, they could not overcome many of the difficulties associated with single-pole tripping and their use was limited. By taking advantage of today's microprocessor-based recloser controls and the versatile reclosers available, solutions to these problems can be found. Over the past 5 years, Alabama Power Company has begun implementing single-pole tripping with microprocessor-based recloser controls on distribution feeder circuits. The goal for this initiative is to improve reliability and service to customers. By isolating the smallest portion of the system possible to clear a fault, Alabama Power can minimize the number of customers affected, and the improvement to reliability numbers will be significant. Additionally, customer satisfaction will improve as customers on unfaulted phases no longer experience momentary outages for faults on other phases.

1.5.3.2 Benefits of Single-Pole Tripping

SLG faults are the most common type of fault on distribution systems. Some studies show that only 2–3% of distribution faults are three-phase faults. For conservative estimates of reliability impact, Alabama Power has estimated that 60% of faults can be cleared with single-pole tripping, 25% involve two phases, and 15% require three-pole tripping to clear. When we consider phase selectivity for tripping to clear a fault, there are obvious benefits to isolating only the affected phase(s) over tripping a three-pole device for all faults. With three-phase tripping devices, the number of customers seeing an outage, whether momentary or sustained, is the same regardless of the type of fault. Single-phase tripping schemes, however, can open only the phase(s) involved in the fault and can reduce outage numbers by two-thirds for line-to-ground faults and one-third for faults involving two phases. Given that the vast majority of faults do not involve all three phases, single-pole tripping offers significant improvement to reliability numbers.

1.5.3.3 Single-Pole Tripping Concerns and Solutions

Although single-pole tripping holds obvious benefits over fixed three-pole tripping, there are some concerns that we must address to see the advantages without sacrificing protection. Some of these concerns are described in the following text.

1. Tripping/Reclosing Modes: With traditional independent reclosers used for single-pole tripping, each unit operates independently of the others for tripping and reclosing. Because of this fixed mode operation, their use is limited to locations that can always accommodate single-pole tripping. Microprocessor recloser controls that monitor all three phases and can direct tripping and closing from a central unit allow flexibility in the trip/close modes. Three basic operation modes are possible when we use the common control along with single-phase trip/close capable reclosers. (1) The first mode, Single-Pole Tripping/Single-Pole Lockout (SPTSPLO), allows tripping, reclosing, and lockout of each phase independently of each other. This mode results in the lowest outage impact for a single-phase fault. You cannot use this mode when sustained single phasing of three-phase loads or sustained load unbalance is unacceptable. (2) The second mode, Single-Pole Tripping/Three-Pole Lockout (SPT3PLO), allows for independent pole tripping and reclosing, but if a pole trips to lockout, the other poles open and lock out as well. Use this mode when the system cannot tolerate extended periods of unbalanced current resulting from an unbalanced lockout condition. (3) The final mode, Three-Pole Tripping/Three-Pole Lockout (3PT3PLO), allows the recloser and control to operate as a traditional three-phase recloser. In this mode, the recloser trips, recloses, and locks out all three phases as necessary to clear any type of fault. Use this mode when load levels prohibit any incremental load unbalance that a momentary single-phase interruption would cause and when you must avoid even short intervals of single-phasing

of a three-phase load. With a recloser control that offers flexible logic programming, variations and enhancements beyond these basic operating modes are possible. Additionally, with this ability, you can make the control adapt to system conditions. This allows the control to change the operation as necessary and maximize opportunities to take advantage of single-pole tripping, while avoiding problems when you require three-pole tripping only.

2. Ground-Fault Sensitivity: In the past, when individual single-phase hydraulic reclosers have been grouped together to achieve single-pole tripping for feeder faults, the trip value of the phase recloser limited the sensitivity for ground faults. The single-phase hydraulic recloser is an individual phase-sensing element, so load-carrying capability determines the sensitivity of the element. Often, the required sensitivity of the overcurrent element dictates the load ability of the feeder or the placement and type of the protective devices used. With three-phase protection, we typically use ground time-overcurrent elements operating from the residual of the phase currents to improve protection by increasing sensitivity. While individual hydraulic reclosers provide the desired phase selectivity because they do not offer the sensitivity of a ground-sensing protection element, we might be unable to use these where the required loading and restricted earth fault detection are in conflict.

   A microprocessor-based common control operating three independent tripping/closing recloser poles offers improved functionality. With a common protective device, the control can modify operation and protection according to conditions on all three phases. Additionally, the common control can calculate or measure a ground residual current value from the three phases. An overcurrent element operating on this value offers improved sensitivity for ground faults compared to the phase protection element. This protection is not possible with independent single-phase units. Providing this improved sensitivity offers a solution to the load ability versus sensitivity problem described previously. Typically, we can set the ground-overcurrent element as low as 10%–15% of the phase value. This is a considerable improvement in detecting low-magnitude ground faults. It is important to note that this setting depends on the presence of typical load unbalance.

3. Load Unbalance Following Single- or Two-Pole Trip: Figure 1.55 shows an elementary diagram of a feeder with balanced operation and Figure 1.56 shows the unbalanced operation resulting from a single pole open. Figure 1.57 illustrate the normal load currents on each phase. Figures 1.58 through 1.60 show the unbalance resulting from single-pole, double-pole and three-pole trips. Typically, there will be some load unbalance present, and this will show up as a low-level ground current. Normally, we can set ground elements to easily accommodate this normal unbalance and provide sensitivity for low-magnitude down-line faults. During times when one or two poles are open, however, even with only normal load current flowing in the closed poles, the unbalance of the circuit will appear as significant ground current. We must set ground protection locally at the single-pole tripping recloser and on any protective devices upline to accommodate this unbalance current. Depending on load levels, this ground current due to a pole-open condition can be large compared to the typical unbalance and protection sensitivity may have to be sacrificed to allow single-pole tripping. Since loading can vary substantially on a feeder depending on various factors such as season, time of day, weather, etc., we must consider the worst-case loading conditions when setting the ground protection.

4. Possible Solutions

   a. Raise Pickup: An obvious solution would be to raise the pickup setting of the ground element that detects the unbalance current. We would need to increase the pickup above the anticipated unbalance current, but we may be unable to do this where fault levels are lower than the possible current unbalance. If we use a fixed higher setting, we can accommodate the problem of the unbalance current, but the sensitivity advantage of the ground element is compromised or lost. We can take advantage of the logic capabilities of the microprocessor recloser control and configure it to adapt the overcurrent pickup to system conditions. With the recloser control metering loading values as well as monitoring status of each pole of the recloser, we can modify the ground-overcurrent setting as necessary in the control to avoid tripping for unbalance conditions. For operations where all poles are closed, this setting allows the overcurrent element sensitivity to be at a maximum. We can adjust it as necessary for changing conditions. To illustrate this scenario, consider the case of a feeder with a maximum expected phase current of 600 A primary. Typical phase-overcurrent settings would need to carry this loading along with a margin to allow for emergency loading conditions. If we consider a 25% margin acceptable, we would set the phase overcurrent to $600 \times 1.25 = 750$ A primary. With traditional per-phase protection, this setting would limit the sensitivity for ground faults to the 750 A pickup. If a ground-overcurrent element is available, we can set the pickup to accommodate a typical unbalance of 15% ($750 \times 15\% = 112.5$ A). This is a substantial increase in sensitivity for ground faults.

If one or two phases open either for a fault or other condition, we will see ground current in the recloser control resulting from the unbalance. The resulting unbalance current is the 3I0 current, which we can calculate by summing the phase currents. For the example given, considering the three-phase current angles to be separated by 120 degrees.

Because the 600 A ground current is more than five times the pickup of the ground element, the ground-overcurrent element would trip for this condition. Because the control has a status indication from each of the recloser poles, we can use these elements to supervise the ground overcurrent element. We would use this logic to prevent operation of the ground element if all three poles are not closed. 52aA 52aB 52aC 51GTC 51G Enable of Figure 1.61 depicts the ground enable logic. We must consider disabling the ground element for this condition when performing coordination studies. Any upline device must be set to coordinate with the downline device, considering that the ground curve is not active in the downline device. We would lose the overcurrent sensitivity advantage of the ground element for this condition. Because we can set the ground element significantly lower than the phase elements, there may be instances where we need tripping for faults that are below phase pickup. Logic functions were designed to allow the recloser to trip all three phases for faults that are below phase pickup but are detected by the ground element. With the ground element enabled (all three poles closed), unbalanced faults would typically have an asserted phase overcurrent element and ground-overcurrent element. For these instances, the particular phase involved would be tripped regardless of whether the particular phase element or the ground element timed out first. For low magnitude faults where the faulted phase cannot be determined (ground element timed out, no phase element picked up), we trip all three poles to ensure the fault is cleared. Figure 1.62 shows the phase trip selection logic for ground faults.

b. Coordination with Upline Three-Phase Devices: Coordination between a single-pole tripping recloser and an upline three-phase device can present problems. Often, the three-phase tripping device will use a maximum phase overcurrent element in which the time-overcurrent element uses the highest phase current that the relay detects. For multiphase faults, the upline maximum phase-overcurrent element will be timing against multiple single-phase overcurrent elements. The currents the elements detect on the involved phases may be different, so the maximum time overcurrent element upline may not be timing for corresponding currents with the single-phase downline devices involved. Once the first single-phase device trips downline, the maximum phase-overcurrent element may see a lower current corresponding to the second single-phase overcurrent element that might remain timing. Because the current that the three-phase device uses initially was higher prior to the first single-phase element tripping, the time overcurrent element in the three-phase device will have advanced further than for the current corresponding to the fault on the remaining phase and will in effect have a head start. This may lead to a potential mis-coordination between devices if sufficient margin does not exist between the overcurrent elements involved. Evolving faults present a similar problem. Consider the case where a fault evolves from a single-phase-to-ground to phase-to-phase-to-ground. The single-pole tripping device is using independent overcurrent elements for each phase, so the first overcurrent element corresponding to the initial faulted phase will begin timing along with the upline three-phase device.

Once the fault evolves, the second phase will begin timing from the reset position. The three-phase device, however, will continue timing without regard to another phase being involved. The fault current the three-phase device detects changes phases but does not go away when the first downline device operates, so the upline device continues to time with a head start of the time between onset of the initial fault and the time the second phase became involved. The worst-case scenario for this would be for the fault to evolve just prior to the first involved phase clearing. This would require that the upline overcurrent device be set at a minimum of two times the operate time of the downline device. Different fault current magnitudes between the phases could further complicate the situation just described. Using a microprocessor recloser control, we can design a solution to aid with both of these potential difficulties. Because the recloser control monitors currents on all three phases, we can design logic so that we can trip all phases in the process of timing when the first time-overcurrent element expires. Using this logic, the upline device only has to coordinate with the fastest operating element in the downline recloser. Operating in this manner does have the drawback of potentially affecting coordination with other downline devices, because the tripping times on additional phases may not correspond to the current seen on the respective phase. Figure 1.63 shows the multi-phase false trip selection.

c. Single-Pole Reclosing: Another complication to properly coordinating a maximum time-overcurrent element with a single-pole tripping device is reclosing. As discussed previously, because the upline device is seeing the highest current of the three phases while the single-pole tripping device times individually on each phase for potentially different magnitudes of current, we must take additional care to provide a sufficient margin. The onset of the fault may not have occurred simultaneously on the phases involved, so there can be a discrepancy in operating times of the individual phase overcurrents even for the same fault magnitudes. These are valid concerns for the initial trip of the event, but when reclosing is added to the coordination equation, the problem can be magnified as the reclosing sequence moves along. If the single-pole tripping device operates independently for both tripping and reclosing, the reclose counter at the single-pole tripping device can become unsynchronized between phases, resulting in numerous operations that the upline three-phase device must coordinate with. If the upline device is an electromechanical induction disk type or induction disk emulating microprocessor type, we must account for disk buildup or ratcheting that may occur during the reclose operation sequence.

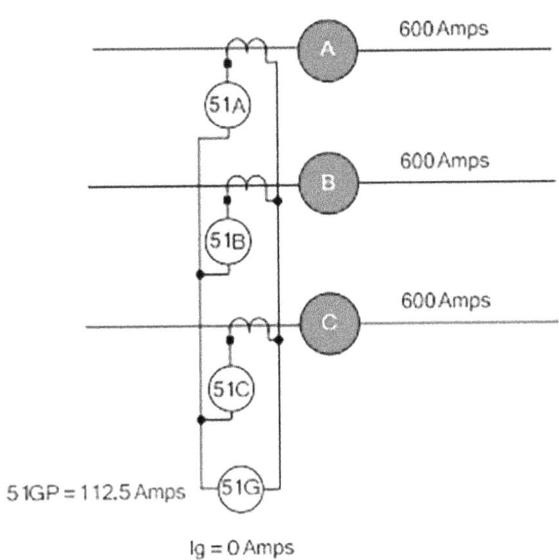

FIGURE 1.55 Balanced operation.

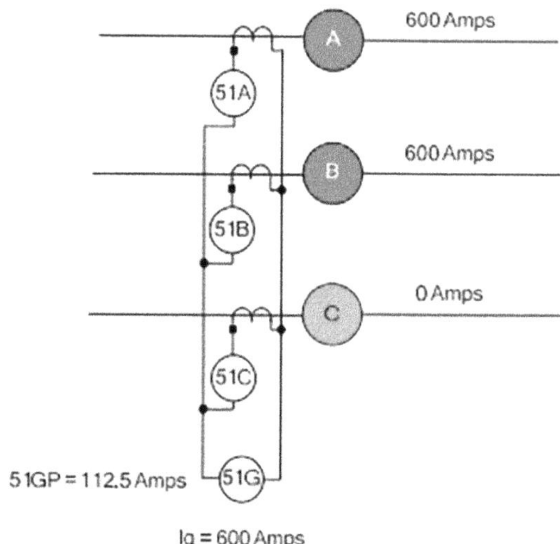

FIGURE 1.56 Unbalance resulting from a single-pole open.

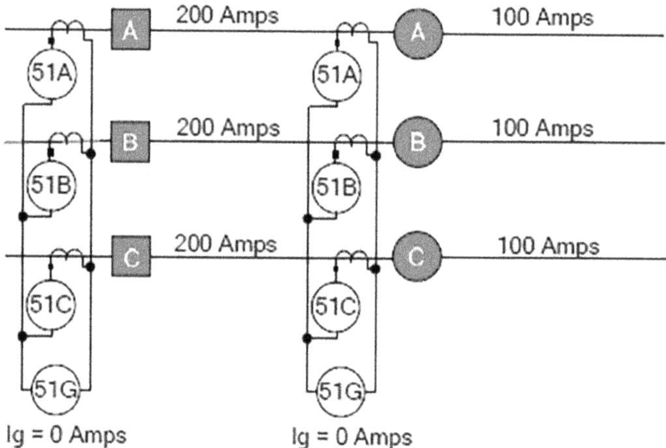

FIGURE 1.57 Typical feeder with balanced load.

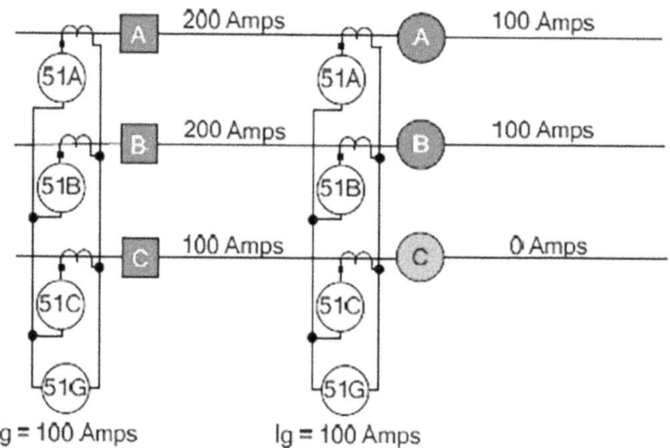

FIGURE 1.58 Unbalance resulting from single-pole trip.

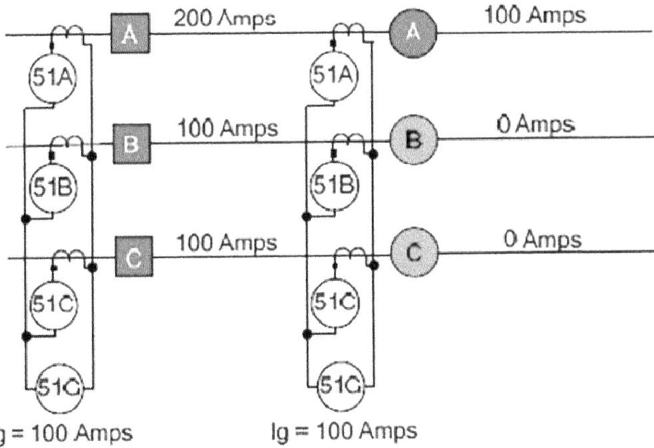

FIGURE 1.59 Unbalance resulting from two-pole trip.

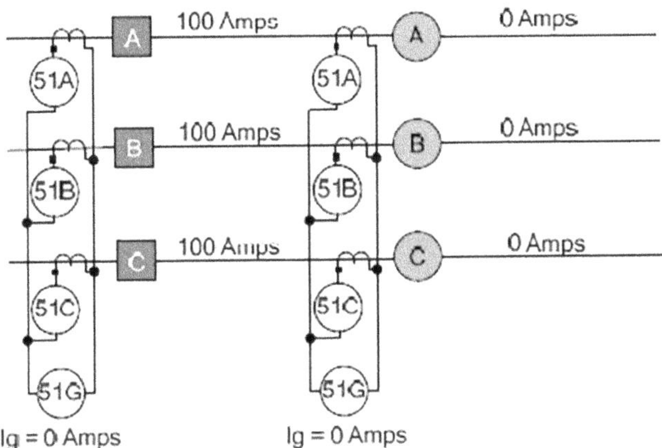

FIGURE 1.60 Unbalance from three-pole trip.

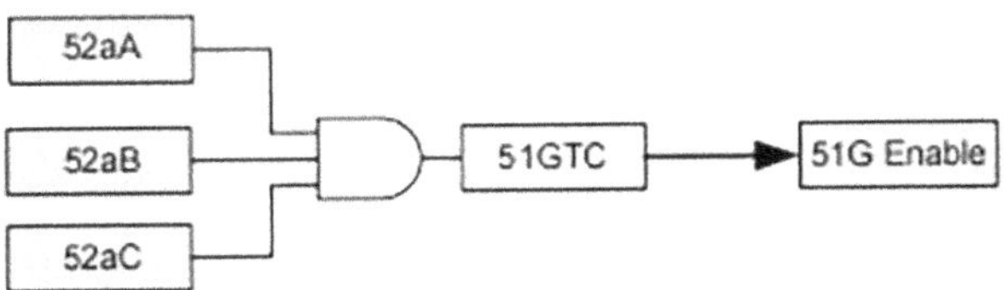

FIGURE 1.61 Ground-enabled logic.

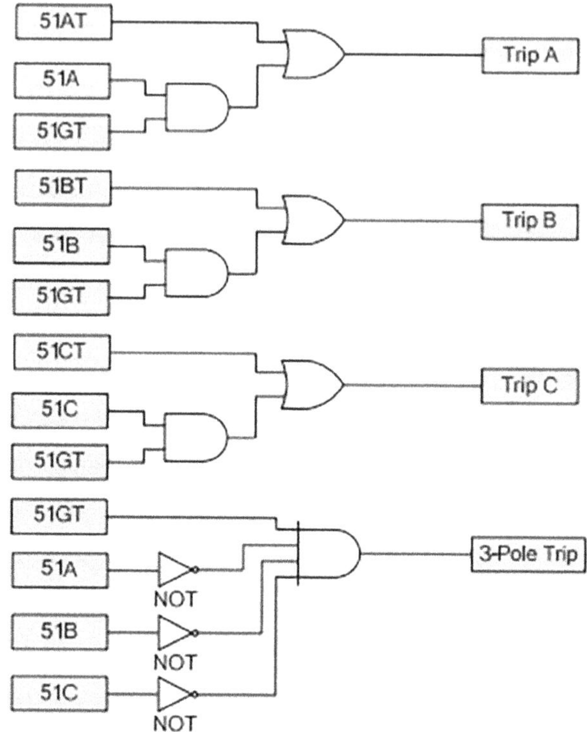

FIGURE 1.62 Phase-trip selection logic for ground faults.

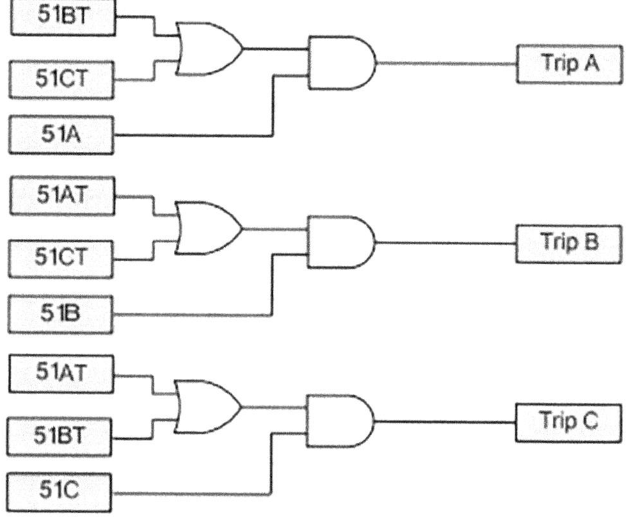

FIGURE 1.63 Multi-phase false trip selection.

1.6 Description of Long-Duration Power Quality Issues

1.6.1 Transients

Transients are short-duration, high-amplitude pulses superimposed on a normal voltage waveform. They can vary widely from twice the normal voltage to several thousand volts and last from less than a microsecond up to a few hundredths of a second. Transients can be classified as impulsive transients and oscillatory transients. Impulse transients are mainly caused by the impact of lightning strikes to the power system. The typical causes of oscillatory transients are capacitor or transformer energization and converter switching. While an impulsive transient is sudden and has non-power frequency change in voltage and current with a fast rise and decaying time, an oscillatory transient has one or more sinusoidal components with frequencies in the range from power frequency (50 Hz) to 500 kHz and decays in time.

1.6.2 Short-Duration Voltage Variations

Short Duration Voltage Variations are defined as the variations in the supply voltage for durations not exceeding 1 minute and caused by faults, energization of large loads that have large inrush currents, or rapidly varying large reactive power demands of the loads. These are further classified as voltage sags, voltage swells, and interruption.

1.6.3 Long-Duration Voltage Variations

Long-Duration Voltage Variations are defined as the rms variations in the supply voltage at fundamental frequency exceeding one minute, such as overvoltage, undervoltage, and sustained interruption. The causes of overvoltage (or undervoltage) may be the switching off (or on) of a large load having poor power factor, or the energization of a large capacitor bank or reactors.

1.6.4 Voltage Unbalance

Voltage Unbalance is the condition in which three phase voltages of the supply are not equal in magnitude and may not be equally displaced in time. The primary causes are the single-phase loads, open circuit in any one phase of a balanced 3-phase loads, and unequal loads connected in each phase of a polyphase system.

1.6.5 Waveform Distortion

Waveform Distortion is defined as steady-state deviation in the voltage or current waveform from an ideal sine wave. These distortions are classified as DC-offset, harmonics, and notching. The causes of DC offsets in power systems are geomagnetic disturbances, especially at higher altitudes and half-wave rectifications. These may increase the peak value of the flux in the transformer, pushing it into saturation and resulting in heating in the transformer. Power electronics equipment like UPS and adjustable-speed drives inject harmonics in the power systems. Notching is a periodic voltage distortion due to the operation of power converters when current commutates from one phase to another.

1.6.6 Voltage Fluctuations

Voltage Fluctuations are defined as rapid, systematic, and random variations in the supply voltage. This is also called as "Voltage Flicker" and is caused by rapid and large variations in current magnitude of loads having poor power factor such as arc furnaces. These large variations in load current cause severe dip in the supply voltage unless the supply bus is very stiff.

1.6.7 Power Frequency Variations

Power Frequency Variations are the variations that are caused by rapid changes in the load connected to the system, such as the operation of draglines connected to a comparatively low-inertia system. Since the frequency is directly related to rotational speeds of the generators, large variations in power frequency may reduce the life span of turbine blades on the shaft connected to the generator.

Although the above terms are not new, customer awareness on power quality has increased. In recent times, power quality issues and custom solutions have generated tremendous amount of interest among power system authorities and engineers. The International Electro Technical Commission (IEC) and Institute of Electrical and Electronics Engineers (IEEE) have proposed various standards on power quality. This led to more stringent regulations and limits imposed by electricity authorities although they differ from one country to another in a limited extend. Although terms of power quality are valid for transmission and distribution systems, their approach to power quality has different concerns. An engineer of transmission system deals with the control of active and reactive power flow in order to maximize both the loading capability and stability limits of the transmission system. On the other hand, an engineer of distribution system deals with load compensation either by means of individual or group compensation in order to maintain power quality for each load in the distribution system. The utilization of power electronic-based power conditioning devices brought the solution for these power quality issues in distribution system.

2

Mitigation Techniques

2.1 Introduction

In recent years, many multinational software companies and automobile industries have established their units in India. In turn, this initiates many other small industries to supply their needs. The growth of these industries is found to be very fast and it pollutes the power system by injecting harmonics into it. These industries need electrical power for their operation. Establishing a new power generation unit is not so easy in India due to the initial cost. In addition, it has many constraints like fuel, political, economical and technological requirements. This spurs researchers to think of an alternate solution for the scarcity of power by improving the quality of existing power. Reducing the wastages and improving the quality of available power is equivalent to generation of power. To improve the reliability and deliver energy at the lowest possible cost with improved power quality (PQ), power supply industries must increase flexibility in the transmission and distribution systems. The power industries are handling these challenges with the power electronics-based technology of flexible AC transmission systems (FACTS). This term covers the whole family of power electronic controllers, some of which may have achieved maturity within the industries, while some others are yet in the design stage. As Higorani et al. (1999) described, the various VSC-based FACTS controllers are available for PQ improvement.

FACT's has been defined by the IEEE as follows:

> "Power electronics-based system and other static equipment that provide control of one or more AC transmission system parameters to enhance controllability and increase power transfer capability."

In general, FACTs controllers can be classified as follows:

1. Series controllers
2. Shunt controllers
3. Combined series and shunt controllers
4. Combined shunt and series controllers

Based on the power electronic devices used in the controller, the FACTS controllers can be classified as:

1. Variable impedance-type FACTS controller
2. Voltage source converter (VSC)-based FACTS controller

The variable impedance-type controllers include:

1. Shunt connected—static var compensator (SVC)
2. Series connected—thyristor-controlled series capacitor or compensator (TCSC)
3. Combined shunt and series connected—thyristor-controlled phase-shifting transformer (TCPST) of static PST

The VSC-based FACTS controllers are:

1. Static synchronous compensator (STATCOM) (shunt connected)
2. Static synchronous series compensator (SSSC) (series connected)
3. Interline power flow controller (IPFC) (combined series-series)
4. Unified power flow controller (UPFC) (combined shunt-series)

The VSC-based FACTS controllers have several advantages over the variable impedance type. VSC-based STATCOM response is much faster than a variable impedance-type SVC. STATCOM requires less space than SVC for same rating. It can supply required reactive power even at low values of the bus voltage. In addition, a STATCOM can supply active power if it has an energy source or large energy storage at its DC terminals. It can also be designed to have in built, short-term overload capability. The only drawback with VSC-based controllers is that they require self-commutating power semiconductor switches such as gate turn-off (GTO) thyristors, insulated gate bipolar transistors (IGBT), and integrated gate commutated thyristors (IGCT). However, the VSC-based controllers built with emerging power semiconductor devices using silicon carbide technology will lead to the widespread use of VSC-based controllers in future.

Among FACTs controllers, the shunt controllers have shown feasibility in terms of cost effectiveness in a wide range of problem solving from transmission to distribution levels. For more than a decade, it has been recognized that the transmittable power through transmission lines could be increased and the voltage profile along the transmission line could be controlled by an appropriate amount of compensated reactive power. Moreover, the shunt controller can improve transient stability and can damp power oscillation during a post-fault event. Using a high-speed power converter, the shunt controller can further alleviate the flicker problem caused by electrical arc furnaces.

2.1.1 Series Controllers

Static SSSCs are series reactive power compensation devices used at the transmission level. The series compensation is obtained by controlling the equivalent impedance of a transmission line, to regulate the power flow through the line. The SSSC can be considered as a static synchronous generator that acts as a series compensator whose output voltage is fully controllable, independent of line current and kept in quadrature with it, with the aim of increasing or decreasing the voltage drop across the line, thus controlling the power flow.

The SSSC injects a voltage V_q in quadrature with line current. It can provide either capacitive compensation if V_q leads the line current by Π/2 rad or inductive compensation if V_q lags line current by Π/2 rad. A relatively small active power exchange is required to compensate for coupling transformer and switching losses, and maintain the required DC voltage.

2.1.2 Shunt Controllers—STATCOM

In principle, all shunt-type controllers inject additional current into the system at the point of common coupling (PCC). VSC uses charged capacitors as the input DC source and produces a 3Φ AC voltage output in synchronism and in phase with the AC systems. The converter is connected in shunt to a bus by means of the impedance of a coupling transformer. A control on the output voltage of this converter is either lower or higher than the connecting bus voltage and controls the reactive power drawn from or supplied to the connected bus. The impedance of the shunt controller, which is connected to the line, causes a variable current to flow and hence represents an injection of current into the line. As long as the injected current is in phase quadrature with the line voltage, the shunt controller can either supply or consume variable reactive power.

With a 6-pulse VSC with suitable controller, the phase angle and the magnitude of the AC voltage injected by the VSC can be controlled. The phase lock loop (PLL) ensures that the sinusoidal component of the injected voltage is synchronized (matching in frequency and required phase angle) with the AC bus voltage to which VSC is connected through a coupling inductor. Often, the leakage impedance of the interconnecting transformer serves as the coupling inductor. It also serves as harmonic filter for the voltage injected by the VSC. The injection of harmonic voltages can also be minimized by multi-pulse (12, 24 or 48) and/or multilevel convertors. At low power levels, the pulse width modulation (PWM) technique is sufficient to control the magnitude of the fundamental component of the injected voltage. The high-voltage IGBT devices can be switched at high frequency (2 kHz and above) of sinusoidal modulation, which enables the use of simple LC low-pass filters to reduce harmonic components.

2.1.3 Combined Shunt and Series Controllers

2.1.3.1 Unified Power Flow Controller

The UPFC is the most versatile FACTS controller for the regulation of voltage and power flow in a transmission line. It consists of two VSCs, with one connected in shunt and the other one connected in series. The shunt-connected converters work as STATCOM and control the reactive current injected into the line. Series-connected converters work as SSSC and control reactive voltage injected series with the line. The combination of these two converters enables the exchange of active power flow between the two converters. The series-connected converter can supply or absorb the active power.

The controllable power source on the DC side of the series-connected converter results in the control of both real and reactive power flow in the line at the receiving end of the line. The shunt-connected converter provides the required reactive power and injects the reactive current at the converter bus. Thus, a UPFC has three degrees of freedom whereas other FACTS controllers have only one degree of freedom or control variable. The concept of combining two or more converters can be extended to provide flexibility and additional degrees of freedom. A generalized UPFC refers to the use of three or more converters of which one is shunt connected while the remaining converters are series connected.

2.1.3.2 Interline Power Flow Controller

An IPFC refers to the configuration of two or more series-connected VSCs sharing a common DC bus. The IPFC is used in reactive (series) compensation of each individual line. In addition to this, the IPFC is capable of exchanging real power between the two or more compensated lines. To achieve this, the AC side of the series-connected VSCs are connected in different lines and on the DC side, while all the DC capacitors of individual converters are connected in parallel. This is possible because all the series converters are located inside the substation in close proximity.

An IPFC is similar to a UPFC in that the magnitude and phase angle of the injected voltage in the line (main system) can be controlled by exchanging real power with the second line (support system) in which a series converter is connected. The basic difference with a UPFC is that the support system in the UPFC is the shunt converter instead of a series converter. The series converter associated with the main system of one IPFC is termed as the master converter while the series converter associated with the support system is termed as the slave converter. The master converter controls both active and reactive voltage within limits while the slave converter controls the DC voltage across the capacitor and the reactive voltage magnitude.

2.2 Application of FACTS Controllers in Distribution Systems

Although the concept of FACTS was developed originally for transmission networks, later on this has been extended since last decade for improvement of PQ in distribution systems operating at low or medium voltages. In the early days, the PQ referred primarily to the uninterrupted power supply at acceptable voltage and frequency. In the modern context, a PQ problem is defined as any problem manifested in voltage, current, or frequency deviations that result in failure or malfunctioning of customer equipment. However, the increase in the use of computers, microprocessors, and power electronic systems has resulted in PQ issues involving transient disturbances in voltage magnitude, waveform, and frequency. The nonlinear loads not only cause PQ problems but also are very sensitive to the voltage deviations. The unbalanced load in the distribution systems like single-phase railway loading creates PQ problem at the distribution level. The highly inductive loads like arc furnaces are a major source of PQ problems in distribution networks.

2.3 Introduction to Long-Duration Voltage Variations

Electrical appliances become more complex due to complicated functions and mutual interactions. There are many advantages of automated manufactory systems and variable-speed drives (VSDs) in industry, as well as information systems and fluorescent lights at the public and domestic consumers. Therefore, development of these systems and equipment is very fast and their usage is growing. Most electrical equipment nowadays is more sensitive to deviations from sinusoidal supply voltage. At the same time, the same or other equipment causes modifications to the characteristics of the supply voltage.

2.3.1 Observation of System Performance

Long-duration overvoltages are close cousins to voltage swells, except they last longer. Like voltage swells, they are rms voltage variations that exceed 110% of the nominal voltage. Unlike swells, they last longer than a minute.

Several types of initiating events cause overvoltages. The major cause of overvoltages is capacitor switching. This is because a capacitor is a charging device. When a capacitor is switched on, it adds voltage to the utility's system. Another cause of overvoltage is the dropping of load. Light load conditions in the evening also cause overvoltages on high voltage systems. Another common cause of overvoltage is the mis-setting of voltage taps on transformers. Extended overvoltages shorten the life of lighting filaments and motors. Solutions to overvoltages include using inductors during light load conditions and correctly setting transformer taps. Figure 2.1 shows a plot of overvoltage versus time.

2.3.2 Principle of Regulating Voltage

A voltage regulator is a voltage stabilizer that is designed to automatically stabilize a constant voltage level. A voltage regulator circuit is also used to change or stabilize the voltage level according to the necessity of the circuit. Thus, a voltage regulator is used for two reasons:

- To regulate or vary the output voltage of the circuit.
- To keep the output voltage constant at the desired value in-spite of variations in the supply voltage or in the load current.

Voltage regulators find their applications in computers, alternators, and power generator plants where the circuit is used to control the output of the plant. Voltage regulators may be classified as electromechanical or electronic. They can also be classified as AC regulators or DC regulators.

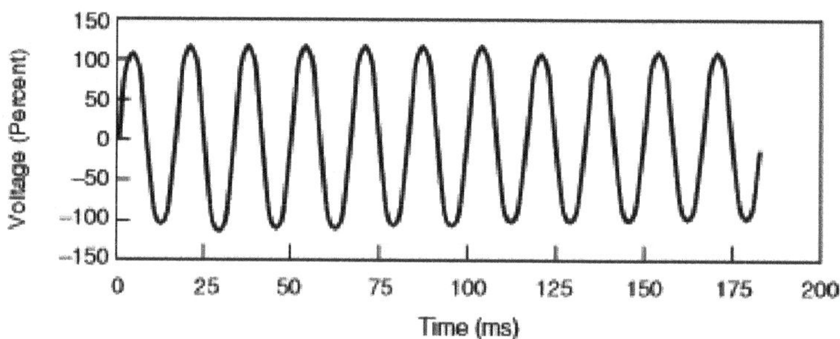

FIGURE 2.1 Voltage swell plot.

2.4 Devices for Voltage Regulation

2.4.1 Electronic Voltage Regulator

All electronic voltage regulators will have a stable voltage reference source that is provided by the reverse breakdown voltage operating diode called Zener diode. The main reason to use a voltage regulator is to maintain a constant DC output voltage. It also blocks the AC ripple voltage that cannot be blocked by the filter. A good voltage regulator may also include additional circuits for protection against problems like short circuits, current limiting circuit, thermal shutdown, and overvoltage. Electronic voltage regulators are designed with any combination of the three regulators given below.

2.4.2 Zener-Controlled Transistor Voltage Regulator

A Zener-controlled voltage regulator is used when the efficiency of a regulated power supply becomes very low due to high current. There are two kinds of Zener-controlled transistor voltage regulators.

2.4.3 Zener-Controlled Transistor Series Voltage Regulator

Such a circuit is also named an emitter-follower voltage regulator. It is so called because the transistor used is connected in an emitter-follower configuration. The circuit consists of an N-P-N transistor and a Zener diode. As shown in Figure 2.2, the collector and emitter terminals of the transistor are in series with the load. Thus, this regulator has the name series in it. The transistor used is a series-pass transistor.

The output of the rectifier that is filtered is then given to the input terminals and regulated output voltage V_{load} is obtained across the load resistor R load. The reference voltage is provided by the Zener diode and the transistor acts as a variable resistor, whose resistance varies with the operating conditions of base current, I_{base}.

The main principle behind the working of such a regulator is that a large proportion of the change in supply or input voltage appears across the transistor and thus the output voltage tends to remain constant.

The output voltage can thus be written as

$$V_{out} = V_{zener} - V_{be} \tag{2.1}$$

The transistor base voltage V_{base} and the Zener diode voltage V_{zener} are equal and thus the value of V_{base} remains almost constant.

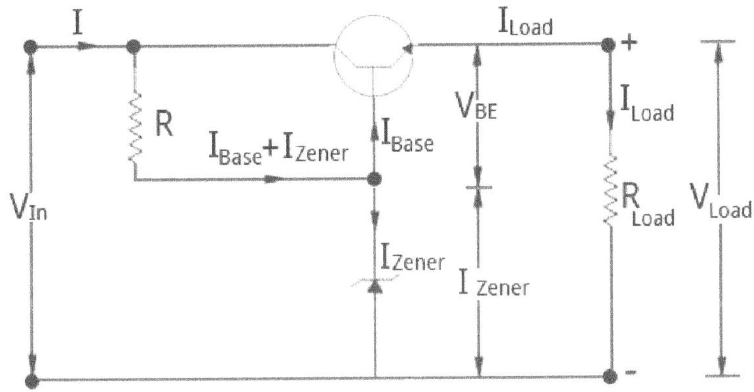

FIGURE 2.2 Zener-controlled transistor series voltage regulator.

2.4.3.1 Operation

When the input supply voltage V_{in} increases the output voltage V_{load} also increases. This increase in V_{load} will cause a reduced voltage of the transistor base emitter voltage V_{be} as the Zener voltage V_{zener} is constant. This reduction in V_{be} causes a decrease in the level of conduction, which will further increase the collector-emitter resistance of the transistor and thus cause an increase in the transistor collector-emitter voltage and all of this causes the output voltage V_{out} to reduce. Thus, the output voltage remains constant. The operation is similar when the input supply voltage decreases.

The next condition would be the effect of the output load change in regard to the output voltage. Let us consider a case where the current is increased by the decrease in load resistance R_L. This causes a decrease in the value of output voltage and thus causes the transistor base emitter voltage to increase. This causes the collector emitter resistance value to decrease due to an increase in the conduction level of the transistor. This causes the input current to increase slightly and thus compensates for the decrease in the load resistance R_L.

The biggest advantage of this circuit is that the changes in the Zener current are reduced by a factor β and thus the Zener effect is greatly reduced and a much more stabilized output is obtained.

The output voltage of the series regulator is $V_{out} = V_{zener} - V_{be}$. The load current I_{load} of the circuit will be the maximum emitter current that the transistor can pass. For a normal transistor like the 2N3055, the load current can go up to 15 A. If the load current is zero or has no value, then the current drawn from the supply can be written as $I_{zener} + I_{c(min)}$. Such an emitter follower voltage regulator is more efficient than a normal Zener regulator. A normal Zener regulator that has only a resistor and a Zener diode has to supply the base current of the transistor.

2.4.3.2 Limitations

The limitations listed below has proved the use of this series voltage regulator only suitable for low output voltages. With the increase in room temperature, the values of V_{be} and V_{zener} tend to decrease. Thus, the output voltage cannot be maintained a constant. This will further increase the transistor base emitter voltage and thus the load.

1. There is no option to change the output voltage in the circuit.
2. Due to the small amplification process provided by only one transistor, the circuit cannot provide good regulation at high currents.
3. When compared to other regulators, this regulator has poor regulation and ripple suppression with respect to input variations.
4. The power dissipation of a pass transistor is large because it is equal to V_{cc} I_c and almost all variation appears at V_{ce} and the load current is approximately equal to collector current. Thus, for heavy load currents pass transistor has to dissipate a lot of power and, therefore, becomes hot.

2.4.4 Zener-Controlled Transistor Shunt Voltage Regulator

Figure 2.3 shows the circuit diagram of a shunt voltage regulator. The circuit consists of an NPN transistor and a Zener diode along with a series resistor R_{series} that is connected in series with the input supply. The Zener diode is connected across the base and the collector of the transistor, which is connected across the output.

2.4.4.1 Operation

As there is a voltage drop in the series resistance R_{series} the unregulated voltage is also decreased along with it. The amount of voltage drop depends on the current supplied to the load R_L. The value of the voltage across the load depends on the Zener diode and the transistor base emitter voltage V_{be}.

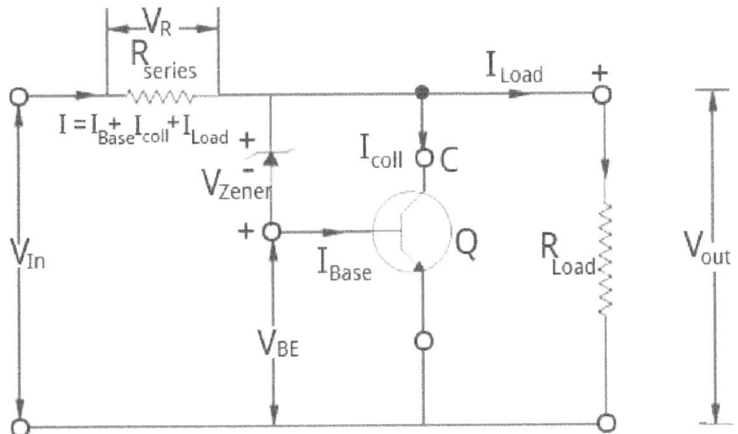

FIGURE 2.3 Zener-controlled transistor shunt voltage regulator.

Thus, the output voltage can be written as

$$V_{\text{out}} = V_{\text{zener}} + V_{\text{be}} = V_{\text{in}} - I \cdot R_{\text{series}}$$

The output remains nearly a constant as the values of V_{zener} and V_{be} are nearly constant. This condition is explained below.

When the supply voltage increases, the output voltage and base emitter voltage of transistor increases and thus increases the base current I_{base} and therefore causes an increase in the collector current I_{coll} ($I_{coll} = \beta \times I_{base}$).

Thus, the supply voltage increases causing an increase in supply current, which in turn causes a voltage drop i the series resistance R_{series} and thereby decreases the output voltage. This decrease will be more than enough to compensate for the initial increase in output voltage. Thus, the output remains nearly a constant. The working explained above happens in reverse if the supply voltage decreases.

When the load resistance R_L decreases, the load current I_L increases due to the decrease in currents through base and collector I_b and I_{coll}. Thus, there will not be any voltage drop across R_{series} and the input current remains constant. Thus, the output voltage will remain constant and will be the difference of the supply voltage and the voltage drop in the series resistance. It happens in reverse if there is an increase in load resistance.

2.4.4.2 Limitations

1. The series resistor causes a huge amount of power loss.
2. The supply current flow will be more through the transistor than it is through the load.
3. The circuit may have problems regarding overvoltage mishaps.

2.4.5 Discrete Transistor Voltage Regulator

Discrete transistor voltage regulators can be classified into two. They are explained below. These two circuits are able to produce a regulated output DC voltage that is regulated or maintained at a predetermined value even if the input voltage varies or the load connected to the output terminal changes.

1. Discrete Transistor Series Voltage Regulator

   The block diagram of a discrete transistor type voltage regulator is given Figure 2.4. A control element is placed to collect the unregulated input, which controls the magnitude of the input

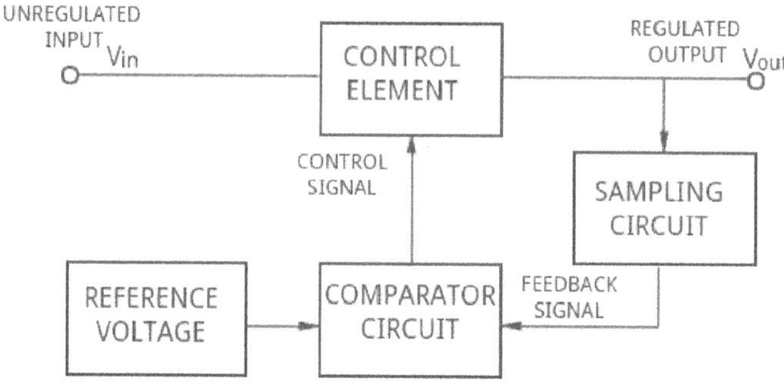

FIGURE 2.4 Discrete transistor series voltage regulator.

voltage and passes it to the output. The output voltage is then fed back to a sampling circuit and then compared with a reference voltage and sent back to the output. Thus, if the output voltage tends to increase, the comparator circuit provides a control signal to cause the control element to reduce the magnitude of the output voltage by passing it through the sampling circuit and comparing it, thereby maintaining a constant and steady output voltage. Suppose the output voltage tends to decrease, the comparator circuit provides a control signal that causes the series control element to increase the magnitude of output voltage, thus maintaining the steadiness.

2. Discrete Transistor Shunt Voltage Regulator

The block diagram of a discrete transistor shunt voltage regulator is given in Figure 2.5. As the name says the voltage regulation is provided by shunting the current away from the load. The control element shunts a part of the current that is produced as a result of the input unregulated voltage that is given to the load. Thus, the voltage is regulated across the load. Due to the change in load, if there is a change in the output voltage, it will be corrected by giving a feedback signal to the comparator circuit, which compares with a reference voltage and gives the output control signal to the control element to correct the magnitude of the signal required to shunt the current away from the load.

If the output voltage increases, the shunt current increases and thus produces less load current and maintains a regulated output voltage. If the output voltage reduces, the shunt current reduces and thus produces more load current and maintains a regulated constant output voltage. In both cases, the sampling circuit, comparator circuit, and control element play an important role.

2.4.5.1 Limitations of Transistor Voltage Regulators

The steady and stabilized output voltage that is obtained from the regulator is limited to a voltage range of (30–40) V. This is because of the small value of the maximum collector emitter voltage of transistor (50 V). This puts a limit to the use of transistorized power supplies.

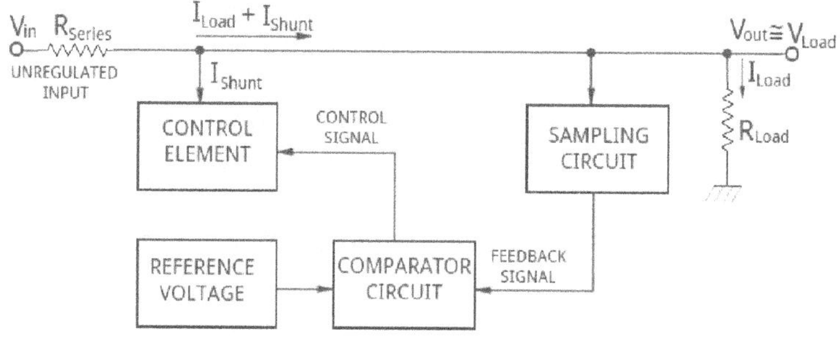

FIGURE 2.5 Discrete transistor shunt voltage regulator.

2.4.6 Electromechanical Regulator

As the name implies it is a regulator with the combination of electrical and mechanical characteristics. The voltage regulation process is carried out by the coiled sensing wire to perform as an electromagnet. A magnetic field is produced by the solenoid according to the current that passes through it. This magnetic field attracts a moving ferrous core material that is connected to a spring tension or gravitational pull. When the voltage increases, the current strengthen the magnetic field, so the core is attracted towards the solenoid. The magnet is physically connected to a mechanical switch. When the voltage decreases, the magnetic field produced by the core reduces, so the spring tension causes the core to retract. This closes the mechanical switch and allows the power to flow.

If the mechanical regulator design is sensitive to small voltages variations, a selector switch can be added to the solenoid across the range of the resistances or transformer winding to gradually step the output voltage up and down, or to rotate the position of a moving coil AC regulator.

Earlier automobile generators and alternators contained mechanical regulators. In these kinds of regulators, the process is carried out by one, two, or three relays and various resistors to establish the generator output slightly more than 6 or 12 V; this process is independent of the rpm of the engine or the load varying on the vehicle's electrical system. The relays are used to carry out the PWM to regulate the output of the generator and control the field current passing through the generator.

The regulator used for DC generators will disconnect from the generator when it was not working, to prevent the reverse flow of electricity from the battery to the generator. Otherwise it will work as a motor.

2.4.7 Automatic Voltage Regulator

This active system regulator is mostly used to regulate the voltage output of very huge generators that are commonly used in ships, oil rigs, big buildings, and so on. The circuitry of an automatic voltage regulator (AVR) is complex and consists of all active and passive elements along with microcontrollers. The basic principle of the AVR is the same as an ordinary voltage regulator. The input voltage of the generator exciter is controlled by the AVR and when the generator voltage is increased or decreased, the output voltage of the generator automatically increases or decreases. There will be a predetermined set point on which the AVR sorts out the amount of voltage that is to be passed on to the exciter every millisecond. Thus, the output voltage is regulated. The same operation becomes more complex when only one AVR is used to regulate multi generators that are connected in parallel.

2.4.8 Constant Voltage Transformer

In some cases, a constant voltage transformer (CVT) is also used as a voltage regulator. A CVT consists of a high-voltage resonant winding and a capacitor that produces a regulated output voltage for any type of input varying current. Like a regular transformer, the CVT has a primary and a secondary. The primary is on the side of a magnetic shunt and the secondary is on the opposite side with a tuned coil circuit. The regulation is maintained by the magnetic saturation in the secondary coils.

2.4.9 Utility Voltage Regulator Application

Step-voltage regulators applied in the utility's distribution systems are generally medium-voltage mechanical AVR. It should be noted that there are two distinct types of AC AVRs: medium-voltage (mechanical) and low-voltage (mechanical or electronic). The difference in their operation and design clearly demonstrates that their applications are not the same. The latter is intended to protect end-user devices from overvoltage and undervoltage conditions. Nonetheless, this post will focus on the medium-voltage regulators, which are primarily used by the electric utility to compensate for the voltage drop in the feeders or distribution systems. In addition, the term step-voltage regulator is often used to refer to utility AVR, which is shown in Figure 2.6.

FIGURE 2.6 Step-voltage regulators for utility applications.

2.5 Step-Voltage Regulator Basic Operation

The step-voltage regulator is basically a transformer that has its high-voltage winding (shunt) and low-voltage winding (series) connected to either aid or oppose their respective voltages. Subsequently, the output voltage could be the sum or difference between the winding voltages. For example, if the transformer has a turns ratio of 10:1 with 1000 V applied in the primary, then the secondary voltage will be 100 V. Adding or subtracting by using the connection mentioned above, the output voltage would be 1100 V or 900 V, respectively, as shown in Figures 2.7 and 2.8. Thus, the transformer becomes an autotransformer with the capability to boost (raise/step-up) or buck (lower/step-down) the system voltage by 10%.

In other words, by switching the location of the physical connection from the shunt to the series winding (reversing switch) and with the turns ratio made variable through automatic tap-changing, the system voltage is adjusted to the required level, which is shown in Figure 2.9. This is made possible since the AVR includes microprocessor-based and/or mechanical controls that tell the unit when and how to change taps. Moreover, modern controllers are equipped with data acquisition and communication capabilities for remote applications. Utility step-voltage regulators usually allow a maximum voltage regulation range of ±10% of the incoming line voltage in 32 steps of 5/8% or 0.625%. That makes 16 steps each for buck and boost—5/8% × 16 steps = 10%. Utility AVRs can be installed out on the feeders or at the substation bus. The voltage regulator units could either be single-phase or three-phase. However, on

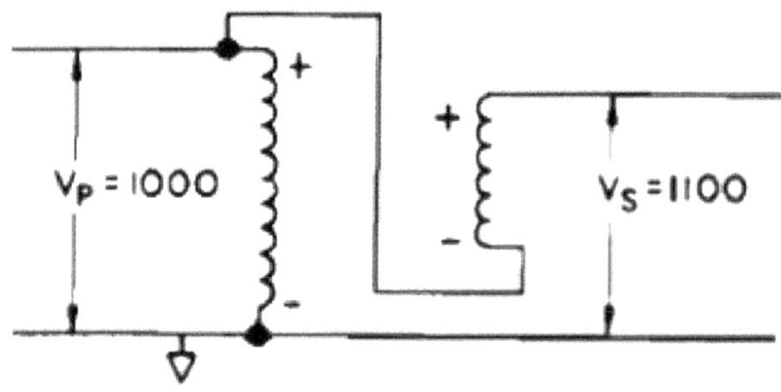

FIGURE 2.7 Step-up autotransformer (boost).

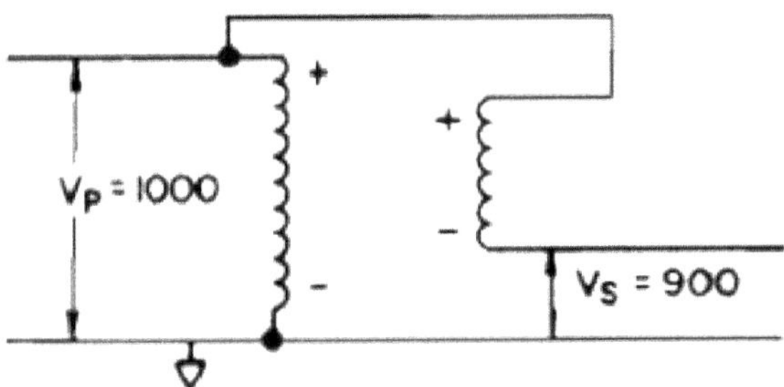

FIGURE 2.8 Step-down autotransformer (buck).

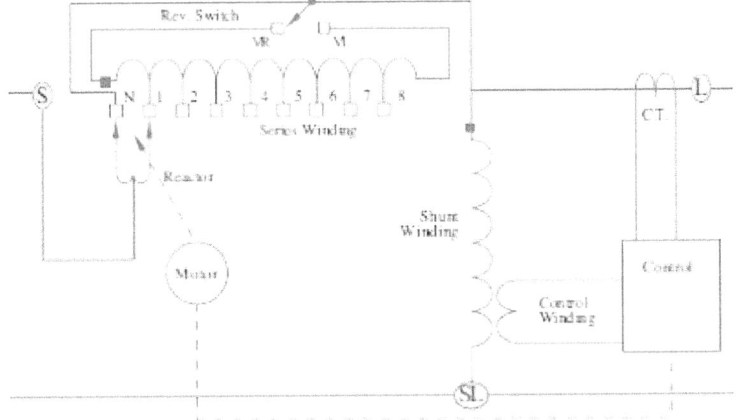

FIGURE 2.9 Step voltage regulator schematic diagram.

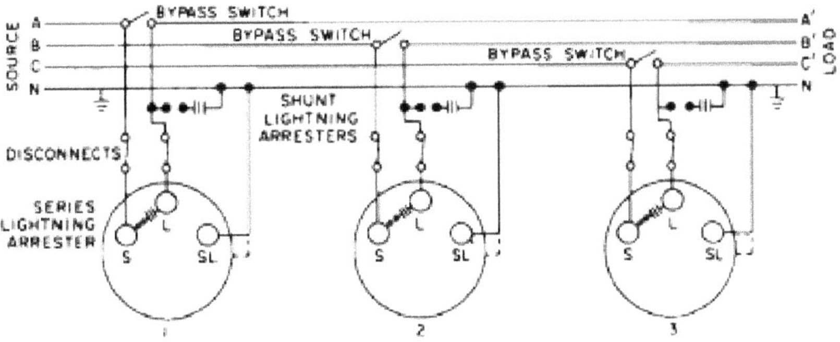

FIGURE 2.10 Voltage regulator grounded-wye connection.

a three-phase feeder, it is more common in utility applications to use single-phase units connected in banks of three (e.g., wye-grounded, closed delta shown in Figures 2.10 and 2.11). This is because utility distribution lines are typically unbalanced in their construction, added with single-phase loads that create significant unbalance in the line currents.

Thus, three independently controlled regulators may very well yield better balance between the phase voltages than a single three-phase unit or ganged operation. Also, there are many installations of open-delta regulator banks on lightly loaded three-phase feeders, which require only two regulators and are less costly than a full three-phase bank.

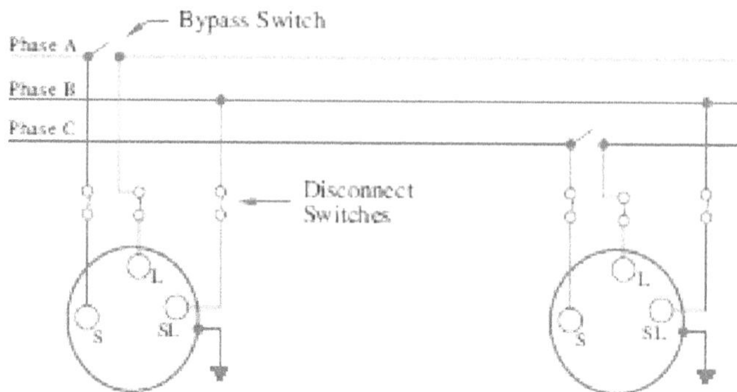

FIGURE 2.11 Voltage regulator closed-delta connection.

2.5.1 Voltage Regulator Applications

Step-voltage regulators are typically installed on the following:

1. Existing feeders—before the point where the voltage drop problem starts with heavy load
2. Important laterals
3. To serve a remotely located load

Voltage regulators in the utility distribution systems are relatively slow. These AVRs have a time delay of at least 15 s. Therefore, they are not suitable to applications where voltages may vary in cycles or seconds. Utility step-voltage regulators are primarily used for boosting voltage on long feeders where the load is changing slowly over several minutes or hours. The voltage band typically ranges from 1.5 to 3.0 V on a 120-V base. The control can be set to maintain voltage at some point down the line from the feeder by using the line drop compensator capability. This results in a more level average voltage response and helps prevent overvoltages on customers near the regulator.

2.5.2 Voltage Regulator Sizing and Connection

These are the basic steps in determining the size and connection type of the voltage regulator for utility applications:

1. Determine the system configuration (i.e., three-phase, 4-wire multi grounded wye or 3-phase, 3-wire delta). This will be the basis for AVR connection type.
2. Establish the amount of voltage regulation needed (e.g., ±5%, ±10%).
3. Determine the system phase voltage on which the AVRs will be connected. Remember that the phase voltage is affected by the system configuration (1).
4. Calculate the maximum load current of the feeder or line.
5. Multiply the voltage regulation (2), system phase voltage (3), and maximum line current (4) to get the required kVA size of the AVR.

(For example, compute for the step-voltage regulator size needed by a three-phase, 4-wire multi-grounded feeder with a system voltage of 13800Y/7970 V. The required voltage regulation is 10% and the peak connected load is 6.0 MVA.)

1. System configuration is three-phase, 4-wire, multi grounded wye—meaning that the voltage regulators shall be connected grounded wye.
2. Voltage regulation = 10%.
3. Phase voltage is the line-to-neutral voltage = 7.97 kV (since it is a 4-wire multi grounded wye feeder).

4. Load current = 6.0 MVA/(1.732 × 13.8 kV) = 251 A.
5. Voltage Regulator kVA Size = 10% × 7.97 kV × 251 A = 200 kVA.

Use three 32-step voltage regulators, each with a standard rating of 250 kVA, 7970 V, ± 10% regulation.

2.5.3 Capacitor Selection Is Key to Good Voltage Regulator Design

Modular DC–DC switching voltage converters (or voltage regulators) are fully integrated devices that take away most of the complexity of power supply design—but not all. One of the key areas that are still left to the design engineer's discretion is the choice of components for, and layout of, the energy storage and filtering circuits. In principle, these look like simple circuits comprising a few resistors, capacitors, and the energy-storage element usually an inductor. However, this simplicity belies the fact that the vast majority of problems associated with switching regulators have nothing to do with the module itself, rather the improper use of capacitors in the input filter and energy storage and filtering circuitry. Choosing the wrong type of capacitor, getting the required capacitance just slightly incorrect, or misplacing the passive device can cause an otherwise perfectly functioning voltage regulator module to generate excessive electromagnetic interference (EMI). In a worse-case scenario, poor capacitor selection can result in a good voltage regulator becoming unstable and failing prematurely.

2.5.4 Dealing with EMI

Switching voltage regulators have become popular because they are efficient and flexible. However, there is a trade-off. The chips are more expensive, take up more space, have a relatively slow transient response to load variances, and, because of the switching operation, it can be difficult to prevent EMI from the switching circuit radiating out into both input and output sides of the device.

The best way to tackle EMI is to target the source, which in a switching regulator is typically the power FET, particularly when it turns off. Figure 2.12 shows a basic step-down ("buck") regulator. When the power FET ("Q1") is on, current flows around the circuit as indicated by the arrows. When

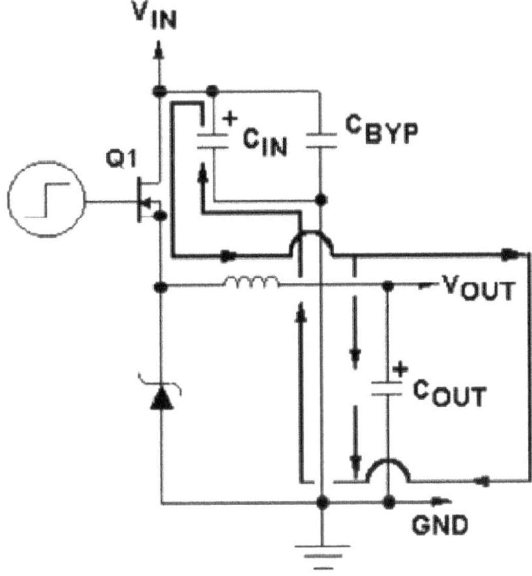

FIGURE 2.12 Basic buck-switching voltage regulator circuit showing current flow when Q1 is ON.

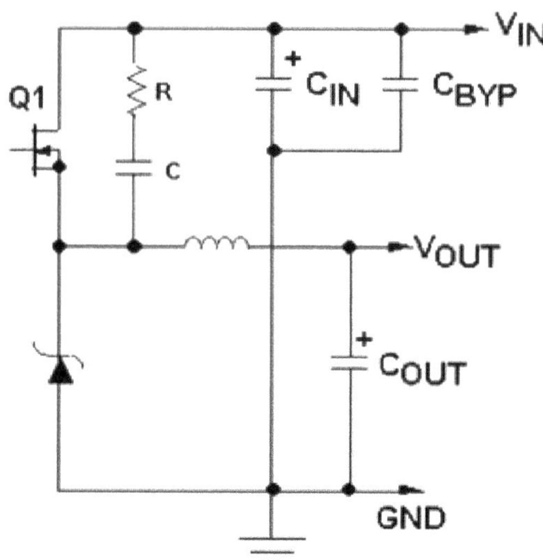

FIGURE 2.13 R-C snubber circuit helps to reduce EMI radiating from power FET ("Q1").

Q1 is turned OFF, the voltage on the end of the inductor closest to Q1 is forced to swing negative until the diode responds by turning on to hold the current through the inductor constant. However, because the diode cannot switch on in zero time, the voltage can overshoot beyond the diode drop and then ripple (or "ring") as the diode eventually turns on. Even if a Schottky diode is used, parasitic trace inductance will still cause some rippling to occur. This rippling generates EMI in the 20–100 MHz range. A proven way to reduce EMI caused by ringing is to add an R-C "snubber," comprising a ceramic capacitor plus carbon film resistor, close to the FET as shown in Figure 2.13. The snubber works because when the FET switches off and stops sourcing current, the snubber capacitor sources enough current such that the rate of current change through the inductor is slowed, lowering the incidence of rippling in the diode. Providing the component values of the snubber circuit are appropriately selected, they will damp any rippling that does occur.

The R-C snubber circuit can reduce overall voltage regulator efficiency by a few percentage points. The loss of efficiency shows up most at light loads because the power expended in the snubber is relatively constant and independent of output-load current. (Some newer switching voltage regulator modules integrate the R-C snubber into the chip, saving the design engineer the task of coming up with a suitable circuit shown in Figure 2.13.)

The value of the input capacitance is not the main consideration when selecting this capacitor; rather it is the RMS current and the voltage rating. The RMS current of the input capacitor (CIN) can be calculated from:

$$I_{CIN,RMS} = I_{OUT} \times \sqrt{(D - D^2)} \tag{2.2}$$

where D is the PWM square-wave duty cycle.

The worst case occurs at $D = 50\%$ (i.e., $V_{IN} = 2 \times V_{OUT}$), which yields $I_{IN,\ RMS} = I_{OUT}/2$.

The Equivalent Series Resistance (ESR) of the bulk-input capacitor causes the device to heat up under this RMS current. Higher ESR increases heating, so a good place to start is to specify a device with as low an ESR as possible within the constraints of the budget.

Ceramic and tantalum capacitors are both suitable as input capacitors for switching voltage regulator circuits. Choose ceramic capacitors with a voltage rating of at least 1.5 times the maximum-input voltage. If tantalum capacitors are selected, they should be chosen with a voltage rating of at least twice the maximum-input voltage.

A small ceramic capacitor in parallel to the bulk capacitor is recommended for high-frequency decoupling.

2.5.5 The L-C Output Filter

Perhaps the most important capacitor choice a power supply design engineer can make is the selection of the component for the voltage regulator's L-C output filter. The job of this circuit is to filter the voltage square wave at its input (the switching node) to produce a constant regulated voltage at its output (V_{OUT}).

In a buck converter under steady-state conditions, the average current in the inductor (I_L) is equal to the output current I_{OUT}. Due to the input being a square wave, the inductor current is not constant but ripples between a maximum and a minimum value as the input voltage switches on and off. The difference between the maximum and minimum (ΔI_L) is called the peak-to-peak inductor current ripple (Figure 2.14). In the L-C output filter, the role of the capacitor is to maintain a constant output voltage (V_{OUT}) and limit any voltage spikes

The design engineer should specify the maximum current and voltage ripple for the circuit depending on the application. The inductor is typically selected to keep the ripple current to less than 20 to 30% of the rated DC current. Sensitive modern silicon circuits require tighter control of voltage ripple typically in a 5–100 mV range. Figure 2.15 illustrates acceptable voltage ripple profiles for a buck regulator delivering a 2.0 V output.

Selecting the best capacitor for a switching voltage regulator's output filter is not a trivial task. However, a good starting point is to estimate the maximum ESR and minimum capacitance for a given output voltage ripple. The ESR can be calculated from the formula:

$$\text{ESR} = \frac{\Delta V_{\text{OUT}}}{\Delta I_L} \tag{2.3}$$

And the minimum output capacitance (C_{OUT}) can be estimated from the following equation:

$$C_{OUT} = \frac{L \times (I_{L,\text{max}})^2}{(V_{\text{OUT}} + \Delta V_{\text{OUT,overshoot}})^2 = V_{\text{OUT}^2}} \tag{2.4}$$

where $\Delta V_{\text{OUT, overshoot}}$ is the maximum voltage overshoot allowed on the output and I_L, max is the maximum inductor current.

However, there is a little more to the output capacitor selection than that. Even a device with an apparently correct ESR can suffer overheating if the input "ripple current rating" is not taken into consideration. Prolonged exposure to high temperatures can cause a gradual evaporation of the device's liquid

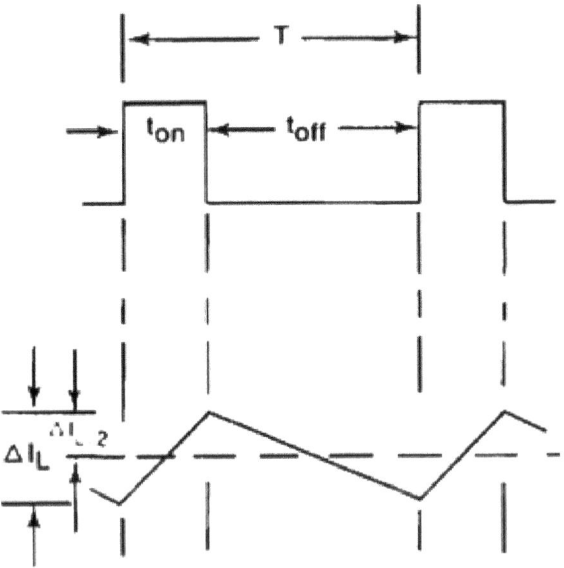

FIGURE 2.14 Inductor current ripples in response to voltage-input square wave.

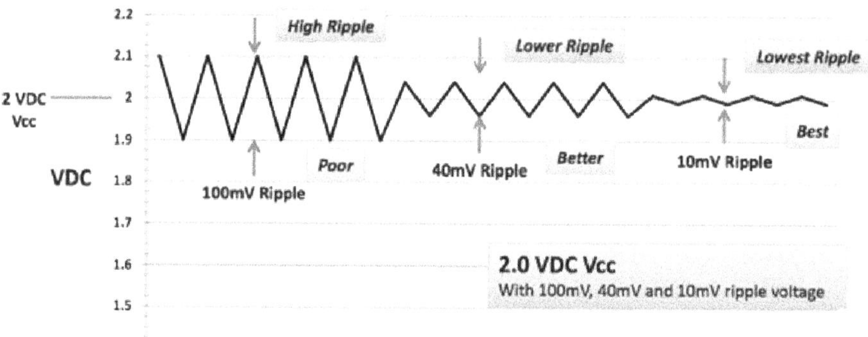

FIGURE 2.15 Lowering voltage ripple improves downstream circuit performance.

electrolyte through the seal, raising ESR and causing even more heating in a vicious cycle that leads to capacitance drop and ultimately failure. The core temperature rise should be limited to 5°C–10°C to prevent such damage. The power dissipation (P_{CAP}) of the output filter circuit capacitor is:

$$P_{CAP} = I_{RMS^2} \times ESR \tag{2.5}$$

where I_{RMS} is the input ripple current.

In other words, for a given ESR, the internal temperature rise is proportional to the square of the ripple current. It is important to note that ripple current rating can vary widely between technologies, manufacturer, and voltage for a given capacitance. High-ripple current-rating capacitors tend to have low ESR, greater surface area, and a high-heat transfer constant. Tall capacitors of equal volume to short, fat devices dissipate heat better because the heat is transferred from the core to the case more easily. Nonetheless, capacitor size compared to ESR rating also varies widely for different capacitor technologies and even between the same technologies from different manufacturers. The design engineer is advised to carefully familiarize themselves with the spec sheet for a shortlisted device before making a selection.

Another important factor is placement of the output capacitor on the PCB. It is a good idea to keep traces short for high-frequency switching circuits to minimize EMI. It is also a good idea to make sure that the capacitor is not sited too close to sources of heat, such as diodes, as this is likely to raise the internal temperature of the device even higher.

2.5.6 Seeking Guidance

Help is at hand for design engineers looking to source the correct external capacitors to work with the many switching voltage regulator modules on the market. The spec sheets that accompany and describe the module maker's products typically include applications circuits that suggest capacitor values for the input- and output-filter circuits for the devices.

Consider Texas Instruments' (TI) TPS53318 where the chip is a synchronous buck-switching voltage regulator that features an adjustable switching frequency from 250 kHz–1 MHz. The device features a wide conversion input voltage range (4.5–25 V), very low external component count, auto-skip-mode operation, internal soft-start control, and no need for compensation. The TPS51462 offers an output voltage range of 0.6–5.5 V (at up to 8 A output current) and is available in a 5 × 6 mm, 22-pin, QFN package.

The data sheet for TI's TPS53318 provides an application circuit diagram and the company helpfully suggests typical capacitance values for the output filter circuit. Further on, the data sheet includes a very detailed design procedure that describes how to select the inductor, set the output voltage, determine the output capacitance, and check the power supply's output stability for a given output capacitor.

Linear Technology's data sheet for its LTC3549 switching buck voltage regulator includes an application circuit advising on values of input and output capacitors (Figure 2.16).

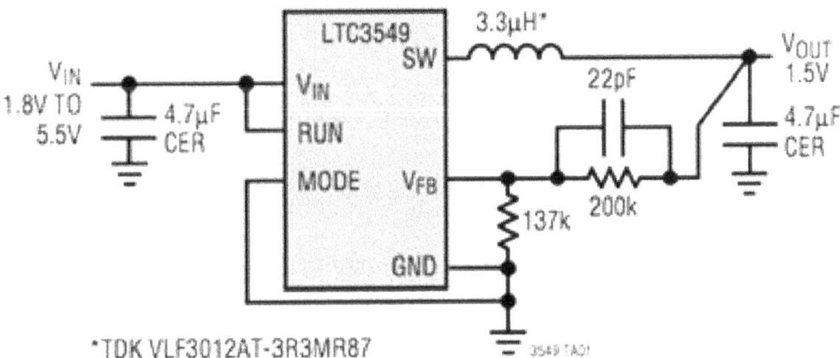

FIGURE 2.16 An application circuit for Linear Technology's LTC3549 advises values for input and output capacitors.

The data sheet also includes some handy design tips on external component selection such as pointing out that ripple-current ratings are often based on just 2,000 h of life, so it can make sense to specify a capacitor with a higher temperature capability than required. In addition, Linear Technology's spec sheet advises that because the LTC3549's control loop does not depend on the capacitor's ESR for stable operation, ceramic devices can be used (with care) to achieve very-low-output ripple and small circuit size.

The LTC3549 employs a fixed operational frequency of 2.25 MHz. The 1.6–5.5 V input voltage range makes the device ideally suited for single-cell Li-Ion, Li-Metal, Alkaline, NiCd or NiMH battery-powered applications. Burst mode operation can be user-enabled, increasing efficiency at light loads, further extending battery life.

For its part, STMicroelectronics explains that the input capacitor must have an RMS current rating higher than the maximum RMS input current and an ESR value compliant with the expected efficiency in its data sheet for the ST1S31 synchronous-buck switching regulator. Further, the data sheet advises that on the output side for a ceramic (MLCC) capacitor, the capacitive component of the voltage ripple dominates the resistive one, while for an electrolytic capacitor the opposite is true. The company's data sheet includes all the relevant formulas for calculating capacitor values and provides advice on which manufacturers supply suitable components.

The ST1S31 is an internally compensated, 1.5 MHz synchronous-buck voltage regulator that operates from a 2.8–5.5 V input, regulating an output voltage from V_{IN} to as low as 0.8 V. The chip features a peak current mode control with internal compensation and is claimed to deliver a very compact solution with a minimum component count. The device is available in 3 × 3 mm, 8-lead VFDFPN, and SO8 packages.

2.5.7 A Critical Part of Power Supply Design

Considering the sophistication of contemporary switching voltage converter modules, the choice of input and output capacitors for the chip's filter- and energy-storage circuits would seem to be a trivial part of power supply design. However, nothing could be further from the truth.

An incorrect choice of technology, dimensions, capacitance, or ESR can see the lowly capacitor at best ruin the circuit's efficiency and at worst lead to premature product failure. The recommended approach is to use the wealth of information supplied in the vendors' datasheets to calculate initial capacitance and ESR values and then test the prototype circuit with several alternative devices across a range of technologies.

2.5.8 End-User Capacitor Application

Capacitors have many uses in electronic and electrical systems. They are so ubiquitous that it is rare that an electrical product does not include at least one for some purpose.

2.5.9 Energy Storage Device

A capacitor can store electric energy when it is connected to its charging circuit. When it is disconnected from its charging circuit, it can dissipate that stored energy, so it can be used like a temporary battery.

FIGURE 2.17 Vintage paper in wax capacitor often found in antique tube radio circuits.

Capacitors are commonly used in electronic devices to maintain power supply while batteries are being changed. (This prevents loss of information in volatile memory.)

Conventional electrostatic capacitors provide less than 360 joules per kilogram of energy density, while capacitors using developing technology can provide more than 2.52 kilojoules per kilogram.

In car audio systems, large capacitors store energy for the amplifier to use on demand. Figure 2.17 shows the vintage paper-in-wax capacitor often found in antique tube radio circuits. An uninterruptible power supply (UPS) can be equipped with maintenance-free capacitors to extend service life.

2.5.10 Pulsed Power and Weapons

Groups of large, specially constructed, low-inductance high-voltage capacitors (capacitor banks) are used to supply huge pulses of current for many pulsed power applications. These include electromagnetic forming, Marx generators, pulsed lasers (especially TEA lasers), pulse-forming networks, fusion research, and particle accelerators.

Large capacitor banks (reservoirs) are used as energy sources for the exploding-bridge wire detonators or slapper detonators in nuclear weapons and other specialty weapons. Experimental work is under way using banks of capacitors as power sources for electromagnetic armor and electromagnetic rail guns or coil guns.

2.5.11 Power Conditioning

Reservoir capacitors are used in power supplies where they smooth the output of a full- or half-wave rectifier. They can also be used in charge pump circuits as the energy storage element in the generation of higher voltages than the input voltage.

Capacitors are connected in parallel with the DC power circuits of most electronic devices to smooth current fluctuations for signal or control circuits. Audio equipment, for example, uses several capacitors in this way, to shunt away power line hum before it gets into the signal circuitry. The capacitors act as a local reserve for the DC power source and bypass AC currents from the power supply. This is used in car audio applications, when a stiffening capacitor compensates for the inductance and resistance of the leads to the lead-acid car battery.

2.5.12 Power Factor Correction

In electric power distribution, capacitors are used for power factor correction. Such capacitors often come as three capacitors connected as a three-phase load. Usually, the values of these capacitors are given not in farads but rather as a reactive power in volt-amperes reactive (VAr). The purpose is to counteract inductive loading from devices like electric motors and transmission lines to make the load appear to be mostly resistive. Individual motor or lamp loads may have capacitors for power factor correction, or larger sets of capacitors (usually with automatic switching devices) may be installed at a load center within a building or in a large utility substation. In high-voltage direct current transmission systems, power factor correction capacitors may have tuning inductors to suppress harmonic currents that would otherwise be injected into the AC power system.

2.5.13 Motor Starters

In single-phase squirrel cage motors, the primary winding within the motor housing is not capable of starting a rotational motion on the rotor, but is capable of sustaining one. To start the motor, a secondary winding is used in series with a non-polarized *starting capacitor* to introduce a lag in the sinusoidal current through the starting winding. When the secondary winding is placed at an angle with respect to the primary winding, a rotating electric field is created. The force of the rotational field is not constant, but is sufficient to start the rotor spinning. When the rotor comes close to operating speed, a centrifugal switch (or current-sensitive relay in series with the main winding) disconnects the capacitor. The start capacitor is typically mounted to the side of the motor housing. These are called capacitor-start motors, and have relatively high starting torque.

There are also capacitor-run induction motors that have a permanently connected phase-shifting capacitor in series with a second winding. The motor is much like a two-phase induction motor.

Motor-starting capacitors are typically non-polarized electrolytic types, while running capacitors are conventional paper or plastic film dielectric types.

2.5.14 Signal Processing

The energy stored in capacitor can be used to represent information, either in binary form, as in DRAMs, or in analog form, as in analog sampled filters and CCDs. Capacitors can be used in analog circuits as components of integrators or more complex filters and in negative feedback loop stabilization. Signal processing circuits also use capacitors to integrate a current signal.

2.5.15 Tuned Circuits

Capacitors and inductors are applied together in tuned circuits to select information in particular frequency bands. For example, radio receivers rely on variable capacitors to tune the station frequency. Speakers use passive analog crossovers, and analog equalizers use capacitors to select different audio bands.

The resonant frequency f of a tuned circuit is a function of the inductance (L) and capacitance (C) in series, and is given by:

$$f = \frac{1}{2\pi\sqrt{LC}} \tag{2.6}$$

where L is in henry and C is in farads.

2.5.16 Regulating Utility Voltage with Distributed Resources

It is becoming more popular for utility distribution planners to consider distributed generation (DG) and storage devices to defer investments in substations and transmission lines until the load has grown to a sufficient size to warrant the larger investment. This concept is particularly useful when there are relatively few hours each year when the load approaches the system capacity limits. The movement toward utility deregulation in recent years has created renewed interest in distributed resources, and many of the issues related to PQ are addressed. Here, we will restrict our discussion to the potential of using distributed generators for distribution feeder voltage regulation.

Most of the utility-owned installations have been located in utility distribution substations. This offers load relief for the substation and transmission facilities, but contributes little else to the quality of power for the distribution feeder. Now, many distribution engineers are considering the benefits of moving the devices out onto the feeder to gain additional system capacity, loss reduction, improved reliability, and voltage regulation. These generators will often be owned by end users, but could be contracted to operate for utility system benefits as well. While this option may be too expensive to consider for voltage regulation alone, it is a useful side effect of dispersed sources justified on the basis of deferment of capital expansion. While few utility distribution planners will rely on customer-owned generation for base capacity, it is more palatable to employ them to help cover contingencies. One example is illustrated in Figure 2.18.

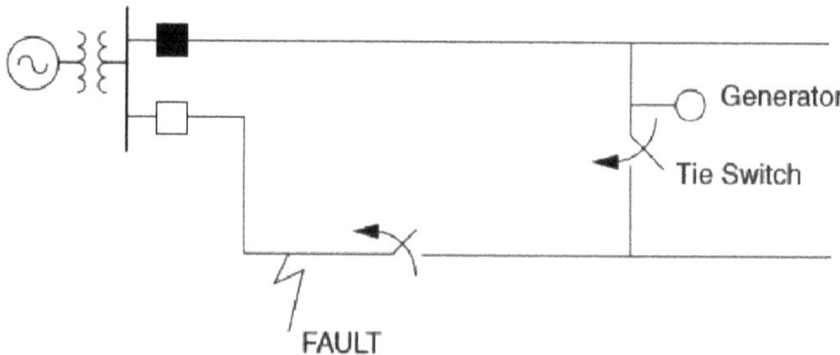

FIGURE 2.18 Using a generator to support restoration of service to the unfaulted portion of a feeder.

Utilities usually have sectionalizing switches installed so that portions of a distribution feeder can be served from different feeders or substations during emergencies. If the fault occurs at the time of peak load, it may be impossible to pick up any more loads from other feeders in the normal manner simply by closing a switch. However, a generator located near the switch tie point can potentially provide enough power to support the additional load at a satisfactory voltage. If the generator is of sufficient size, it could be employed to help regulate the voltage. One advantage of using a generator to regulate the voltage is that its controls respond much faster and more smoothly than discrete tap changing devices like regulators and LTCs. While this potential exists, most utility distribution protection engineers are reluctant to allow this type of operation without careful study and costly control equipment. The issue is that operating with automotive voltage regulation makes the system more susceptible to sustaining an inadvertent island. Therefore, some means of direct transfer trip is generally required to ensure that the generator disconnects from the system when certain utility breakers operate. A more normal connection of DG is to use power and power factor control. This minimizes the risk of islanding. Although the DG no longer attempts to regulate the voltage, it is still useful for voltage regulation purposes during constrained loading conditions by displacing some active and reactive power. Alternatively, customer-owned DG may be exploited simply by operating off-grid and supporting part or the customer's entire load off-line. This avoids interconnection issues and provides some assistance to voltage regulation by reducing the load. The controls of distributed sources must be carefully coordinated with existing line regulators and substation LTCs. Reverse power flow can sometimes fool voltage regulators into moving the tap changer in the wrong direction. Also, it is possible for the generator to cause regulators to change taps constantly, causing early failure of the tap-changing mechanism. Fortunately, some regulator manufacturers have anticipated these problems and now provide sophisticated microcomputer-based regulator controls that are able to compensate. To exploit dispersed sources for voltage regulation, one is limited in options to the types of devices with steady, controllable outputs such as reciprocating engines, combustion turbines, fuel cells, and battery storage. Randomly varying sources such as wind turbines and photovoltaics are unsatisfactory for this role and often must be placed on a relatively stiff part of the system or have special regulation to avoid voltage regulation difficulties. DG used for voltage regulation must also be large enough to accomplish the task. Not all technologies are suitable for regulating voltage. They must be capable of producing a controlled amount of reactive power. Manufacturers of devices requiring inverters for interconnection sometimes program the inverter controls to operate only at unity power factor while grid-connected. Simple induction generators consume reactive power like an induction motor, which can cause low voltage.

2.5.17 Flicker

Loads that can exhibit continuous, rapid variations in the load current magnitude can cause voltage variations that are often referred to as flicker. The term flicker is derived from the impact of the voltage fluctuation on lamps such that they are perceived by the human eye to flicker. To be technically correct, voltage fluctuation is an electromagnetic phenomenon while flicker is an undesirable result of the voltage fluctuation in some loads. However, the two terms are often linked together in standards. Therefore, we will also use the common term voltage flicker to describe such voltage fluctuations. An example of

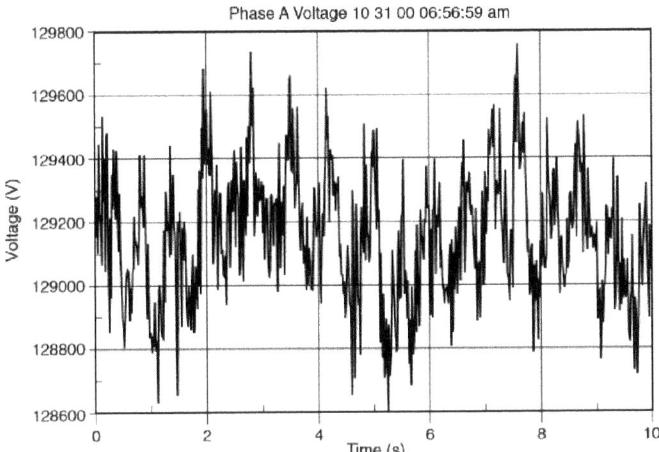

FIGURE 2.19 Example of voltage fluctuation caused by arc furnace operation.

a voltage waveform that produces flicker is shown in Figure 2.19. This is caused by an arc furnace, one of the most common causes of voltage fluctuations on utility transmission and distribution systems. The flicker signal is defined by its rms magnitude expressed as a percent of the fundamental. Voltage flicker is measured with respect to the sensitivity of the human eye. Typically, magnitudes as low as 0.5% can result in perceptible lamp flicker if the frequencies are in the range of 6 to 8 Hz. IEC 61000-4-15 defines the methodology and specifications of instrumentation for measuring flicker. The IEEE Voltage Flicker Working Group has recently agreed to adopt this standard as amended for 60 Hz power systems for use in North America. This standard devises a simple means of describing the potential for visible light flicker through voltage measurements. The measurement method simulates the lamp/eye/brain transfer function and produces a fundamental metric called short-term flicker sensation (P_{st}). This value is normalized to 1.0 to represent the level of voltage fluctuations sufficient to cause noticeable flicker to 50% of a sample observing group. Another measure called long-term flicker sensation (P_{lt}) is often used for the purpose of verifying compliance with compatibility levels established by standards bodies and used in utility power contracts. This value is a longer-term average of P_{st} samples.

Figure 2.19 illustrates a trend of P_{st} measurements taken at a 161-kV substation bus serving an arc furnace load. P_{st} samples are normally reported at 10-min intervals. A statistical evaluation process defined in the measurement standard processes instantaneous flicker measurements to produce the P_{st} value.

2.5.17.1 Standards and Regulation

The primary United States standards for voltage fluctuation are contained in IEEE 519-1992, Recommended Practices and Requirements for Harmonic Control in Electric Power Systems, and IEEE 141-1995, Recommended Practice for Electric Power Distribution for Industrial Plants (The Red Book). There are no United States standards for measuring flicker at this time. The international standard for measuring flicker is IEC 1000-4-15, Flicker meter—Functional and Design Specifications (formerly IEC 868), and for setting flicker limits for individual appliances is IEC 1000-3-3, Disturbances in Supply Systems Caused by Household Appliances and Similar Electrical Equipment (formerly IEC 555-3). All these standards attempt to limit the lighting flicker so that it does not irritate a person seeing it. Because of the subjective nature of light flicker, standards organizations have had difficulty correlating voltage fluctuation standards to perceptible light flicker that is irritating to the observer.

IEEE 519-1992, Recommended Practices and Requirements for Harmonic Control in Electric Power Systems, page 80, says “sources of flicker in industrial power distribution systems can be, for instance, the somewhat random variations of load typified by an arc furnace melting scrap steel or an elevator motor’s starts and stops. A flicker source may be nearly periodic, as in the case of jogging or manual spot-welding. A source may also be periodic, as in the case of an automatic spot-welder." It mentions that static volt-amperes reactive (VAR) compensators at the flicker source keep the voltage steady under varying load conditions and therefore solve the flicker problem. A person experiencing irritating flicker can sometimes get rid of the flicker by changing the light bulb type. IEEE-5191992 not only sets flicker standards but is the basis for setting harmonic standards in the United States.

2.6 Introduction to Voltage Sag

In electrical engineering, spikes are fast, short duration electrical transients in voltage (voltage spikes), current (current spikes), or transferred energy (energy spikes) in an electrical circuit. Fast, short duration electrical transients (overvoltage) in the electric potential of a circuit are typically caused by

- Lightning strikes
- Power outages
- Tripped circuit breakers
- Short circuits
- Power transitions in other large equipment on the same power line
- Malfunctions caused by the power company
- Electromagnetic pulses (EMP) with electromagnetic energy distributed typically up to the 100 kHz and 1 MHz frequency range
- Inductive spikes

In the design of critical infrastructure and military hardware, one concern is of pulses produced by nuclear explosions, whose nuclear EMPs distribute large energies in frequencies from 1 kHz into the gigahertz range through the atmosphere. The effect of a voltage spike is to produce a corresponding increase in current (current spike). However, some voltage spikes may be created by current sources. Voltage would increase as necessary so that a constant current will flow. Current from a discharging inductor is one example. For sensitive electronics, excessive current can flow if this voltage spike exceeds a material's breakdown voltage, or if it causes avalanche breakdown. In semiconductor junctions, excessive electric current may destroy or severely weaken that device. An avalanche diode, transient voltage suppression diode, voltage crowbar, or a range of other overvoltage protective devices can divert (shunt) this transient current thereby minimizing voltage. While generally referred to as a voltage spike, the phenomenon in question is actually an energy spike, in that it is measured not in volts but in joules; a transient response defined by a mathematical product of voltage, current, and time. Voltage spikes may be created by a rapid buildup or decay of a magnetic field, which may induce energy into the associated circuit. However, voltage spikes can also have more mundane causes such as a fault in a transformer or higher-voltage (primary circuit) power wires falling onto lower-voltage (secondary circuit) power wires as a result of accident or storm damage. Voltage spikes may be longitudinal (common) mode or metallic (normal or differential) mode. Some equipment damage from surges and spikes can be prevented by use of surge protection equipment. Each type of spike requires selective use of protective equipment. For example, a common mode voltage spike may not even be detected by a protector installed for normal mode transients.

2.6.1 Voltage Sag

Voltage sags are short duration reductions in RMS voltage, mainly caused by short circuits and starting or large motors. The large interest in voltage sags is due to the problems they cause on several types of equipment; in particular, adjustable-speed drives (ASDs), process-control equipment, and computers are notorious for their sensitivity. Some pieces of equipment trip when the RMS voltage drops below 90% for longer than one or two cycles. Such a piece of equipment will trip tens of times a year. If this is the process-control equipment of a paper mill, one can imagine that the damage due to voltage sags can be

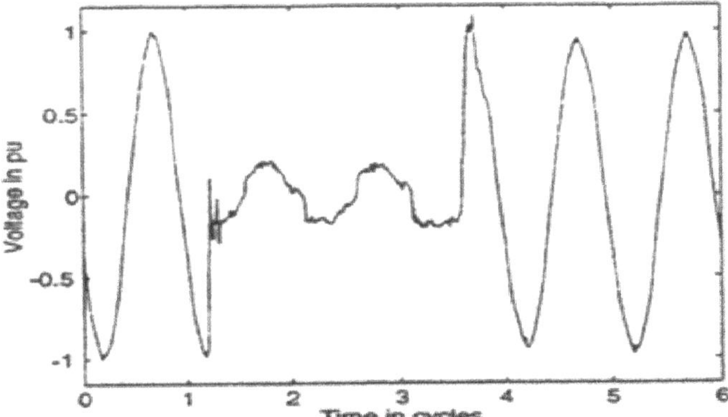

FIGURE 2.20 A voltage sag voltage in one phase in time domain.

enormous. Of course, voltage sag is not as damaging to industry as a (long or short) interruption. But as there are far more voltage sags than interruptions the total damage due to sags is still larger. Another important aspect of voltage sags is that they are rather hard to mitigate. Short interruptions and many long interruptions can be prevented via rather simple, although relatively expensive, measures in the local distribution network. However, voltage sags at equipment terminals can be due to short circuit faults hundreds of kilometers away in the transmission system. It will be clear that there is no simple method to prevent them. An example of voltage sag in shown in Figure 2.20. We see that the voltage amplitude drops to a value of about 20% of the pre-event voltage for about two cycles. After these two cycles the voltage comes back to about the pre-sag voltage. This magnitude and duration are the main characteristics of voltage sag. Both will be discussed in more detail in the forthcoming sections. We can also conclude from Figure 2.20 that magnitude and duration do not completely characterize the sag. The during-sag voltage contains a rather large amount of higher frequency components. Also, the voltage shows a small overshoot immediately after the sag. To what degree these higher frequency components are of any influence on the equipment behavior due to sags remains a point of discussion. Voltage sags are mainly caused by short circuits. The sag in Figure 2.20 is due to a short circuit. In addition, however, the starting of large load can lead to voltage sag. Large induction motors are the typical load that causes voltage sags. Voltage sags due to induction motor starting last longer than those due to short circuits. Typical durations are seconds to tens of seconds. The remainder of this chapter will concentrate on voltage sags due to short circuits.

2.6.2 Voltage Sag Magnitude

The magnitude of voltage sag can be determined in a number of ways. At the moment there appears general agreement that the magnitude should be determined from the rms voltages. As voltage sags are initially recorded as sampled points in time, the rms voltage will have to be calculated. This has been done for the sag shown in Figure 2.20, resulting in Figure 2.21. The rms voltage has been calculated over a window of one cycle, which is 256 samples for the recording used. Each point in Figure 2.21 is the rms voltage over the preceding 256 points (the first 255 rms values have been made equal to the value for sample 256):

$$V_{\text{rms}}(k) = \sqrt{\frac{1}{N}\sum\nolimits_{i=k-N+1}^{i=k}(v)^2} \tag{2.7}$$

with $N = 256$ and V_i the sampled voltage in time domain. We see that the rms voltage does not immediately drop to a lower value but takes one cycle for the transition. We also see that the rms value during the sag is not completely constant and that the voltage does not immediately recover after the fault.

A surprising observation is that the rms voltage immediately after the fault is only about 90% of the pre-sag voltage. From Figure 2.21 one can see that the voltage in time domain shows a small overvoltage instead. In

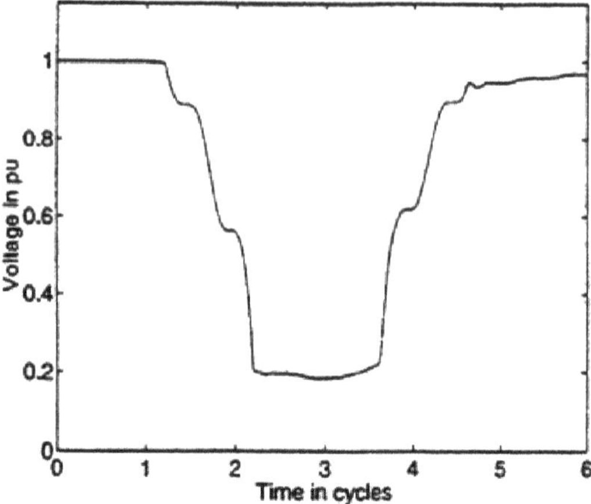

FIGURE 2.21 One-cycle rms voltage for the voltage sag.

this example the rms voltage has been calculated after each sample. In PQ monitors, this calculation is typically only made once a cycle. It is thus likely that the monitor will give one value with an intermediate magnitude before its rms voltage value settles down. In the US the general practice is to characterize the sag through the remaining voltage during the sag. This is then given as a percentage of the nominal voltage. Thus a 70% of sag in a 120 Volt system means that the voltage dropped to 84 V. The confusion with this terminology is clear. One could be tricked into thinking that 70% sag refers to a drop of 70%, thus a remaining voltage of 30%. The recommendation is therefore to use the phrase sag down to 70%. The IEC has solved this ambiguity by characterizing the sag through the actual drop. This has somewhat become common practice in Europe. Characterizing sag through its drop-in voltage does not solve all problems, however, because the next question will be: what is the reference voltage? There are arguments in favor of using the pre-fault voltage and there are arguments in favor of using the nominal voltage. The International Union of Producers and Distributors of Electrical Energy (Union Internationale des Producteurs et Distributeurs d'Energie Electrique, UNIPEDE) recommends to use the nominal voltage as a reference. In the remainder of these chapter, we will use the term "magnitude" in the meaning of the remaining voltage during the fault.

2.6.3 Voltage Sag Duration

We have seen that the drop-in voltage during sag is due to a short circuit being present in the system. The moment the short circuit fault is cleared by the protection, the voltage can return to its original value. The duration of sag is thus determined by the fault-clearing time. However, the duration of sag is normally longer than the fault-clearing time. Generally speaking faults in transmission systems are cleared faster than faults in distribution systems. In transmission systems the critical fault-clearing time is rather small. Thus, fast protection and fast circuit breakers are essential. Also, transmission and sub-transmission systems are normally operated as a grid, requiring distance protection or differential protection, both of which are rather fast. The principle form of protection in distribution systems is over-current protection. This requires often some time-grading, which increases the fault-clearing time. An exception is systems in which current-limiting fuses are used. These have the ability to clear a fault within one half-cycle. The sag duration will be longer when sag originates at a lower voltage level. This is due to the fault-clearing time typically becoming shorter for higher voltage levels. We saw before that faults in distribution systems will lead to deep sags if they are at the same voltage level as the point-of-common coupling (pee) and to shallow sags if they are at lower voltage level than the pee. We also saw that transmission system faults lead to shorter duration sags than distribution system faults and that they cover the whole range of sag magnitude. Current-limiting fuses allow very short fault-clearing times, they are only found in low-voltage and distribution systems.

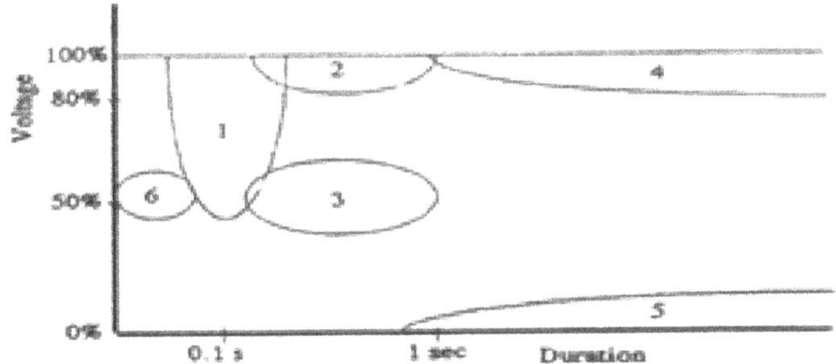

FIGURE 2.22 Sags of different origin in magnitude-duration plot.

In a magnitude versus duration plot we can now distinguish a number of areas. This is shown in Figure 2.22. The numbers in the figure refer to the following sag origins:

- Transmission system faults
- Remote distribution system faults
- Local distribution system faults
- Starting of large motors
- Short interruptions
- Fuses

The magnitude-duration plot is an often-used tool to show the quality of supply at a certain location or the average quality of supply of a number of locations. Measurement of sag duration is much less trivial than it might appear from the previous section. For sag like in Figure 2.20 it is obvious that the duration is about 2~ cycles. However, to come up with an automatic way for a PQ monitor to obtain the sag duration is no longer straightforward. The commonly used definition of sag duration is the number of cycles during which the rms voltage is below a given threshold. This threshold will be somewhat different for each monitor, but typical values are around 90%. A PQ monitor will typically calculate the rms value once every cycle. This gives an overestimation of the sag duration as shown in Figure 2.22. The normal situation is shown in the upper figure. The rms calculation is performed at regular instants in time and the voltage sag starts somewhere in between two of those instants. As there is no correlation between the calculation instants and the sag commencement, this is the most likely situation.

We see that the rms value is low for three samples in a row. The sag duration according to the monitor will be three cycles. Here it is assumed that the sag is deep enough for the intermediate rms value to be below the threshold. For shallow sags both intermediate values might be above the threshold and the monitor will record a one-cycle sag. The bottom curve shows the rare situation where the sag commencement almost coincides with one of the instants on which the voltage is calculated. In that case the monitor gives the correct sag duration. The one cycle or one half-cycle error in sag duration is only significant for short-duration sags. For longer sags it does not really matter. But for longer sags the so-called post-fault sag will give a more serious error in sag duration. When the fault is cleared the voltage does not recover immediately. The rms voltage after the sag is slightly lower than before the sag. The effect can be especially severe for sags due to three-phase faults. Due to the drop-in voltage during the sag, induction motors will slow down. The torque produced by an induction motor is proportional to the square of the voltage, so even a rather small drop in voltage can already produce a large drop in torque and thus in speed. The moment the fault is cleared and the voltage comes back, the induction motors start to draw a large current: up to 10 times their nominal current. Immediately after the sag, the air-gap field will have to be built up again. In other words: the induction motor behaves like a short-circuited transformer. After the flux has come back into the air gap, the motor can start reaccelerating, which also requires a rather large current. It is this post-fault

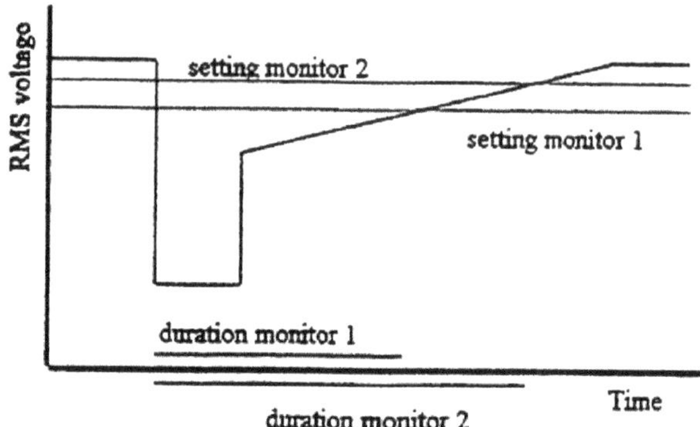

FIGURE 2.23 Monitoring of post fault sags.

inrush current of induction motors that leads to an extended sag. This post-fault sag can last for several seconds, much the larger than actual sag.

The post-fault sag can cause a serious error on the sag duration as obtained by a PQ monitor. An even more serious problem is that different monitors can give different results. This is shown schematically in Figure 2.23. Assume that monitor 1 has a setting as indicated and monitor 2 a slightly higher setting. Both monitors will record a sag duration much longer than the fault-clearing time. The fault-clearing time can be estimated from the duration of the deep part of the sag. We see that monitor 2 will record a significantly longer duration than monitor 1.

2.6.3.1 Three-Phase Unbalance

For each type of fault, expressions can be derived for the voltages at the pcc. But this voltage is not equal to the voltage at the equipment terminals. Equipment is normally connected at a lower voltage level than the level at which the fault occurs. The voltages at the equipment terminals therefore not only depend on the voltages at the pcc but also on the winding connection of the transformers between the pcc and the equipment terminals. The voltages at the equipment terminals further depend on the load connection. Three-phase load is normally connected in delta but star-connection is also used. Single-phase load is normally connected in star (i.e., between one phase and neutral) but sometimes in delta (between two phases). In this section we will derive a classification for three-phase unbalanced voltage sags, based on the following assumptions:

- The zero-sequence component of the voltage does not propagate down to the equipment terminals, so that we can consider phase-to-neutral voltages.
- Load currents, before, during and after the fault can be neglected.
- Positive- and negative-sequence source impedances are equal.
- Faults are single-phase, phase-to-phase, or three-phase.

2.6.3.2 Phase Angle Jumps

A short circuit in a system not only causes a drop-in voltage magnitude but also a jump in the phase angle of the voltage. In a 50 or 60 Hz system, voltage is a complex quantity (a phasor) that has magnitude and phase-angle. A change in the system, like a short circuit, causes a change in voltage. This change is not limited to the magnitude of the phasor but can equally well include a change in phase-angle. The phase-angle jump manifests itself as a shift in voltage zero crossing compared to a synchronous voltage, e.g., as obtained by using a phase-locked loop (PLL). Phase-angle jumps are not of concern for most equipment. But power electronics converters using phase-angle information for their firing instants could easily get disturbed. The concept of phase-angle jump will be introduced by means of a three-phase fault,

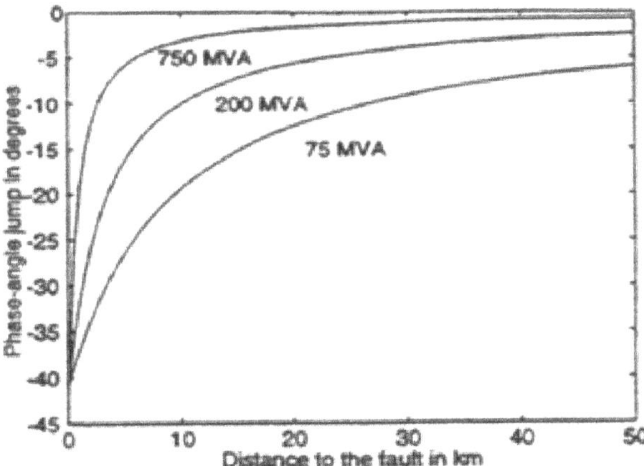

FIGURE 2.24 Phase angle jumps vs. distance for faults on a 150 mm[2] 11 kV overhead header with different source strength.

as that enables us to use the single-phase model. Phase-angle jumps during three phase faults are due to the difference in *X*/*R* ratio between the source and the feeder. A second cause of phase-angle jumps is the transformation of sags to lower voltage levels. This phenomenon has already been mentioned when three-phase unbalanced sags were discussed before. To understand the origin of phase-angle jumps associated with voltage sags, the single-phase voltage divider model of Figure 2.24 can be used, with the difference that Z_s and Z_p are complex quantities that we will denote as Z_s and Z_p.

The (complex) voltage at the pee during the fault is:

$$V_{\text{sag}} = \frac{Z_f}{Z_s + Z_f} \tag{2.8}$$

Let $Z_s = R_s + jX_s$ and $Z_p = R_p + jX_f$. The argument of V_{sag}, the phase-angle jump in the voltage, is given by the following expression:

$$\Delta\varnothing = \arg(V_{\text{sag}}) \tag{2.9}$$

$$= \arctan\frac{X_f}{R_f} - \arctan\left(\frac{X_s + X_f}{R_s + R_f}\right) \tag{2.10}$$

2.6.3.3 Magnitude and Phase-Angle Jumps for Three-Phase Unbalanced Sags

The magnitude of voltage sag was defined as the rms value of the voltage during the fault. For single-phase loads this is an implementable definition, as despite the problems with actually obtaining the rms value three-phase unbalanced sags the problem becomes more complicated as there are now three rms values to choose from. The most commonly used definition is: The magnitude of a three-phase sag is the rms value of the lowest of the three voltages. Alternatives suggested are to use the average of the three rms values, or the lowest value but one. Based on the classification of three-phase unbalanced sags we distinguish between three different kinds of magnitude and phase-angle jump. In all cases magnitude and phase-angle jump are absolute value and argument, respectively, of a complex voltage.

- The initial complex voltage is the voltage at the point-of-common coupling at the faulted voltage level. For a single-phase-to-ground fault the initial complex voltage is the voltage between the faulted phase and ground at the pee. For a phase-to-phase fault the initial complex voltage is the voltage between the two faulted phases. For a two-phase-to-ground or a three-phase fault it can be either the voltage in one of the faulted phases or between two faulted phases (as long as pu values are used). The initial sag magnitude is the absolute value of the complex initial voltage; the initial phase-angle jump is the argument of the complex initial voltage.

- We will give an alternative interpretation of the characteristic complex voltage later on. The characteristic sag magnitude is the absolute value of the characteristic complex voltage. The characteristic phase-angle jump is the argument of the characteristic complex voltage. These can be viewed as generalized definitions of magnitude and phase angle jumps for three-phase unbalanced sags.
- The sag magnitude and phase-angle jump at the equipment terminals are absolute value and argument, respectively, of the complex voltages at the equipment terminals. For single-phase equipment these are simply sag magnitude and phase-angle jump as previously defined for single-phase voltage sags.

Before, we have introduced three types of sags together with their characteristic complex voltage, a mathematically elegant method to obtain the characteristic complex voltage from the sampled *voltages*. Here we will give a simple method for obtaining the sag magnitude. For type D the magnitude is the rms value of the lowest of the three voltages. For type C it is the rms value of the difference between the two lowest voltages (in pu). For type A, either definition holds. This leads to the following way of determining the characteristic magnitude of a three-phase sag from the voltages measured at the equipment terminals:

- Obtain the three voltages as the function of the time $V_a(t)$, $V_b(t)$, and $V_c(t)$.
- Determine the zero-sequence voltage.
- Determine the remaining voltage after subtracting the zero-sequence voltage.
- Determine the rms voltage of the $V_a^f(t)$, $V_b^f(t)$, and $V_c^f(t)$.
- Determine the three-voltage difference.
- Determine the rms voltage of the three V_{ab}, V_{bc}, and V_{ca}.
- The magnitude of the three-phase sag is the lowest of the six rms values.

This procedure has been applied to the voltage sag shown in Figure 2.25. At first the rms values have been determined for the three measured phase-to-ground voltages, resulting in Figure 2.25. The rms value has been determined each half cycle over the preceding 128 samples (one half-cycle). We see the behavior typical for a single-phase fault on an overhead feeder: a drop-in voltage in one phase and a rise in voltage in the two remaining voltages. After subtraction of the zero-sequence component, all three

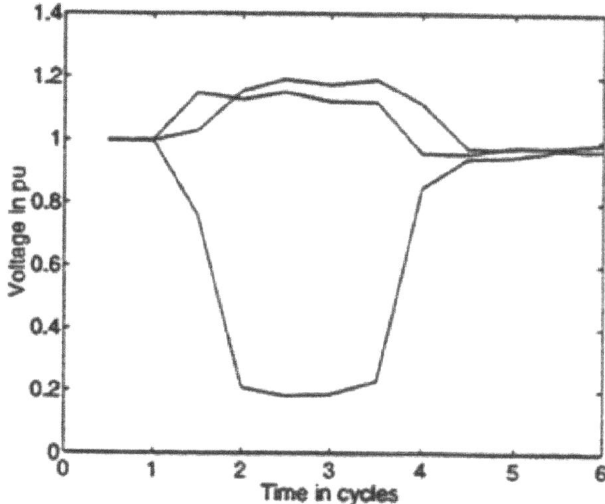

FIGURE 2.25 RMS values for the phase to ground voltages for the sag is shown.

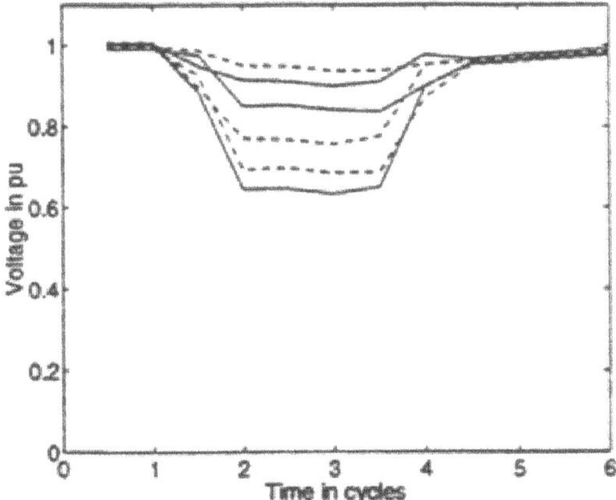

FIGURE 2.26 RMS values of phase to phase (dashed line) and phase to neutral voltages after removal of the zero-sequence component for the sag is shown.

voltages show a drop-in magnitude. This is shown in Figure 2.26. The phase-to-ground voltages minus the zero sequence are indicated through solid lines, the phase-to-phase voltages through dotted lines. The lowest rms value is reached for a phase-to-ground voltage, which indicates sag of type D. This is not surprising as the original sag was of type B (albeit with a larger than normal zero-sequence component). After removal of the zero-sequence, voltage sag of type D remains. The characteristic magnitude of this three-phase unbalanced sag is 63%.

2.6.3.4 Other Characteristics of Voltage Sags

Some devices are affected by other sag characteristics such as the phase unbalance during the sag event, the point in the wave at which the sag is initiated, or any transient oscillations occurring during the disturbance. These characteristics are subtler than magnitude and duration, and their impacts are much more difficult to generalize. As a result, the rms variation performance indices defined here are focused on the more common magnitude and duration characteristics.

2.6.3.5 Load Influence on Voltage Sags

Consider an example: A transformer failure can be the initiating event that causes a fault on the utility power system that results in voltage sag. These faults draw energy from the power system. Voltage sag occurs while the fault is on the utility's power system. As soon as a breaker or recloser clears the fault, the voltage returns to normal. Transmission faults cause voltage sags that last about 6 cycles, or 0.10 s. Distribution faults last longer than transmission faults, while large motor loads can cause voltage sag on the utility's and end user's power systems. Compared to other PQ problems affecting industrial and commercial end users, voltage sags occur most frequently. They reduce the energy being delivered to the end user and cause computers to fail, ASDs to shut down, and motors to stall and overheat.

2.6.3.5.1 Voltage Sags due to Induction Motor Starting

Since induction motors are balanced 3φ loads, voltage sags due to their starting are symmetrical. Each phase draws approximately the same in-rush current. The magnitude of voltage sag depends on:

- Characteristics of the induction motor
- Strength of the system at the point where motor is connected

Figure 2.27 represents the shape of the voltage sag on the three phases (A, B, and C) due to voltage sags.

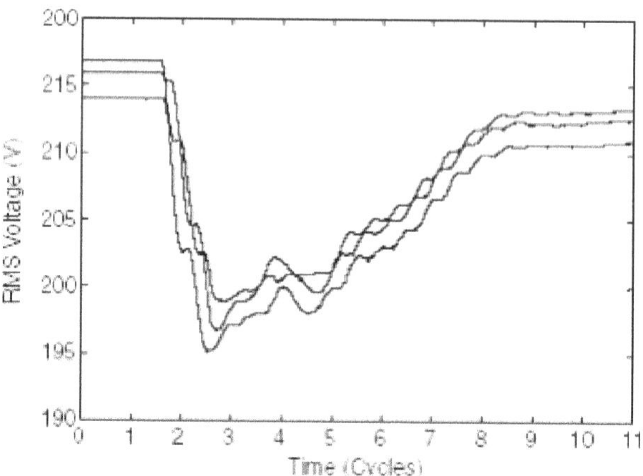

FIGURE 2.27 Voltage sag due to motor starting.

2.6.4 Equipment Behavior

2.6.4.1 Voltage-Tolerance Curves

Generally speaking, electrical equipment prefers a constant rms voltage. That is what the equipment has been designed for and that is what it will operate best for. The other extreme is no voltage for a longer period of time. In that case the equipment will simply completely stop operating. No piece of electrical equipment can operate indefinitely without electricity. Some equipment will stop within one second like most desktop computers. Other equipment can withstand a supply interruption much longer, like a laptop computer, which is designed to withstand (intentional) power interruptions. But even a laptop computer's battery only contains enough energy for typically a few hours. For each piece of equipment, it is possible to determine how long it will continue to operate after the supply becomes interrupted. A rather simple test would give the answer. The same test can be done for a voltage of 10% (of nominal), for a voltage of 20%, etc. If the voltage becomes high enough, the equipment will be able to operate on it indefinitely. Connecting the points obtained by performing these tests results in the so-called "voltage-tolerance curve." An example of a voltage-tolerance curve is shown in Figure 2.28. Strictly speaking one can claim that this is not a voltage-tolerance curve, but a requirement for the voltage tolerance; in this case the voltage tolerance of power stations connected to the Swedish National Grid. One could refer to this as a voltage-tolerance requirement and to the result of equipment tests as a voltage-tolerance performance. We will refer to both the measured curve and the required curve as voltage-tolerance curves. It will be clear from the context whether one refers to the voltage-tolerance requirement or the voltage-tolerance performance. We see in Figure 2.28 that a Swedish power station has to withstand voltage sag down to 25% of nominal for 250 ms, and that the power station should be able to operate normally for any voltage of 95% or higher. The concept of voltage-tolerance curve was introduced in 1978 by Thomas Key. When studying the reliability of the power supply to military installations, he realized that voltage sags and their resulting tripping of mainframe computers could be a greater threat than complete interruptions of the supply. He therefore contacted some manufacturers for their design criteria and performed some tests himself. The resulting voltage-tolerance curve became known as the "CBEMA curve" several years later.

2.6.4.2 Voltage-Tolerance Tests

The only standard that describes how to obtain voltage tolerance of equipment is IEC 61000-4-11. This standard does not, however, mention the term voltage-tolerance curve. Instead it defines a number of preferred magnitudes and durations of sags for which the equipment has to be tested. (Note: The standard

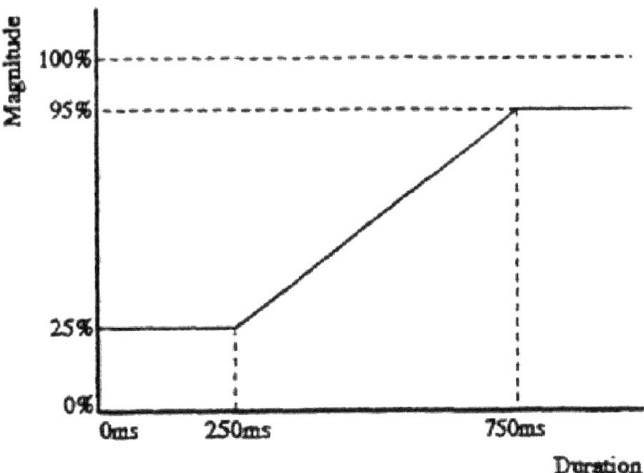

FIGURE 2.28 Voltage-tolerance requirement for the power station in Sweden.

uses the term "test levels," which refers to the remaining voltage during the sag). The equipment does not need to be tested for all these values, but one or more of the magnitudes and durations may be chosen. The preferred magnitudes and duration are shown in Table 2.1. The IEC standard also allows the choice of one sag duration outside of the list of preferred durations.

The standard in its current form does not set any voltage-tolerance requirements. It only defines the way in which the voltage tolerance of equipment shall be obtained. The standard also does not mention any specific testing method. The only requirement is that the transition from the pre-sag to the during-sag and from the during-sag to the post-sag voltage is instantaneous. An informative appendix to the standard does, however, mention two examples of test set-ups:

- Use a transformer with two output voltages. Make one output voltage equal to 100% and the other to the required during-sag magnitude value. Switch very fast between the two outputs.
- Generate the sag by using a wave-form generator in cascade with a power amplifier.

Both methods are only aimed at testing one piece of equipment at a time. To make a whole installation tolerate a certain voltage sag, each piece needs to be tested hoping that their interconnection does not cause any unexpected deterioration in performance. A method for testing a whole installation is presented. A three-phase diesel generator is used to power the installation under test. Voltage sag is made by reducing the field voltage. It takes about two cycles for the AC voltage to drop after a drop in field voltage, so that this method can only be used for sags of 5 cycles and longer. For equipment testing this is no serious limitation.

TABLE 2.1

Record of Voltage-Tolerance Test

	Duration in Cycles of 50 Hz					
Magnitude	**0.5**	**1**	**5**	**10**	**25**	**50**
70%						
40%						
0%						

2.6.5 Computers and Consumer Electronics

The power supply of a computer and of most consumer electronics equipment normally consists of a diode rectifier followed by some kind of electronic voltage regulator. The power supply of all these low-power electronic devices is similar and so is their sensitivity to voltage sags. The difference is in the consequences of a sag-induced trip. A television will show a black screen for up to a few seconds; a compact disc player will reset itself and start from the beginning of the disc, or just wait for a new command. But the trip of the process-control computer of a chemical plant leads to a complete restart of the plant taking up to several days, plus sometimes a very dangerous situation.

2.6.5.1 Estimation of Computer Voltage Tolerance

In the configuration of the power supply to a computer, the capacitor connected to the non-regulated DC bus reduces the voltage ripple at the input of the voltage regulator. The voltage regulator transforms a non-regulated DC voltage of a few hundred volts into a regulated DC voltage of the order of 10 V. If the AC voltage drops, the voltage on the DC side of the rectifier (the non-regulated DC voltage) drops. The voltage regulator is able to keep its output voltage constant over a certain range of input voltage. If the voltage at the DC bus becomes too low the regulated voltage will also start to drop and ultimately errors will occur in the digital electronics.

DC voltage during sag: When the rms voltage drops suddenly, the maximum AC voltage remains less than the DC voltage for the whole cycle. Thus, the capacitor continues to discharge. This discharging goes on for a number of cycles, until the capacitor voltage drops below the maximum of the AC voltage. After that a new equilibrium will be reached. It is important to realize that the discharging of the capacitor is only determined by the load connected to the DC bus, not by the AC voltage. Thus, all sags will cause the same initial decay in DC voltage. But the duration of the decay is determined by the magnitude of the sag. The deeper the sag the longer it takes before the capacitor has discharged enough to enable charging from the supply.

As long as the absolute value of the AC voltage is less than the DC bus voltage, all electrical energy for the load comes from the energy stored in the capacitor. For a capacitance C, the stored energy, a time t after sag initiation, is equal to $1/2\{CV(t)\}^2$, with $V(t)$ the bus voltage. This energy is equal to the energy at sag initiation minus the energy taken by the load:

$$\frac{1}{2}CV^2 = \frac{1}{2}CV_0^2 - Pt \tag{2.11}$$

With V_0 the DC bus voltage at sag initiation and P the loading of the DC bus. Expression (1) holds as long as the DC bus voltage is higher than the absolute value of the AC voltage. Solving (1) gives an expression for the voltage during this initial decay period.

$$V = \sqrt{\left(V_0^2 - 2\frac{P}{C}t\right)} \tag{2.12}$$

The moment the DC bus voltage drops below the absolute value of the AC voltage, the normal operation mode of the rectifier takes over and the DC bus voltage remains constant, apart from the unavoidable DC voltage ripple. From (2) we can calculate how long it takes for the DC bus voltage to decay to its new steady-state value. But first we obtain an expression for the DC voltage ripple $\in$

$$\in = \frac{PT}{2V_0^2G} \tag{2.13}$$

With T cycle of the fundamental frequency, the ripple is defined as the difference between the maximum and the minimum value of the DC voltage (Table 2.2). The discharge period of the capacitor is assumed

TABLE 2.2
DC Ripple Value for Various DC Voltages

Min Bus DC Voltage	5% DC Ripple	1% DC Ripple
0	5 cycles	25 cycles
50%	4 cycles	19 cycles
70%	2.5 cycles	13 cycles
90%	1 cycle	5 cycles

to be equal to one half cycle. Inserting the expression for the DC voltage ripple (3) in (2) gives an expression for the DC voltage during the discharge period, thus during the initial cycles of a voltage sag:

$$V(t) = V_0 \sqrt{\left(1 - 4 \in \frac{t}{T}\right)} \tag{2.14}$$

Thus, if a computer trips at 50% DC bus voltage, and as the normal operation DC voltage ripple is 5%, sag of less than 4 cycles in duration will not cause a mal-trip. Any sag below 50% for more than 4 cycles will trip the computer. Of course, sag to a voltage above 50% can be withstood permanently by this computer. This results in what is called a "rectangular voltage tolerance curve," shown in Figure 2.29.

2.6.6 Adjustable AC Drive System

ASDs are fed through either a three-phase diode rectifier or a three-phase controlled rectifier. Generally speaking, the first type is found in AC motor drives, the second in DC drives and in large AC drives. The configuration of most AC drives is as shown in Figure 2.30. The three voltages are fed to a three-phase diode rectifier. The output voltage of the rectifier is smoothened by means of a DC capacitor. The inductance present in some drives aims at smoothing the DC link current and so reduces the harmonic distortion in the current taken from the supply. Here we will only consider the effect of the capacitor. The DC voltage is inverted to an AC voltage of variable frequency and magnitude. The motor speed is controlled through the magnitude and frequency of the output voltage of the VSC. For AC motors, the rotational speed is mainly determined by the frequency of the stator voltages. Thus, by changing the frequency an easy method of speed control is obtained. ASDs are often very sensitive to voltage sags. Typical AC drives topology.

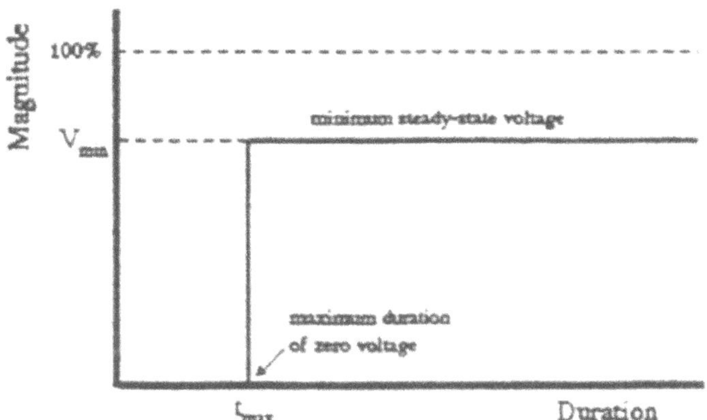

FIGURE 2.29 Voltage tolerance curve of a computer.

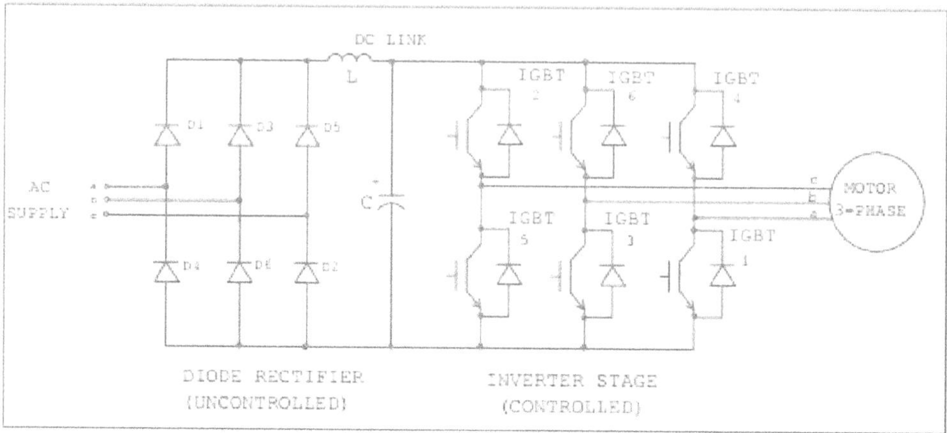

FIGURE 2.30 AC drive configuration.

Tripping of ASDs occurs due to several phenomena:

- The drive controller or protection will detect the sudden change in operating conditions and trip the drive to prevent damage to the power electronic components. Tripping of the drive is mainly on DC bus undervoltage, sometimes on AC bus undervoltage, on DC voltage ripple, or on missing pulses through the rectifier diodes.

  The increased AC currents during the sag or the post sag over currents charging the DC capacitor will cause an over current trip or blowing of fuses protecting the power electronics components. This effect is normally considered in the drive design by setting the DC bus undervoltage protection such that the drive will trip before a dangerous over current can occur.
- The process driven by the motor will not be able to tolerate the drop in speed or the torque variations due to the sag.
- During unbalanced sag, the currents through the rectifier diodes become unbalanced as well. Already a small unbalance in voltage can lead to a large unbalance in current, with one current twice as large as number and another current zero. The large current may lead to component damage and tripping of the over current protection.

Most of the existing drives still trip on DC bus undervoltage. Some of the more modern drives restart immediately when the voltage comes back; others restart after a certain delay time or only after a manual restart. The various automatic restart options are only relevant when the process tolerates a certain level of speed and torque variations.

2.6.7 Adjustable DC Drives

The DC drive behaves much differently than the typical AC drive due to its topology and its lack of inherent energy storage. A typical DC drive layout is shown in Figure 2.31. The AC source is coupled to the armature of the DC motor via a controlled rectifier. Although it has many variations, this 6-pulse six-SCR topology is the most common. During normal operation, an SCR is fired every 1/6 of a cycle in a sequence determined by the phase rotation of the AC supply, with each SCR conducting for 2/6 of a cycle. The gate timing of the SCRs is precisely controlled relative to the supply voltage waveform to yield an output with the desired average voltage. Usually a feedback loop (voltage or speed) controls this firing angle.

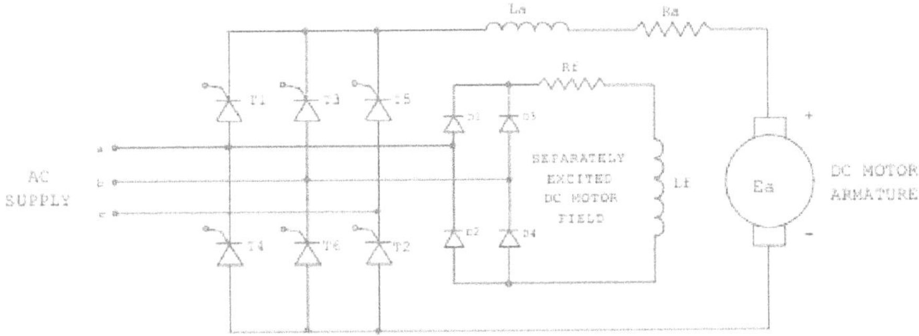

FIGURE 2.31 Typical DC drive topology.

Modern DC drives incorporate timing circuitry that normally synchronizes to the zero crossing of one of the line voltages. A PLL stabilizes the timing circuitry. Phase rotation is determined automatically, thus the firing sequence is determined by what the drive's internal sensing circuitry. Most drives monitor the rms value of the incoming waveform (usually through a peak detecting circuit) and will trip for undervoltage, overvoltage, phase loss, and excessive current, either line current and/or armature current. A voltage sag can create havoc with the timing circuitry for a DC drive. The firing circuit timing requires a solid zero crossing and looks for this zero crossing to occur at a regular interval. During an asymmetrical sag, the phase jump and/or sharp discontinuity in the voltage waveform can fool the drive and cause it to lose synchronization with the source. The PLL circuit can usually hold the synchronization for a short time; often the sag will recover within a few cycles. Too often, however, the drive will experience confusion in the timing circuit. Some drives will initially disable firing of the SCRs during the anomaly and allow the motor to coast. Other times, the phase jump will cause a phase-loss detector to trip the drive. In an extreme case, the phase sequence detector will sense a reversal of the phase rotation of the source and begin firing the SCRs in the incorrect sequence. In this case, the drive will then usually trip due to over current or fuses will blow.

2.6.7.1 Other Sensitive Loads

Intermixing loads can cause PQ problems in any facility. When nonsensitive and sensitive loads are connected to the same circuit, they often interact with one another. For example, when a large motor on an elevator or an air conditioner starts, it causes a large inrush current that can cause voltage sag. The voltage sag inside a facility has the same effect that voltage sag has outside of the facility. It causes lights to dim and computer equipment to malfunction. The solution is to not connect nonsensitive loads that will interact with sensitive loads. Wiring sensitive loads to separate circuits connected to the main electrical service panel separates sensitive loads from nonsensitive loads.

2.7 Stochastic Assessment of Voltage Sag

2.7.1 Compatibility between Equipment and Supply

Stochastic assessment of voltage sags is needed to find out whether a piece of equipment is compatible with the supply. A study of the worst-case scenario is not feasible as the worst-case voltage disturbance is a very long interruption. In some cases, a kind of "likely-worst-case scenario" is chosen, e.g., a fault close to the equipment terminals, cleared by the primary protection, not leading to an interruption. But that will not give any information about the likelihood of an equipment trip. To obtain information like that, a "stochastic compatibility assessment" is required. Such a study typically consists of three steps:

1. Obtain system performance. Information must be obtained on the expected number of voltage sags with different characteristics for the supply point. There are various ways to obtain this information: contacting the utility, monitoring the supply for several months or years, or doing a stochastic prediction study. Both voltage sag monitoring and stochastic prediction will be discussed in detail in this chapter.
2. Obtain equipment voltage tolerance. Information has to be obtained on the behavior of the piece of equipment for various voltage sags. This information can be obtained from the equipment manufacturer, by doing equipment tests, or simply by taking typical values for the voltage tolerance.
3. Determine expected impact. If the two types of information are available in an appropriate format, it is possible to estimate how often the piece of equipment is expected to trip per year, and what the (e.g., financial) impact of that will be. Based on the outcome of this study one can decide to opt for a better supply, for better equipment, or to be satisfied with the situation.

An example of a stochastic compatibility assessment will be given, based on Figure 2.32. The aim is to compare two supply alternatives and two equipment tolerance. The two supply alternatives are indicated in Figure 2.32 through the expected number of sags as a function of the sag severity: supply I is indicated through a solid line and supply II through a dashed line. We further assume the following costs to be associated with the two supply alternatives and the two devices (in arbitrary units):

Supply I 200 units
Supply II 500 units
Device A 100 units
Device B 200 units

We also assume that the costs of an equipment trip are 10 units. One can read the number of spurious trips per year, for each of the four design options, at the intersection between the supply curve and the device (vertical) line. For device A and supply I, we find 72.6 spurious equipment trips per year, etc. The results are shown in Tables 2.3 and 2.4. Knowing the number of trips per year, the annual costs of each

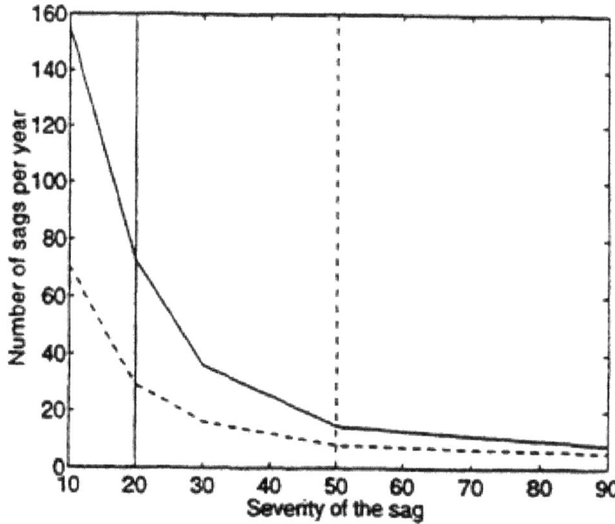

FIGURE 2.32 Comparison of two supply alternatives and two equipment tolerances.

TABLE 2.3

Number of Spurious Trips Per Year for Four Design Alternatives

	Supply 1	Supply 2
Device 1	72.6	29.1
Device 2	14.6	7.9

TABLE 2.4

Total Cost Per Year of Four Design Alternatives

	Supply 1	Supply 2
Device A	1026	891
Device B	546	779

of the four design options, and the costs per spurious trip, it is easy to calculate the total annual costs. For the combination of device A and supply I these costs are: $72.6 \times 10 + 100 + 200 = 1026$.

In practical cases, two additional problems have to be solved before the actual comparison can be made. At first one needs to obtain the data, both about the supply performance and about the equipment voltage tolerance. Methods for obtaining the system performance are discussed further on in this part. For obtaining the data, a customer often needs co-operation from the utility and from the equipment manufacturer. The second problem that has to be solved is the presentation of the data. System performance and equipment immunity are normally not one-dimensional, as suggested in the above example. We already saw that for voltage sags both magnitude and duration play a role, and possibly also unbalance and phase angle jump. The data has to be presented in such a way that a compatibility study can be made. Some suggestions for this are given in the next section.

2.7.1.1 Presentation of Results: Voltage Sag Co-ordination Chart

1. **The Scatter Diagram**

   When the supply is monitored for a certain period of time, a number of sags will occur. Each sag can be characterized with its own magnitude and duration and be plotted as a point in the magnitude duration plane. An example of a resulting scatter diagram is shown in Figure 2.33. The scatter diagram is obtained from one year of monitoring at an industrial site. For a large PQ survey, the scatter diagrams of all the sites can be merged.

2. **The Sag Density Table**

   A straightforward way of quantifying the number of sags is through a table with magnitude and duration ranges. The length of each bar is now proportional to the number of sags in the corresponding range. The bar chart is easier to get an impression of the distribution of the sag characteristics, but it is less useful to get numerical values. In this case we see from Figure 2.34 that the majority of sags have a magnitude above 80% and duration less than 200 ms. There is also a concentration of short interruptions with durations of 800 ms and over.

3. **The Cumulative Table**

   The interest to the customer is not so much the number of sags in a given magnitude and duration range, but the number of times that a certain piece of equipment will trip due to sag. It therefore makes sense to show the number of sags worse than a given magnitude and duration. For this a so-called "cumulative sag table" can be calculated.

4. **The Voltage Sag Co-ordination Chart**

   The values in the cumulative table belong to a continuous monotone function: the value increases towards the left-rear corner in Figure 2.35. The values shown in Table 2.4 can thus be

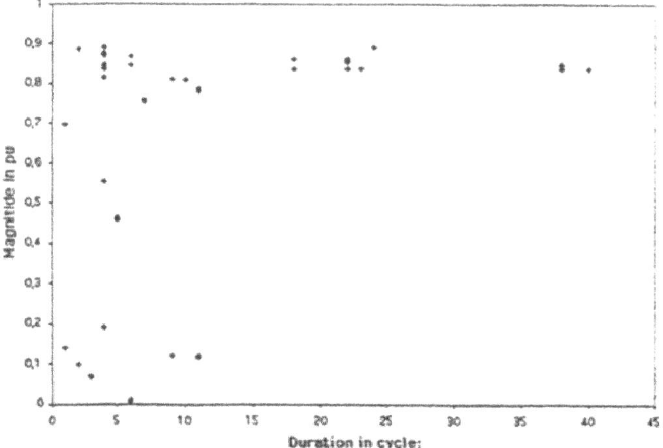

FIGURE 2.33 Scatter diagram.

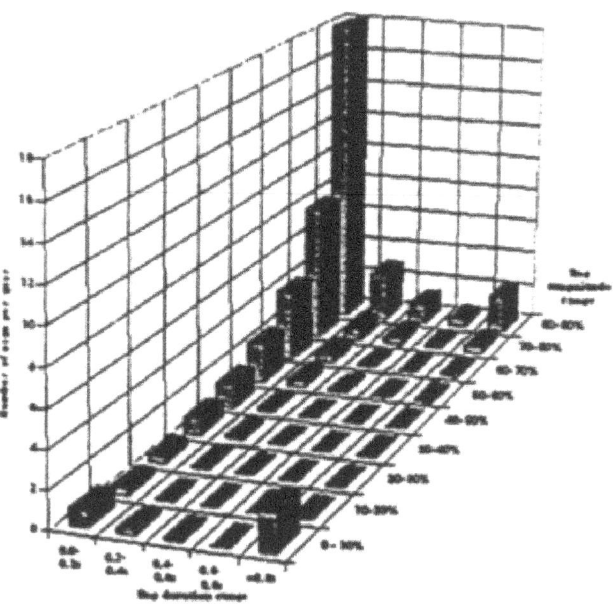

FIGURE 2.34 Two-dimensional bar chart of sag density.

seen as a two-dimensional function of number of sags versus magnitude and duration. A common way of presenting a two-dimensional function is through a contour chart. This was done by Conrad for the two-dimensional cumulative sag function, resulting in Figure 2.36. The contour chart is recommended as a "voltage sag co-ordination chart" in IEEE Standard 493 [1,2] and in IEEE Standard 1346.

In a voltage sag co-ordination chart the contour chart of the supply is combined with the equipment voltage tolerance curve to estimate the number of times the equipment will trip. Drawn in the chart are two equipment voltage-tolerance curves. Both curves are rectangular; i.e., the equipment trips when the voltage drops below a certain voltage for longer than a given duration.

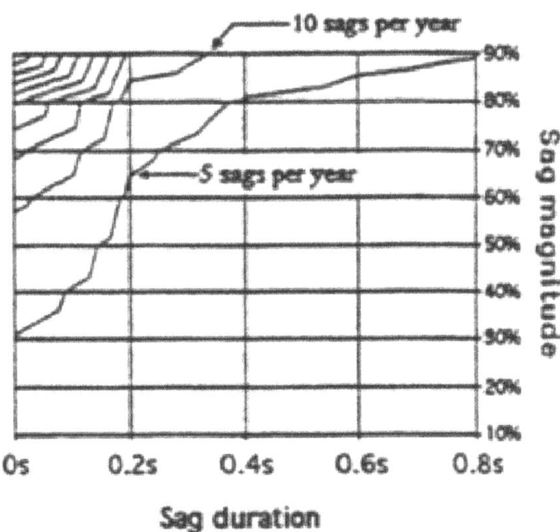

FIGURE 2.35 Counter chart for cumulative sag.

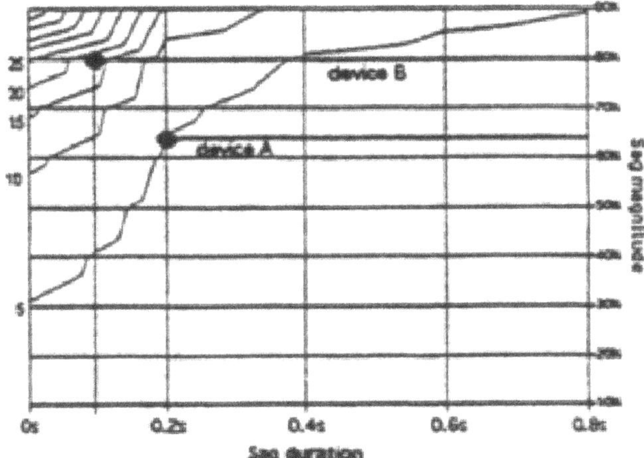

FIGURE 2.36 Voltage chart co-ordination chart.

2.8 Mitigation of Voltage Sag

2.8.1 From the Fault to Trip

To understand the various ways of mitigation, the mechanism leading to an equipment trip needs to be understood. The equipment trip is what makes the event a problem; if there were no equipment trips, there would be no voltage quality problem. The underlying event of the equipment trip is a short-circuit fault: a low impedance connection between two or more phases, or between one or more phases and ground. At the fault position the voltage drops to zero, or to a very low value. This zero voltage is changed into an event of a certain magnitude and duration at the interface between the equipment and the power system. The short-circuit fault will always cause a voltage sag for some customers. If the fault takes place in a radial part of the system, the protection intervention clearing the fault will also lead to

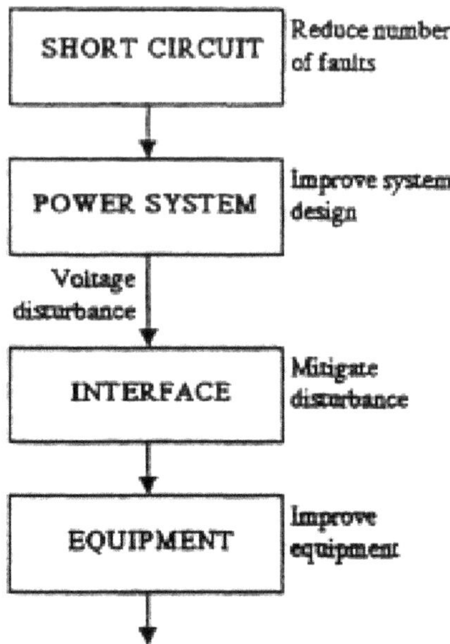

FIGURE 2.37 Voltage quality problem and ways of mitigation.

an interruption. If there is sufficient redundancy present, the short circuit will only lead to voltage sag. If the resulting event exceeds a certain severity, it will cause an equipment trip. Figure 2.37 enables us to distinguish between the various mitigation methods:

- Reducing the number of short-circuit faults.
- Reducing the fault-clearing time.
- Changing the system such that short-circuits faults result in less severe events at the equipment terminals or at the customer interface.
- Reducing the number of faults improves system design.
- Connecting mitigation equipment between the sensitive equipment and the supply.
- Improving the immunity of the equipment.

2.8.2 Reducing the Number of Faults

Reducing the number of short-circuit faults in a system reduces not only the sag frequency but also the frequency of sustained interruptions. This is thus a very effective way of improving the quality of supply and many customers suggest this as the obvious solution when a voltage sag or short interruption problem occurs. The solution is unfortunately most of the time not that obvious. A short circuit not only leads to a voltage sag or interruption at the customer interface but also causes damage to utility equipment and plant. Therefore, most utilities will already have reduced the fault frequency as far as economically feasible. In individual cases there could still be room for improvement, e.g., when the majority of trips is due to faults on one or two distribution lines. Some examples of fault mitigation are:

- Replace overhead lines by underground cables.
- Use special wires for overhead lines.
- Implement a strict policy of tree trimming.

- Install additional shielding wires.
- Increase the insulation level.
- Increase maintenance and inspection frequencies.

One has to keep in mind however that these measures can be very expensive and that its costs have to be weighed against the consequences of the equipment trips.

2.8.3 Reducing the Fault-Clearing Time

Reducing the fault-clearing time does not reduce the number of events but only their duration. The ultimate reduction of fault-clearing time is achieved by using current-limiting fuses (a proven technology) or static circuit breakers (an emerging technology). These devices are able to clear a fault within one half-cycle, thus ensuring that no voltage sag lasts longer. Additionally, several types of fault current limiters have been proposed that not so much clear the fault, but significantly reduce the fault current magnitude within one or two cycles. But the fault-clearing time is not only the time needed to open the breaker, also the time needed for the protection to make a decision. Here we need to consider two significantly different types of distribution networks, both shown in Figure 2.38. The top drawing in Figure 2.38 shows a system with one circuit breaker protecting the whole feeder. The protection relay with the breaker has a certain current setting. This setting is such that it will be exceeded for any fault on the feeder but not exceeded for any fault elsewhere in the system or for any loading situation. The moment the current value exceeds the setting the relay gives a trip signal to the breaker and the breaker opens within a few cycles. Typical fault-clearing times in these systems are around 100 ms. To limit the number of long interruptions for the customers, reclosing is used in combination with (slow) expulsion fuses in the laterals or in combination with interrupters along the feeder.

This type of protection is commonly used in overhead systems. Reducing the fault-clearing time mainly requires a faster breaker. The static circuit breaker or several of the other current limiters would be good options for these systems. A current-limiting fuse to protect the whole feeder is not

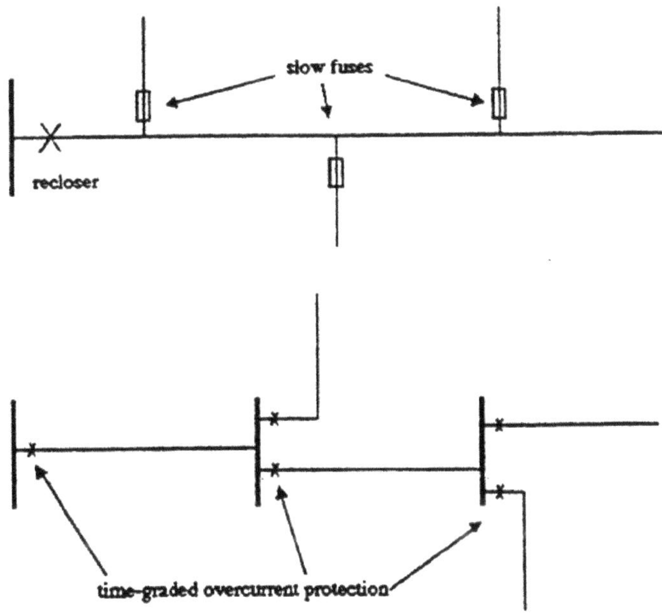

FIGURE 2.38 Distribution system with one circuit breaker protecting the whole feeder and with number of substations.

suitable as it makes fast reclosing more complicated. Current-limiting fuses can also not be used for the protection of the laterals because they would start arcing before the main breaker opens. Using a faster clearing with the main breaker enables faster clearing in the laterals as well. The network in the bottom drawing of Figure 2.38 consists of a number of distribution substations in cascade. To achieve selectivity, time grading of the over current relay is used. The relays furthest away from the source trip instantaneously on over current. When moving closer to the source, the tripping delay increases each time with typically 500 ms. In the example in Figure 2.38, the delay times would be 1000, 500 ms, and zero (from left to right). Close to the source, fault-clearing times can be up to several seconds. These kinds of systems are used in underground networks and in industrial distribution systems. The fault-clearing time can be somewhat reduced by using inverse-time over current protection where the delay time decreases for increasing fault current. But even with these schemes, fault-clearing times above one second are possible. The various techniques for reducing the fault clearing time without losing selectivity are discussed in various publications on power system protection. To achieve a serious reduction in fault-clearing time one needs to reduce the grading margin, thereby allowing a certain loss of selectivity. The setting rules described in most publications are based on preventing incorrect trips. Future protection settings need to be based on a maximum fault-clearing time. The latter curve can be used in combination with relay curves to obtain the various settings. The opening time of the downstream breaker is an important term in the expression for the grading margin. By using faster breakers, the grading margin can be significantly reduced, thus leading to a significant reduction in fault-clearing time. The impact of static circuit breakers might be bigger in these systems that in the ones with one breaker protecting the whole feeder. In transmission systems the fault-clearing time is already short, so further reduction is much more difficult. The fault-clearing time is often limited by transient-stability constraints.

In some cases, faster circuit breakers could be of help. This again not only limits the fault-clearing time directly, but it also limits the grading margin for distance protection. One should realize however that faster circuit breakers could be very expensive.

- A certain reduction in grading margin is probably possible. This will not so much reduce the fault clearing time in normal situations, but in case the protection fails and a backup relay have to intervene. When reducing the grading margin, one should realize that loss of selectivity is unacceptable in most transmission systems as it leads to the loss of two or more components at the same time.
- Faster back-up protection is one of the few effective means of reducing fault-clearing time in transmission systems. Possible options are to use intercropping for distance protection and breaker-failure protection.

2.8.4 Including Changes in Power System

By implementing changes in the supply system, the severity of the event can be reduced. Here again the costs can become very high, especially for transmission and sub-transmission voltage levels. Some examples of mitigation methods especially directed towards voltage sags are:

1. Install a generator near the sensitive load. The generators will keep up the voltage during remote sag. The reduction in voltage drop is equal to the percentage contribution of the generator station to the fault current. In case a combined-heat-and-power station is planned, it is worth to consider the position of its electrical connection to the supply.
2. Split buses or substations in the supply path to limit the number of feeders in the exposed area.
3. Install current-limiting coils at strategic places in the system to increase the "electrical distance" to the fault. One should realize that this can make the sag worse for other customers.

4. Feed the bus with the sensitive equipment from two or more substations. A voltage sag in one substation will be mitigated by the indeed from the other substations. The more independent the substations are the more the mitigation effect. The best mitigation effect is by feeding from two different transmission substations. Introducing the second in feed increases the number of sags, but reduces their severity. The number of short interruptions can be prevented by correcting fewer customers to one re-closer (thus by installing more recloses), or by getting rid of the re-closure scheme altogether. Short as well as long interruptions are considerably reduced in frequency by installing additional redundancy in the system. The costs for this are only justified for large industrial and commercial customers. Intermediate solutions reduce the duration of (long) interruptions by having a level of redundancy available within a certain time.

2.8.5 Installing Mitigation Equipment

The most commonly applied method of mitigation is the installation of additional equipment at the system equipment interface. Also, recent developments point towards a continued interest in this way of mitigation. The popularity of mitigation equipment is explained by it being the only place where the customer has control over the situation. Both changes in the supply as well as improvement of the equipment are often completely outside of the control of the end-user. Some examples of mitigation equipment are

- Dynamic voltage restorer (DVR). This device uses modern power electronic components to insert a series voltage source between the supply and the load. The voltage source compensates for the voltage drop due to the sag. Some devices use internal energy storage to make up for the drop in active power supplied by the system. They can only mitigate sags up to a maximum duration. Other devices take the same amount of active power from the supply by increasing the current. These can only mitigate sags down to a minimum magnitude. This is also a rather new and promising technology, available both for medium voltage and low voltage levels. Also, a number of alternative configurations have been suggested, some more promising than others. For low voltage equipment this new technology may not add much above a UPS, for medium voltage load, this may prove a very expensive but the only feasible solution. Motor-generator sets are the classical solution for sag and interruption mitigation with large equipment. They are obviously not suitable for an office environment but the noise and the maintenance requirements are often no objection in an industrial environment. Some manufacturers combine the MG-set with a backup generator; others combine it with power-electronic converters to obtain a longer ride-through time.

2.8.6 Improvising Equipment Immunity

Improvement of equipment immunity is probably the most effective solution against equipment trips due to voltage sags. But as a short-time solution it is often not suitable. A customer often only finds out about equipment immunity after the equipment has been installed. For consumer electronics it is very hard for a customer to find out about immunity of the equipment as he is not in direct contact with the manufacturer. Even most ASDs have become off-the-shelf equipment where the customer has no influence on the specifications. Only large industrial equipment is custom-made for a certain application, which enables the incorporation of voltage tolerance requirements.

For short interruptions equipment immunity is very hard to achieve; for long interruptions it is impossible to achieve. The equipment should in so far as possible be immune to interruptions that no damage is caused and no dangerous situation arises. This is especially important when considering a complete installation.

2.9 Different Events and Mitigation Methods

For different events different mitigation strategies apply

- Sags due to short-circuit faults in the transmission and sub-transmission system are characterized by a short duration, typically up to 100 ms. These sags are very hard to mitigate at the source and also improvements in the system are seldom feasible. The only way of mitigating these sags is by improvement of the equipment or, where this turns out to be unfeasible, installing mitigation equipment. For low-power equipment a UPS is a straightforward solution; for high-power equipment and for complete installations several competing tools are emerging.
- As we saw before the duration of sags due to distribution system faults depends on the type of protection used. Ranging from less than a cycle for current-limiting fuses up to several seconds for over current relays in underground or industrial distribution systems. The long sag duration makes that ion feeders fed from another HV/MV substation. For deep long-duration sags, equipment improvement becomes more difficult and system improvement easier. The latter could well become the preferred solution, although a critical assessment of the various options is certainly needed.
- Sags due to faults in remote distribution systems and sags due to motor starting should not lead to equipment tripping for sags down to 85%. If there are problems the equipment needs to be improved. If equipment trips occur for long-duration sags in the 70%–80% magnitude range, changes in the system have to be considered as an option.
- For interruptions, especially the longer ones, equipment improvement is no longer feasible. System improvements or a UPS in combination with an emergency generator are possible solutions here.

2.10 Voltage Imbalance and Voltage Fluctuation

2.10.1 Voltage Imbalance

Electrical equipment, especially motors and their controllers, will not operate reliably on unbalanced voltages in a three-phase system. Generally, the difference between the highest and the lowest voltages should not exceed 4% of the lowest voltage. Greater imbalances may cause overheating of components and result in intermittent shutdown of motor controllers. Motors operated on unbalanced voltages will overheat, and many overload relays can't sense the overheating. In addition, many solid-state motor controllers and inverters include components that are sensitive to voltage imbalances.

Consider a three-phase circuit with the lowest voltage on first phase is 230 V, while another phase is 235 V, and the third phase is 240 V. To determine the voltage imbalance, apply the 4% rule.

Four percent of the lowest voltage (230 V) is 9.2 V. The difference between the highest voltage (240 V) and the lowest voltage (230 V) is 10 V. The difference of 10 V is greater than 4% of the lowest voltage (9.2 V), which makes it a greater imbalance. The procedure for calculating voltage imbalance with precision as follows: The first step is to calculate the average voltage by adding all three phases and dividing by 3. In our example, the average is 235 V. The sum of voltage 705 V, which is divided by the number of phases; this makes 235 V. Next, add up the absolute differences between each phase voltage and the average voltage. In this case, the difference between the average voltage and 230 V is 5 V. The difference between the average and itself is 0 V; and the difference between the average and 240 V is 5 V. Adding up the differences, we get 10 V. And that 10 V is what we call the total imbalance. Now divide the total imbalance in half to get an adjusted imbalance. Half of 10 V is 5 V. Finally, divide the adjusted imbalance into the average voltage to get a percentage imbalance. In this case, 5 V/235 V give 0.021. That is 2.1% imbalance.

Reliable, long-term operation of most electrical equipment requires a voltage imbalance of less than 2%. This means the example considered has too much imbalance.

If voltage imbalances are identified in the locality, check for electrical distribution systems in which one leg of a three-phase supply powers both single-phase and three-phase loads. You may find single phase loads not evenly balanced across the phases. Or, look for in-line reactors installed to correct imbalances. These reactors usually have taps for adjustment, and somebody may have adjusted them. Or, the imbalance they originally corrected may have shifted over time. Circuits with tapped reactors rarely stay in balance indefinitely.

2.10.2 Voltage Fluctuation

Voltage fluctuation is defined as a series of rms voltage changes or a cyclic variation of the voltage waveform envelope. The defining characteristics of voltage fluctuations are:

- The amplitude of voltage change (difference of maximum and minimum rms or peak voltage value occurring during the disturbance)
- The number of voltage changes over a specified unit of time
- The consequential effects (such as flicker) of voltage changes associated with the disturbances

Voltage fluctuations in power systems can cause a number of harmful technical effects, resulting in disruption to production processes and substantial costs. But flicker, with its negative physiological results, can affect worker safety as well as productivity.

Humans can be sensitive to light flicker caused by voltage fluctuations. Flicker can significantly impair our vision and cause general discomfort and fatigue. In some situations, it can even result in workplace accidents because it affects the ergonomics of the production environment by causing operator fatigue and reduced concentration levels.

Flicker is the impression of unsteadiness of visual sensation induced by a light source whose luminance or spectral distribution fluctuates with time. Usually, it applies to cyclic variation of light intensity resulting from fluctuation of the supply voltage, which, in turn, can be caused by disturbances introduced during power generation, transmission, or distribution. However, flicker is typically caused by the use of large loads having rapidly fluctuating active and reactive power demand. The phenomenon of flickering in light sources has been an issue since the beginning of power distribution systems. However, with the increase in number of customers and installed power, the problem of flicker has grown rapidly. Study reveals voltage fluctuations as a cause of flicker, what prompts fluctuations, methods of their mitigation, and applicable standards regarding flicker levels.

2.10.2.1 Causes of Voltage Fluctuations

Theoretically, for any supply line, the voltage at the load end is different from that at the source. We can demonstrate this with a per-phase equivalent circuit, as shown in Figure 2.39. Here, E is the source voltage, R_S is the equivalent line resistance, X_S is the equivalent line reactance, Z_S is the equivalent line

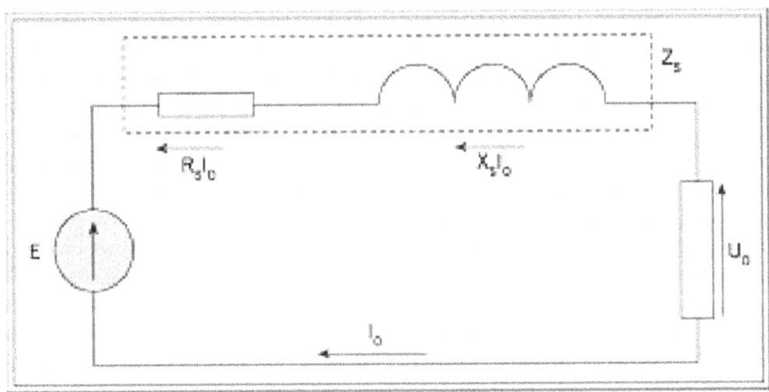

FIGURE 2.39 Per-phase equivalent circuit.

impedance, I_O is the current, and U_O is the voltage at the load terminals. Depending on its cause, a voltage change can take the form of a voltage drop having a constant value over a long time interval, a slow or rapid voltage change, or a voltage fluctuation.

Voltage fluctuations in power systems and at the load terminals were characterized using factors associated with the peak-to-peak rms voltage change in the power system. The energy of voltage fluctuations, their power spectrum (also called the energy spectrum of voltage fluctuations), and their duration were considered when assessing voltage fluctuations. Currently, the basic parameters that determine voltage fluctuations are short-term flicker severity, called the P_{ST} index, and long-term flicker severity, called the P_{LT} index. These parameters refer to voltage fluctuation effects on lighting and their influence on humans.

2.10.2.2 Sources of Voltage Fluctuations

1. The primary cause of voltage changes is the time variability of the reactive power component of fluctuating loads. Such loads include arc furnaces, rolling mill drives, and main winders—all of which are loads with a high rate of change of power with respect to the short-circuit capacity at the PCC.
2. Small power loads, such as starting of induction motors, welders, boilers, power regulators, electric saws and hammers, pumps and compressors, cranes, and elevators also can be sources of flicker.
3. Other causes are capacitor switching and on-load transformer tap changers, which can change the inductive component of the source impedance. Variations in generation capacity of wind turbines, for example, also can have an effect. Sometimes, voltage fluctuations are caused by low-frequency voltage inter-harmonics.

2.10.2.3 Mitigation of Voltage Fluctuations in Power Systems

The effects of voltage fluctuations depend on (i) amplitude, which is influenced by the characteristics of the power system, and (ii) rate of their occurrence, which is determined by the type of load and character of its operation. Usually, mitigation measures focus on limiting the amplitude of the voltage fluctuations. The technological process is seldom influenced.

Examples of mitigation methods for various types of equipment include:

1. Arc furnaces—Incorporate series reactors (or variable saturation), which ensure proper functioning of the electrode control system, segregate and provide preliminary heating of charge.
2. Welding plants—Supply the plant from a dedicated transformer, connect single-phase welders to a three-phase network for balanced load distribution between phases; connect single-phase welding machines to different phases from those powering lighting equipment.
3. Adjustable speed drives—Use soft-start devices.

Another way to reduce the amplitude of voltage fluctuations is to increase the short-circuit power, with respect to the load power, at the PCC to which a fluctuating load is connected. This can be done by:

- Connecting the load at a higher nominal voltage level
- Supplying this category of loads from dedicated lines
- Separating supplies to fluctuating loads from steady loads by using separate windings of a three-winding transformer
- Increasing the rated power of the transformer supplying the fluctuating load
- Installing series capacitors

2.10.3 Voltage Stabilization Solutions

Another way of reducing the amplitude of voltage fluctuations is to reduce the changes of reactive power in the supply system. This is implemented by installing dynamic voltage stabilizers. Their effectiveness depends mainly on their rated power and speed of reaction. By drawing reactive power at the fundamental frequency, dynamic voltage stabilizers produce voltage drops on the supply network impedances. Depending on whether the reactive power is inductive or capacitive, the rms voltage value at the PCC can be increased or reduced.

2.11 Waveform Distortion

Any undesired changes in the waveform of a signal passing through a circuit, device, or transmission medium are called distortion. Such distortion includes amplitude, harmonic, inter modulation, and phase distortion.

1. **Typical causes**
   Voltage waveform distortion typically relates to electronic equipment, which has an internal switch-mode power supply (SMPS) that draws a nonlinear current waveform. Whereas a linear load produces a sine wave current, the SMPS draws current pulses at only one portion of the applied voltage waveform. This nonlinear current increases with each added device. The nonlinear current, combined with the impedance of the circuit conductors and the power source, creates a voltage drop you see at the peak portions of the voltage waveform. The heavier the loading, the greater the root-means-square (rms) voltage drop. Recent changes in SMPS design and efficiency decreased the amplitude of current and, therefore, the amount of voltage waveform distortion.
2. **Effects on equipment**
   Most equipment can operate properly with little voltage waveform distortion and still maintain a regulated DC output. However, with greater distortion, efficiency of the power system and loads decreases. Feeder and branch-circuit lengths also affect the amount of distortion. Overloaded branch circuits and associated wiring can't accommodate increased current flow.
3. **Investigation techniques**
   Voltage waveform distortion can cause equipment malfunction. AC electrical panelboard feeding the affected equipment is to be investigated. The following are the precautions to be ensured during investigation.
   a. Check for tight connections.
   b. Perform true rms current measurements around each of the phase conductors. Make sure there are no overloaded conductors. Verify there is not a significant imbalance between any of the phase conductors. Also, make sure the individual phase conductor currents do not exceed 20% of the average phase current.
   c. In any instances where significant waveform distortion exists, the neutral conductor current is equal to or greater than the average phase current. Verify the neutral conductor current is at least 15% less than the average phase current.
   d. Compare the phase-neutral voltages at the panel board with phase-neutral voltage measurements made at the branch circuit. If the voltage drop is greater than 3% across the branch circuit, then it's possible to overload a particular circuit, or a loose connection exists on the hot or neutral conductors.
   e. Always make sure and measure the voltage across the circuit breaker contacts. This is a good indication of the unit's integrity. If a voltage drop of more than 50 mV exists across the contacts, replace the breaker.

f. Use a harmonic analyzer to determine the total harmonic distortion (THD) for the various areas of the facility. Using IEEE Std. 519 as a guide, verify the voltage THD does not exceed 5%. Also make sure the phase currents do not exceed 20% THD.
g. Even though Std. 519 deals specifically with the PCC, it's good practice to carry over the same logic to the secondary of isolation transformers.

4. **Solutions**

Consider the following approaches to address and troubleshoot voltage waveform distortion problems.

a. Reduce transformer loading

Since the transformer provides power to nonlinear loads, it should not be loaded more than 60% of its kVA nameplate rating. This practice reduces distortion and prevents overheating.

b. Increase the size of the neutral conductor

This reduces the impedance of the neutral conductor path. Lower neutral conductor impedance decreases the voltage drop and reduces waveform distortion.

c. Limit the length of feeder circuits to less than 200 ft. This reduces the overall impedance of the power system.
d. Limit the length of branch circuits to less than 50 ft.
e. Eliminate shared neutral conductors and "daisy-chained" circuits

When wiring for electronic equipment, it's best to wire no more than five outlets off of a single circuit breaker.

f. Reduce the amount of load on a branch-circuit panel board.

Voltage waveform distortion exists on every site. If it's a cause of equipment malfunction or overloaded power sources, the above recommendations should improve the efficiency of your system and extend the life of your equipment.

2.11.1 Power Frequency Variation

2.11.1.1 Variation from Rated Voltage

In accordance with NEMA MG 1, 12.44, motors shall operate successfully under running conditions at rated load with variation in the voltage up to the following percentages of rated voltage:

1. Universal motors except fan motors—plus or minus 6% (with rated frequency).
2. Induction motors—plus or minus 10% (with rated frequency).

Performance within these voltage variations will not necessarily be in accordance with the standards established for operation at rated voltage.

2.11.1.2 Variation from Rated Frequency

Alternating-current motors shall operate successfully under running conditions at rated load and at rated voltage with a variation in the frequency up to 5% above or below the rated frequency. Performance within this frequency variation will not necessarily be in accordance with the standards established for operation at rated frequency.

2.11.1.3 Combined Variation of Voltage and Frequency

Alternating-current motors shall operate successfully under running conditions at rated load with a combined variation in the voltage and frequency up to 10% above or below the rated voltage and the rated frequency, provided that the frequency variation does not exceed 5%. Performance within this combined variation will not necessarily be in accordance with the standards established for operation at rated voltage and rated frequency.

2.11.1.4 Effects of Variation of Voltage and Frequency upon the Performance of Induction Motors

1. Induction motors are at times operated on circuits of voltage or frequency other than those for which the motors are rated. Under such conditions, the performance of the motor will vary from the rating. The following is a brief statement of some operating results caused by small variations of voltage and frequency and is indicative of the general changes produced by such variations in operating conditions.
2. With a 10% increase or decrease in voltage from that given on the nameplate, the heating at rated horsepower load may increase. Such operation for extended periods of time may accelerate the deterioration of the insulation system.
3. In a motor of normal characteristics at full rated horsepower load, a 10% increase of voltage above that given on the nameplate would usually result in a decided lowering in power factor. A 10% decrease of voltage below that given on the nameplate would usually give an increase in power factor.
4. The locked-rotor and breakdown torque will be proportional to the square of the voltage applied.
5. An increase of 10% in voltage will result in a decrease of slip of about 17%, while a reduction of 10% will increase the slip about 21%. Thus, if the slip at rated voltage were 5%, it would be increased to 6.05% if the voltage were reduced 10 percent.
6. A frequency higher than the rated frequency usually improves the power factor but decreases locked-rotor torque and increases the speed and friction and windage loss. At a frequency lower than the rated frequency, the speed is decreased, locked-rotor torque is increased, and power factor is decreased. For certain kinds of motor load, such as in textile mills, close frequency regulation is essential.
7. If variation in both voltage and frequency occur simultaneously, the effect will be superimposed. Thus, if the voltage is high and the frequency low, the locked-rotor torque will be greatly increased, but the power factor will be decreased and the temperature rise increased with normal load.
8. The foregoing facts apply particularly to general-purpose motors. They may not always be true in connection with special-purpose motors, built for a particular purpose, or as applied to very small motors.

2.11.1.5 Operation of General-Purpose Alternating-Current Polyphase 2-, 4-, and 8-Pole, 60 Hz Integral-Horsepower Induction Motors Operated on 50 Hz

Induction motors are not designed to operate at 60 Hz ratings on 50 Hz circuits, but they are capable of being operated satisfactorily on 50 Hz circuits if their voltage and horsepower ratings are appropriately reduced. When such 60 Hz motors are operated on 50 Hz circuits, the applied voltage at 50 Hz should be reduced to 5/6 of the 60 Hz horsepower rating of the motor.

When a 60 Hz motor is operated on 50 Hz at 5/6 of the 60 Hz voltage and horsepower ratings, the other performance characteristics for 50 Hz operation are as follows:

1. **Speed**

   The synchronous speed will be 5/6 of the 60 Hz synchronous speed and the slip will be 6/5 of the 60-Hz slip.

2. **Torque**

   The rated load torque in pound-feet will be approximately the same as the 60 Hz rated load torque in pound-feet. The locked-rotor and breakdown torques in pound-feet of 50 Hz motors will be approximately the same as the 60 Hz locked-rotor and breakdown torques in pound-feet.
3. **Locked-Rotor Current**

   The locked-rotor current (ampere) will be approximately 5% less than the 60 Hz locked-rotor current (amperes). The code letter appearing on the motor nameplate to indicate locked-rotor kVA per horsepower applies only to the 60 Hz rating of the motor.
4. **Service Factor**

   The service factor will be 1.0.
5. **Temperature Rise**

   The temperature rise should not exceed 90°C.

2.11.1.6 Effects of Voltages over 600 V on the Performance of Low-Voltage Motors

Polyphase motors are regularly built for voltage ratings of 575 V or less and are expected to operate satisfactorily with a voltage variation of plus or minus 10%. This means that motors of this insulation level may be successfully applied up to an operating voltage of 635 V.

Based on motor manufacturers' high-potential tests and performance in the field, it has been found that where service voltages exceed 635 V, the safety factor of the insulation has been reduced to a level inconsistent with good engineering procedures. Motors of this insulation level should not be applied to power systems either with or without grounded neutral where the voltage exceeds 630 V, regardless of the motor connection employed.

2.11.2 Electrical Noise

Noise, or interference, can be defined as undesirable electrical signals that distort or interfere with an original (or desired) signal. Noise could be transient (temporary) or constant. Unpredictable transient noise is caused, for example, by lightning. Constant noise can be due to the predictable 50 or 60 Hz AC 'hum' from power circuits or harmonic multiples of power frequency close to the data communications cable. This unpredictability makes the design of a data communications system quite challenging.

Noise can be generated from within the system itself (internal noise) or from an outside source (external noise). Examples of these types of noise are:

2.11.2.1 Internal Noise

- Thermal noise (due to electron movement within the electrical circuits)
- Imperfections (in the electrical design).

2.11.2.2 External Noise

- Natural origins (electrostatic interference and electrical storms)
- EMI—from currents in cables
- Radio frequency interference (RFI)—from radio systems radiating signals
- Cross talk (from other cables separated by a small distance).

From a general point of view, there must be three contributing factors before an electrical noise problem can exist.

They are:

1. A source of electrical noise
2. A mechanism coupling the source to the affected circuit
3. A circuit conveying the sensitive communication signals

Typical sources of noise are devices, which produce quick changes (spikes) in voltage or current or harmonics, such as:

- Large electrical motors being switched on
- Fluorescent lighting tubes
- Solid-state converters or drive systems
- Lightning strikes
- High-voltage surges due to electrical faults
- Welding equipment

Figure 2.40 shows a typical noise waveform when superimposed on the power source voltage waveform. Electrical systems are prone to noise due to various reasons. Lightning and switching surges produce very short duration of distortions of the voltage wave. Another common example is "notching," which appears in circuits using silicon-controlled rectifiers (power thyristors). The switching of these devices causes sharp inverted spikes during commutation (transfer of conduction from one phase arm to the next). Figure 2.41 shows the typical waveform with this type of disturbance. Switching of large loads in power circuits to which automatic data processing (ADP) loads are connected can also cause disturbances. Similarly, faults in power systems can cause voltage disturbances.

Apart from these directly communicated disturbances, sparks and arcing generated in power-switching devices and high-frequency harmonic current components can produce EMI in signal circuits, which will require proper shielding or screening to avoid interference.

The following general principles are applicable for reducing the effects of electrical noise:

- Physical segregation of noise sources from noise-sensitive equipment
- Electrical segregation

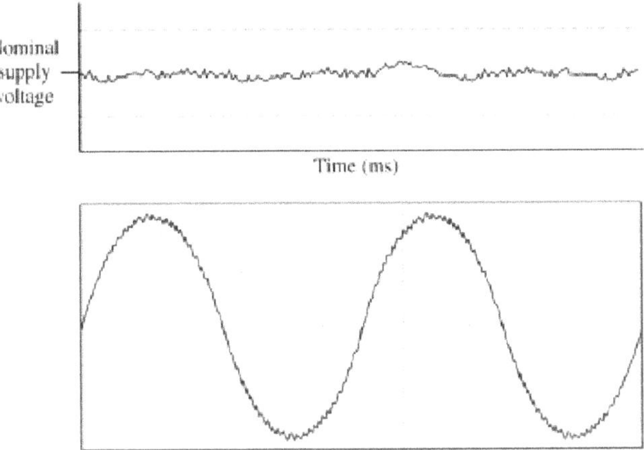

FIGURE 2.40 Noise signal (top) and noise over AC power (bottom).

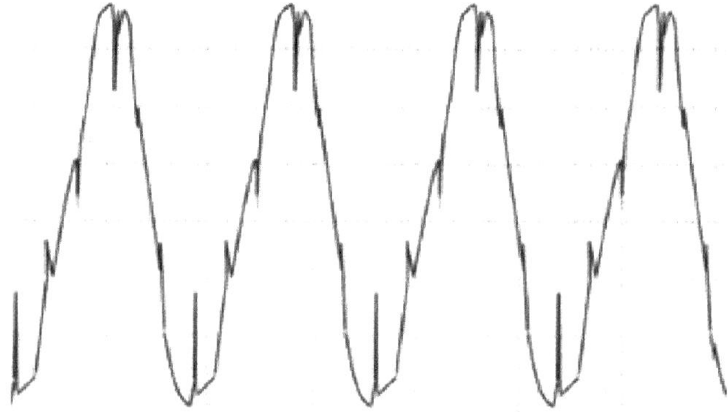

FIGURE 2.41 Waveform distorted by notching.

- Harmonic current control
- Avoiding ground loops, which are a major cause of noise propagation

The power and logic voltages of present-day devices have been drastically reduced and the speed of these devices has increased with propagation times, which are measured in picoseconds. While the speed of the equipment has gone up and the voltage sensitivity has gone down, the noise conditions coming from the power supply side have not reduced at all. Figure 2.42 shows the noise emanating from electrical systems within a facility.

The best illustration to explore noise signals is to consider the signal voltage as shown in Figure 2.43. Signal voltages may have been 30 V or more but since then have steadily been decreasing. As long as the signal voltage was high and the noise voltage was only 1 V, then it is said to be very high signal to noise ratio, 30:1.

Study reveals distinguishing the signal as long as high signal to noise ratio exist finds a stable system. As the electronic equipment industry advanced, the signal strength went down further, below 10 and then below 5. Today we are fighting 1-, 2-, and 3-V signals and still finding ourselves with 1, 2 and 3 V of electrical noise. When this takes place for brief periods of time, the noise signal may be larger than the actual signal. The sensors within the sensitive equipment turn and try to run on the noise signal itself as the predominant voltage.

When this takes place, a parity check or a security check signal is sent out from the sensitive equipment asking if this particular voltage is one of the voltages the sensor should recognize. Usually, this check fails

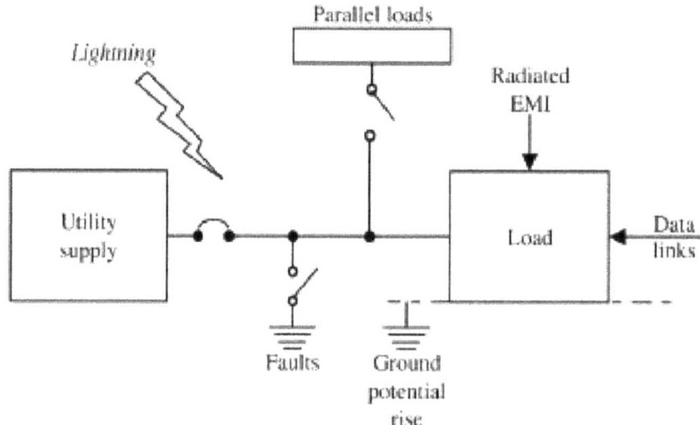

FIGURE 2.42 Noise emanating from electrical systems within a facility.

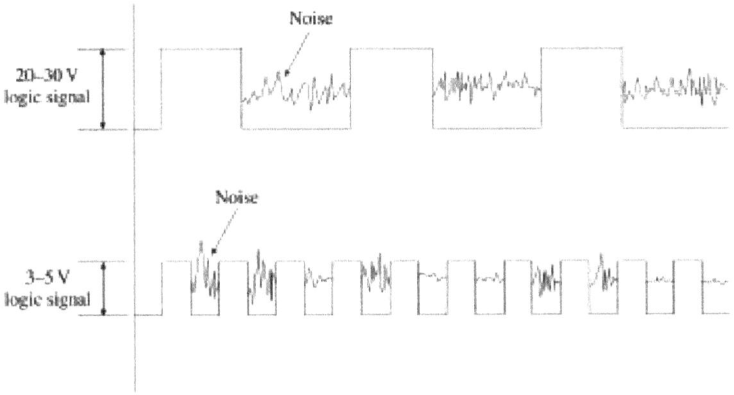

FIGURE 2.43 Relative magnitudes of signal and noise (then and now).

when it is a noise voltage rather than the proper signal that it should be looking at and the equipment shuts down because it has no signal. In other words, the equipment self-protects when there is no signal to keep it operating. When the signal to noise ratio has fallen from a positive direction to a negative direction, the equipment interprets that as the need to turn off so it will not be running on sporadic signals.

From Figure 2.44, the difference between the upper and lower pictures in the graph shows the speed with which the signal was transmitted. In the upper graph, the ONs and OFFs are relatively slow, evidenced by the large spaces between the traces. In the lower graph, the trace is now much faster. There are many more ONs and OFFs jammed into the same space and as such, the erratic noise behavior may now interfere with the actual transmission.

The ratio of the signal voltage to the noise voltage determines the strength of the signal in relation to the noise. In data communications, the signal voltage is relatively stable and is determined by the voltage at the source (transmitter) and the volt drop along the line due to the cable resistance (size and length). The SNR is therefore a measure of the interference on the communication link. The SNR is usually expressed in decibels (dB), which is the logarithmic ratio of the signal voltage (S) to noise voltage (N).

$$\text{SNR} = 10\log(S/N)\text{dB} \tag{2.15}$$

An SNR of 20 dB is considered low (bad), while an SNR of 60 dB is considered high (good). The higher the SNR, the easier it is to provide acceptable performance with simpler circuitry and cheaper cabling.

In data communications, a more relevant performance measurement of the link is the bit error rate (BER). This is a measure of the number of successful bits received compared to bits that are in error.

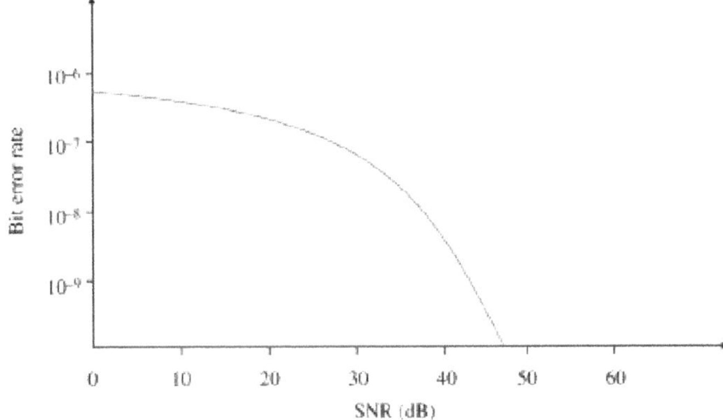

FIGURE 2.44 Relationship between the bit error rate and the signal to noise ratio.

A BER of 10^{-6} means that one bit in a million will be in error and is considered poor performance on a bulk data communications system with high data rates.

A BER of 10^{-12} (one error bit in a million) is considered to be very good. Over industrial systems, with low data requirements, a BER of 10^{-4} could be quite acceptable. There is a relationship between SNR and BER. As the SNR increases, the error rate drops off rapidly as is shown in Figure 2.44. Most of the communications systems start to provide reasonably good BERs when the SNR is above 20 dB.

2.11.2.3 Frequency Analysis of Noise

Another useful way of evaluating the effects of noise is to examine its frequency spectrum. Noise can be seen to fall into three groups:

1. Wideband noise
2. Impulse noise
3. Frequency-specific noise

The three groups are shown in the simplified frequency domain as well as the conventional time domain. In this way, we can appreciate the signal's changing properties as well as viewing the amplitude in the customary time domain.

Wideband noise contains numerous frequency components and amplitude values. These are depicted in the time domain graph shown in Figure 2.45 and in the frequency domain graph shown in Figure 2.46. In the frequency domain, the energy components of wideband noise extend over a wide range of frequencies (frequency spectrum).

Wideband noise will often result in the occasional loss or corruption of a data bit. This occurs at times when the noise signal amplitude is large enough to confuse the system into making a wrong decision on what digital information or character was received. Encoding techniques such as parity checking and block character checking (BCC) are important for wideband error detection so that the receiver can determine when an error has occurred.

Impulse noise is best described as a burst of noise, which may last for a duration of say up to 20 ms and it appears in the time domain as indicated in Figure 2.47.

Figure 2.48 illustrates the frequency domain of this type of noise. It affects a wide bandwidth with decreasing amplitude vs. frequency. Impulse noise is brought about by the transient disturbances in electrical activity such as when an electric motor starts up, or from switching elements within telephone exchanges. Impulse noise swamps the desired signal, thus corrupting a string of data bits. As a result of this effect, synchronization may be lost or the character framing may be disrupted. Noise of this nature usually results in garbled data making messages difficult to decipher. Cyclic redundancy checking (CRC) error detection techniques may be required to detect such corruption.

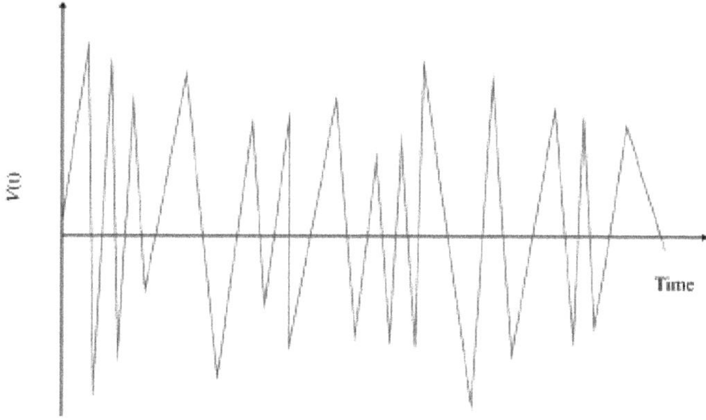

FIGURE 2.45 Time domain plot of wideband noise.

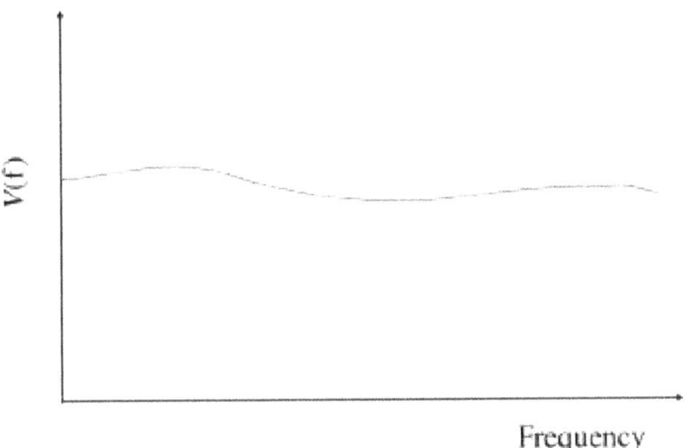

FIGURE 2.46 Frequency domain plot of wideband noise.

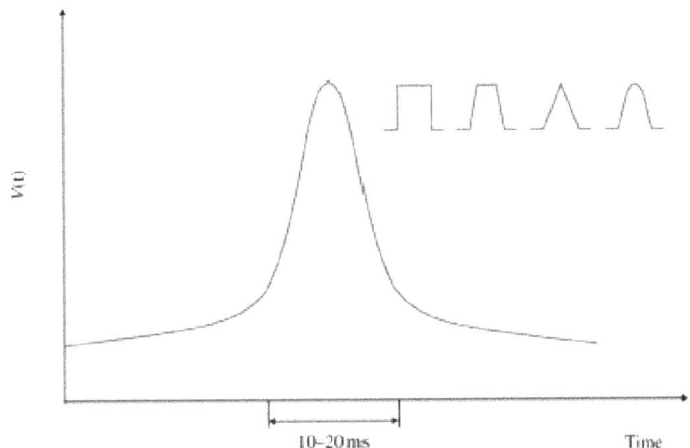

FIGURE 2.47 Time domain plot of impulse noise.

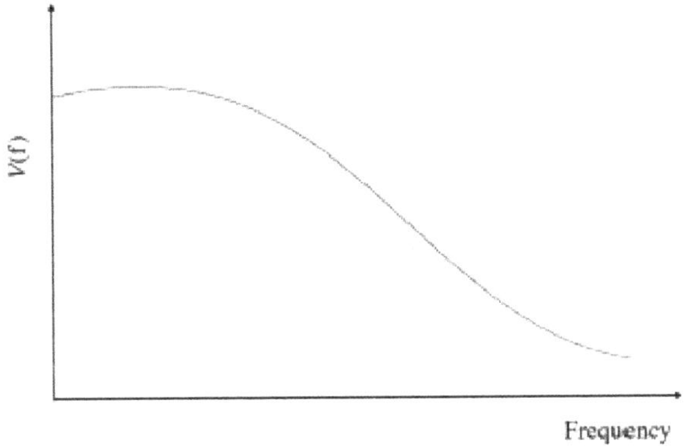

FIGURE 2.48 Frequency domain plot of impulse noise.

Although more damaging than wideband noise, impulse noise is generally less frequent. The time and frequency domain plots for impulse noise will vary depending on the actual shape of the pulse. Pulse shapes may be square, trapezoid, triangular, or sine, for example. In general, the narrower and steeper a pulse, the more energy is placed in the higher-frequency regions.

Frequency-specific noise is characterized by a constant frequency, but its amplitude may vary depending on how far the communication system is from the noise source, the amplitude of the noise signal, and the shielding techniques used. This noise group is typical of AC power systems and can be reduced by separating the data communication system from the power source. As this form of noise has a predictable frequency spectrum, noise resistance is easier to implement within the system design. Filters are typically used to reduce this to an acceptable level.

2.11.3 Overvoltage and Undervoltage

2.11.3.1 Overvoltage

When the voltage in a circuit or part of it is raised above its upper design limit, this is known as overvoltage. The conditions may be hazardous. Depending on its duration, the overvoltage event can be transient (a voltage spike) or permanent, leading to a power surge. An overvoltage is a voltage pulse or wave that is superimposed on the rated voltage of the network, which is shown in Figure 2.49. Overvoltage is characterized in Figure 2.50:

- The rise time *tf* (in μs)
- The gradient *S* (in kV/μs)

An overvoltage disturbs equipment and produces electromagnetic radiation. Moreover, the duration of the overvoltage (*T*) causes an energy peak in the electric circuits, which could destroy equipment.

Four types of overvoltage can disturb electrical installations and loads.

- Switching surges: high-frequency overvoltages or burst disturbance (shown in Figures 2.51 and 2.52) caused by a change in the steady state in an electrical network (during operation of switchgear)
- Power-frequency overvoltage: overvoltage of the same frequency as the network (50, 60, or 400 Hz) caused by a permanent change of state in the network (insulation fault, breakdown of neutral conductor, etc.)

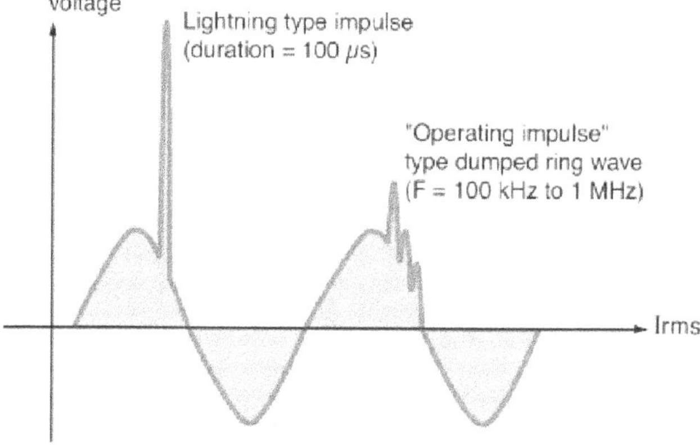

FIGURE 2.49 Examples of overvoltage.

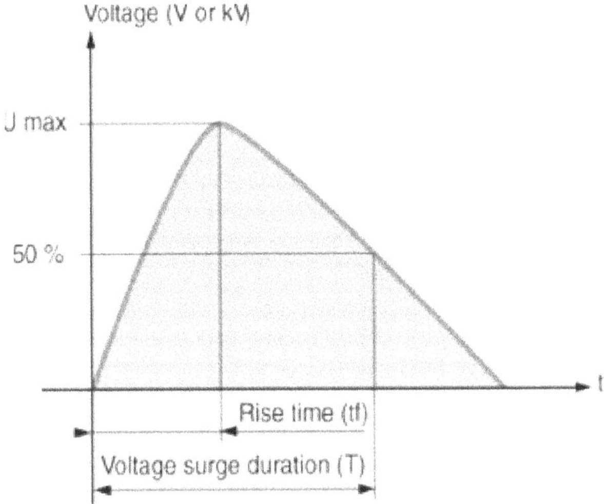

FIGURE 2.50 Main characteristics of an overvoltage.

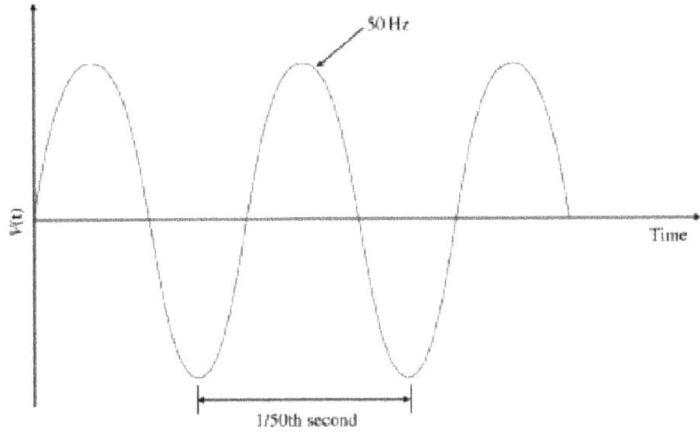

FIGURE 2.51 Time domain plot of constant frequency noise.

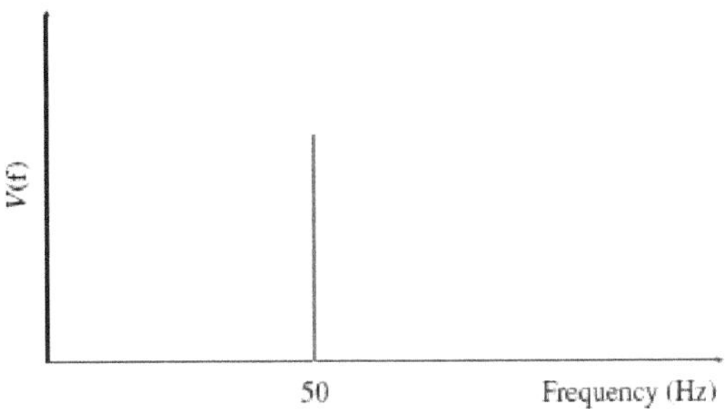

FIGURE 2.52 Frequency domain plot of constant frequency noise.

- Overvoltage caused by electrostatic discharge: very short overvoltage (a few nanoseconds) of very high frequency caused by the discharge of accumulated electric charges (for example, a person walking on a carpet with insulating soles is electrically charged with a voltage of several kilovolts)
- Overvoltage of atmospheric origin

Protecting susceptible electronic equipment from AC power line disturbances is a major concern, whether their business environment is commercial or industrial. Computers are subject to data errors, crashing, and even damage or destruction by voltage transients as a result of an absence or misapplication of protective devices. Similarly, programmable logic controllers (PLCs), solid-state motor controllers, VSDs, and communications equipment all can be damaged from transients.

This damage can be immediate, as when the transient voltage causes a breakdown of solid-state components due to the energy of the transient, or latent, as when insulation or components are severely stressed by a single transient or several transients but not to the point of immediate failure. In the latter case, a transient or other stress that would normally cause no problem affects the weakened insulation or component, causing it to fail without any apparent reason.

Transients can divide into two main categories: those that are derived from natural causes, such as lightning, and those generated through the use of equipment, either on-site or elsewhere.

2.11.3.2 Lightning

A lightning stroke can produce much more energy than we previously thought possible. In fact, a "typical" lightning stroke can carry nearly 3 billion kW at approximately 125 million volts, with an average current of more than 20,000 A. In addition to this prodigious energy production and its dramatic visual display, lightning also produces extremely powerful, short-duration transients on power distribution systems either by a direct strike or a near hit. In most instances, a lightning strike-induced surge on local power distribution lines causes damage to susceptible equipment.

- Lightning strikes can also have secondary effects that cause problems to susceptible electronic equipment in a building. Circuit breakers protecting utility lines can trip and then try to reclose. The resulting voltage sags and outages can cause more problems to computers and other electronic devices than the voltage transients themselves.
- Consider the pad-mounted transformer feeding the building is rated 13.8 kV–480 Y/277 V and is served by a 13.8 kV overhead distribution line. If the overhead line is struck by lightning, lightning arresters (LAs) will protect up to rated spark voltage of approximately 30 kV. This level of protection provides for the full peak capability of the 13.8 kV distribution line while protecting the system from any increased discharge problems. The wiring system and the apparatus connected to it are most likely rated for 95 kV BIL (basic impulse insulation level). With the lightning arresters operating as required, cross arms, wire and insulation, insulators, and the transformer's primary insulation are safely protected.
- While the lightning charge is building up to 30 kV on the overhead power lines, the resultant increasing energy acts as a traveling wave, traversing the overhead lines through Points 1, 2, and 3. At any point where this wave encounters a "discontinuity," it then treats that point as a reflecting point, as indicated by the U-shaped arrow at Point 3. In reflecting back along the line, the reflected traveling wave follows the laws of physics and doubles in voltage. If the wave is traveling into the reflecting point at 30 kV, the reflected traveling wave will travel out of and away from the reflecting point at 60 kV. This doubling effect takes place at dead-ends, open disconnect switches, or the front end of transformer primaries.

2.11.3.3 Surges Induced by Equipment

Arresters at the service entrance will not necessarily eliminate all power line surges within a building. Although greatly reduced in magnitude, those surges that slip by can add to transients that are generated within the facility, causing significant problems to susceptible electronic equipment located there.

Equipment-caused transient voltages result from the basic nature of alternating current. A sudden change in an electrical circuit will generate a transient voltage due to the stored energy contained in any circuit inductances (L) and capacitances (C). The size and duration of the transient depend on the value of L and C and the applied waveform.

a. ***Continuous surges:*** These surges, which can range from 250 V to 1,000 V, can come from operation of electric motors or other inductive loads. Other causes include DC motor drives, the power electronics of VSDs, DC power supply switching, and even portable tools.
b. ***Momentary surges:*** These surges, which can range from 250 V to 3,000 V, can originate from the switching of inductive loads. When you interrupt an inductor's current, a surge voltage will be generated. Its magnitude is described by the equation $e = L \times di/dt$, where the voltage is equal to the inductance times the rate of current shutdown.

The opening and closing of electric motor starters or the use of arc welders and furnace igniters can induce these surges. When the conductors carrying these surge currents are in proximity to conductors of signaling or data circuits, induced voltages will be generated within these circuits. The result is the introduction of electrical noise and loop currents. Deenergizing inductive circuits with air-gap switches such as relays and contactors can generate bursts of high-frequency impulses. Figure 2.53 shows a 15-ms (.015 s) burst composed of impulses having 5-ns (.000000005 s) rise times and 50-ns durations.

A common symptom that directly relates to utility capacitor switching overvoltage is that the resulting oscillatory transients appear at nearly identical times each day. This is because electric utilities, in anticipation of an increase in load, frequently switch their capacitors by time clock. The resulting conditions result in switching include VSD tripping as well as disoperation of other electronically controlled equipment—all without any blinking of lights or other noticeable effects on more conventional loads.

Magnification of the utility capacitor-switching transients can occur when the end-user adds power factor correction capacitors at its facility. These capacitors can magnify the transient overvoltage for certain low-voltage capacitor and step-down transformer sizes. To counter this potential problem, electric utility synchronous closing breakers or switches with reinsertion resistors are used. Also, high-energy surge arresters at the end-user location are used to limit the magnitude of the transient voltage at the end-user bus.

Another possibility is to convert end-user power factor capacitor banks to harmonic filters. Appearing as an inductance in series with capacitors, these filters decrease transient voltages at the end-user bus to acceptable levels. This solution has multiple benefits:

- Correction for displacement power factor
- Control of harmonic distortion levels
- Limitation of magnified capacitor-switching transients

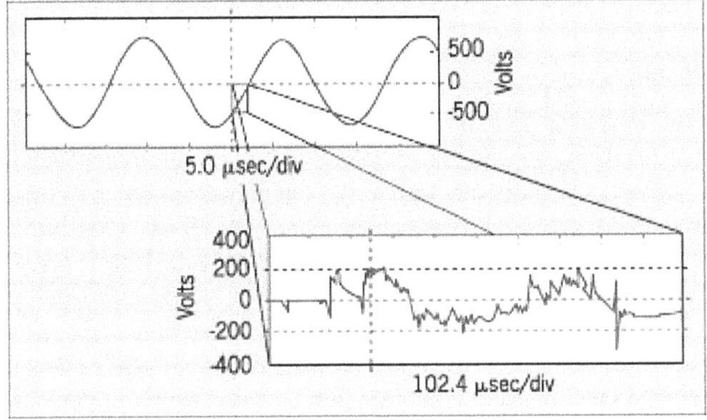

FIGURE 2.53 High frequency impulses.

Overvoltages are caused on power systems due to external and internal influencing factors. The voltage stress caused by overvoltage can damage the lines and equipment connected to the system. Overvoltages arising on a system can be generally classified into two main categories as below:

2.11.3.3.1 Types of Overvoltage

1. External Overvoltages

   This type of overvoltage originates from atmospheric disturbances, such as lightning. This takes the form of a surge and has no direct relationship with the operating voltage of the line. It may be due to any of the following causes:

   a. Direct lightning stroke
   b. Electromagnetically induced overvoltages due to lightning discharge taking place near the line, called "side stroke"
   c. Voltages induced due to atmospheric changes along the length of the line
   d. Electrostatically induced voltages due to presence of charged clouds nearby
   e. Electrostatically induced overvoltages due to the frictional effects of small particles like dust or dry snow in the atmosphere or due to change in the altitude of the line

2. Internal Overvoltages

   These overvoltages are caused by changes in the operating conditions of the power system. These can be divided into two groups as below:

   1. Switching overvoltages or transient over operation voltages of high frequency

      This is caused when switching operation is carried out under normal conditions or when fault occurs in the network. When an unloaded long line is charged, due to Ferranti Effect the receiving end voltage is increased considerably resulting in overvoltage in the system. Similarly, when the primary side of the transformers or reactors is switched on, overvoltage of transient nature occurs.

   2. Temporary overvoltages

These are caused when some major load gets disconnected from the long line under normal or steady state condition.

2.11.3.4 Effects of Overvoltages on Power System

Overvoltage tends to stress the insulation of the electrical equipment and causes damage when it frequently occurs. Overvoltage caused by surges can result in sparkover and flashover between phase and ground at the weakest point in the network, breakdown of gaseous/solid/liquid insulation, and failure of transformers and rotating machines.

2.11.3.5 Undervoltage

Undervoltage is classified as a long-duration voltage variation phenomenon. Long-duration voltage variation is commonly defined as the root-mean-square (RMS) value deviations at power frequencies for longer than one minute. It is important to note the duration of one minute or more as this differentiates undervoltage from short-duration voltage variations such as voltage sags. That is, AC voltage (RMS) decreases, typically to 80%–90% of nominal, at the power frequency for a period of time greater than 1 min.

Undervoltage generally results from low distribution voltage because of heavily loaded circuits that lead to considerable voltage drop, switching on a large load or group of loads, or a capacitor bank switching off. Figure 2.54 shows the voltage profile in utility lines. Undervoltage can expose electrical devices to problems such as overheating, malfunction, premature failure, and shut down, especially for motors (i.e., refrigerators, dryers, and air conditioners). Common symptoms of undervoltage include: motors run hotter than normal and fail prematurely, dim incandescent lighting and batteries fail to recharge properly.

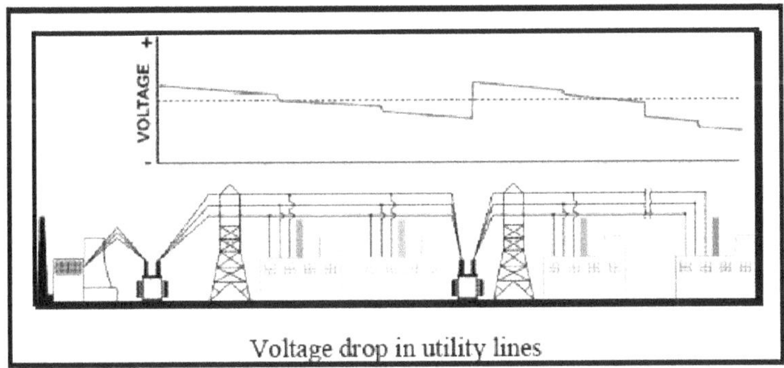

FIGURE 2.54 Voltage profile in utility lines.

Brownout is sometimes used to describe sustained periods of low power frequency voltage initiated as a specific utility dispatch strategy to reduce delivered power. Basically, the disturbance described by brownout has the same meaning as that of undervoltage. Undervoltage problems may be alleviated by:

1. Reducing the system impedance—increase the size of the transformer, reduce the line length, add series capacitors, or increase the size of line conductors.
2. Improving the voltage profile—adjust transformers to the correct tap setting (for manual tap changers) or install voltage regulators or automatic on-load tap changers. Voltage regulators include the mechanical tap changing voltage regulators, electronic tap switching voltage regulators and the ferroresonant transformers.
3. Reducing the line current—de-load the feeder or circuit by transferring some loads to other substations or load centers, add shunt capacitors or SVCs, or upgrade the line to the next voltage level.

2.11.3.6 Outage

Interruptions or outages can be initiated by transients that cause utility breakers and switches to operate.

2.11.4 Harmonics

Harmonics are the major source of sine waveform distortion. The increased use of nonlinear equipment has caused harmonics to become more common. This section provides the sources, effects and causes of harmonics.

2.11.4.1 Harmonic Number (h)

Harmonic number (h) refers to the individual frequency elements that comprise a composite waveform. For example, $h = 5$ refers to the fifth harmonic component with a frequency equal to five times the fundamental frequency. If the fundamental frequency is 60 Hz, then the fifth harmonic frequency is 5×60, or 300 Hz. The harmonic number 6 is a component with a frequency of 360 Hz. Dealing with harmonic numbers and not with harmonic frequencies is done for two reasons. The fundamental frequency varies among individual countries and applications. The fundamental frequency in the U.S. is 60 Hz, whereas in Europe and many Asian countries it is 50 Hz. Also, some applications use frequencies other than 50 or 60 Hz; for example, 400 Hz is a common frequency in the aerospace industry, while some AC systems for electric traction use 25 Hz as the frequency. The inverter part of an AC ASD can operate at any frequency between zero and its full rated maximum frequency, and the fundamental frequency then becomes the frequency at which the motor is operating. The use of harmonic numbers allows us to simplify how we express harmonics. The second reason for using harmonic numbers is the simplification realized in performing mathematical operations involving harmonics.

2.11.4.2 Harmonic Signatures

Many of the loads installed in present-day power systems are harmonic current generators. Combined with the impedance of the electrical system, the loads also produce harmonic voltages. The nonlinear loads may therefore be viewed as both harmonic current generators and harmonic voltage generators. Prior to the 1970s, speed control of AC motors was primarily achieved using belts and pulleys. Now, ASDs perform speed control functions very efficiently. ASDs are generators of large harmonic currents. Fluorescent lights use less electrical energy for the same light output as incandescent lighting but produce substantial harmonic currents in the process. The explosion of personal computer use has resulted in harmonic current proliferation in commercial buildings. This section is devoted to describing, in no particular order, a few of the more common nonlinear loads that surround us in our everyday life.

2.11.4.2.1 Fluorescent Lighting

Current waveform at a distribution panel is primarily comprised of the third and the fifth harmonic frequencies. The waveform also contains slight traces of even harmonics, especially of the higher frequency order. The current waveform is flat topped due to initiation of arc within the gas tube, which causes the voltage across the tube and the current to become essentially unchanged for a portion of each half of a cycle.

2.11.4.2.2 Adjustable-Speed Drives

While several technologies exist for creating a variable voltage and variable frequency power source for the speed control of AC motors, the pulse-width modulation (PWM) drive technology is currently the most widely used. The characteristic double hump for each half cycle of the AC waveform is due to conduction of the input rectifier modules for duration of two 60° periods for each half cycle. As the operating frequency is reduced, the humps become pronounced with a large increase in the THD. The THD of 74.2% for 45-Hz operation is excessive and can produce many deleterious effects, as will be shown in later sections of this chapter.

It was stated earlier that large current distortions can produce significant voltage distortions. In this particular case, the voltage THD is 8.3%, which is higher than levels typically found in most industrial installations. High levels of voltage THD also produce unwanted results. This drive contains line-side inductors that, along with the higher inductance of the motor, produce a current waveform with less distortion. Table 4.6 provides the harmonic frequency distribution for this ASD.

2.11.4.2.3 Personal Computer and Monitor

The predominance of the third and fifth harmonics is evident. The current THD for both devices exceeds 100%, as the result of high levels of individual distortions introduced by the third and fifth harmonics. The total current drawn by a personal computer and its monitor is less than 2 A, but a typical high-rise building can contain several hundred computers and monitors. The net effect of this on the total current harmonic distortion of a facility is not difficult to visualize. So far, we have examined some of the more common harmonic current generators. The examples illustrate that a wide spectrum of harmonic currents is generated. Depending on the size of the power source and the harmonic current makeup, the composite harmonic picture will be different from facility to facility.

2.11.4.3 Effect of Harmonics on Power System Devices

Harmonic effects are so insidious and are not known until failure occurs. Insight into how harmonics can interact within a power system and how they can affect power system components is important for preventing failures. In this section, we will look at the effect of harmonics on some common power system devices.

2.11.4.3.1 Transformers

Harmonics can affect transformers primarily in two ways. Voltage harmonics produce additional losses in the transformer core as the higher frequency harmonic voltages set up hysteresis loops, which superimpose on the fundamental loop. Each loop represents higher magnetization power requirements and higher core losses. A second and a more serious effect of harmonics is due to harmonic frequency currents in the transformer windings. The harmonic currents increase the net RMS current flowing in the transformer windings, which results in additional I^2R losses. Winding eddy current losses are also increased. Winding eddy currents are circulating currents induced in the conductors by the leakage magnetic flux. Eddy current concentrations are higher at the ends of the windings due to the crowding effect of the leakage magnetic field at the coil extremities. The winding eddy current losses increase as the square of the harmonic current and the square of the frequency of the current. Thus, the eddy loss (EC) is proportional to $I_h^2 \times h^2$, where I_h is the RMS value of the harmonic current of order h, and h is the harmonic frequency order or number. Eddy currents due to harmonics can significantly increase the transformer winding temperature. Transformers that are required to supply large nonlinear loads must be de-rated to handle the harmonics. This derating factor is based on the percentage of the harmonic currents in the load and the rated winding eddy current losses. One method by which transformers may be rated for suitability to handle harmonic loads is by *k* factor ratings. The *k* factor is equal to the sum of the square of the harmonic frequency currents (expressed as a ratio of the total RMS current) multiplied by the square of the harmonic frequency numbers:

$$k = I_1^2(1)^2 + I_2^2(2)^2 + I_3^2(3)^2 + I_4^2(4)^2 + \ldots + I_n^2(n)^2 \tag{2.16}$$

Where, I_1 is the ratio between the fundamental current and the total RMS current. I_2 is the ratio between the second harmonic current and the total RMS current. I_3 is the ratio between the third harmonic current and the total RMS current.

Equation (i) can be rewritten as:

$$k = \sum I_n^2 h^2 (h = 1,2,3,\ldots n) \tag{2.17}$$

2.11.4.3.2 AC Motors

Application of distorted voltage to a motor results in additional losses in the magnetic core of the motor. Hysteresis and eddy current losses in the core increase as higher frequency harmonic voltages are impressed on the motor windings. Hysteresis losses increase with frequency and eddy current losses increase as the square of the frequency. Also, harmonic currents produce additional I^2R losses in the motor windings that must be accounted for.

Another effect, and perhaps a more serious one, is torsional oscillations due to harmonics. Two of the more prominent harmonics found in a typical power system are the fifth and seventh harmonics. The fifth harmonic is a negative sequence harmonic, and the resulting magnetic field revolves in a direction opposite to that of the fundamental field at a speed five times the fundamental. The seventh harmonic is a positive sequence harmonic with a resulting magnetic field revolving in the same direction as the fundamental field at a speed seven times the fundamental. The net effect is a magnetic field that revolves at a relative speed of six times the speed of the rotor. This induces currents in the rotor bars at a frequency of six times the fundamental frequency. The resulting interaction between the magnetic fields and the rotor-induced currents produces torsional oscillations of the motor shaft. If the frequency of the oscillation coincides with the natural frequency of the motor rotating members, severe damage to the motor can result. Excessive vibration and noise in a motor operating in a harmonic environment should be investigated to prevent failures.

Motors intended for operation in a severe harmonic environment must be specially designed for the application. Motor manufacturers provide motors for operation with ASD units. If the harmonic levels become excessive, filters may be applied at the motor terminals to keep the harmonic currents from the motor windings. Large motors supplied from ASDs are usually provided with harmonic filters to prevent motor damage due to harmonics.

2.11.4.3.3 Capacitor Banks

Capacitor banks are commonly found in commercial and industrial power systems to correct for low-power factor conditions. Capacitor banks are designed to operate at a maximum voltage of 110% of their rated voltages and at 135% of their rated kVARs. When large levels of voltage and current harmonics are present, the ratings are quite often exceeded, resulting in failures. Because the reactance of a capacitor bank is inversely proportional to frequency, harmonic currents can find their way into a capacitor bank. The capacitor bank acts as a sink, absorbing stray harmonic currents and causing overloads and subsequent failure of the bank. A more serious condition with potential for substantial damage occurs due to a phenomenon called harmonic resonance. Resonance conditions are created when the inductive and capacitive reactances become equal at one of the harmonic frequencies. The two types of resonances are series and parallel. In general, series resonance produces voltage amplification and parallel resonance results in current multiplication. Resonance will not be analyzed in this book, but many textbooks on electrical circuit theory are available that can be consulted for further explanation. In a harmonic-rich environment, both series and parallel resonance may be present. If a high level of harmonic voltage or current corresponding to the resonance frequency exists in a power system, considerable damage to the capacitor bank as well as other power system devices can result.

2.11.4.3.4 Cables

Current flowing in a cable produces I^2R losses. When the load current contains harmonic content, additional losses are introduced. To compound the problem, the effective resistance of the cable increases with frequency because of the phenomenon known as skin effect. Skin effect is due to unequal flux linkage across the cross section of the conductor, which causes AC currents to flow only on the outer periphery of the conductor. This has the effect of increasing the resistance of the conductor for AC currents. The higher the frequency of the current, the greater the tendency of the current to crowd at the outer periphery of the conductor and the greater the effective resistance for that frequency. The capacity of a cable to carry nonlinear loads may be determined as follows. The skin effect factor is calculated first. The skin effect factor depends on the skin depth, which is an indicator of the penetration of the current in a conductor. Skin depth (δ) is inversely proportional to the square root of the frequency:

$$\delta = \frac{S}{\sqrt{f}} \tag{2.18}$$

where S is a proportionality constant based on the physical characteristics of the cable and its magnetic permeability and f is the frequency of the current.

2.11.4.3.5 Bus Ways

Most commercial multistory installations contain busways that serve as the primary source of electrical power to various floors. Busways that incorporate sandwiched busbars are susceptible to nonlinear loading, especially if the neutral bus carries large levels of triplen harmonic currents (third, ninth, etc.). Under the worst possible conditions, the neutral bus may be forced to carry a current equal to 173% of the phase currents. In cases where substantial neutral currents are expected, the busways must be suitably derated. The data are shown for busways with neutral busbars that are 100% and 200% in size.

2.11.4.3.6 Protective Devices

Harmonic currents influence the operation of protective devices. Fuses and motor thermal overload devices are prone to nuisance operation when subjected to nonlinear currents. This factor should be given due consideration when sizing protective devices for use in a harmonic environment. Electromechanical relays are also affected by harmonics. Depending on the design, an electromechanical relay may operate faster or slower than the expected times for operation at the fundamental frequency alone. Such factors should be carefully considered prior to placing the relays in service.

2.11.4.4 Guidelines for Harmonic Voltage and Current Limitation

So far, we have discussed the adverse effects of harmonics on power system operation. It is important, therefore, that attempts be made to limit the harmonic distortion that a facility might produce. There are two reasons for this. First, the lower the harmonic currents produced in an electrical system, the better the equipment within the confinement of the system will perform. Also, lower harmonic currents produce less of an impact on other power users sharing the same power lines of the harmonic generating power system. The IEEE 519 standard provides guidelines for harmonic current limits at the PCC between the facility and the utility. The rationale behind the use of the PCC as the reference location is simple. It is a given fact that within a particular power use environment, harmonic currents will be generated and propagated. Harmonic current injection at the PCC determines how one facility might affect other power users and the utility that supplies the power. Table 2.5 (per IEEE 519) lists harmonic current limits based on the size of the power user.

I_{SC} is the maximum short-circuit current at PCC; I_L is the maximum fundamental frequency demand load current at PCC (average current of the maximum demand for the preceding 12 months); h is the individual harmonic order; THD is the total harmonic distortion based on the maximum demand load current. The table applies to odd harmonics, even harmonics are limited to 25% of the odd harmonic limits shown in Table 2.5.

As the ratio between the maximum available short circuit current at the PCC and the maximum demand load current increases, the percentage of the harmonic currents that are allowed also increases. This means that larger power users are allowed to inject into the system only a minimal amount of harmonic current (as a percentage of the fundamental current). Such a scheme tends to equalize the amounts of harmonic currents that large and small users of power are allowed to inject into the power system at the PCC. IEEE 519 also provides guidelines for maximum voltage distortion at the PCC (see Table 2.6). Limiting the voltage distortion at the PCC is the concern of the utility. It can be expected that as long as a facility's harmonic current contribution is within the IEEE 519 limits the voltage distortion at the PCC will also be within the specified limits.

When the IEEE 519 harmonic limits are used as guidelines within a facility, the PCC is the common junction between the harmonic generating loads and other electrical equipment in the power system. It is expected that applying IEEE guidelines renders power system operation more reliable. In the future, more and more utilities might require facilities to limit their harmonic current injection to levels stipulated by IEEE 519. The following section contains information on how harmonic mitigation might be achieved.

TABLE 2.5

Harmonic Current Limits for General Distribution Systems (120–69,000 V)

I_{SC}/I_L	$h < 11$	$11 \leq h < 17$	$17 \leq h < 23$	$23 \leq h < 35$	$35 \leq h$	THD
<20	4.0	2.0	1.5	0.6	0.3	5.0
20–50	7.0	3.5	2.5	1.0	0.5	8.0
50–100	10.0	4.5	4.0	1.5	0.7	12.0
100–1000	12.0	5.5	5.0	2.0	1.0	15.0
>1000	15.0	7.0	6.0	2.5	1.4	20.0

TABLE 2.6

Voltage Harmonic Distortion Limits

Bus Voltage at PCC	Individual Voltage Distortion (%)	Total Voltage Distortion THD (%)
69 kV and below	3.0	5.0
69.001 kV through 161 kV	1.5	2.5
161.001 kV and above	1.0	1.5

Note: PCC = point of common coupling; THD = total harmonic distortion.

2.11.4.5 Harmonic Current Cancellation

Transformer connections employing phase shift are sometimes used to effect cancellation of harmonic currents in a power system. Triplen harmonic (3rd, 9th, 15th, etc.) currents are a set of currents that can be effectively trapped using a special transformer configuration called the zigzag connection. In power systems, triplen harmonics add in the neutral circuit, as these currents are in phase. Using a zigzag connection, the triplens can be effectively kept away from the source. The transformer phase-shifting principle is also used to achieve cancellation of the 5th and the 7th harmonic currents. Using a Δ–Δ and a Δ–*Y* transformer to supply harmonic producing loads in parallel, the 5th and the 7th harmonics are canceled at the point of common connection. This is due to the 30° phase shift between the two transformer connections. As the result of this, the source does not see any significant amount of the 5th and 7th harmonics. If the nonlinear loads supplied by the two transformers are identical, then maximum harmonic current cancellation takes place; otherwise, some 5th and 7th harmonic currents would still be present. Other phase-shifting methods may be used to cancel higher harmonics if they are found to be a problem. Some transformer manufacturers offer multiple phase-shifting connections in a single package that saves cost and space compared to using individual transformers.

2.11.4.6 Harmonic Filters

There are several ways to reduce or eliminate harmonics. The most common way is to add filters to the electrical power system. Harmonic filters or chokes reduce electrical harmonics just as shock absorbers reduce mechanical harmonics. Filters contain capacitors and inductors in series. Filters siphon off the harmonic currents to ground. They prevent the harmonic currents from getting onto the utility's or end user's distribution system and doing damage to the utility's and other end users' equipment. There are two types of filters: static and active. Static filters do not change their value. Active filters change their value to fit the harmonic being filtered. Other ways of reducing or eliminating harmonics include using isolation transformers and detuning capacitors and designing the source of the harmonics to change the type of harmonics.

Nonlinear loads produce harmonic currents that can travel to other locations in the power system and eventually back to the source. As we saw earlier, harmonic currents can produce a variety of effects that are harmful to the power system. Harmonic currents are a result of the characteristics of particular loads. As long as we choose to employ those loads, we must deal with the reality that harmonic currents will exist to a degree dependent upon the loads. One means of ensuring that harmonic currents produced by a nonlinear current source will not unduly interfere with the rest of the power system is to filter out the harmonics. Application of harmonic filters helps to accomplish this.

Figure 2.55 is a typical series-tuned filter. Here the values of the inductor and the capacitor are chosen to present low impedance to the harmonic frequency that is to be filtered out. Due to the lower impedance of the filter in comparison to the impedance of the source, the harmonic frequency current will circulate between the load and the filter. This keeps the harmonic current of the desired frequency away from the source and other loads in the power system. If other harmonic frequencies are to be filtered out, additional tuned filters are applied in parallel. Applications such as arc furnaces require multiple harmonic filters, as they generate large quantities of harmonic currents at several frequencies. Applying harmonic filters requires careful consideration. Series-tuned filters appear to be of low impedance to harmonic currents but they also form a parallel resonance circuit with the source impedance. In some

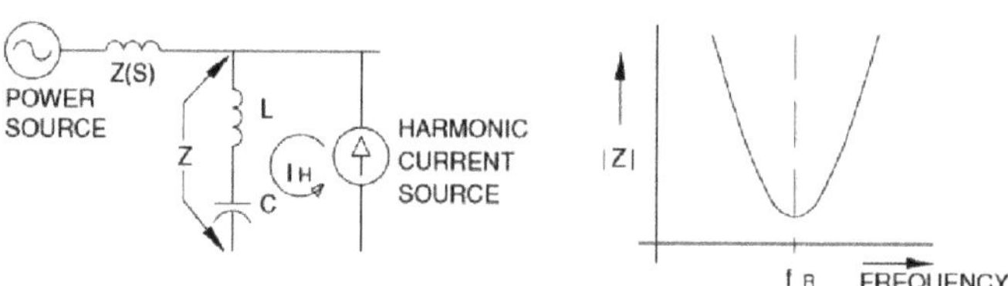

FIGURE 2.55 Series-tuned filter and filter frequency response.

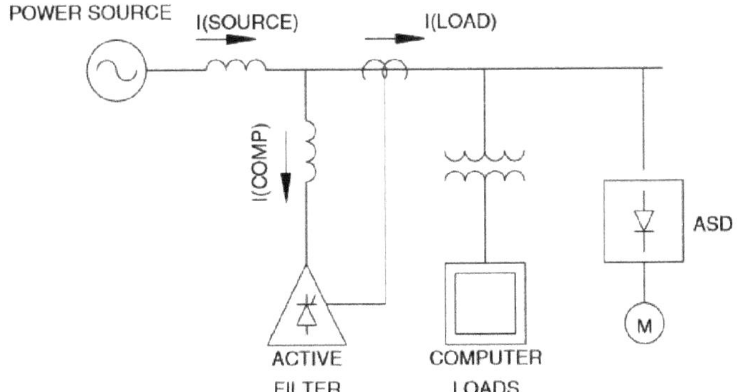

FIGURE 2.56 Active filter to cancel harmonic currents.

instances, a situation can be created that is worse than the condition being corrected. It is imperative that computer simulations of the entire power system be performed prior to applying harmonic filters. As a first step in the computer simulation, the power system is modeled to indicate the locations of the harmonic sources, then hypothetical harmonic filters are placed in the model and the response of the power system to the filter is examined. If unacceptable results are obtained, the location and values of the filter parameters are changed until the results are satisfactory.

When applying harmonic filters, the units are almost never tuned to the exact harmonic frequency. For example, the 5th harmonic frequency may be designed for resonance at the 7th harmonic frequency. By not creating a resonance circuit at precisely the 5th harmonic frequency, we can minimize the possibility of the filter resonating with other loads or the source, thus forming a parallel resonance circuit at the 5th harmonic.

The 7th harmonic filter would still be effective in filtering out the 5th harmonic currents. This is evident from the series-tuned frequency vs. impedance curve shown in Figure 2.55. Sometimes, tuned filters are configured to provide power factor correction for a facility as well as harmonic current filtering. In such cases the filter would be designed to carry the resonant harmonic frequency current and also the normal frequency current at the fundamental frequency. In either case, a power system harmonic study is paramount to ensure that no ill effects would be produced by the application of the power factor correction/filter circuit. Active filters use active conditioning to compensate for harmonic currents in a power system. Figure 2.56 shows an active filter applied in a harmonic environment. The filter samples the distorted current and, using power electronic switching devices, draws a current from the source of such magnitude, frequency composition, and phase shift to cancel the harmonics in the load. The result is that the current drawn from the source is free of harmonics. An advantage of active filters over passive filters is that the active filters can respond to changing load and harmonic conditions, whereas passive filters are fixed in their harmonic response. As we saw earlier, application of passive filters requires careful analysis. Active filters have no serious ill effects associated with them. However, active filters are expensive and not suited for application in small facilities.

2.11.4.7 Cures for Low-Frequency Disturbances

Power-frequency or low-frequency disturbances are slow phenomena caused by switching events related to the power frequency. Such disturbances are dispersed with time once the incident causing the disturbance is removed. This allows the power system to return to normal operation. Low-frequency disturbances also reveal themselves more readily. For example, dimming of lights accompanies voltage sag on the system; when the voltage rises, lights shine brighter. While low-frequency disturbances are easily detected or measured, they are not easily corrected. Transients, on the other hand, are not easily detected or measured but are cured with much more ease than a low-frequency event. Measures available to deal with low-frequency disturbances are discussed.

2.11.4.8 Isolation Transformers

Isolation transformers, as their name indicates, have primary and secondary windings, which are separated by an insulating or isolating medium. Isolation transformers do not help in curing voltage sags or swells; they merely transform the voltage from a primary level to a secondary level to enable power transfer from one winding to the other. However, if the problem is due to common mode noise, isolation transformers help to minimize noise coupling, and shielded isolation transformers can help to a greater degree. Common mode noise is equally present in the line and the neutral circuits with respect to ground. Common mode noise may be converted to transverse mode noise (noise between the line and the neutral) in electrical circuits, which is troublesome for sensitive data and signal circuits. Shielded isolation transformers can limit the amount of common mode noise converted to transverse mode noise. The effectiveness with which a transformer limits common mode noise is called attenuation (A) and is expressed in decibels (dB):

$$A = 20\log(V_1 / V_2) \tag{2.19}$$

where V_1 is the common mode noise voltage at the transformer primary and V_2 is the differential mode noise at the transformer secondary. Figure 2.57 shows how common mode noise attenuation is obtained by the use of a shielded isolation transformer.

The presence of a shield between the primary and secondary windings reduces inter-winding capacitance and thereby reduces noise coupling between the two windings. Example: Find the attenuation of a transformer that can limit 1 V common mode noise to 10 mV of transverse mode noise at the secondary:

$$A = 20\log(1 / 0.01) = 40\,\text{dB} \tag{2.20}$$

Isolation transformers using a single shield provide attenuation in a range of 40 to 60 dB. Higher attenuation may be obtained by specially designed isolation transformers using multiple shields configured to form a continuous enclosure around the secondary winding. Attenuation of the order of 100 dB may be realized with such techniques.

2.11.4.9 Voltage Regulators

Voltage regulators are devices that can maintain a constant voltage (within tolerance) for voltage changes of predetermined limits above and below the nominal value. A switching voltage regulator

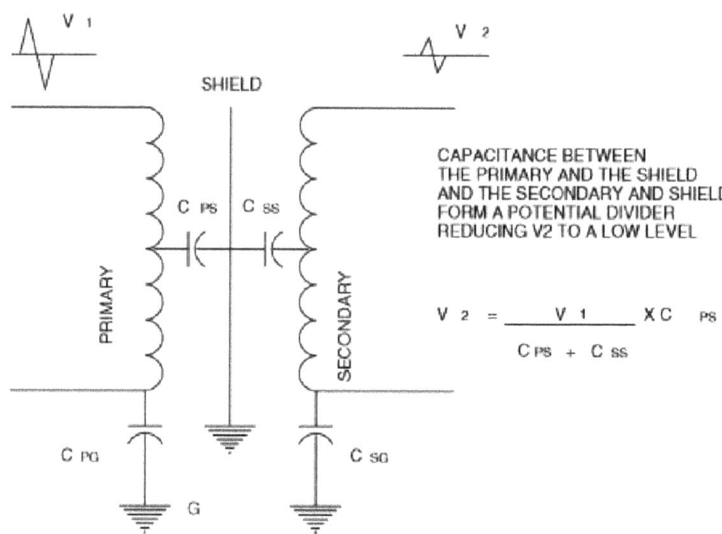

FIGURE 2.57 Common mode noise attenuation by shielded isolation transformer.

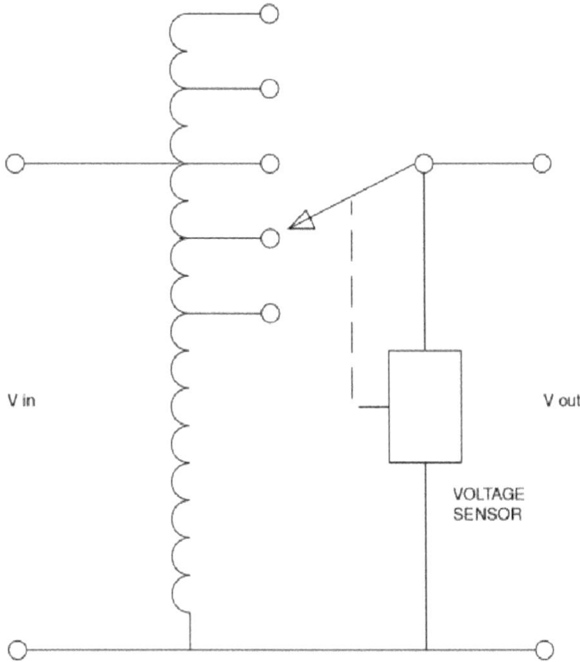

FIGURE 2.58 Tap-changer voltage regulator.

maintains constant output voltage by switching the taps of an autotransformer in response to changes in the system voltage, as shown in Figure 2.58. The electronic switch responds to a signal from the voltage-sensing circuitry and switches to the tap connection necessary to maintain the output voltage constant. The switching is typically accomplished within half of a cycle, which is within the ride-through capability of most sensitive devices. Ferroresonant voltage regulators are static devices that have no moving components. They operate on the principle that, in a transformer, when the secondary magnetic circuit is operating in the saturation region the secondary winding is decoupled from the primary and therefore is not sensitive to voltage changes in the primary. The secondary winding has a capacitor connected across its terminals that form a parallel resonant circuit with the inductance of the secondary winding. Large magnetic fields are created in the magnetic core surrounding the secondary windings, thereby decoupling the secondary winding from the primary. Typically, ferroresonant transformer regulators can maintain secondary voltage to within ±0.5% for changes in the primary voltages of ± 20%. Figure 2.59 contains the schematic of a ferroresonance transformer type of voltage regulator. Ferroresonance transformers are sensitive to loads above their rated current. In extreme cases of overload, secondary windings can become detuned, at which point the output of the transformer becomes very low. Voltage sags far below the rated level can also have a detuning effect on the transformer. Within the rated voltage and load limits, however, the ferroresonance transformer regulators are very effective in maintaining fairly constant voltage levels.

2.11.4.10 Static Uninterruptible Power Source Systems

Static UPSs have no rotating parts, such as motors or generators. These are devices that maintain power to the loads during loss of normal power for a duration that is a function of the individual UPS system. All UPS units have an input rectifier to convert the AC voltage into DC voltage, a battery system to provide power to loads during loss of normal power, and an inverter that converts the DC voltage of the battery to an AC voltage suitable for the load being supplied. Depending on the UPS unit, these three main components are configured differently.

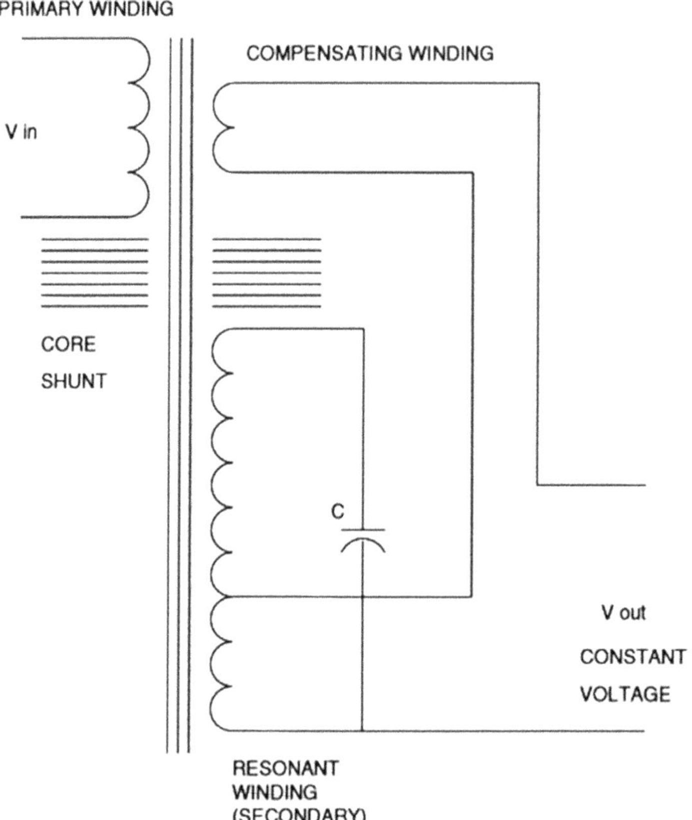

FIGURE 2.59 Ferroresonant transformer.

Static UPS systems may be broadly classified into offline and online units. In the offline units, the loads are normally supplied from the primary electrical source directly. The primary electrical source may be utility power or an in-house generator. If the primary power source fails or falls outside preset parameters, the power to the loads is switched to the batteries and the inverter. The switching is accomplished within half of a cycle in most UPS units, thereby allowing critical loads to continue to receive power. During power transfer from the normal power to the batteries, the loads might be subjected to transients. Once the loads are transferred to the batteries, the length of time for which the loads would continue to receive power depends on the capacity of the batteries and the amount of load. UPS units usually can supply power for 15 to 30 min, at which time the batteries become depleted to a level insufficient to supply the loads, and the UPS unit shuts down. Some offline UPS system manufacturers provide optional battery packs to enhance the time of operation of the units after loss of normal power.

In online UPS units, normal power is rectified into DC power and in turn inverted to AC power to supply the loads. The loads are continuously supplied from the DC bus even during times when the normal power is available. A battery system is also connected to the DC bus of the UPS unit and kept charged from the normal source. When normal power fails, the DC bus is supplied from the battery system. No actual power transfer occurs during this time, as the batteries are already connected to the DC bus. Online units can be equipped with options such as manual and static bypass switches to circumvent the UPS and supply power to the loads directly from the normal source or an

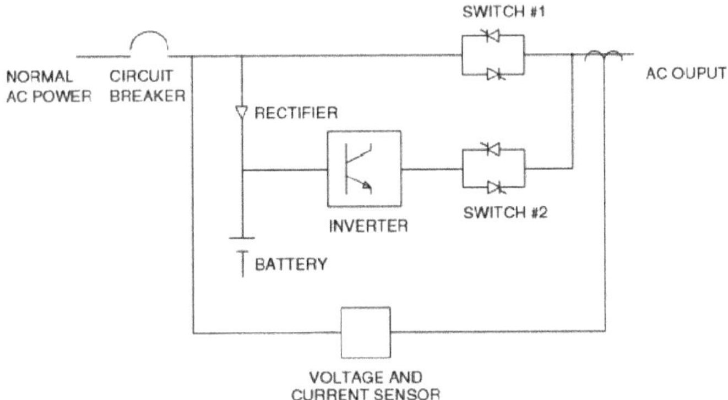

FIGURE 2.60 Offline uninterruptible power source (UPS) system.

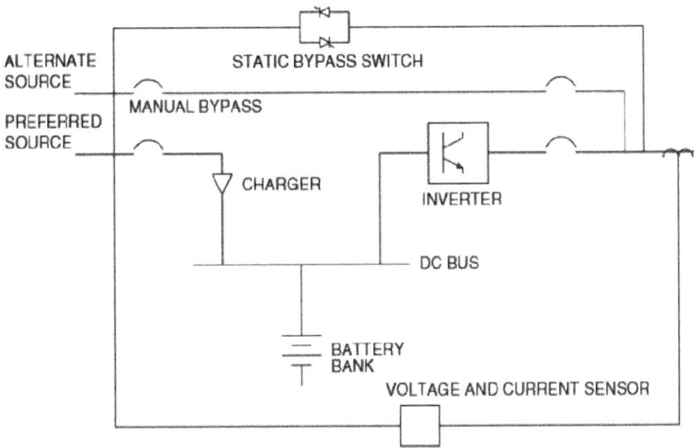

FIGURE 2.61 Online uninterruptible power source (UPS) system.

alternate source such as a standby generator. An offline unit is shown in Figure 2.60, and an online unit in Figure 2.61. Two important advantages of online UPS units are because: (1) power is normally supplied from the DC bus, the UPS unit in effect isolates the loads from the source which keeps power system disturbances and transients from interacting with the loads, and (2) since power to the loads is not switched during loss of normal power, no switching transients are produced. As might be expected, online UPS systems cost considerably more than offline units. The output voltage of static UPS units tends to contain waveform distortions higher than those for normal power derived from the utility or a generator. This is due to the presence of the inverter in the output section of the UPS system. For some lower-priced UPS units, the distortion can be substantial, with the waveform resembling a square wave. Figure 2.62 shows the output waveform of a UPS unit commonly used in offices to supply computer workstations. More expensive units use higher order inverter sections to improve the waveform of the output voltage, as shown in Figure 2.63. It is important to take into consideration the level of susceptibility of the loads to waveform distortion. Problems attributed to excessive voltage distortion have been noticed in some applications involving medical electronics and voice communication.

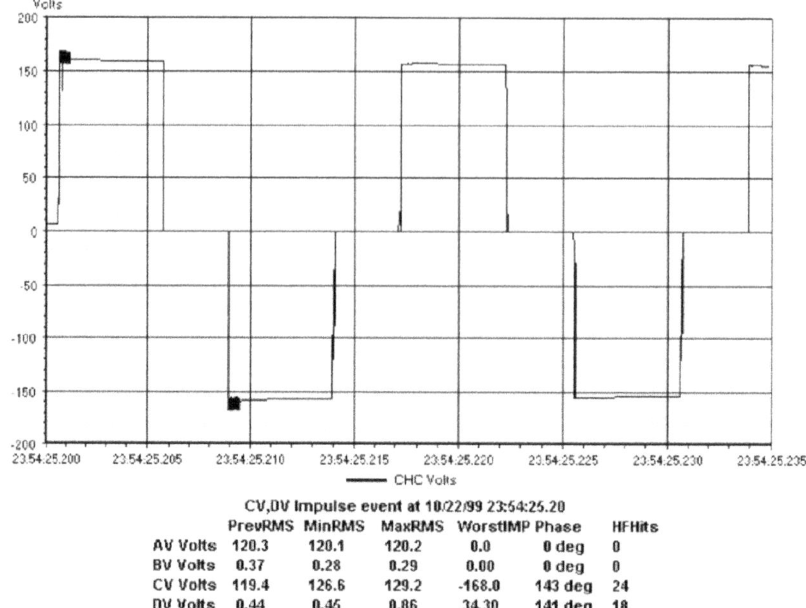

FIGURE 2.62 Output voltage waveform from an offline uninterruptible power source (UPS) system.

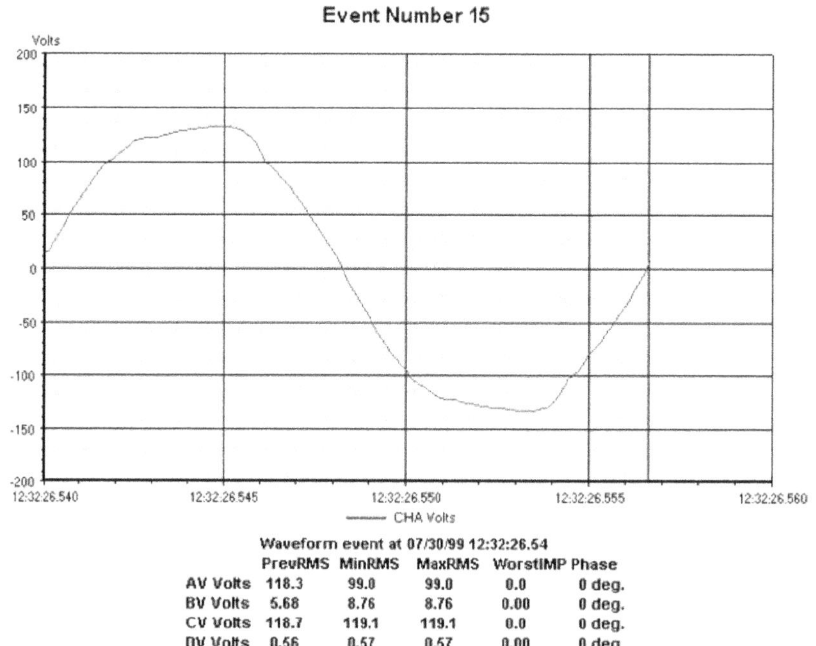

FIGURE 2.63 Voltage waveform from an online uninterruptible power source (UPS) system.

2.11.4.11 Rotary Uninterruptible Power Source Units

Rotary UPS (RUPS) units utilize rotating members to provide uninterrupted power to loads, as shown in Figure 2.64. In this configuration, an AC induction motor drives an AC generator, which supplies power to critical loads. The motor operates from normal utility power. A diesel engine or other type of prime mover is coupled to the same shaft as the motor and the generator. During normal operation, the diesel engine is decoupled from the common shaft by an electric clutch. If the utility power fails, the prime mover shaft is coupled to the generator shaft and the generator gets its mechanical power from the prime mover. The motor shaft is attached to a flywheel, and the total inertia of the system is sufficient to maintain power to the loads until the prime mover comes up to full speed. Once the normal power returns, the induction motor becomes the primary source of mechanical power and the prime mover is decoupled from the shaft.

In a different type of RUPS system, during loss of normal power the AC motor is supplied from a battery bank by means of an inverter (Figure 2.65). The batteries are kept charged by the normal power source. The motor is powered from the batteries until the batteries become depleted. In some applications, standby generators are used to supply the battery bank in case of loss of normal power. Other combinations are used to provide uninterrupted power to critical loads, but we will not attempt to review all the available technologies. It is sufficient to point out that low-frequency disturbances are effectively mitigated using one of this means.

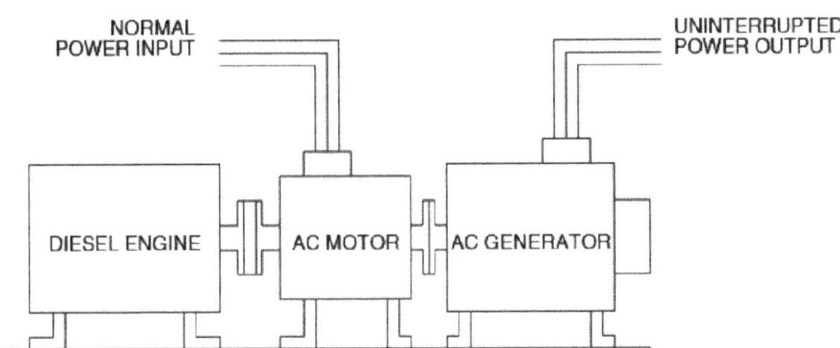

FIGURE 2.64 Rotary uninterruptible power source (RUPS) system using a diesel engine, AC motor, and AC generator to supply uninterrupted power to critical loads.

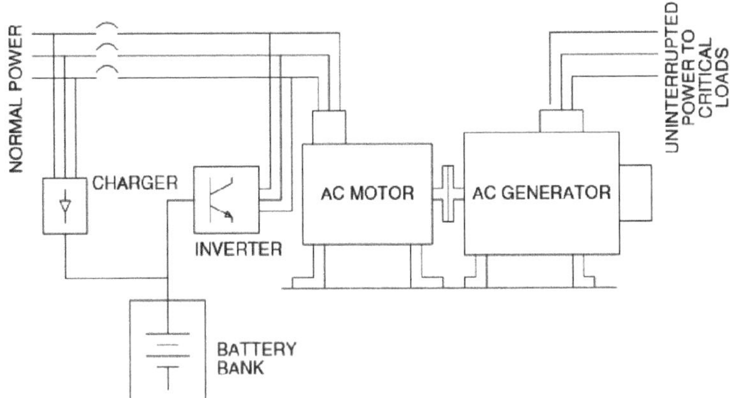

FIGURE 2.65 Rotary uninterruptible power source (RUPS) system using a battery bank, AC motor, and AC generator to provide uninterrupted power to critical loads.

2.11.4.12 Voltage Tolerance Criteria

Manufacturers of computers and data-processing equipment do not generally publish data informing the user of the voltage tolerance limits for their equipment. An agency known as the Information Technology Industry Council (ITIC) has published a graph that provides guidelines as to the voltage tolerance limits within which information technology equipment should function satisfactorily (Figure 2.66). The ordinate (y-axis) represents the voltage as a percentage of the nominal voltage. The abscissa (x-axis) is the time duration in seconds (or cycles). The graph contains three regions. The area within the graph is the voltage tolerance envelope, in which equipment should operate satisfactorily. The area above the graph is the prohibited region, in which equipment damage might result. The area below the graph is the region where the equipment might not function satisfactorily but no damage to the equipment should result. Several types of events fall within the regions bounded by the ITIC graph, as described as follows:

- **Steady-State Tolerance:** The steady-state range describes an RMS voltage that is either slowly varying or is constant. The subject range is ±10% from the nominal voltage. Any voltage in this range may be present for an indefinite period and is a function of the normal loading and losses in the distribution system.
- **Line Voltage Swell:** This region describes a voltage swell having RMS amplitude up to 120% of the nominal voltage, with duration of up to 0.5 sec. This transient may occur when large loads are removed from the system or when voltage is applied from sources other than the utility.
- **Low-Frequency Decaying Ring Wave:** This region describes a decaying ring wave transient that typically results from the connection of power factor correction capacitors to an AC power

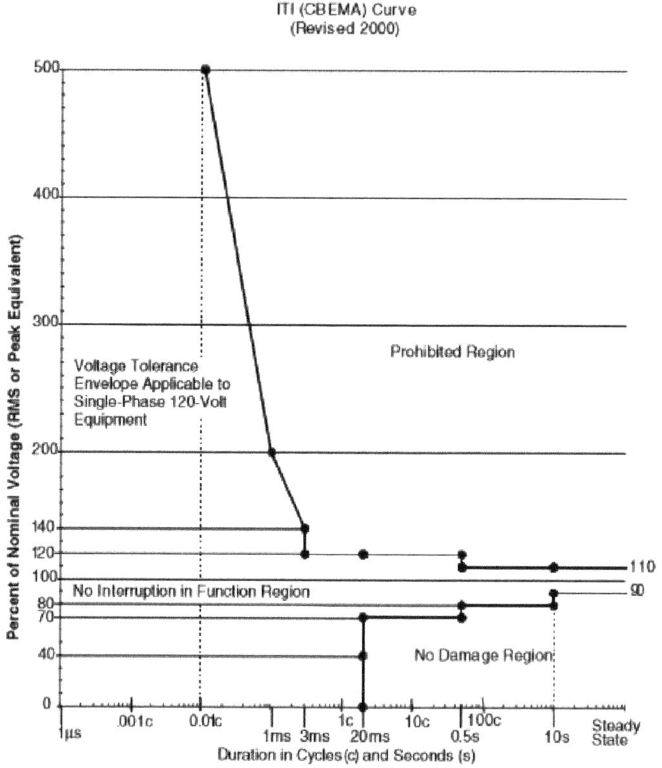

FIGURE 2.66 Information Technology Industry Council (ITIC) graphs.

distribution system. The frequency of this transient may vary from 200 Hz to 5 kHz, depending on the resonant frequency of the AC distribution system. The magnitude of the transient is expressed as a percentage of the peak 60 Hz nominal (not the RMS). The transient is assumed to be completely decayed by the end of the half-cycle in which it occurs. The transient is assumed to occur near the peak of the nominal voltage waveform. The amplitude of the transient varies from 140% for 200-Hz ring waves to 200% for 5-kHz ring waves, with a linear increase in amplitude with frequency.

Figure 2.66 provides guidelines of voltage tolerance limits within which information technology equipment should function satisfactorily.

- **High-Frequency Impulse Ring Wave:** This region describes the transients that typically occur as the result of lightning strikes. Wave shapes applicable to this transient and general test conditions are described in the ANSI/IEEE C62.41 standard. This region of the curve deals with amplitude and duration (energy) rather than RMS amplitude. The intent is to provide 80 J of minimum transient immunity.
- **Voltage Sags:** Two different RMS voltage sags are described. Generally, the transients result from application of heavy loads as well as fault conditions at various points in the AC power distribution system. Sags to 80% of nominal are assumed to have a typical duration of up to 10 s and sags to 70% of nominal are assumed to have a duration of up to 0.5 s.
- **Drop Out:** Voltage drop out includes both severe RMS voltage sags and complete interruption of the applied voltage followed by immediate reapplication of the nominal voltage. The interruption may last up to 20 ms. The transient typically results from the occurrence and subsequent clearing of the faults in the distribution system.
- **No Damage Region:** Events in this region include sags and drop outs that are more severe than those specified in the preceding paragraphs and continuously applied voltages that are less than the lower limit of the steady-state tolerance range. A normal functional state of the information technology equipment is not expected during these conditions, but no damage to equipment should result.
- **Prohibited Region:** This region includes any surge or swell that exceeds the upper limit of the envelope. If information technology equipment is subjected to such conditions damage might result.

The ITIC graph applies to 120-V circuits obtained from 120-V, 120/240-V, and 120/208-V distribution systems. Other nominal voltages and frequencies are not specifically considered, but their applicability may be determined in each case. The curve is useful in determining if problems could be expected under particular power system voltage conditions.

2.11.5 Harmonic Distortion

The PQ of distribution systems has a drastic effect on power regulation and consumption. Johan Lundquist of the Chalmers University of Technology in Goteborg, Sweden put it best, stating "The phrase 'power quality' has been widely used during the last decade and includes all aspects of events in the system that deviates from normal operation." This has been especially true after the second half of the 20th century when new types of electronic power sources caused distortion in waveforms of the power system.

Power sources act as non-linear loads, drawing a distorted waveform that contains harmonics. These harmonics can cause problems ranging from telephone transmission interference to degradation of conductors and insulating material in motors and transformers. Therefore, it is important to gauge the total effect of these harmonics. The summation of all harmonics in a system is known as THD. This paper will attempt to explain the concept of THD and its effects on electrical equipment. It will also outline the low THD of the Associated Power Technologies (APT) line of programmable sources and how these can be used to more effectively test equipment.

2.11.5.1 Total Harmonic Distortion

Total harmonic distortion is a complex and often confusing concept to grasp. However, when broken down into the basic definitions of harmonics and distortion, it becomes much easier to understand. Imagine a power system with an AC source and an electrical load (Figure 2.67).

Now imagine that this load is going to take on one of two basic types: linear or non-linear. The type of load is going to affect the PQ of the system. This is due to the current draw of each type of load. Linear loads draw current that is sinusoidal in nature so they generally do not distort the waveform (Figure 2.68). Most household appliances are categorized as linear loads. Non-linear loads, however, can draw current that is not perfectly sinusoidal (Figure 2.69). Since the current waveform deviates from a sine wave, voltage waveform distortions are created.

As can be observed from the waveform in Figure 2.69, waveform distortions can drastically alter the shape of the sinusoid. However, no matter the level of complexity of the fundamental wave, it is actually just a composite of multiple waveforms called harmonics. Harmonics have frequencies that are integer multiples of the waveform's fundamental frequency. For example, given a 60 Hz fundamental waveform, the 2nd, 3rd, 4th, and 5th harmonic components will be at 120 Hz, 180 Hz, 240 Hz, and 300 Hz, respectively. Thus, harmonic distortion is the degree to which a waveform deviates from its pure sinusoidal values as a result of the summation of all these harmonic elements. The ideal sine wave has zero harmonic components. In that case, there is nothing to distort this perfect wave.

Total harmonic distortion, or THD, is the summation of all harmonic components of the voltage or current waveform compared against the fundamental component of the voltage or current wave:

$$THD = \frac{\sqrt{(V_2^2 + V_3^2 + \ldots + V_n^2)}}{V_1} \times 100\% \tag{2.21}$$

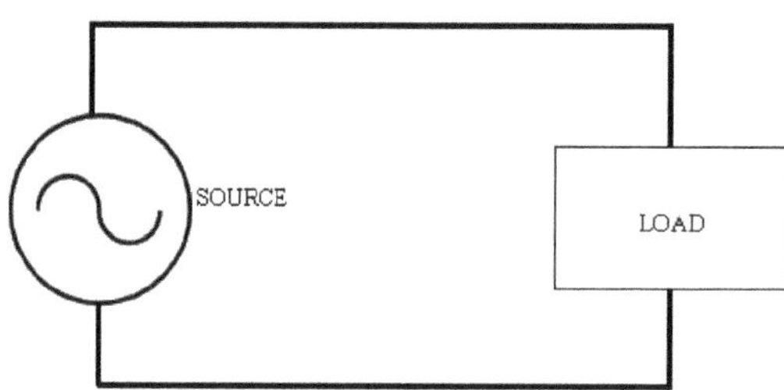

FIGURE 2.67 Power system with AC source and electrical load.

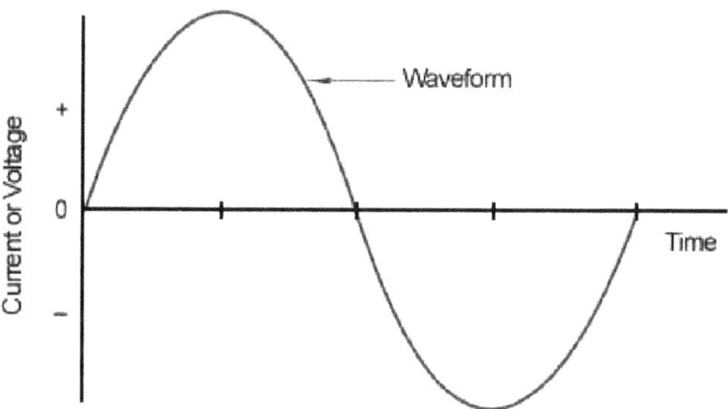

FIGURE 2.68 Ideal sine wave.

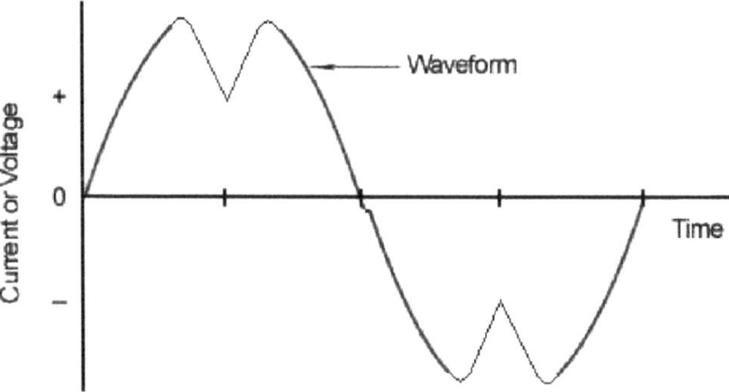

FIGURE 2.69 Distorted waveform.

The formula above shows the calculation for THD on a voltage signal. The end result is a percentage comparing the harmonic components to the fundamental component of a signal. The higher the percentage, the more distortion that is present on the mains signal.

2.11.5.2 The Usual Suspects

Harmonics have existed on power systems from the time of the very first generators. However, the harmonic components were so small that their effects on systems were negligible. This was due to the lack of non-linear loads before the 1960s. As associated professor of the University of Wollongong, V.J. Gosbell, wrote, "Harmonic distortion is not generally due to the operation of the power system, and was largely absent before the 1960s. At about this time, a different type of customer load with electronic power supplies became popular." This was the beginning of the era of non-linear loads, which now include electronic ballasts, computer power supplies, fax machines, arc furnaces, and variable frequency drives (VFDs).

Harmonic distortion can have detrimental effects on electrical equipment. Unwanted distortion can increase the current in power systems, which results in higher temperatures in neutral conductors and distribution transformers. Higher frequency harmonics cause additional core loss in motors, which results in excessive heating of the motor core. These higher order harmonics can also interfere with communication transmission lines since they oscillate at the same frequencies as the transmit frequency. If left unchecked, increased temperatures and interference can greatly shorten the life of electronic equipment and cause damage to power systems.

2.11.5.3 Importance of Mitigating THD

While there is no national standard dictating THD limits on systems, there are recommended values for acceptable harmonic distortion. IEEE Std 519, "Recommended Practices and Requirements for Harmonic Control in Electrical Power Systems," provides suggested harmonic values for power system "Computers and allied equipment, such as programmable controllers, frequently require AC sources that have no more than 5% harmonic voltage distortion factor [THD], with the largest single harmonic being no more than 3% of the fundamental voltage. Higher levels of harmonics result in erratic, sometimes subtle, malfunctions of the equipment that can, in some cases, have serious consequences."

The limits on voltage harmonics are thus set at 5% for THD and 3% for any single harmonic. It is important to note that the suggestions and values given in this standard are purely voluntary. However, keeping low THD values on a system will further ensure proper operation of equipment and a longer equipment life span.

2.11.5.4 Voltage vs. Current Distortion

An electrical system supplies power to loads by delivering current at the fundamental frequency. Only fundamental frequency current can provide real power. Current delivered at harmonic frequencies doesn't deliver any real power to the load. When current of a single frequency is present in a system, you can use the measured values in Ohm's Law and power calculations. However, when currents of more than one frequency are present, direct addition of the current values leads to a summed value that doesn't correctly represent the total effect of the multiple currents. Instead, you need to add the currents in a manner known as the "root mean square" summation. The equation for the rms addition of currents is as follows:

$$I_{\text{rms}} = \sqrt{I_1^2 + I_2^2 + I_3^2 + \ldots + I_n^2} \tag{2.22}$$

So, if a system carries 70 A of fundamental current, 18 A of 5th harmonic current, 14 A of 7th harmonic current, and 11 A of 11th harmonic current, the effective current would be 74.3 A rms, not the arithmetic sum of 113 A. This 74.3 A rms value would be the correct value to use in all power calculations. The same is true for harmonic voltages. To obtain the effective voltage for a system in which voltages of several frequencies are present, you must add the voltages in an rms fashion.

2.11.5.5 Current Measurement with Harmonics

Figure 2.70 illustrates the measurement of current drawn by a nonlinear load. The rms current contains both the fundamental and harmonics. Note that the value of current at each harmonic as well as that for the rms current are the same at each measuring point, just as in a system containing only fundamental current. The term "distortion rms" is used to denote the rms value of harmonic current with the fundamental left out of the summation. The rms current is basically the total effective load current. Calculate THD by using the value of distortion rms current in the following equation:

$$\text{THD}_{\%\text{fundamental}} = \left(\frac{I_{\text{rms(distortion)}}}{I_{\text{fundamental}}} \right) \times 100 \tag{2.23}$$

The current THD is the same at each measuring point.

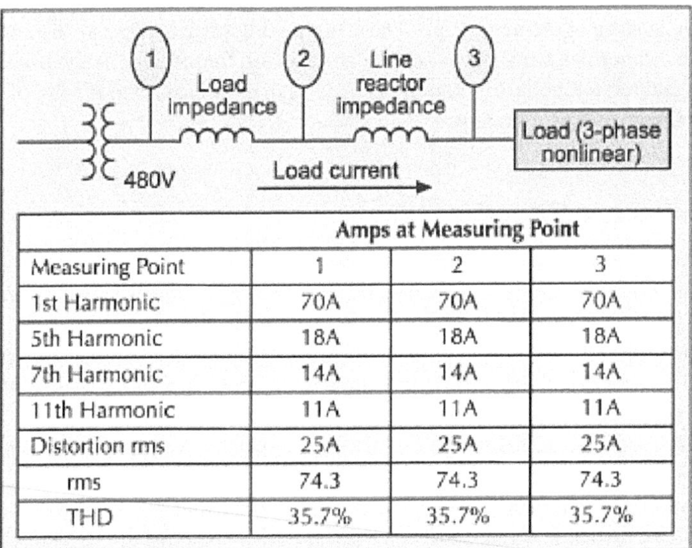

	Amps at Measuring Point		
Measuring Point	1	2	3
1st Harmonic	70A	70A	70A
5th Harmonic	18A	18A	18A
7th Harmonic	14A	14A	14A
11th Harmonic	11A	11A	11A
Distortion rms	25A	25A	25A
rms	74.3	74.3	74.3
THD	35.7%	35.7%	35.7%

FIGURE 2.70 Current measurement with harmonics.

2.11.5.6 *Voltage Measurement with Harmonics*

Figure 2.71 at right illustrates the measurement of voltage in a system powering a nonlinear load. The voltage contains both the fundamental and harmonics. The farther you measure from the source, the higher the harmonic voltage generated. The increased impedance, through which the harmonic current must flow, results in higher harmonic voltage generation. This is the opposite effect of impedance on the fundamental voltage, whereas fundamental voltage causes fundamental current flow, and harmonic current flow causes harmonic voltage.

Looking at Figure 2.71 again, we see that although the fundamental voltage has dropped from 480 V at Point 1 to 465 V at Point 3, the increase in harmonic voltage causes the rms voltage to be higher at each measuring point than the fundamental voltage alone. Distortion rms is the rms value of the harmonic voltage with the fundamental left out of the summation, and it increases the farther away you measure from the source. Calculate THD using the same equation used for current distortion.

2.11.5.7 *Effects of Current Distortion*

Since operation of nonlinear loads causes the distorted current, which is path dependent, the effect of current distortion on loads within a facility is minimal. Therefore, harmonic currents can't flow into equipment other than the nonlinear loads that caused them. However, the effect of current distortion on distribution systems can be serious, primarily because of the increased current flowing in the system.

All distribution systems are rms current-limited. In a system powering three-phase loads, for example, a 1,000 kV A, 480 V transformer is rated to deliver 1,200 A rms. But the more harmonic current this transformer has to supply, the less fundamental current it can provide for powering loads. In other words, because the harmonic current doesn't deliver any power, its presence simply uses up system capacity and reduces the number of loads that can be powered.

Harmonic currents also increase I^2Z heat losses in transformers and wiring. Since transformer impedance is frequency dependent, increasing with harmonic number, the impedance at the 5th harmonic is five times that of the fundamental frequency. So, each ampere of 5th harmonic current causes five times as much heating as an ampere of fundamental current.

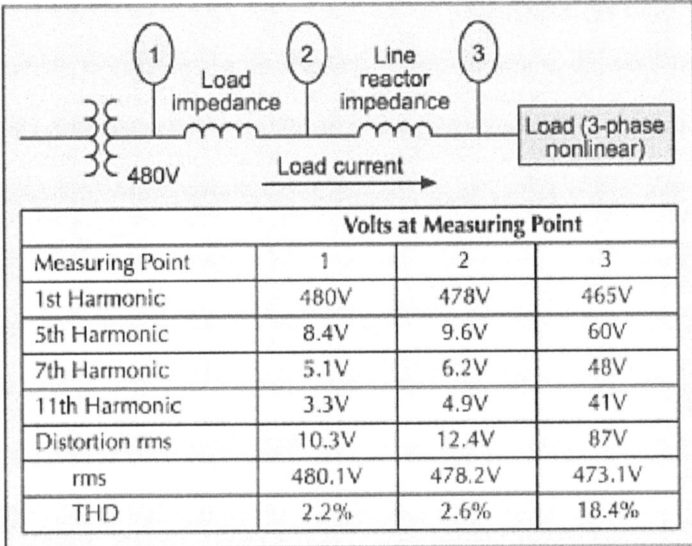

	Volts at Measuring Point		
Measuring Point	1	2	3
1st Harmonic	480V	478V	465V
5th Harmonic	8.4V	9.6V	60V
7th Harmonic	5.1V	6.2V	48V
11th Harmonic	3.3V	4.9V	41V
Distortion rms	10.3V	12.4V	87V
rms	480.1V	478.2V	473.1V
THD	2.2%	2.6%	18.4%

FIGURE 2.71 Voltage measurement with harmonics.

In a system powering phase-to-neutral connected loads, detrimental effects are again due to the harmonic currents using up system capacity and reducing the number of useful loads that can be powered. Third harmonic currents cause a further detriment, because they're additive in the neutral conductor. When many computers, which are nonlinear loads, are connected, the neutral current—primarily 3rd harmonic—can be larger than any of the phase currents. These 3rd harmonic currents circulate in the primary (delta) of the distribution transformer that serves the portion of the system powering the computers and are dissipated as heat.

2.11.5.8 Effects of Voltage Distortion

Besides overheating, the other major effect of current distortion on an electrical system is the creation of voltage distortion. This distortion will have minimal effect on a distribution system, but unlike current distortion, it isn't path dependent. So harmonic voltages generated in one part of a facility will appear on common buses within that facility. High-voltage distortion at the terminals of a nonlinear load doesn't mean high distortion will be present throughout the system. In fact, the voltage distortion becomes lower the closer a bus is located to the service transformer. However, if excessive voltage distortion does exist at the transformer, it can pass through the unit and appear in facilities distant from the origin.

The effect on loads within the facility could be detrimental in certain cases. For example, extreme voltage distortion can cause multiple zero crossings for the voltage wave. For equipment where proper sequencing of operations depends on a zero crossing for timing, voltage distortion can cause misoperation. Most modern electrical equipment uses an internal clock for timing sequencing so it's unaffected by multiple zero crossings.

Voltage distortion appears to have little effect on operation of nonlinear loads connected either phase-to-phase or phase-to-neutral.

On the other hand, 5th harmonic voltage distortion can cause serious problems for three-phase motors. The 5th harmonic is a negative sequence harmonic, and when supplied to an induction motor it produces a negative torque. In other words, it attempts to drive the motor in a reverse direction and slows down its rotation. So, the motor draws more 60-Hz current to offset the reverse torque and regain its normal operating speed. The result is overcurrent in the motor, which either causes protective devices to open or the motor to overheat and fail. For this reason, removing 5th harmonic current from systems powering three-phase loads is often a high priority in industrial facilities.

2.11.5.9 Harmonics vs. Transients

Generally, harmonic analysis is a steady-state calculation or measurement of the various frequencies present in the system and the system sensitivity to those frequencies. Transient analysis is a time-based measurement, calculation, or simulation of the response of the system to an event. There is some crossover when it comes to dealing with actual systems, because a harmonic analysis (calculation) can be used to predict the damped oscillatory response of the system to a transient event. In power systems, we use a harmonic analysis to determine if the voltages and currents are within the allowable steady-state limits across the frequency range of interest. We use the transient analysis to determine what a single event (e.g., lightning stroke) will do to the equipment.

2.11.5.10 Sources of Current Harmonics

A pure sinusoidal voltage is a conceptual quantity produced by an ideal AC generator built with finely distributed stator and field windings that operate in a uniform magnetic field. Since neither the winding distribution nor the magnetic field are uniform in a working AC machine, voltage waveform distortions are created, and the voltage-time relationship deviates from the pure sine function. The distortion at the point of generation is very small (about 1% to 2%), but nonetheless it exists. Because this is a deviation from a pure sine wave, the deviation is in the form of a periodic function, and by definition, the voltage distortion contains harmonics.

When a sinusoidal voltage is applied to a certain type of load, the current drawn by the load is determined by the voltage and impedance and follows the voltage waveform. These loads are referred to as linear loads; examples of linear loads are resistive heaters, incandescent lamps, and constant speed induction and synchronous motors.

In contrast, some loads cause the current to vary disproportionately with the voltage during each cyclic period. These are classified as nonlinear loads, and the current taken by them has a nonsinusoidal waveform.

When there is significant impedance in the path from the power source to a nonlinear load, these current distortions will also produce distortions in the voltage waveform at the load. However, in most cases where the power delivery system is functioning correctly under normal conditions, the voltage distortions will be quite small and can usually be ignored.

Waveform distortion can be mathematically analyzed to show that it is equivalent to superimposing additional frequency components onto a pure sinewave. These frequencies are harmonics (integer multiples) of the fundamental frequency, and can sometimes propagate outwards from nonlinear loads, causing problems elsewhere on the power system.

The classic example of a non-linear load is a rectifier with a capacitor input filter, where the rectifier diode only allows current to pass to the load during the time that the applied voltage exceeds the voltage stored in the capacitor, which might be a relatively small portion of the incoming voltage cycle.

Other examples of nonlinear loads are battery chargers, electronic ballasts, VFDs, and switching mode power supplies.

2.11.5.11 Voltage and Current Harmonics

Current harmonics are caused by non-linear loads such as thyristor drives, induction furnaces, etc. The effect of these loads is the distortion of the fundamental sinusoidal current waveform alternating at 50 Hz. Current harmonics affects the system by loading the distribution system as the waveforms of the other frequencies use up capacity without contributing any power to the load. They also contribute the copper losses I^2Z experiences in the system.

Besides, harmonic currents load the power sources such as transformers and alternators. However, current harmonics do not affect the remainder of the loads in the system which is linear. They only impact the loads that are causing them, i.e., non-linear loads.

Voltage harmonics are caused by the current harmonics that distort the voltage waveform. These voltage harmonics affect the entire system not just the loads that are causing them. Their impact depends on the distance of the load causing the harmonics from the power source. If other harmless loads are connected between the source and harmonics causing loads, these innocent loads will also be affected by the harmonics.

Hence, one way of mitigating the effect of harmonics is by connecting the harmonics-causing loads as close to the source as possible in a separate feeder. Another method is by using an isolating transformer between the problem loads and the rest of the distribution system.

2.11.6 Harmonic Indices

2.11.6.1 Single Site Indices

The harmonic distortion can be described by several indices (individual harmonic distortion, THD, the peak factor, and so on). In the next, without loss of generality, we will refer to the THD and peak factor (k_p); on the other hand, they are very useful indices and also easily transferable in the probabilistic field, since only the marginal probability density function is generally required to fully describe their random properties. Moreover, the Joint Working Group CIGRE C4.07/CIRED has collected available measurement data and existing probabilistic indices for MV, HV, and EHV systems and recommends a set of internationally relevant harmonic indices and objectives.

The THD of Y (current or voltage) is given by:

$$THDY = \frac{\sqrt{\sum_{h=2}^{\infty} Y_h^2}}{Y_1}, \tag{2.24}$$

with Y_h harmonic superimposed to the fundamental Y_l.

The peak factor is defined as:

$$k_{pY} = \frac{Y_p}{Y_{pln}}, \tag{2.25}$$

where Y_p is the peak value of the distorted waveform Y, and Y_{pln} the rated value of the fundamental. The peak factor can be separated into two aliquots k_{psY} and k_{phY}, the first related to the effects of fundamental variations and the second one related to the effects of harmonics superimposed to fundamental Y_{pl}; it results:

$$k_{pY} = \frac{Y_{pl}}{Y_{pln}}\frac{Y_p}{Y_{pl}} = k_{psY}k_{phY}. \tag{2.26}$$

Both THD and peak value can be applied to voltage and current distortions. The peak factor of voltage given by (2) and its aliquots given by (3) have been considered for translation in the probabilistic field while the current indices have been considered.

The probability density functions (pdf's) of the previously mentioned indices exhibit different statistical characteristics, as clearly revealed by the analysis of several on-site measurements and simulation results.

First of all, the pdfs of all above indices can have several shapes and there is not a general rule to choose the probability density function that best fits one of them; in fact, various basic pdfs (Gaussian, uniform, linear increasing, and so on) have been tested to ascertain the best approximating the harmonic index pdf behavior, but any of them can't be generally applied.

From the analysis of several daily or weekly on-site measurements and of numerical simulations on low voltage and medium voltage distribution systems, the following considerations on their probabilistic behavior can be provided.

The pdfs of the THD of the voltage (THD_V) and current (THD_I) can have both monomodal and multimodal (usually bimodal) pattern. Moreover, mainly when the harmonic data refer to a week, they are generally characterized by significant values of standard deviations, due to the variability of harmonics in large ranges.

As an example, Figure 2.72a and b report the pdfs of the THD_V (a) and of the THD_I (b) obtained processing over one week of recorded data referred to a medium voltage distribution system.

Starting from the considerations about both the pdf shapes and statistical measures of THD_V and THD_I, it follows that to impose reasonable probabilistic limits on these indices is very difficult. In fact, to

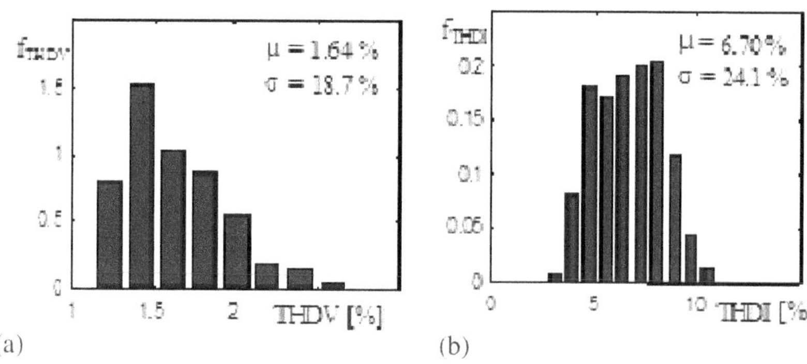

FIGURE 2.72 Probability density functions: (a) THD_V; (b) THD_I.

directly set limits on the whole pdfs is neither realistic nor a reasonable solution, as a consequence of the obvious difficulties in the successive implementation in the standards. On the other hand, to impose limits on one or more statistical measures (minimum value, maximum value, mean value, standard deviation, 95th percentile, and so on) can be practical but can lead to unfair decisions. In fact, if we refer to the 95th percentile or maximum value, as proposed by the IEC 61000-3-6 in its assessment procedure, and we account for all the harmonic effects, disagreeable consequences can follow, since THD_V and THD_I pdfs with the same 95th percentile and maximum value can have very different effects on electrical component behavior. As a very easy example, we can refer to the joule losses in a conductor (in percent of the losses at fundamental frequency) associated to the two different THD_I pdfs of Figure 2.73, characterized by the same 95th percentile and maximum. Assuming the conductor resistance independent on the frequency, the joule loss expected values in the two cases are equal to 0.095 (Figure 2.73a) and 0.035 (Figure 2.73b), that are far each other.

The pdfs of the voltage peak factor (k_{pV}) are characterized, in the most general case, by very low values of standard deviations. In all the results reported in literature and in our experience, the standard deviation shows values at least of very few percent, because the voltage harmonics have values lower than fundamental voltage that ranges in no particularly large intervals for system operation safety. Due to the very low values of standard deviations, it is useless to talk about the shape of the pdf (monomodal or multimodal). A similar behavior has been experimented with reference to the aliquot k_{phV}, related to the effects of harmonics superimposed to the fundamental.

As an example, Figure 2.74 shows the pdfs of the k_{pV} (a) and of k_{phV} (b) obtained processing over one week of recorded data referred to a medium voltage distribution system. From the analysis of Figure 2.74b and from all the measurements on distribution systems we analyzed, it was also noted that the voltage

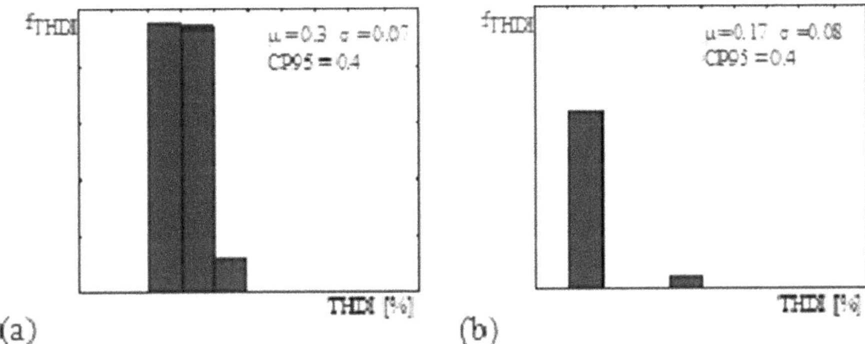

FIGURE 2.73 Different THD_I pdfs with same 95th percentile and maximum values.

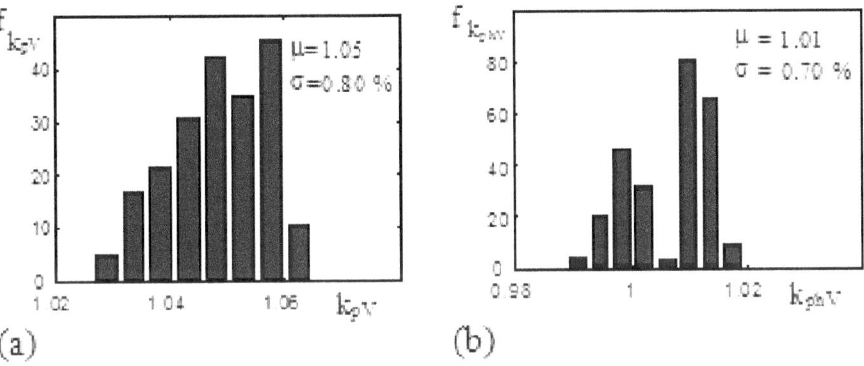

FIGURE 2.74 Probability density functions: (a) k_{pV}; (b) k_{phV}.

harmonics in several cases cause increasing of the peak voltage with respect to the value that would have been in presence of only the fundamental component.

Starting from the considerations about the k_{pV} and k_{phV} pdf shapes and their statistical measures, it follows that to impose probabilistic limits on these indices is now possible and can be a reasonable choice. In fact, the very low value of pdf standard deviations allows us to say that it can be enough to assign a limit on only the mean value of the pdf. Moreover, in practice only the mean value of the voltage peak factor[1] k_{pV} seems to be suitable for inclusion in Standards or in PQ contracts. The aliquot k_{phV} does not allow to compute all the harmonic effects on distribution system components, because to distinguish the effects between fundamental and harmonics is a very difficult matter[2].

As an example of setting limits on the mean value of peak factor, let's consider the case of low voltage industrial systems. The limits can be assigned making use of the diagram in Figure 2.75, where the normalized expected value of life is plotted versus the mean value of peak factor; each point of Figure 2.75 is obtained averaging the expected values of normalized useful life for different distribution system components (cables, transformers, capacitors, and AC motors).

The pdfs of the current peak factor (k_{pI}) and of its aliquot k_{phI}, related to the effects of harmonics superimposed to the fundamental can have both mono modal and multimodal pattern, once again usually bimodal. The variance of the k_{pI} pdf is characterized, in the most general case, by significant values of standard deviations; on the contrary, reduced variances of k_{phI} pdfs were experimented.

As an example, Figure 2.76a and b show the pdfs of the k_{pI} (a) and of the k_{phI} (b) obtained processing over one week of recorded data referred to an actual medium voltage distribution system.

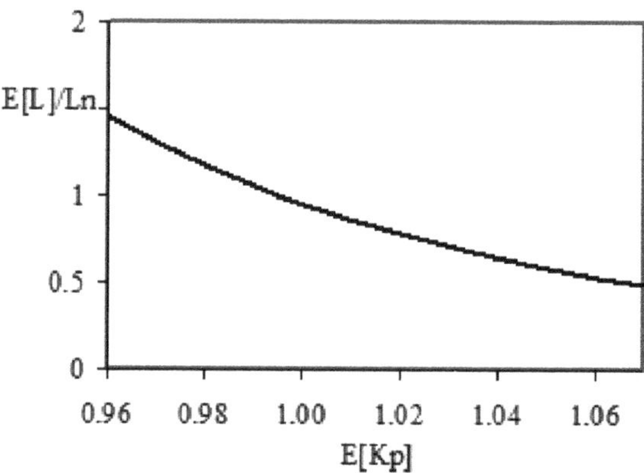

FIGURE 2.75 Expected value of normalized useful life versus expected value of peak factor averaged for cables, capacitors, transformers, and AC motors.

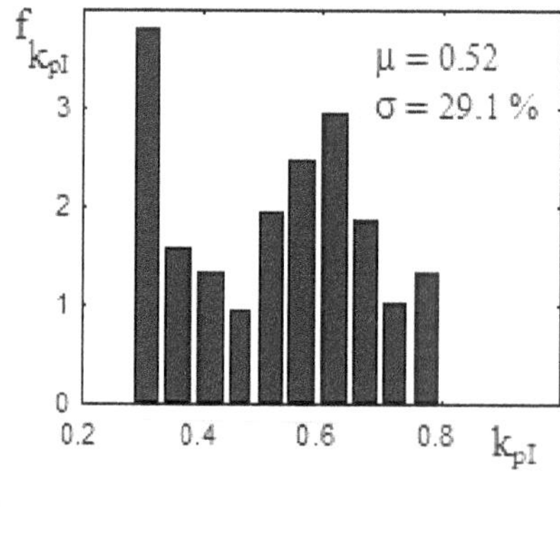

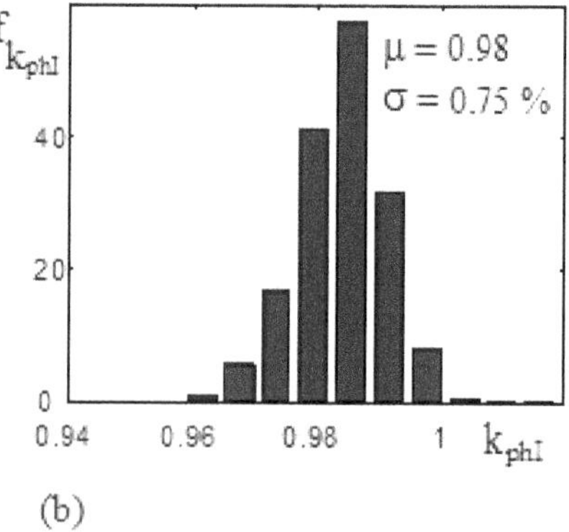

FIGURE 2.76 Probability density functions: (a) k_{pI}; (b) k_{phI}.

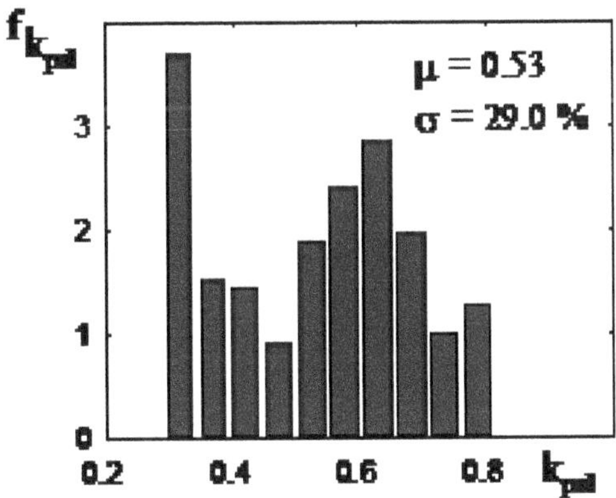

FIGURE 2.77 Probability density function of k_{psI}.

From the analysis of Figure 2.76a and from all the measurements we analyzed on distribution systems, it was also noted that the high variability of current peak factor k_{pI} is due to the wide variation range of the fundamental component, as clearly highlighted in Figure 2.77, which reports pdf of the aliquot k_{psI} corresponding to the same measurements of Figure 2.76. This is a reasonable result, for the very large variability of connected loads.

From the analysis of Figure 2.76b, it appears that the current harmonics usually cause decreasing of the peak current with respect to the value that would have been in presence of only the fundamental component. However, this is not a general rule, because in other on-site measurements on distribution network it was experimented an increasing of the peak current, in some cases significant.

Finally, starting from the considerations about both the k_{pI} pdf shapes and their statistical measures, it follows that to impose reasonable probabilistic limits on this index is a difficult task, for the same reasons evidenced in the case of THD_V.

On the contrary, if the trend evidenced by our measurements were confirmed, to impose probabilistic limits on k_{phI} index would be possible and also a reasonable choice. In fact, the very low value of pdf standard deviation allows to say, once again, that it is enough to assign a limit only on the mean value. A possible criterion for assigning limits to the mean value of k_{phI} could be to set the fundamental current at rated value and derive the limits on the basis of the component effects.

2.11.6.2 System Indices

The system indices serve as metrics and are not intended as an exact representation of the quality of service provided to each customer served from the assessed system. They can be used as a benchmark against which quality levels of parts of the distribution system and of different distribution systems can be compared.

The system indices here considered only refer to voltage distortion; they are:

- the System Total Harmonic Distortion 95th percentile, *STHD95*
- the System Average Total Harmonic Distortion (Peak Factor) *SATHD* (SAk_p)
- the System Average Excessive Total Harmonic Distortion (Peak Factor) Ratio Index, $SAETHDRI_{THD*}$, ($SAEk_pRI_{KP*}$)

Let's refer to a distribution system with N busbars. Even if the system indices can also refer to small segments of the entire distribution system, such as a single feeder or a single busbar, in the following, without loss of generality, we consider indices referred to the entire distribution system.

The *STHD95* index is defined as the 95th percentile value of a weighted distribution; this weighted distribution is obtained collecting the 95th percentile values of the *M* individual index distributions, each distribution obtained with the measurements recorded at a monitoring site. The weights can be linked to the connected powers or the number of customers served from the area that monitor data represents, and so on; in the next, without loss of generality, we assume as weights the connected powers. Under these assumptions, the following relations compute *STHD95*:

$$\frac{\sum_{-\infty}^{STHD95} f_t\left(CP95_{THDs}\right)L_s}{\sum_{-\infty}^{+\infty} f_t\left(CP95_{THDs}\right)L_s} = 0.95, \tag{2.27}$$

with $CP95_{THDs}$ calculated via the succeeding expression:

$$\frac{\sum_{-\infty}^{CP95\ THD,s} f_{THD,s}\left(X_{THD_{s,i}}\right)}{\sum_{-\infty}^{+\infty} f_{THD,s}\left(X_{THD\ _{s,i}}\right)} = 0.95 \tag{2.28}$$

where s is the monitoring site; $X_{THDs,\,i}$ = ith is the steady state *THD* measurement at the site s; L_s is the connected kVA served from the system segment where there is the monitoring site s; f_{THDs} $(X_{THDs,\,i})$ is the probability distribution function of the sampled *THD* values for monitoring site s; $CP95_{THDs}$ is the 95th cumulative probability value of *THD* for the monitoring site s; and $f_t(CP95_{THDs})$ is the probability distribution function of the individual monitoring site $CP95_{THDs}$ values.

The system index *STHD95* allows us to summarize the measurements both temporally and spatially handling measurements at *M* sites of the system in a defined time period, assumed as significant for the characterization of system service conditions.

As an example, the value of *STHD95* obtained processing over one week of recorded data referred to the same medium voltage distribution system considered to 2.2%.

The System Average Total Harmonic Distortion (Peak Factor) is based on the mean value of the distributions rather than the CP95 value:

$$SATHD = \frac{\sum_{s=1}^{M} L_s \mu\left(THD_s\right)}{L_T} \tag{2.29}$$

$$SAk_p = \frac{\sum_{s=1}^{M} L_s \mu\left(k_{p,s}\right)}{L_T} \tag{2.30}$$

being:

- (THD_s) = mean value of sampled *THD* values for monitoring site s
- $(k_{p,\,s})$ = mean value of sampled k_p values for monitoring site s
- L_T = total connected kVA served from the system

SATHD and SAk_p give average indications on the system voltage quality and, thanks to the introduced weights, allow us to assign different importance to the various sections of the entire distribution system.

The values of *SATHD* and of SAk_p, again computed on the same distribution system before referred to, are equal to 1.16% and 1.05, respectively. It is worthwhile to note that *SATHD* is far from *STHD95* due to the not negligible value of standard deviation.

The System Average Excessive Total Harmonic Distortion (Peak Factor) Ratio Index, $SAETHDRI_{THD*}$ $(SAEk_pRI_{kp*})$ is related to the number of steady-state measurements that exhibit a *THD* (k_p) value exceeding the *THD** $(k_p{*})$ threshold. It is computed, for each monitoring site of the system, counting the measurements exceeding the THD^* (k_p^*) value and normalizing this number to the total number of the measurements effected at site s. They are defined by the following relations:

$$SAETHDRI_{THD^*} = \frac{\sum_{s=1}^{M} L_s \left(N_{THD^*s} / N_{Tot,s} \right)}{L_T} \tag{2.31}$$

$$SAEk_pRI_{k_p^*} = \frac{\sum_{s=1}^{M} L_s \left(N_{k_p^*s} / N_{Tot,s} \right)}{L_T} \tag{2.32}$$

where N_{THD^*s} (N_{kp^*s}) is the number of steady state measurements at monitoring site s that exhibit a THD (k_p) value that exceeds the specified threshold THD^* (k_p^*); $N_{Tot,\,s}$ is the total number of steady state measurements effected at monitoring site s over the assessed period of time.

$SAETHDRI_{THD^*}$ and $SAEk_pRI_{kp^*}$ are linked to the measure of the total portion of the time that the system exceeds the specified THD^* and k_p^* thresholds, respectively.

Table 2.7 shows the values of $SAETHDRI_{THD^*}$ and $SAEk_pRI_{kp^*}$ computed on the same distribution system already referred to and with two thresholds.

Summarizing, the introduced indices provide different global measures to qualify the system in terms of service quality. The availability of several indices for a phenomenon can be useful to give evidence to different aspects of the same disturbance. No general rule can be established to suggest the most adequate index to be used; it will depend on which aspect is judged more relevant.

For example, in the considered real distribution system, $SATHD$ equal to 1.16% indicates not only the average level of distortion, but also considers the sites serving larger loads. This can be more evident if we consider that the average value of THD on the same distribution system, without any weights, is greater than about 25%.

Further indication on the same system and on the same measurements concerns the highest levels of distortion. $STHD95$ value indicates the weighted 95th percentile of the 95th percentiles of the system sites and $SAETHDRI_{THD^*}$ values give more details about the weighted relative frequencies of the records above specified threshold. Analogous considerations can be developed concerning the system indices referred to the peak factor.

An alternative way to compute the aforementioned indices concerns the statistical analysis of all the measurements effected at the M sites of the system. Assuming that centralized handling all the measurements of the M sites is not hard, this method seems to be interesting, since any statistical measure of a pdf is estimated with an accuracy that depends on the amount of processed data. This extensive procedure gives indication on the global behavior of the system; on the contrary the first recalled procedure furnishes the indication on the behavior of the sites.

For example, with reference to THD, for each site s we compute the relative frequencies of the measured THD and we weight them with the ratio L_s/L_T. Starting from the weighted relative frequencies at each site s, we evaluate the system weighted relative frequencies WF_{THD} and, then, compute the mean value and the 95th percentile that coincide with the indices $SATHD$ and $STHD95$, respectively. The system weighted relative frequencies of THD can be also used to evaluate the indices $SAETHDRI_{THD^*}$.

An analogous procedure can be implemented to calculate the indices SAk_p and $SAEk_pRI_{kp^*}$ referred to the peak factor.

TABLE 2.7

Values of $SAETHDRI_{THD^*}$ and $SAEk_pRI_{KP^*}$

SAETHDRI5%	0
SAETHDRI2%	0.016
AEkpRI1	0.97
SAEkpRI1	0.41

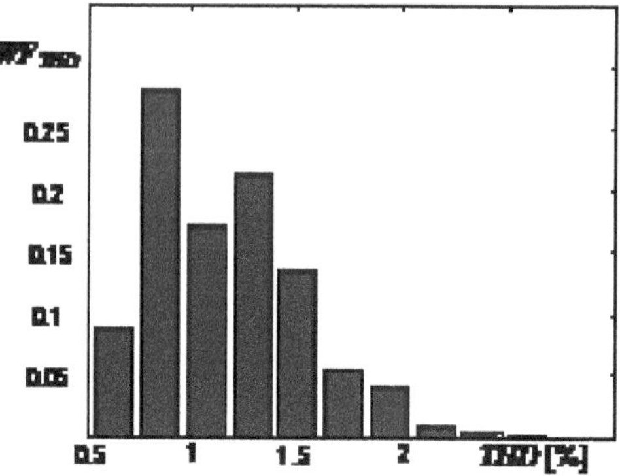

FIGURE 2.78 Histogram of system-weighted relative frequencies for *THD*.

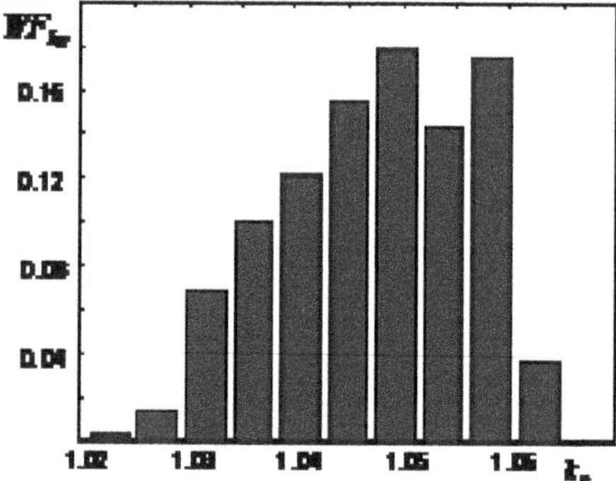

FIGURE 2.79 Histogram of system-weighted relative frequencies for k_p.

Figures 2.78 and 2.79 show the histograms of system weighted relative frequencies of *THD* and $k_p(WF_{THD}$ and $WF_{kp})$, respectively, with reference to the distribution system under consideration.

Table 2.8 shows the results in terms of the system indices obtained with the aforementioned procedure.

Comparing the results of the indices computed via the two procedures, it is important to evidence that the obtained values of *SATHD*, $SAETHDRI_{THD*}$, SAk_p, and $SAEk_pRI_{kp*}$ coincide, as expected.

On the contrary, the values of *STHD95* are different. This is due to the fact that in the first procedure, the weighted distribution cannot consider the values greater than the 95th percentiles. These values are instead considered, for the second procedure, in the construction of the system weighted relative frequency histogram, so that they can determine a lower value of the 95th percentile.

TABLE 2.8

Values of System Indices

STHD95	1.94%
THD	1.16%
SAkp	1.05
SAETHDRI2%	0.016
SAEkpRI1	0.97

2.11.6.3 Harmonic Sources from Commercial Loads

Commercial facilities such as office complexes, department stores, hospitals, and Internet data centers are dominated with high-efficiency fluorescent lighting with electronic ballasts, ASDs for the heating, ventilation, and air conditioning (HVAC) loads, elevator drives, and sensitive electronic equipment supplied by single-phase switch-mode power supplies.

Commercial loads are characterized by a large number of small harmonic-producing loads. Depending on the diversity of the different load types, these small harmonic currents may add in phase or cancel each other. The voltage distortion levels depend on both the circuit impedances and the overall harmonic current distortion. Since power factor correction capacitors are not typically used in commercial facilities, the circuit impedance is dominated by the service entrance transformers and conductor impedances.

Therefore, the voltage distortion can be estimated simply by multiplying the current by the impedance adjusted for frequency.

2.11.6.3.1 Single-Phase Power Supplies

Electronic power converter loads with their capacity for producing harmonic currents now constitute the most important class of nonlinear loads in the power system. Advances in semiconductor device technology have fueled a revolution in power electronics over the past decade, and there is every indication that this trend will continue. Equipment includes adjustable-speed motor drives, electronic power supplies, DC motor drives, battery chargers, electronic ballasts, and many other rectifier and inverter applications.

A major concern in commercial buildings is that power supplies for single-phase electronic equipment will produce too much harmonic current for the wiring. DC power for modern electronic and microprocessor-based office equipment is commonly derived from single-phase full-wave diode bridge rectifiers. The percentage of load that contains electronic power supplies is increasing at a dramatic pace, with the increased utilization of personal computers in every commercial sector.

There are two common types of single-phase power supplies. Older technologies use AC-side voltage control methods, such as transformers, to reduce voltages to the level required for the DC bus. The inductance of the transformer provides a beneficial side effect by smoothing the input current waveform, reducing harmonic content.

Newer technology switch mode power supplies (see Figure 2.80) use DC-to-DC conversion techniques to achieve a smooth DC output with small, lightweight components. The input diode bridge is directly connected to the AC line, eliminating the transformer. This results in a coarsely regulated DC voltage on the capacitor. This direct current is then converted back to alternating current at a very high frequency by the switcher and subsequently rectified again. Personal computers, printers, copiers, and most other single-phase electronic equipment now almost universally employ switch-mode power supplies. The key advantages are the lightweight, compact size, efficient operation, and lack of need for a transformer. Switch-mode power supplies can usually tolerate large variations in input voltage.

Because there is no large AC-side inductance, the input current to the power supply comes in very short pulses as the capacitor *C*1 regains its charge on each half cycle. Figure 2.81 illustrates the current waveform and spectrum for an entire circuit supplying a variety of electronic equipment with switch-mode power supplies.

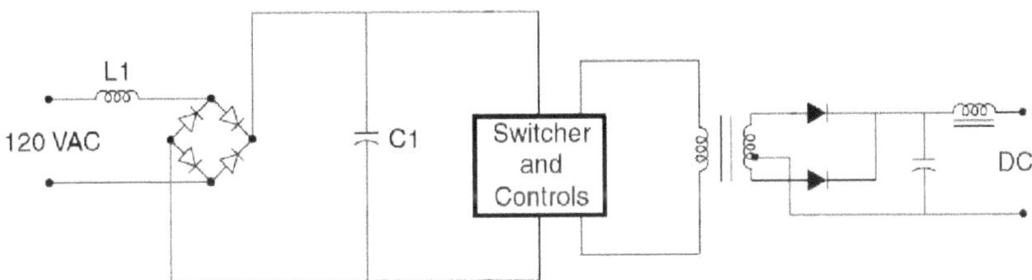

FIGURE 2.80 Switch-mode power supply.

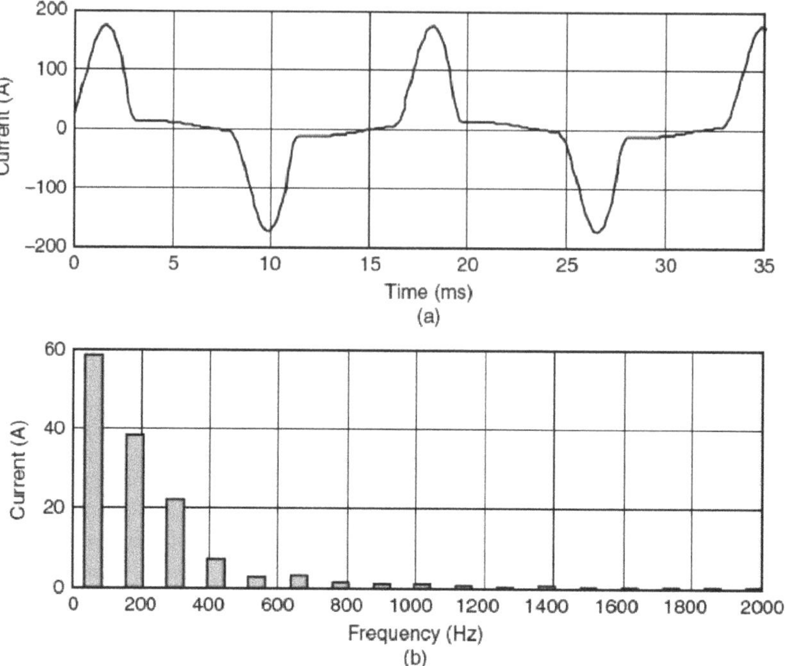

FIGURE 2.81 SMPS current and harmonic spectrum.

A distinctive characteristic of switch-mode power supplies is a very high third-harmonic content in the current. Since third-harmonic current components are additive in the neutral of a three-phase system, the increasing application of switch-mode power supplies causes concern for overloading of neutral conductors, especially in older buildings where an undersized neutral may have been installed. There is also a concern for transformer overheating due to a combination of harmonic content of the current, stray flux, and high neutral currents.

2.11.6.3.2 Fluorescent Lighting

Lighting typically accounts for 40% to 60% of a commercial building load. Fluorescent lights are discharge lamps; thus, they require a ballast to provide a high initial voltage to initiate the discharge for the electric current to flow between two electrodes in the fluorescent tube. Once the discharge is established, the voltage decreases as the arc current increases.

It is essentially a short circuit between the two electrodes, and the ballast has to quickly reduce the current to a level to maintain the specified lumen output. Thus, ballast is also a current-limiting device in lighting applications. There are two types of ballasts, magnetic and electronic. Standard magnetic ballast is simply made up of an iron-core transformer with a capacitor encased in an insulating material. Single magnetic ballast can drive one or two fluorescent lamps, and it operates at the line fundamental frequency, i.e., 50 or 60 Hz. The iron-core magnetic ballast contributes additional heat losses, which makes it inefficient compared to electronic ballast. Electronic ballast employs a switch-mode-type power supply to convert the incoming fundamental frequency voltage to a much higher frequency voltage typically in the range of 25 to 40 kHz. This high frequency has two advantages. First, a small inductor is sufficient to limit reduces the 100- or 120-Hz flicker associated with an iron-core magnetic ballast. Single electronic ballast typically can drive up to four fluorescent lamps.

Standard magnetic ballasts are usually rather benign sources of additional harmonics themselves since the main harmonic distortion comes from the behavior of the arc. The current THD is a moderate 15%. As a comparison, electronic ballasts, which employ switch-mode power supplies, can produce double or triple the standard magnetic ballast harmonic output.

Other electronic ballasts have been specifically designed to minimize harmonics and may actually produce less harmonic distortion than the normal magnetic ballast-lamp combination. Electronic ballasts typically produce current THDs in the range of between 10% and 32%. A current THD greater than 32% is considered excessive according to ANSI C82.11-1993, *High-Frequency Fluorescent Lamp Ballasts*. Most electronic ballasts are equipped with passive filtering to reduce the input current harmonic distortion to less than 20%.

Since fluorescent lamps are a significant source of harmonics in commercial buildings, they are usually distributed among the phases in a nearly balanced manner. With a delta-connected supply transformer, this reduces the amount of triplen harmonic currents flowing onto the power supply system.

2.11.6.3.3 Adjustable-Speed Drives for HVAC and Elevators

Common applications of ASDs in commercial loads can be found in elevator motors and in pumps and fans in HVAC systems. An ASD consists of an electronic power converter that converts AC voltage and frequency into variable voltage and frequency. The variable voltage and frequency allow the ASD to control motor speed to match the application requirement such as slowing a pump or fan. ASDs also find many applications in industrial loads.

2.11.6.3.4 Harmonic Sources from Industrial Loads

- In modern industries, nonlinear loads are unavoidable today. They are injecting harmonics in to the system.
- Nonlinear loads are operating at low power factor. Therefore, power factor correction strategies are applied to the system. Widely using power factor correction capacitors magnify the harmonics.
- High voltage distortions are experienced at LV side (Capacitor side).
- At resonance condition, motor and transformer overheating and misoperation of sensitive electronic equipment occur.
- There are three categories of nonlinear industrial loads:
  - Three-phase power converters
  - Arcing devices
  - Saturable devices

2.11.6.3.4.1 Three-Phase Power Converters All equipment containing static converters, as variable-speed controllers, UPS units and AC/DC converters in general, are based on a three-phase bridge, also known as a 6-pulse bridge because there are six voltage pulses per cycle (one per half cycle per phase) on the DC output.

This bridge produces in supply networks current harmonics of order $6n\pm1$, which means one more and one less than each multiple of six. In theory, the magnitude of each harmonic should be equal to the reciprocal of the harmonic number, so there would be 20% of the 5th harmonic and 9% of the 11th harmonic, etc. Figure 2.82 shows an example waveform of a thyristor bridge current against the phase voltage. Commutation notches are clearly visible in the voltage waveform (the source of high-frequency distorting components). The magnitude of the harmonics is significantly reduced by the use of a 12-pulse converter.

2.11.6.3.4.2 Arcing Devices The following are the arcing devices:

- Arc furnaces
- Arc welders and discharge-type
- Lighting (fluorescent, sodium vapor, mercury vapor) with magnetic ballasts

The voltage-current characteristics of electric arcs are nonlinear. Following arc ignition, the voltage decreases as the arc current increases, limited only by the impedance of the power system.

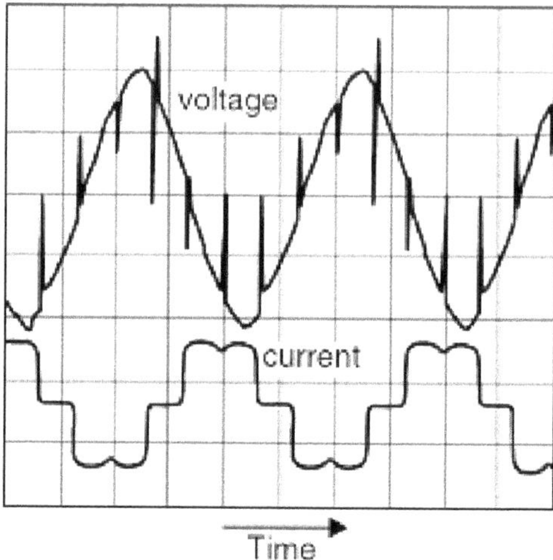

FIGURE 2.82 Example waveforms of the supply voltage and current of a 6-pulse thyristor bridge with DC side reactor.

In electric arc furnace applications, the limiting impedance is primarily the furnace cable and leads with some contribution from the power system and furnace transformer. Currents in excess of 60,000 A are common.

The electric arc itself is actually best represented as a source of voltage harmonics. Its magnitude is largely a function of the length of the arc. The arcing load thus appears to be a relatively stable harmonic current source, which is adequate for most analyses. The exception occurs when the system is near resonance and a Thevenin-equivalent model using the arc voltage waveform gives more realistic answers.

Three phase arcing devices can be arranged to cancel the triplen harmonics through the transformer connection. However, this cancellation may not work in three-phase arc furnaces because of the frequent unbalanced operation during the melting phase. During the refining stage when the arc is more constant, the cancellation is better.

2.11.6.3.4.3 Saturable Devices Equipment in this category includes transformers and other electromagnetic devices with a steel core, including motors. Harmonics are generated due to the nonlinear magnetizing characteristics of the steel.

Power transformers are designed to operate below the "knee point" of the magnetic saturation characteristics. Selection of the operating point depends on,

- Steel cost
- No-load losses
- Noise
- Other factors

Some transformers are purposefully operated in the saturated region. One example is a triplen transformer used to generate 180 Hz for induction furnaces.

Motors also exhibit some distortion in the current when overexcited, although it is generally of little consequence. There are, however, some fractional horsepower, single-phase motors that have a nearly triangular waveform with significant third-harmonic currents.

The waveform shown in Figure 2.83 is for single-phase or wye-grounded three-phase transformers. The current obviously contains a large amount of third harmonic. Delta connections and ungrounded-wye connections prevent the flow of zero-sequence harmonic, which triplens tend to be. Thus, the line current will be void of these harmonics unless there is an imbalance in the system.

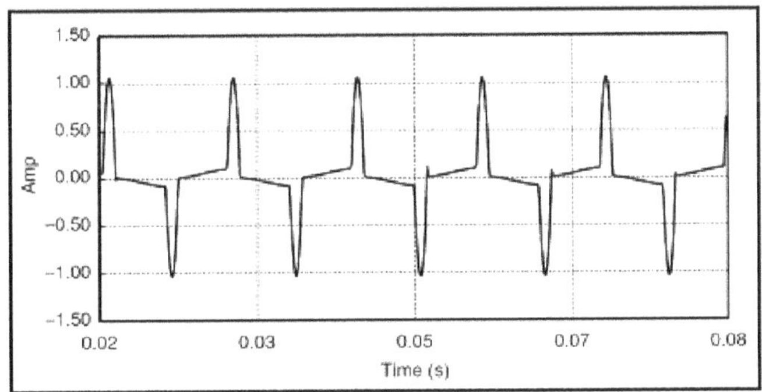

FIGURE 2.83 Transformer magnetizing current and harmonic spectrum.

2.11.6.3.5 Locating Harmonic Sources

On radial utility distribution feeders and industrial plant power systems, the main tendency is for the harmonic currents to flow from the harmonic-producing load to the power system source. This is illustrated in Figure 2.84. The impedance of the power system is normally the lowest impedance seen by the harmonic currents. Thus, the bulk of the current flows into the source.

This general tendency of harmonic current flows can be used to locate sources of harmonics. Using a PQ monitor capable of reporting the harmonic content of the current, simply measure the harmonic currents in each branch starting at the beginning of the circuit and trace the harmonics to the source.

Power factor correction capacitors can alter this flow pattern for at least one of the harmonics. For example, adding a capacitor to the previous circuit as shown in Figure 2.85 may draw a large amount of harmonic current into that portion of the circuit. In such a situation, following the path of the harmonic current will lead to a capacitor bank instead of the actual harmonic source. Thus, it is generally necessary to temporarily disconnect all capacitors to reliably locate the sources of harmonics.

It is usually straightforward to differentiate harmonic currents due to actual sources from harmonic currents that are strictly due to resonance involving a capacitor bank. A resonance current typically has only one dominant harmonic riding on top of the fundamental sine wave.

They all produce more than one single harmonic frequency. Waveforms of these harmonic sources have somewhat arbitrary wave shapes depending on the distorting phenomena, but they contain several harmonics in significant quantities. A single, large, significant harmonic nearly always signifies resonance.

Another method to locate harmonic sources is by correlating the time variations of the voltage distortion with specific customer and load characteristics. Patterns from the harmonic distortion measurements can be compared to particular types of loads, such as arc furnaces, mill drives, and mass transits that appear intermittently. Correlating the time from the measurements and the actual operation time can identify the harmonic source.

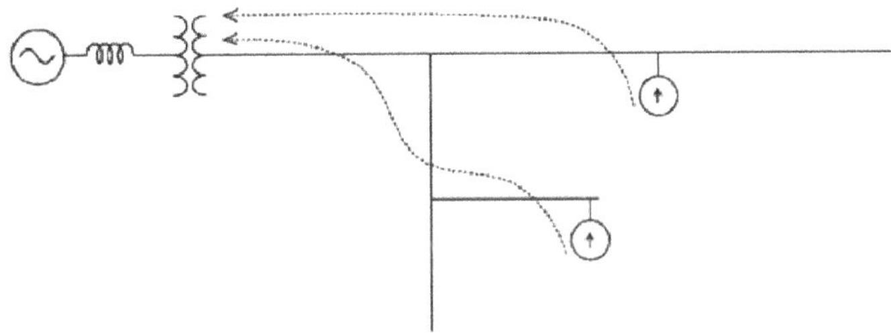

FIGURE 2.84 General flow of harmonic currents in a radial power system.

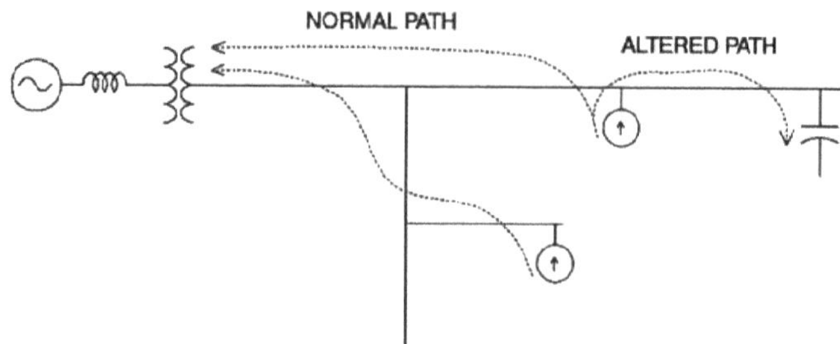

FIGURE 2.85 Power factor capacitors can alter the direction of flow of one of the harmonic components of the current.

2.11.6.3.6 System Response Characteristics

In power systems, the response of the system is equally as important as the sources of harmonics. In fact, power systems are quite tolerant of the currents injected by harmonic-producing loads unless there is some adverse interaction with the impedance of the system. Identifying the sources is only half the job of harmonic analysis. The response of the power system at each harmonic frequency determines the true impact of the nonlinear load on harmonic voltage distortion. There are three primary variables affecting the system response characteristics, i.e., the system impedance, the presence of a capacitor bank, and the number of resistive loads in the system.

2.11.6.3.7 Effects of Harmonic Distortion

A pure sinusoidal voltage is a conceptual quantity produced by an ideal AC generator built with finely distributed stator and field windings that operate in a uniform magnetic field. Since neither the winding distribution nor the magnetic field is uniform in a working AC machine, voltage waveform distortions are created, and the voltage-time relationship deviates from the pure sine function. The distortion at the point of generation is very small (about 1% to 2%), but nonetheless it exists. Because this is a deviation from a pure sine wave, the deviation is in the form of a periodic function, and by definition, the voltage distortion contains harmonics.

When a sinusoidal voltage is applied to a certain type of load, the current drawn by the load is proportional to the voltage and impedance and follows the envelope of the voltage waveform. These loads are referred to as linear loads (loads where the voltage and current follow one another without any distortion to their pure sine waves). Examples of linear loads are resistive heaters, incandescent lamps, and constant speed induction and synchronous motors.

In contrast, some loads cause the current to vary disproportionately with the voltage during each half cycle. These loads are classified as nonlinear loads, and the current and voltage have waveforms that are nonsinusoidal, containing distortions, whereby the 60-Hz waveform has numerous additional waveforms superimposed upon it, creating multiple frequencies within the normal 60-Hz sine wave. The multiple frequencies are harmonics of the fundamental frequency.

Normally, current distortions produce voltage distortions. However, when there is a stiff sinusoidal voltage source (when there is a low impedance path from the power source, which has sufficient capacity so that loads placed upon it will not affect the voltage), one need not be concerned about current distortions producing voltage distortions.

Examples of nonlinear loads are battery chargers, electronic ballasts, VFDs, and switching mode power supplies. As nonlinear currents flow through a facility's electrical system and the distribution-transmission lines, additional voltage distortions are produced due to the impedance associated with the electrical network. Thus, as electrical power is generated, distributed, and utilized, voltage and current waveform distortions are produced.

Power systems designed to function at the fundamental frequency, which is 60 Hz in the United States, are prone to unsatisfactory operation and, at times, failure when subjected to voltages and currents that contain substantial harmonic frequency elements. Very often, the operation of electrical equipment may seem normal, but under a certain combination of conditions, the impact of harmonics is enhanced, with damaging results.

2.11.6.3.7.1 Motors There is an increasing use of VFDs that power electric motors. The voltages and currents emanating from a VFD that goes to a motor are rich in harmonic frequency components. Voltage supplied to a motor sets up magnetic fields in the core, which create iron losses in the magnetic frame of the motor. Hysteresis and eddy current losses are part of iron losses that are produced in the core due to the alternating magnetic field. Hysteresis losses are proportional to frequency, and eddy current losses vary as the square of the frequency. Therefore, higher frequency voltage components produce additional losses in the core of AC motors, which in turn, increase the operating temperature of the core and the windings surrounding in the core. Application of non-sinusoidal voltages to motors results in harmonic current circulation in the windings of motors. The net rms current is

$$I_{rms} = \sqrt{[(I_1)^2 + (I_2)^2 + (I_3)^2 + \ldots]}$$

where the subscripts 1, 2, 3, etc. represent the different harmonic currents. The I^2R losses in the motor windings vary as the square of the rms current. Due to skin effect, actual losses would be slightly higher than calculated values. Stray motor losses, which include winding eddy current losses, high frequency rotor and stator surface losses, and tooth pulsation losses, also increase due to harmonic voltages and currents.

The phenomenon of torsional oscillation of the motor shaft due to harmonics is not clearly understood, and this condition is often disregarded by plant personnel. Torque in AC motors is produced by the interaction between the air gap magnetic field and the rotor-induced currents. When a motor is supplied non-sinusoidal voltages and currents, the air gap magnetic fields and the rotor currents contain harmonic frequency components.

The harmonics are grouped into positive (+), negative (–), and zero (0) sequence components. Positive sequence harmonics (harmonic numbers 1, 4, 7, 10, 13, etc.) produce magnetic fields and currents rotating in the same direction as the fundamental frequency harmonic. Negative sequence harmonics (harmonic numbers 2, 5, 8, 11, 14, etc.) develop magnetic fields and currents that rotate in a direction opposite to the positive frequency set. Zero sequence harmonics (harmonic numbers 3, 9, 15, 21, etc.) do not develop usable torque, but produce additional losses in the machine. The interaction between the positive and negative sequence magnetic fields and currents produces torsional oscillations of the motor shaft. These oscillations result in shaft vibrations. If the frequency of oscillations coincides with the natural mechanical frequency of the shaft, the vibrations are amplified and severe damage to the motor shaft may occur. It is important that for large VFD motor installations, harmonic analyses be performed to determine the levels of harmonic distortions and assess their impact on the motor.

2.11.6.3.7.2 Transformers The harmful effects of harmonic voltages and currents on transformer performance often go unnoticed until an actual failure occurs. In some instances, transformers that have operated satisfactorily for long periods have failed in a relatively short time when plant loads were changed or a facility's electrical system was reconfigured. Changes could include installation of VFDs, electronic ballasts, power factor improvement capacitors, arc furnaces, and the addition or removal of large motors.

Application of nonsinusoidal excitation voltages to transformers increases the iron losses in the magnetic core of the transformer in much the same way as in a motor. A more serious effect of harmonic loads served by transformers is due to an increase in winding eddy current losses. Eddy currents are circulating currents in the conductors induced by the sweeping action of the leakage magnetic field on the conductors. Eddy current concentrations are higher at the ends of the transformer windings due to the crowding effect of the leakage magnetic fields at the coil extremities. The eddy current losses increase as the square of the current in the conductor and the square of its frequency. The increase in transformer eddy current loss due to harmonics has a significant effect on the operating temperature of the transformer. Transformers that are required to supply power to nonlinear loads must be derated based on the percentages of harmonic components in the load current and the rated winding eddy current loss.

One method of determining the capability of transformers to handle harmonic loads is by *k* factor ratings. The *k* factor is equal to the sum of the square of the harmonic currents multiplied by the square of the frequencies.

k = [([*I*.sub.1]).sup.2]([1.sup.2])+[([*I*.sub.2]).sup.2]([2.sup.2])+
[([*I*.sub.3]).sup.2]([3.sup.2]) + … + [([*I*.sub.*n*]).sup.2]([*n*.sup.2]) (2.33)

where [*I*.sub.1] = ratio of fundamental current to total rms current, [*I*.sub.2] = ratio of second harmonic current to total rms current, [*I*.sub.3] = ratio of third harmonic current to total rms current, etc., and 1,2,3,... *n* are harmonic frequency numbers. The total rms current is the square root of the sum of square of the individual currents.

By providing additional capacity (larger-size or multiple winding conductors), *k* factor rated transformers are capable of safely withstanding additional winding eddy current losses equal to *k* times the rated eddy current loss. Also, due to the additive nature of triplen harmonic (3, 9, 15, etc.) currents flowing in the neutral conductor, *k* rated transformers are provided with a neutral terminal that is sized at least twice as large as the phase terminals.

Example: A transformer is required to supply a nonlinear load comprised of 200 A of fundamental (60 Hz), 30 A of 3rd harmonic, 48 A of 5th harmonic, and 79 A of 7th harmonic. Find the required *k* factor rating of the transformer:

Total rms current, *I* = [square root of [([*I*.sub.1]).sup.2] + [([*I*.sub.3]).sup.2] +
[([*I*.sub.5]).sup.2] + [([*I*.sub.7]).sup.2]]
Total rms current, *I* = [square root of [(200).sup.2] + [(30).sup.2] + [(48).sup.2] +
[(79).sup.2]] = 222.4 A
[*I*.sub.1] = 200 / 222.4 = 0.899
[*I*.sub.3] = 30 / 222.4 = 0.135
[*I*.sub.5] = 48 / 222.4 = 0.216
[*I*.sub.7] = 79 / 222.4 = 0.355
k = [(0.899).sup.2][(1).sup.2] + [(0.135).sup.2] [(3).sup.2] + [(0.216).sup.2][(5).sup.2] +
[(0.355).sup.2][(7).sup.2] = 8.31

To address the harmonic loading in this example, you should specify a transformer capable of supplying a minimum of 222.4 A with a *k* rating of 9. Of course, it would be best to consider possible load growth and adjust the minimum capacity accordingly.

2.11.6.3.7.3 Capacitor Banks Many industrial and commercial electrical systems have capacitors installed to offset the effect of low power factor. Most capacitors are designed to operate at a maximum of 110% of rated voltage and at 135% of their kVAR ratings. In a power system characterized by large voltage or current harmonics, these limitations are frequently exceeded, resulting in capacitor bank failures. Since capacitive reactance is inversely proportional to frequency, unfiltered harmonic currents in the power system find their way into capacitor banks. These banks act like a sink, attracting harmonic currents, thereby becoming overloaded.

A more serious condition, with potential for substantial damage, occurs as a result of harmonic resonance. Resonant conditions are created when the inductive and capacitive reactances become equal in an electrical system. Resonance in a power system may be classified as series or parallel resonance, depending on the configuration of the resonance circuit. Series resonance produces voltage amplification and parallel resonance causes current multiplication within an electrical system. In a harmonic-rich environment, both types of resonance are present. During resonant conditions, if the amplitude of the offending frequency is large, considerable damage to capacitor banks would result. And, there is a high probability that other electrical equipment on the system would also be damaged.

A typical power system incorporates a distribution transformer ([*T*.sub.1]) and two VFDs, each serving a 500 hp induction motor. Assume that transformer [*T*.sub.1] is rated 3 MVA, 13.8 kV/480 V, 7.0% leakage reactance. With a 1,000 kVAR capacitor bank installed on the 480 V bus, the following calculations examine the power system for resonance. Where the secondary current of the 3 MVA transformer is based at a potential of 480 V, and neglecting utility source impedance, the transformer reactance at 7% results in an inductive reactance ([*X*.sub.*L*]) of 0.0161 ohms as determined from the following calculations, based upon a delta electrical configuration.

Transformer line current ([*I*.sub.*L*]) = [VA transformer rating] / [([square root of 3])([*V*.sub.*L*])]

([*I*.sub.*L*]) = [(3)[(10).sup.6]] / [([square root of 3])(480)] = 3608 A

NOTE: Impedance values are calculated using the actual winding current ([*I*.sub.*w*]) and winding voltage ([*V*.sub.*w*]).

[*I*.sub.*w*] = [*I*.sub.*L*] / [square root of 3] = 3608 / [square root of 3] = 2083 A

Winding voltage ([*V*.sub.*w*]) = line voltage ([*V*.sub.*L*]) = 480 V

Percent reactance (7%) = ([*I*.sub.*w*])([*X*.sub.*L*]) / ([*V*.sub.*w*])

Inductive reactance ([*X*.sub.*L*]) = (.07)([*V*.sub.*w*]) / ([*I*.sub.*w*]) = (.07)(480) / (2083) [*X*.sub.*L*] = 0.0161 ohms

Inductance (*L*) = [*X*.sub.*L*] / 2[Pi]*f* = 0.0161 / (2)(3.14)(60) = (0.428)[(10).sup.–4] henry

For a delta-connected capacitor, the following calculations are applicable:

Line current to capacitor bank ([*I*.sub.*L*]) = (capacity in var) / ([square root of 3])([*V*.sub.*L*]) [*I*.sub.*L*] = (1000)[(10).sup.3] / ([square root of 3])(480) = 1203 A

Capacitor current ([*I*.sub.*c*]) = [*I*.sub.*L*] / [square root of 3] = 1203 / 1.732 = 694.6 A

Capacitive reactance ([*X*.sub.*c*]) = [*V*.sub.*L*] / [*I*.sub.*c*] = 480 / 694.4 = 0.691 ohm Capacitance (*C*) = 1 / 2[Pi]*f*[*X*.sub.*c*] = 1 / (2)(3.14)(60)(0.691)= (38.4)[(10).sup.–4] farad

Resonance frequency ([*f*.sub.*R*]) = 1 / 2[Pi][square root of (*L*)(*C*)]

([*f*.sub.*R*])= 1 / (2)(3.14) [[square root of (0.428)[(10).sup.–4] (38.4)[(10.)sup.–4]]]

([*f*.sub.*R*]) = 1 / (6.28) [[square root of (0.428)(38.4)[(10).sup.–8]]] = 393 Hz

A different derivation must be carried out when using a wye-connected transformer and a wye-connected capacitor bank. The wye-connected arrangement is the one normally used when a secondary neutral is required. The following equations are applicable for wye configurations.

For the transformer:

Transformer winding voltage ([*V*.sub.*w*]) = line voltage ([*V*.sub.*L*]) / [square root of 3] = 480 / [square root of 3] = 277 V

Winding current ([*I*.sub.*w*]) = transformer capacity (VA) / ([*V*.sub.*L*])([square root of 3])

[*I*.sub.*w*] = (3)[(10).sup.6] / (480)([square root of 3])= 3608 A

Inductive reactance ([*X*.sub.*L*]) = (.07)([*V*.sub.*w*]) / ([*I*.sub.*w*]) = (.07)(277) / (3608)

[*X*.sub.*L*] = 0.00537 ohms

Inductance (*L*) = [*X*.sub.*L*] / 2[Pi]*f* = 0.00537 / (2)(3.14)(60) = (14.3)[(10).sup.–6] henry

For the capacitor bank:

Capacitor bank current flow ([*I*.sub.*c*]) = (capacity in var) / ([square root of 3])([*V*.sub.*L*])

[*I*.sub.*c*] = (1000)[(10).sup.3] / ([square root of 3])(480) = 1203 A

Capacitor voltage ([*V*.sub.*c*]) = line voltage ([*V*.sub.*L*]) / [square root of 3] = 480 / [square root of 3] = 277 V

Capacitive reactance ([*X*.sub.*c*]) = [*V*.sub.*c*] / [*I*.sub.*c*] = 277 / 1203 = 0.23 ohm

Capacitance (*C*) = 1 / 2[Pi]*f*[*X*.sub.*c*] = 1 / (2)(3.14)(60)(0.23) = 0.0115 farad

Resonance frequency ([*f*.sub.*R*]) = 1 / 2[Pi][square root of (*L*)(*C*)]

([*f*.sub.*R*]) = 1 / (2)(3.14)[[square root of (14.3)[(10).sup.–6]] (0.0115)]
([*f*.sub.*R*]) = 1 / (6.28)[[square root of (0.16445)[(10).sup.–6]]] = 393 Hz

NOTE: that the resonance frequency remains the same, whether for a delta-type circuit or for a wye-type circuit. However, this situation would change should the transformer be one type circuit and the capacitor another type circuit.

The system would therefore be in resonance at a frequency corresponding to the 6.6th harmonic (393/60 = 6.55). This is dangerously close to the 7th harmonic voltage and current produced in VFDs.

The two 500-hp drives draw a combined line current of 1100 A (a typical value assuming motor efficiency of 90% and a.9PF). If the current of the 7th harmonic component is assumed to be 1/7 of the fundamental current (typical in drive applications), then [*I*.sub.7] = 1100 / 7 = 157 A. If the source resistance (*R*) for the transformer and the conductors causes a 1.2% voltage drop based on a 3 MVA load flow, then *R* = (0.92)([10.sup.–3]) ohms. This is because the determination of the inductive reactance ([*X*.sub.*L*]) for the wye-connected transformer was 0.00537 ohms. Thus, *R* = (0.00537)(1.2%) / 7% (transformer leakage reactance) = (0.92)([10.sup.–3]) ohms.

The "Q" or "quality factor" of an electrical system is a measure of the energy stored in the capacitors and inductors in the system. The current amplification factor (CAF) in a parallel resonant circuit (such as where a transformer and a capacitor are in a parallel configuration) is approximately equal to *Q*. Actually, *Q* = (2)([Pi]) (maximum energy storage) / (energy dissipation/cycle) as follows:

Q = [(2)([Pi])][(1/2)(*L*)[([*I*.sub.*M*]).sup.2] / [(*I*).sup.2] (*R*/*f*)]

where [*I*.sub.*M*] (maximum current) = ([square root of 2])(*I*), thus,

Q = (2)([Pi])(*f*)(*L*) / *R* = [*X*.sub.*L*] / *R*

where CAF can be considered *Q* or [*X*.sub.*L*] / *R*.

For the example, with the two 500-hp drives, CAF equals (7)([*X*.sub.*L*]) / *R*, where 7 is a multiplication factor representing the 7th harmonic (or 7 times the fundamental 60 Hz); [*X*.sub.*L*] is the reactive impedance at 0.00537; and *R* = (0.92)([10.sup.–3]) ohms. Thus:

CAF = (7)(.00537) / (0.92)([10.sup.–3]) = 40.86

The resonant current ([*I*.sub.*R*]) equals (CAF)([*I*.sub.7]) = (40.86)(157 A)= 6415 A. This current circulates between the source and the capacitor bank. The net current in the capacitor bank ([*I*.sub.*Q*] is equal to 6527 A, which is derived as follows:

([*I*.sub.*Q*]) = [square root of [([*I*.sub.*R*]).sup.2] + [([*I*.sub.*C*]).sup.2]]
= [square root of [(6415).sup.2] + [(1203).sup.2]] = 6527 A

The value of [*I*.sub.*Q*] will seriously overload the capacitors. If the protective device does not operate to protect the capacitor bank, serious damage will occur.

The transformer and the capacitor bank may also form a series resonance circuit and cause large voltage distortions and overvoltage conditions at the 480 V bus. Prior to installation of a power factor improvement capacitor bank, a harmonic analysis must be performed to ensure that resonance frequencies do not coincide with prominent harmonic components contained in the voltages and currents.

2.11.6.3.7.4 Cables The flow of normal 60-Hz current in a cable produces [*I*.sup.2]*R* losses and current distortion introduces additional losses in the conductor. Also, the effective resistance of the cable increases with frequency due to skin effect, where unequal flux linkages across the cross section of the cable causes the AC current to flow on the outer periphery of the conductor. The higher the frequency of the AC current, the greater this tendency. Because of both the fundamental and the harmonic currents that can flow in a conductor, it is important to make sure a cable is rated for the proper current flow.

A set of calculations should be carried out to determine a cable's ampacity level. To do so, the first thing is to evaluate the skin effect. Skin depth relates to the penetration of the current in a conductor and varies inversely as the square root of the frequency, as follows:

Skin depth ([Delta]) = *S* / [square root of *f*]

where "S" is a proportionality constant based on the physical characteristics of the conductor and its magnetic permeability and "f" is the frequency.

If [*R*.sub.dc] is the DC resistance of a conductor, the AC resistance ([*R*.sub.*f*]) at frequency "f" is given by the expression,

[*R*.sub.*f*] = (*K*)([*R*.sub.dc])

The value of *K* is determined from the table. Its value corresponds to the calculated value of the skin effect resistance parameter (*X*), where *X* can be calculated as follows:

X = 0.0636 [square root of *f*[Mu] / [*R*.sub.dc]]

For this calculation, 0.0636 is a constant for copper conductors, "f" is the frequency, [*R*.sub.dc] is the DC resistance per mile of the conductor, and [Mu] is the permeability of the conducting material. The permeability for nonmagnetic materials, such as copper, is approximately equal to 1 and this is the value used. Tables or graphs that contain values of *X* and *K* are normally available from conductor manufacturers. The value of *K* is a multiplying factor that is to be multiplied by the normal cable resistance.

Example: Find the 60-Hz and 300-Hz AC resistances of a 4/0 copper conductor that has a DC resistance ([*R*.sub.dc]) of 0.276 ohm per mile. Using the following equation

X = 0.0636[square root of f[Mu] /[*R*.sub.DC]],

We find that [*X*.sub.60] = (.0636)[[square root of (60)(1) /.276]] = 0.938. And, the value of *K* from the table, when [*X*.sub.60] = 0.938, is approximately 1.004. Thus, the conductor resistance per mile at 60 Hz = (1.004)(0.276) = 0.277 ohm.

For 300 Hz, [*X*.sub.300] = (.0636) [[square root of (300)(1) / .276]] = 2.097. For this condition, the value of *K*, based on [*X*.sub.300] = 2.097 from the table, is approximately 1.092. And, the conductor resistance per mile at 300 Hz = (1.092)(0.276) = 0.301 ohm.

The ratio of resistance, which is also called the skin effect ratio (E), based on the 300 Hz resistance to the 60 Hz resistance =.301 / .277 = 1.09. As can be seen; *E* = [*X*.sub.*n*] / [*X*.sub.60]

A conservative expression for the current rating factor (*q*) for cables that carry harmonic currents is derived by adding the [*I*.sup.2]*R* losses produced by each harmonic frequency current component at the equivalent 60 Hz level, as follows:

q = [[*I*.sub.[1.sup.2]][*E*.sub.1] + [*I*.sub.[2.sup.2]][*E*.sub.2] + [*I*.sub.[3.sup.2]][*E*.sub.3] +... [*I*.sub.[*n*.sup.2][*E*.sub.*N*]

Where [*I*.sub.1], [*I*.sub.2], [*I*.sub.3]... [*I*.sub.*n*] are the ratios of the harmonic currents to the fundamental frequency current and [*E*.sub.1], [*E*.sub.2], [*E*.sub.3],... [*E*.sub.*E*] are skin effect ratios. (ratio of the effective resistance of the cable at the harmonic frequency to the resistance at the fundamental frequency).

Example: Determine the current rating factor (*q*) for a 60-Hz cable required to carry a nonlinear load with the following harmonic characteristics: fundamental current = 190 A, 5th harmonic current = 50 A, 7th harmonic current = 40 A, 11th harmonic current = 15 A, and 13th harmonic current = 10 A.

The skin effect ratios are as follows:

[*E*.sub.1] = 1.0; [*E*.sub.5] = 1.09; [*E*.sub.7] = 1.17; [*E*.sub.11] = 1.35; [*E*.sub.13] = 1.44.

As previously mentioned, the skin effect ratio (*E*), also called the ratio of resistance, equals [*X*.sub.*n*] / [*X*.sub.60]. As an example, the skin effect ratio for E5 is based on the ratio of the 300 Hz resistance to the 60 Hz resistance, which is 0.301 / 0.277 = 1.09.

The harmonic current ratios are as follows:

[*I*.sub.1] = 190/190 = 1.0 [*I*.sub.5] = 50/190 = 0.263 [*I*.sub.7] = 40/190 = 0.210 [*I*.sub.11] = 15/190 = 0.079 [*I*.sub.13] = 10/190 = 0.053 *q* = [(1.0).sup.2](1.0) + [(0.263).sup.2](1.09) + [(0.210).sup.2](1.17) + [(0.079).sup.2](1.35) + [(0.053).sup.2](1.44)

$$q = 1.14$$

Because the cable must be able to handle both the fundamental and the harmonic loads, based upon the *q* factor, the cable must be rated for a minimum current of (1.14)(190) = 217 A at 60 Hz.

2.11.7 Interharmonics

2.11.7.1 Description of the Phenomenon

IEC-61000-2-1 [1] defines interharmonic as follows:

Between the harmonics of the power frequency voltage and current, further frequencies can be observed that are not an integer of the fundamental. They can appear as discrete frequencies or as a wide-band spectrum.

Harmonics and interharmonics of a waveform can be defined in terms of its spectral components in the quasi-steady state over a range of frequencies. The following table provides a simple, yet effective, mathematical definition:

Harmonic $f = h * f_1$ where h is an integer > 0
DC $f = 0$ Hz ($f = h * f_1$ where $h = 0$)
Interharmonic $f \neq h * f_1$ where h is an integer > 0
Sub-harmonic $f > 0$ Hz and $f < f_1$

Where f_1 is the fundamental power system frequency

The term sub-harmonic does not have any official definition but is simply a special case of interharmonic for frequency components less than the power system frequency. The term has appeared in several references and is in general use in the engineering community so it is mentioned here for completeness. Use of the term sub-synchronous frequency component is preferred, as it is more descriptive of the phenomenon.

2.11.7.1.1 Sources

Chief among interharmonic sources is the cycloconverter. Cycloconverters are well-established, reliable units used in a variety of applications from rolling-mill and linear motor drives to static-var generators. Larger mill drives using cycloconverters quoting ranges up to 8 MVA appeared in the 1970s in the cement and mining industries. As recently as 1995, rolling mill drives as large as 56 MVA have been successfully installed. They have also appeared in 25 Hz railroad traction power applications where they are replacing 25 Hz generation and older rotary frequency converters.

The currents injected into the power system by cycloconverters have a unique type of spectrum. Unlike the p-pulse rectifier, which generates characteristic harmonics:

$$f_i = (p \bullet n \pm 1)f \tag{2.34}$$

where f is the power frequency, $n = 1, 2, 3, \ldots$(integers)

cycloconverters have characteristic frequencies of

$$f_i = (p_1 \bullet m \pm 1)f \pm p_2 \bullet n \bullet f_o \tag{2.35}$$

where p_1 = pulse number of the rectifier section, p_2 = pulse number of the output section, $m = 0, 1, 2, 3, \ldots$ (integers), $n = 0, 1, 2, 3, \ldots$ (integers) (m and n not simultaneously equal to 0), and f_o = output frequency of the cycloconverter.

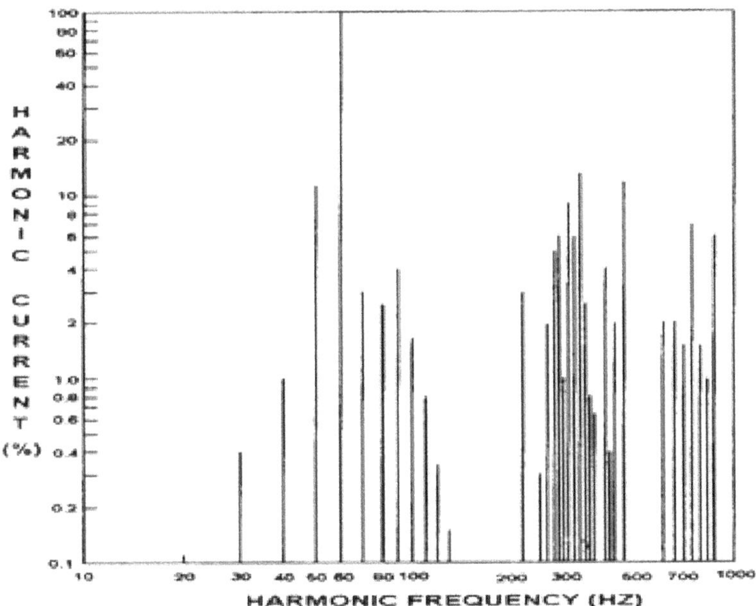

FIGURE 2.86 Typical cycloconverter current spectrum (60 Hz power system).

Figure 2.86 illustrates a typical input current spectrum of a 6-pulse cycloconverter with 5 Hz output frequency. Cycloconverters can be thought of as a special case of a more general class of power electronic device—the static frequency converter. Static frequency converters transform the supply voltage into AC voltage of frequency lower or higher that the supply frequency. They consist of two parts, the AC–DC rectifier and a DC–AC inverter. The DC voltage is modulated by the output frequency of the converter and as a result, interharmonic currents appear in the input current according to equations 1 and 2 causing interharmonic voltages to be generated in the supply voltage.

The magnitude of these frequency components depends on the topology of the power electronics and the degree of coupling and filtering between the rectifier and inverter sections. The cycloconverter is generally the most severe of these devices due to the direct connection between rectifier and inverter common in typical cycloconverter designs, but modern ASDs may also be of concern.

Another common source of interharmonic currents is an arcing load. This includes arc welders and arc furnaces. These types of loads are typically associated with low frequency voltage fluctuations and the resulting light flicker. These voltage fluctuations can be thought of as low frequency interharmonic components. In addition to these components however, arcing loads also exhibit higher frequency interharmonic components across a wide frequency band.

DC arc furnaces do not normally produce significant interharmonics, except when instability occurs due to interactions between the control system and the filters.

Other sources of interharmonics include:

1. Induction motors (wound rotor and sub synchronous converter cascade)
2. Induction furnaces [ref to be determined]
3. Integral cycle control (heating applications)
4. Low frequency power line carrier (ripple control)

It should be noted that most of these sources have interharmonic characteristics that vary in magnitude and frequency over time. This should be considered when characterizing sources.

2.11.7.1.2 Harmonic Current Mitigation

Users of ASDs and other three-phase (rectified) non-linear loads have many choices available when it comes to harmonic mitigation. In the consideration of various alternatives, much depends on the user's objectives as well as the severity of harmonics contributed by internal loads. The typical list of alternative three phase harmonic mitigation equipment includes:

- Line reactors
- Isolation
- Transformers
- *K*-Factor transformers
- Tuned harmonic filters (fixed capacity or automatic switched multiple banks), IGBT-based fast switched harmonic filters,
- Low-pass harmonic filters, 12- and 18-pulse rectifiers, Phase shifting transformers, and active harmonic filters

2.11.7.1.2.1 Line Reactors Line reactors are the simplest and lowest cost means of attenuating harmonics. They are connected in series with an individual non-linear load such as an ASD. By inserting series inductive reactance into the circuit, they attenuate harmonics as well as absorb voltage transients that may otherwise cause a voltage source ASD to trip on overvoltage. The magnitude of harmonic distortion and the actual spectrum of harmonics depend on the effective impedance that the reactor represents in relation to the load. The input harmonic current distortion (both magnitude and phase angle) depends on the rectifier type being used and the effective source impedance. While the nameplates of most main supply transformers and line reactors include an impedance rating, that rating indicates the per-unit impedance relative to its rated full load current. The effective impedance reduces proportionately with the reduction in actual load current. This means that a transformer or line reactor rated at 5% impedance, appears as an impedance of only 2.5% for a load that draws only 50% of the rated current.

In many cases, the main transformer is many times larger in capacity than the individual loads to which it supplies power. This means that relative to the individual loads, the transformer impedance is often very low. As an example, if a 500 kVA, 5% impedance transformer were feeding several 50 kVA loads to each individual load it would actually represent only 0.5% impedance, or one-tenth of its nameplate impedance rating. If the load current is reduced, it will appear as yet a lower percent impedance. Because transformer impedance is usually very small relative to the connected loads, line reactors are applied to each individual non-linear load to increase the effective impedance. Normal and desired levels of impedance range from about 3% to 6% for PQ purposes. This also limits the full load voltage drop to reasonable levels.

In this case, if the load was a 6-pulse rectifier type voltage source drive, the input harmonic current distortion would be as high as 100% total harmonic current distortion (THD-I). If a line reactor, of 5% impedance, based on the actual load current, is added in series to the input of the rectifiers, the harmonic distortion reduces to less than 35% THD-I. If this drive is operated at a lighter load causing a reduction in ASD input current, then the input circuit effective percent impedance appears lower, the input circuit voltage drop is reduced, but the input harmonic current distortion increases.

The chart in Table 2.9 indicates the approximate total harmonic current distortion for a 6-pulse rectified voltage source drive, based on total input circuit effective percent impedance, relative to load current and system voltage.

Line reactors offer the advantage of low cost and they can achieve a significant reduction in harmonics when the appropriate percent impedance is utilized. For reasonable harmonic attenuation, a 5% impedance line reactor should be installed ahead of the motor drive or other 6-pulse non-linear load. Their disadvantages are that they cause a voltage drop, increase system losses and normal impedance values do not achieve current distortion levels much below 35% THD-I. Additionally, the harmonic mitigation capabilities of the reactor reduce as load current is reduced because the reactor's effective percent impedance is reduced. In the range of 100 hp down to 20 hp, they can cost between \$10 and \$30 per horsepower depending on rating, impedance, and enclosure type.

TABLE 2.9

Harmonic Distortion for 6-pulse Rectifiers Based on Total Effective Input Circuit Impedance

	Percent Total Input Impedance						
Harmonic	**0.5%**	**1%**	**1.5%**	**2%**	**3%**	**4%**	**5%**
5th	78%	60%	51%	46%	39%	35%	32%
7th	58%	36%	28%	23%	17.5%	14.5%	12.5%
11th	18%	13%	11%	9%	7.5%	6.5%	6%
13th	10%	8%	6.5%	6%	5%	4.3%	4%
17th	7.5%	5%	4%	3.6%	3%	2.5%	2.3%
19th	6%	4%	3.3%	3%	2.3%	2%	1.8%
23rd	5%	3%	2.6%	2%	1.5%	1.3%	1.1%
25th	2.3%	2%	1.6%	1.3%	1.1%	1%	0.9%
%THD-I	100%	72%	60%	55%	44%	39%	35%

2.11.7.1.2.2 Isolation Transformers Since input circuit reactance is a major determining factor for the magnitude of harmonics that will be present and flowing to an individual load, isolation transformers can be used effectively to reduce harmonic distortion. The leakage inductance of isolation transformers can offer appropriate values of circuit impedance so that harmonics are attenuated. The typical configuration of isolation transformer, for PQ purposes, is delta primary and wye secondary.

Like a reactor, the inductive reactance is low enough at the fundamental frequency to easily pass fundamental current, but increases proportionately for harmonic frequencies and can achieve performance similar to that of a line reactor. Additionally, the isolation transformer can be supplied with an electrostatic shield between the primary and secondary windings. Due to the capacitive coupling between each winding and the shield, a low impedance path is created to attenuate noise, transients, and zero sequence currents. The shield helps to mitigate the common mode disturbances to their originating side (primary or secondary) of the transformer. The wye secondary transformer has the capability of providing a new electrical ground for the load circuit.

Similar to line reactors, the effective percent impedance of the transformer is typically lower than stated on the nameplate because the connected loads are smaller in rating than full load transformer capacity.

The transformer leakage reactance (0.0092 ohms) is precisely 6% of the rated minimum load impedance (0.1536), which correspondingly draws maximum rated load current. Any load that draws less current, does so because it has higher impedance, so the effective percent impedance of the transformer relative to the smaller load, will be proportionately lower. Although the transformer reactance (ohms) does not change, its effective percent impedance depends on the actual connected load.

If a transformer (or reactor) is rated at 5% impedance, but the load is only drawing 60% of the transformer's or reactor's rated current, then the effective impedance will be only 3% impedance (5% × 0.60 = 3%). Now instead of the expected 35% THD-I, associated with 5% impedance, the load will actually draw current with distortion of 44%. For the most effective attenuation of harmonics, transformers or line reactors should be sized as close as possible to the rated load current and have their inductance (impedance) based on rated load current and voltage.

Isolation transformers can achieve the same harmonic attenuation as for line reactors, provided they are sized properly. De-rating a transformer or line reactor, results in lower effective percent impedance and thus higher harmonic current distortion. The advantage of an isolation transformer is that it can reduce both common mode (when an electrostatic shield is used) and normal mode disturbances as well as provide circuit isolation. Disadvantages are physical size, circuit losses, and cost. The typical cost of an isolation transformer for loads of 100 hp down to 20 hp can be $50 to $150 per horsepower.

2.11.7.1.2.3 K-Factor Transformers Transformer losses, which cause transformer heating, are composed of both I^2R losses (based on the rms current) and eddy current losses, which are proportional to both the current and frequency squared. The *K*-factor transformer is designed to accommodate the temperature

TABLE 2.10

Typical *K*-Factor vs. Effective Impedance (6-pulse load)

%impedance	0.5	1	1.5	2	3	4	5
K-Factor	21	14	11	9	6.8	5.6	4.8

rise caused by current harmonics in the transformer windings, in addition to the fundamental frequency losses. *K*-factor is a constant that specifies the ability of the transformer to handle harmonic heating, as a multiple of the normal eddy current losses developed by a sinusoidal current in the transformer windings.

If *K*-factor is 9, the transformer can handle the heat associated with eddy current losses which are 9 times greater than a non-*K*-factor transformer. The *K*-factor transformer has a special design that is not usually found in ordinary dry–type transformers. The neutral bus, in the secondary side, has double the current carrying capacity of the phase conductors. The windings are typically made with multiple insulated conductors that are transposed to reduce the skin effect that current harmonics cause in the coils. Additionally, the magnetic core is designed with a lower flux density than non-*K*-factor transformers.

The *K*-factor can be computed using the information contained in a waveform harmonic spectrum. The *K*-factor is a function of two variables: magnitude of harmonic current and harmonic order. A low THD is not a guarantee of low *K*-factor, especially when the THD contains mostly high order harmonics, such as 12- and 18-pulse rectifiers. Whenever possible, a measurement should be performed if high order harmonics are suspected in the current waveform. For typical values of input circuit effective percent impedance, the *K*-factors associated with the corresponding waveforms and harmonic spectra are as shown in the following chart (Table 2.10).

A transformer with a *K*-factor rating of 9 could be used to supply an individual non- linear load when the effective impedance is between 2% and 5%. When the load contains a mix of linear and non-linear loads, the *K*-factor requirement will often be lower than when the transformer only supplies non-linear loads.

When the transformer is supplying single phase non-linear loads which are connected line to neutral, triplen (multiples of three) harmonics will be experienced. If the load current has a high level of third harmonic, the delta connection can mitigate this harmonic in the circuit on the primary side, however triplen harmonics will be present between phase and neutral conductors on the secondary side of the transformer. Because triplen harmonic currents, relative to each phase conductor, are in phase with each other, they will sum algebraically in the neutral conductor.

Let's suppose that the phase currents in the secondary side of the transformer are distorted, and current harmonics in all phases are: third, fifth, seventh, and ninth. The third harmonic currents in phases A, B, and C have the same phase angle. This is what causes them to sum together in the neutral conductor, but is also the key for the mitigation in the delta winding.

The third harmonics are subtracted arithmetically in the delta corners and total current in phases A, B, and C, if the circuit of the primary side does not contain third harmonic.

There is a limit to the mitigation of third harmonic obtained with a delta–wye transformer: the third harmonics in all the phases on the secondary side must be equal in magnitude and phase angle. If this condition is not satisfied, there will be third harmonics in the circuit of the primary side of the transformer. Whereas standard drive isolation transformers may cost from $50 to $150 per horsepower in the range of 100 hp down to 10 hp, the *K*-factor transformer may cost as much as $100 to $250 per hp.

2.11.7.1.2.4 Tuned Harmonic Filters A tuned harmonic filter is a device with two basic elements: inductive and capacitive. These reactive elements are connected in series to form a tuned LC circuit. The tuned harmonic filter is connected as a shunt device to the power system. In many cases, tuned harmonic filters are applied on a facility-wide basis at the service entrance, or at distribution transformers.

The tuned harmonic filter is a resonant circuit at the tuning frequency so its impedance is very low for the tuned harmonic. Due to its low impedance at the tuned harmonic frequency, the tuned filter now becomes the source of the tuned frequency harmonic energy demanded by the loads, rather than the utility. The filter impedance below the tuning frequency resembles a capacitive behavior, while the impedance above the tuning frequency has an inductive behavior and at the tuning harmonic the filter behavior is like a resistor.

As a result of the capacitive behavior at low frequencies (below the tuning frequency), the filter improves the displacement power factor. At the tuning frequency the filter acts like a very low resistance, and a great amount of harmonic current at this frequency flows through the filter and the total harmonic current distortion in the upstream system decreases. Harmonic currents flow between the filter and connected loads.

The decrease in the total harmonic current distortion improves the distortion power factor and the final result is an improved total power factor, because both displacement power factor and distortion power factor are increased. A tuned filter can be designed in different ways: fixed, automatic, and hybrid (a fixed and one or several automatic parts). The final design is a function of the system reactive power requirements and an electrical survey must be performed to choose the best solution.

A fixed tuned filter should be used when the power factor is low and constant and the harmonic that must be mitigated has a constant magnitude. If the power factor is low and fluctuates over time, or if the harmonic that must be mitigated also changes over time, then a hybrid solution must be used. The hybrid solution would consist of both fixed and automatic filters. The fixed portion of the filter will compensate for the continuous reactive power required by the load, and the automatic portion will compensate for the fluctuating changes in reactive power. When both the power factor and harmonics fluctuate then an automatic filter must be used. The capacity of a tuned filter is calculated following the analysis of the results of a PQ survey. The total current that will flow through a tuned filter consists of both fundamental and harmonic components. The filter must be of a capacity large enough to allow the two components to flow without overheating.

A tuned filter is normally used when the total harmonic current distortion is greater than 20% THD-I. In cases where the initial total harmonic current distortion is lower than this amount, the capacitance of the tuned filter may interact with power system reactance and cause a resonance problem. In situations like this, installation of harmonic filters may be considered on the individual loads, which are generating the harmonics.

When a tuned harmonic filter is applied on a system consisting of a mixture of linear and non-linear loads (service entrance or distribution transformer), it is possible to achieve distortion levels in the range from 3% to 12% THD-I. Although the filter is a very low impedance path for the tuning harmonic, there will always be some harmonic current flowing through the electrical system because the main transformer is a parallel path (from the load point of view) and the current will divide according to Ohm's law between the filter and the main transformer. Like virtually any harmonic mitigation technology, the harmonic filter does not eliminate the tuned harmonic, it only mitigates it.

2.11.7.1.2.5 IGBT-Based Fast-Switched Harmonic Filters An automatic filter is very useful when the reactive power changes occasionally over time, without extremely rapid swings in loading conditions. In those situations, involving dynamic loads rapidly changing demands for reactive power, a typical automatic filter will not respond quickly enough to meet the reactive power requirements of the load and to maintain acceptable power factor or harmonic distortion levels.

The solution to supplying reactive power and harmonic mitigation for dynamic loads is the fast-switched harmonic filter. These filters are switched very rapidly IN and OUT of the circuit using isolated gate bipolar transistors (IGBT) instead of contactors. This type of filter is capable of soft-switching the capacitors, so as not to create a voltage spike. It can be switched, without discharging the capacitors, at switching rates up to 60 times per second. The main advantages of this filter are the capability to switch without transients and to respond in real time to dynamically changing load conditions. The performance of the fast-switched filter is similar to the performance that can be expected from a typical tuned filter, a total harmonic current distortion from 3% to 12% (for multiple mixed loads).

2.11.7.1.2.6 Low-Pass Harmonic Filters Low-pass harmonic filters have gained popularity due to their ability to attenuate all harmonic frequencies, achieve low levels of residual harmonic distortion and offer guarantee-able results. Although historically several circuit configurations have been used.

The series elements increase the input circuit effective impedance to reduce overall harmonics, as well as to de-tune the shunt circuit relative to both the supply end and load side of the filter. From either direction, the shunt circuit is detuned away from a harmonic frequency prevent the attraction of harmonics from other sources supplied by the same feeder or transformer as well as minimizing the possibility of resonance. The shunt elements are tuned in such a manner as to remove most of the remaining circuit

TABLE 2.11

THD Value for Different Loadings

Load/H	1	5	7	11	13	17	19	23	25	THD-I
33% Load	100	5.8	4.8	1.4	0.7	0.8	0.3	0.4	0.4	7.76%
50% Load	100	5.5	1.7	2.9	1.2	0.9	0.8	0.7	0.5	6.72%
75% Load	100	4.1	1.6	2.3	1.3	0.5	0.6	0.4	0.4	5.22%
100% Load	100	3.8	1.5	2.0	1.1	0.4	0.5	0.4	0.4	4.75%

harmonics (primarily 5th and 7th). The low-pass filter forms a hybrid combination of series and shunt elements that can be applied without performing system harmonic analysis.

A low-pass harmonic filter connects in series with and ahead of the 6-pulse rectifier load(s). While there is very little distortion at the input stage of this filter, the output stage, where the load is connected, may have significant amounts of current and voltage distortion. Due to the voltage distortion at the output stage of this filter, it is recommended that only non-linear loads be connected. Operating linear loads, such as motors, from a distorted voltage source, can cause increased heating and lower life expectancy for the linear loads.

Due to the series reactance, the typical low-pass harmonic filter will experience voltage drop under loaded conditions. Voltage boosting will occur under no-load conditions, due to the presence of the shunt capacitor and reactor. Typical regulation is about 5% but experience has shown as much as 10% voltage boosting may occur with some types of low-pass filters. Some low-pass harmonic filters may not be suitable for use with silicon-controlled rectifiers (SCRs). Some low-pass filters may require that external reactors be installed either to the line or load side of the filter in order to achieve the expected results.

The waveform and harmonic measurements for a low-pass filter, connected ahead of a 50-amp load, are shown in Table 2.12. Notice the residual harmonics are lowest when the current is near full load. Also included is the waveform collected at 75% load.

Table 2.11—Measured input current waveform and spectrum for low-pass filter (50 amps) feeding a 6-pulse rectifier-type variable frequency motor drive, at various load conditions.

Based on customer experiences with low-pass filters from various manufacturers, distortion levels of 5% to 15% can be expected depending on the application. Pre-existing voltage distortion and unbalanced line voltages will increase the residual harmonic distortion levels.

Normal applications for the low-pass filter are generally fans and pumps (variable torque) applications. Voltage drop can become excessive above filter rated current, requiring derating for constant torque applications. Since low-pass filters achieve lowest harmonic distortion levels at or near full load, when filters are derated, the residual harmonic current distortion may increase.

The advantages of the low-pass filters include relatively low cost, low residual harmonics, predictable and guarantee-able results, no analysis or harmonic studies required. The disadvantages are that they must connect in series with the load, can only be used with non-linear loads, experience low (leading) power factor at light loads, and induce additional system losses. Low-pass filters may cost in the range of $40 per hp to $125 per hp for ratings of 100 hp down to 20 hp.

2.11.7.1.2.7 12- and 18-Pulse Rectifiers This technique involves a special type of rectifier and transformer configuration as illustrated in Figure 2.87a and b. Rather than being an add-on solution like the previous ones, this version must be ordered upfront from the manufacturer of the power electronics equipment, because it involves internal changes to the input rectifier section.

We noticed earlier that a standard 6-pulse rectifier caused a predictable harmonic spectrum consisting of the 5th, 7th, 11th, 13th, 17th, 19th ... harmonics for three-phase power systems and rectifiers.

In each of these cases, the 5th and 7th harmonics, which normally have the largest magnitudes of all of the individual harmonics, are (theoretically) eliminated. In real practice, it is common to see small amounts (2%–3%) of these harmonics present. Having nearly eliminated the normally strong 5th and 7th harmonics, these rectifier schemes can achieve low levels of harmonic distortion.

Harmonic mitigation occurs by phase shifting one bridge rectifier against the other(s), causing specific harmonics from one bridge rectifier to cancel those from the other. Phase shifting is accomplished through

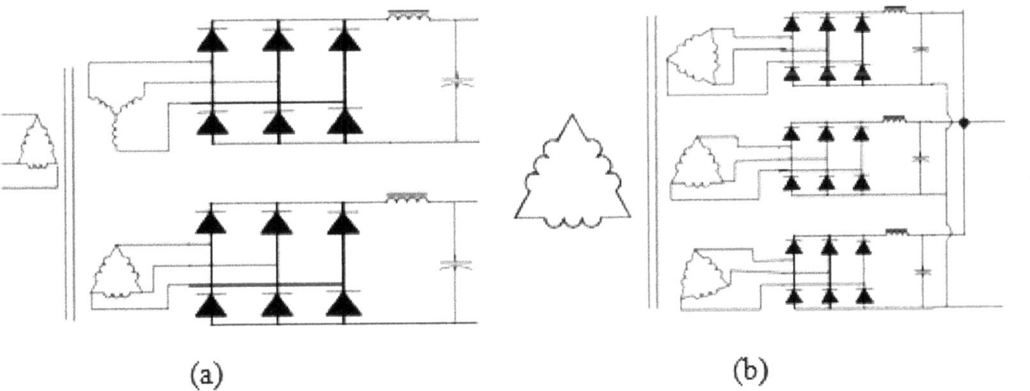

FIGURE 2.87 (a) 12-pulse rectifier, (b) 18-pulse rectifier.

the multiple secondary windings of the transformer. In the case of the 12-pulse system, the transformer has two separate secondary windings—one in wye and one in delta configurations. The 18-pulse system utilizes a transformer with three phase shifted windings. The degrees of phase shift between each secondary winding is 360 divided by the number of rectifier pulses (i.e., 12-pulse = 30° and 18-pulse = 20°).

It is critical to the operation of both 12- and 18-pulse rectifiers that the currents drawn by each rectifier bridge be balanced and that source voltages for all phases are balanced. The presence of unbalanced line voltages can cause triplen harmonics to flow and will increase the residual harmonic current distortion. Likewise, unbalanced rectifier bridge currents will increase the harmonic distortion. For this reason, it is advisable to use inter-phase reactors to cause rectifier bridge currents to be similar. When properly applied under conditions that achieve balanced bridge currents, 12-pulse rectifier systems can achieve input current distortion levels between 10% and 20% THD-I, while 18-pulse systems may achieve between 5% to 10% THD-I. These levels can be expected at full load conditions, but the THD-I will increase as the load is reduced.

While other harmonic mitigation techniques can be installed as aftermarket equipment when necessary, the 12- and 18-pulse rectifiers require special ordering of the drives or other power electronics equipment. The drives must be built with the desired 6-, 12- or 18-pulse rectifier front ends. The 12- and 18-pulse rectifiers may increase the cost of the drive by 50% to 100%, when the cost of the transformer is considered.

2.11.7.1.2.8 Phase Shifting Transformers Quasi 12-pulse methods have also been used to reduce facility harmonic distortion. In these cases, two sets of non-linear loads are fed by two phase shifted transformer windings. It may be a single transformer with two separate windings (i.e., delta and wye), or two transformers, one configured as delta primary/wye secondary and the other configured as delta primary/delta secondary. Similar to a 12-pulse rectifier system, cancellation of the 5th and 7th harmonic can be achieved on the primary side of the transformers to the degree that these currents are balanced in each of the transformer secondary windings.

For only a small premium (5% to 15%) over the price of a drive isolation transformer, facilities can specify phase shifted transformers and improve the input harmonic current distortion. Whenever the load conditions are equal, near cancellation of the 5th and 7th harmonics can be accomplished. When load currents are not matched, the harmonics can be partially cancelled. Based on the effective percent impedance of the transformer, higher order harmonics can also be attenuated.

2.11.7.2.1.8.1 Active Filters The newest technology available for mitigation of harmonics is the active filter. Active filtering techniques can be applied either as a standalone harmonic filter or by incorporating the technology into the rectifier stage of a drive, UPS, or other power electronics equipment.

Typically, active filters will monitor the load currents, filter out the fundamental frequency currents, analyze the frequency and magnitude content of the remainder, and then inject the appropriate inverse currents to cancel the individual harmonics. Active filters will normally cancel harmonics up to about the 50th harmonic and can achieve harmonic distortion levels as low as 5% THD-I or less. To apply active harmonic filters, determine the magnitude of harmonics (by measurement) that you wish to remove from the system, and select an active filter with suitable harmonic current cancellation capacity.

Because active filters utilize fast switching transistors (IGBT), which are connected directly to the facility power circuit phase conductors, switching frequency noise may be present and require additional filtering to prevent interference with other sensitive equipment.

Some active filters may experience lower performance when the power system has high levels of pre-existing voltage distortion. If system harmonics are relatively high, it will be best to use an active filter with immunity to voltage distortion (one that does not require a sinusoidal voltage reference).

Active filters utilize power electronics circuitry and therefore maintenance requirements can be higher than for passive solutions and may be similar to that for a VFD. The losses associated with active filters also tend to be higher than for passive solutions. In terms of harmonic cancellation current, the prices of active filters can range from about $30,000 (50 amps) to $100,000 (300 amps).

2.11.7.1.2.9 Comparison of Harmonic Mitigation Alternatives Table 2.12 shows the approximate costs and typical performance of various *aftermarket* solutions for three-phase harmonic mitigation.

TABLE 2.12

Comparison of Aftermarket Mitigation Alternatives

Harmonic Mitigation Technique	20 hp Price	100 hp Price	400 hp Price	THD-I Non-linear Loads	THD-I Mixed (50–50) Loads
Reactor (5%)	$520	$1100	$3800	35%	17.5%
Isolation Transformer	$2650	$6340	$18,000	35%	17.5%
K-Factor (13) Transformer	$5300	$11000	$48,000	35%	17.5%
Tuned Filter	$2800	$3900	$7000	15%–20%	3%–12%
Low-pass Filter	$2400	$5600	$13,000	8%–15%	n/a
Active Filter	n/a	$27,000	$65,000	5%	5%

Note: 50–50 mixed loads refers to 50% linear and 50% non-linear loads

3

A Voltage-Controlled DSTATCOM for Power Quality Improvement

3.1 Introduction

It has long been recognized that the steady-state transmittable power can be increased and the voltage profile along the line can be improved by appropriate reactive shunt compensation. The purpose of this reactive compensation is to change the natural electrical characteristics of the transmission line to make it more compatible with the prevailing load demand. To overcome the problem related to the power quality custom power device is introduced. A number of power quality problem solutions are provided by custom devices. At present, a wide range of flexible AC controllers that are capitalized on newly available power electronic components are emerging for custom power application. Among these distributions, static compensators are mainly used in the present work. The fast response of the distribution static compensator (DSTATCOM) makes it an efficient solution for improving the power quality in distribution systems.

3.2 DSTATCOM

The DSTATCOM is a voltage source inverter-based static compensator (similar in many respects to the DVR) that is used for the correction of bus voltage sags. Connection (shunt) to the distribution network is via a standard power distribution transformer. The DSTATCOM is capable of generating continuously variable inductive or capacitive shunt compensation at a level up to its maximum MVA rating. The DSTATCOM continuously checks the line waveform with respect to a reference AC signal, and therefore, it can provide the correct amount of leading or lagging reactive current compensation to reduce the amount of voltage fluctuations. The major components of a DSTATCOM are shown in Figure 3.1. It consists of a DC capacitor, one or more inverter modules, an AC filter, a transformer to match the inverter output to the line voltage, and a PWM control strategy. In the DSTATCOM implementation, a voltage-source inverter converts a DC voltage into a three-phase AC voltage that is synchronized with, and connected to, the AC line through a small tie reactor and capacitor (AC filter).

A circuit diagram of a DSTATCOM-compensated distribution system is shown in Figure 3.2. It uses a three-phase, four-wire, two-level, neutral-point-clamped VSI. DSTATCOM is basically a VSC-based FACTS controller. It is employed at distribution level or at load side and also behaves as a shunt active filter. It works as the IEEE-519 standard limit. For the electrical power distribution system, it is very important to balance the supply and demand of active and reactive power. If the balance is lost the frequency and voltage excursion may result in collapse of the power system. Hence balancing is considered as the key of stable power system. The distribution system's power loss and quality problems are increasing due to reactive power. The main application of DSTATCOM is high

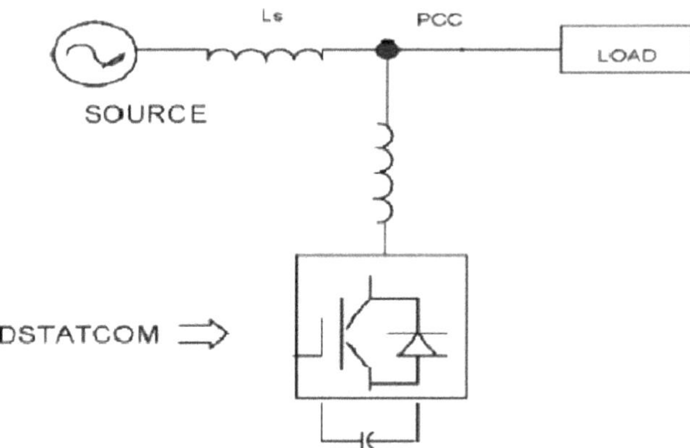

FIGURE 3.1 Block diagram of DSTATCOM.

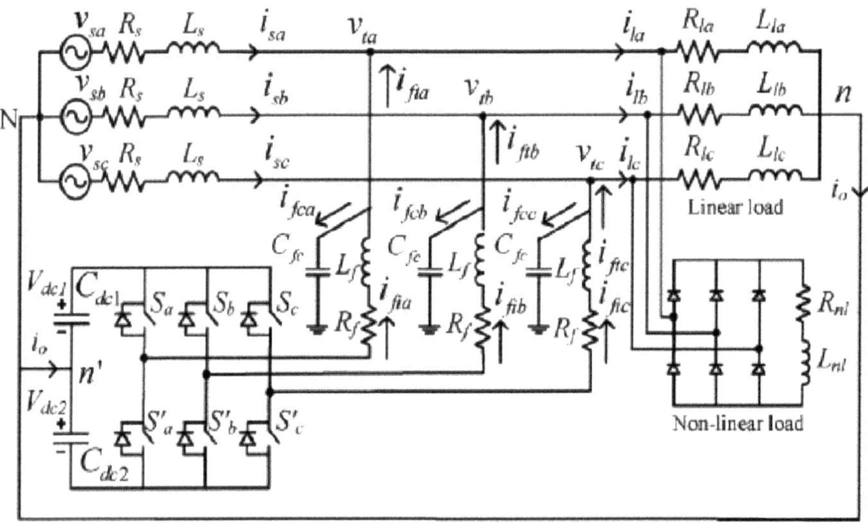

FIGURE 3.2 DSTATCOM-compensated distribution system.

speed control of reactive power to provide voltage stabilization in a power system. This can protect the distribution system from voltage sags and flicker caused by reactive current demand. Adjustment of the phase and magnitude of the DSTATCOM output voltages allows effective control of active and reactive power exchanges between the DSTATCOM and the AC system. Such configuration allows the device to absorb or generate controllable active and reactive power. The VSC connected in shunt with the AC system provides a multifunctional topology that can be used for up to three quite distinct purposes:

- Voltage regulation and compensation of reactive power
- Correction of power factor
- Elimination of current harmonics

3.3 Design of DSTATCOM

Proper selection of VSI components is important for the satisfactory operation of STATCOM. The design procedure is as follows:

1. **Voltage across DC Bus (V_{dc}):** The DC bus voltage is taken as twice the peak of the phase voltage of the source for satisfactory performance. Therefore, for a line voltage of 400 V, the DC bus voltage is maintained at 650 V.
2. **DC Capacitance (C_{dc}):** Values of DC capacitors are chosen based on a period of sag/swell and change in DC bus voltage during transients. The value is given by the equation,

$$\frac{1}{2}C_{dc}\left(V_{dcref}^2 - V_{dc}^2\right) = pST \tag{3.1}$$

   where:

   S is the total load rating and p is the number of cycles
   T is the time period

   Here, S = 10 kVA, 650 V, V_{dcref} = 1, and capacitor values are found to be 2630 microF. The capacitor value 2600 microF is chosen to achieve satisfactory performance during all operating conditions.
3. **Filter Inductance (L_f):** Filter inductance should provide reasonably high switching frequency and a sufficient rate of change of current such that VSI currents follow desired currents. The following equation represents inductor dynamics:

$$L_f \frac{di_{fi}}{dt} = -\upsilon_{fc} - R_f i_{fi} + V_{dc} \tag{3.2}$$

   The inductance is designed to provide good tracking performance at a maximum switching frequency, which is achieved at the zero of the source voltage in the hysteresis controller. Neglecting R_f, L_f is given by

$$L_f = \frac{2V_m}{(2h_c)(2f_{max})} = \frac{0.5V_m}{h_c f_{max}} \tag{3.3}$$

   where $2h_c$ is the ripple in the current. With switching frequency 10 kHz and h_c = 0.75 A (5% of rated current), the value of L_f is found to be 21.8 mH, and 22 mH is used in realizing the filter.
4. **Shunt Capacitor (C_{fc}):** The shunt capacitor should not resonate with feeder inductance at the fundamental frequency. Capacitance, at which resonance will occur, is given as

$$C_{fcr} = \frac{1}{\omega_0^2 L_s}. \tag{3.4}$$

For proper operation, C_{fc} must be very small compared to C_{fcr}. Here, a value of 5 microF is chosen, which provides an impedance of 637 ohm. This does not allow the capacitor to draw significant fundamental reactive current.

3.4 Control Circuit Design and Reference Terminal Voltage Generation

Load angle is calculated by comparing the average DC bus voltage ($[V_{dc1} + V_{dc2}]/2$) with the actual terminal voltage. Then the error is passed through a PI controller. Reference load voltage magnitude is given by,

$$V_t^* = \sqrt{V^2 - \left(\left|\bar{I}_{la1}^+\right| X_s\right)^2} - \left|\bar{I}_{la1}^+\right| R_s. \tag{3.5}$$

Using this reference load, terminal voltage and load angle reference terminal voltage are given by,

$$\begin{aligned} \upsilon_{ta}^*(t) &= \sqrt{2}V_t^* \sin(\omega t - \delta) \\ \upsilon_{tb}^*(t) &= \sqrt{2}V_t^* \sin\left(\omega t - \frac{2\pi}{3} - \delta\right) \\ \upsilon_{tc}^*(t) &= \sqrt{2}V_t^* \sin\left(\omega t + \frac{2\pi}{3} - \delta\right). \end{aligned} \tag{3.6}$$

3.5 Simulation

The control scheme is implemented using MATLAB® software. Simulation parameters are given in Table 3.1. Here simulation of existing and proposed STATCOM models is illustrated.

The Simulink® model shown in Figure 3.3 depicts a three-phase system supplying a linear load and a nonlinear load. The source current and voltage waveforms are as shown in Figure 3.4. The total harmonic distortion of the system is found to be higher, with poor power transfer qualities.

The Simulink model shown in Figure 3.5 gives a three-phase system supplying a linear load and a nonlinear load with compensator reference voltage arbitrarily taken as 1.0 p.u. The source current

TABLE 3.1

Simulation Parameters

System Quantities	Values
Source voltage	400 V rms line to line, 50 Hz
Feeder impedance	$Z_s = 1 + j3.14\Omega$
Linear load	$Z_{la} = 30 + j62.8\Omega$ $Z_{lb} = 40 + j78.5\Omega$ $Z_{lc} = 50 + j50.24\Omega$
Non-linear load	An R-L load of $50 + j62.8\Omega$
VSI parameters	$V_{dc} = 650$ V, $C_{dc} = 2600$ µF, $R_f = 1\Omega$, $L_f = 22$ mH, $C_{fc} = 5$ µF, $I_{rated} = 30$A
PI gains	$K_{p\delta} = 8.5e^{-7}$, $K_{i\delta} = 1.8e^{-6}$
Hysteresis band (h)	1 V

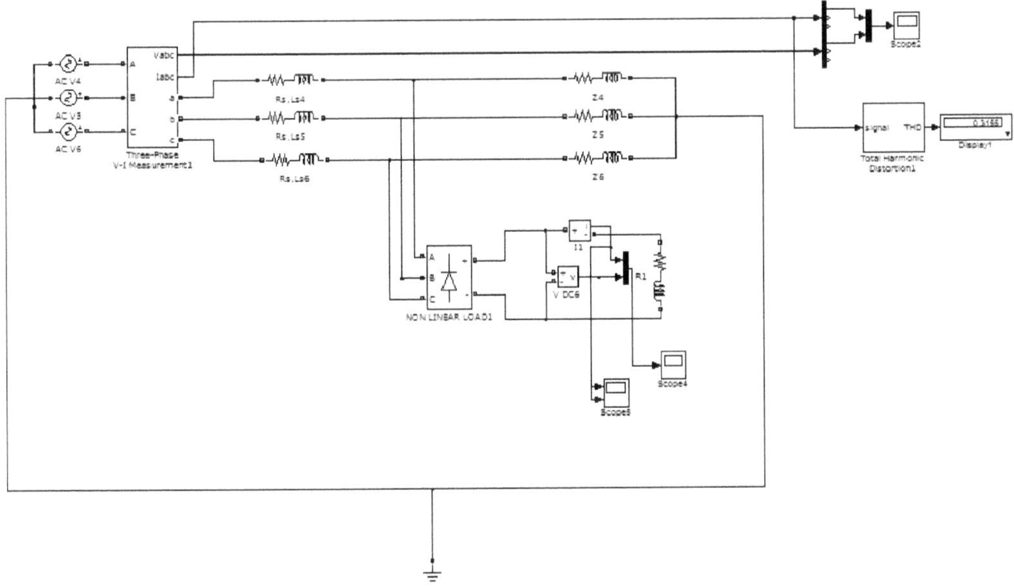

FIGURE 3.3 Simulink model of uncompensated network.

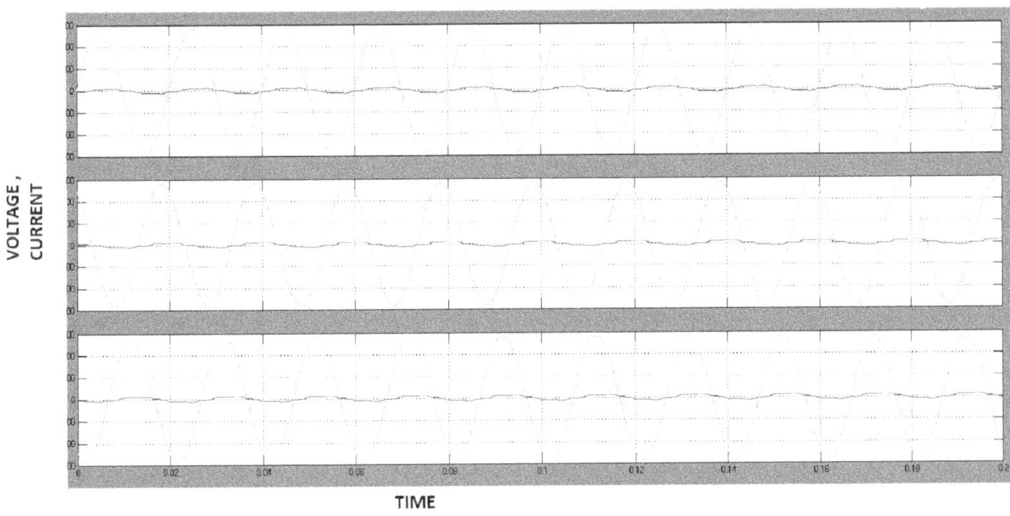

FIGURE 3.4 Source voltage and current waveform of uncompensated network.

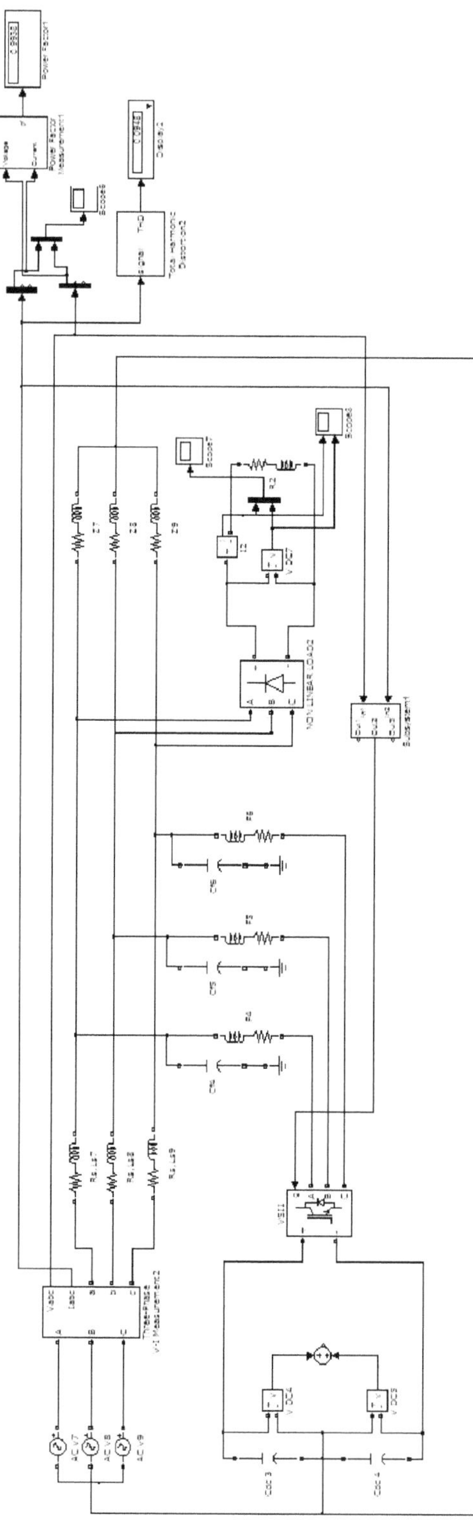

FIGURE 3.5 Simulink model of compensated network (with reference voltage arbitrarily taken as 1.0 p.u.).

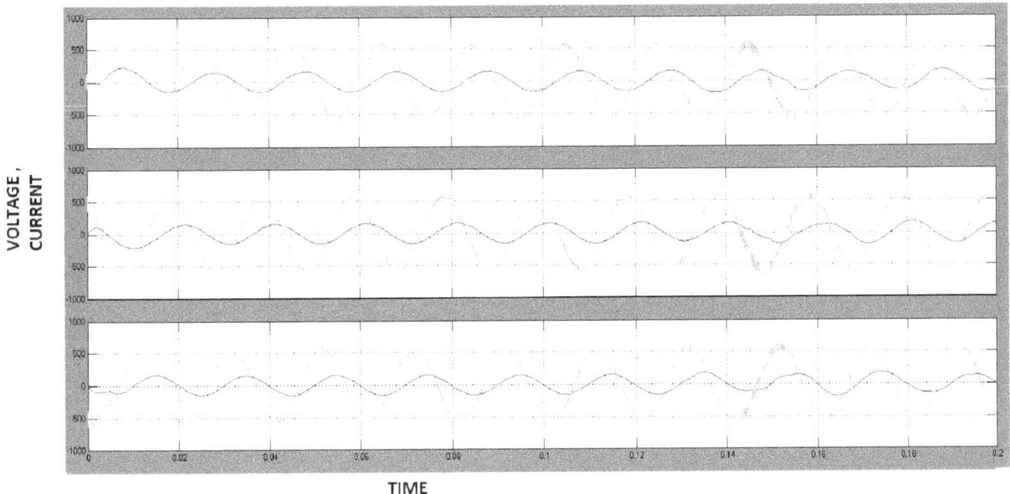

FIGURE 3.6 Source voltage and current waveform of compensated network (with reference voltage arbitrarily taken as 1.0 p.u.).

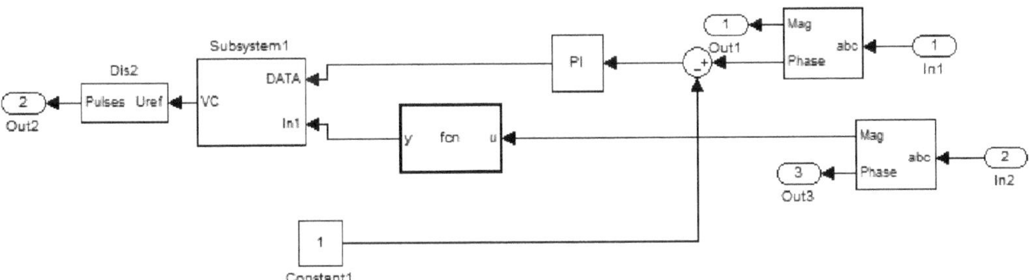

FIGURE 3.7 Control system of compensated network (with reference voltage arbitrarily taken as 1.0 p.u.).

and voltage waveforms are as shown in Figure 3.6. The total harmonic distortion of the system is found to be comparatively lower than the uncompensated network. The control system is as shown in Figure 3.7.

The Simulink model shown in Figure 3.8 shows a three-phase system supplying a linear load and a nonlinear load with compensator and the reference voltage is arbitrarily generated by instantaneous symmetrical component theory. The source current and voltage waveforms are as shown in Figure 3.9. The total harmonic distortion of the system is found to be comparatively lower than the uncompensated network. The control system is as shown in Figure 3.10.

In this chapter, a control algorithm has been proposed for the generation of reference load voltage for a voltage-controlled DSTATCOM. The performance of the proposed scheme is compared with the traditional voltage-controlled DSTATCOM. The proposed method provides the following advantages: (1) at nominal load, the compensator injects reactive and harmonic components of load currents, resulting in

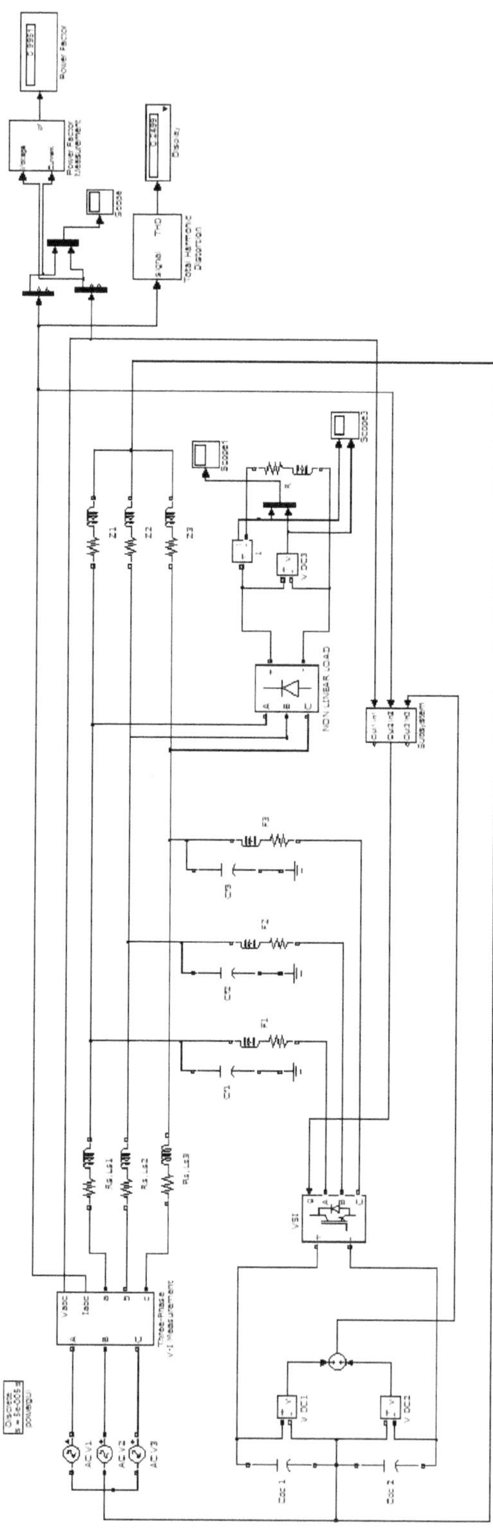

FIGURE 3.8 Simulink model of compensated network (with changing reference voltage).

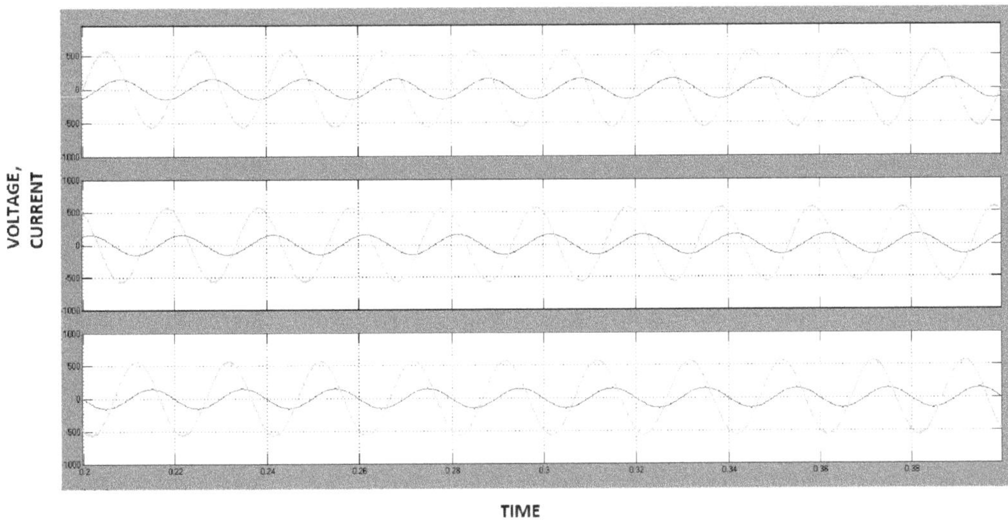

FIGURE 3.9 Source voltage and current waveform of compensated network (with changing reference voltage).

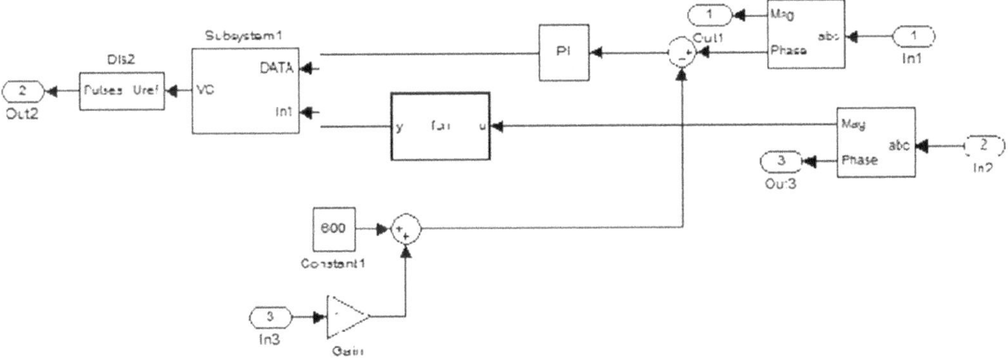

FIGURE 3.10 Control system of compensated network (with changing reference voltage).

UPF; (2) nearly UPF is maintained for a load change; (3) fast voltage regulation has been achieved during voltage disturbances; and (4) losses in the VSI and feeder are reduced considerably and have higher sag supporting capability with the same VSI rating compared to the traditional scheme. The simulation and experimental results show that the proposed scheme provides DSTATCOM a capability to improve several PQ problems (related to voltage and current).

4

Power Quality Issues and Solutions in Renewable Energy Systems

4.1 Introduction

Power quality has been a problem bristling with snags ever since electrical power was invented. It has become a well projected area of interest in recent years because of the electrical appliances (load) it affects. The electric current that the customer's appliances draws from the supply network flows through the impedances of the supply system and causes a voltage drop, which affects the voltage delivered to the customer. Hence both the voltage quality and the current quality are important. The power distribution supplier is responsible for the voltage quality and the customer is accountable for the quality of electric current that they draw from the utility. The categorization of power system electromagnetic phenomena that affects the power quality are transients, short duration variations, long duration variations and waveform distortions. The waveform distortion is defined as the steady state deviation from an ideal sine wave of power frequency principally characterized by the spectral content of the deviation. One of the types of waveform distortion is harmonics. The increased awareness of harmonics in recent years is the result of concerns that harmonic distortion levels are increasing on many electrical power systems.

Power quantity is another major global issue that resulted in increased demand for electrical power. This issue has stimulated the alternate energy sources drastically. It has been estimated that the energy produced by renewable sources is expected to satisfy 50% of the total needs in 2050. Due to fossil fuel exhaustion and environmental problems caused by conventional power generation, renewable energy sources, particularly solar and wind energy, have become very popular and demanding. But the quality of power from these sources has to be improved to protect the loads connected in the system and also to enable the continuity of supply to the consumers without any disturbances. Hence the combination of power quality and power quantity will certainly result in providing clean power from green energy sources.

To meet this objective, a power electronic interface with harmonic reduction capability needs to be connected between the source and load. Unlike the conventional inverters, multilevel inverters (MLI) are recommended as the synthesized multilevel outputs are superior in quality, which results in reduced filter requirements and overall system size. The switching sequence for the MLI is controlled by intelligent techniques that enable the power quality. The input source considered is solar photovoltaic (PV), which intends the power quantity. It is convenient to integrate both functionalities of power generation and power quality improvement using the same hardware structure presented in the distributed generation system.

4.2 Power Quality in Electrical Systems

Any power problem manifested in voltage, current, or frequency deviations that result in failure or misoperation of customer equipment is termed as power quality. The issue of electric power quality is gaining importance because of several reasons. Some of them are: (1) Modern society is becoming increasingly dependent on the electrical supply. A small power outage has a great economic impact on the industrial consumers; (2) the advent of new power electronic equipment; (3) the deregulated environment, which reduces the maintenance and investments into the power system and hence reduces the margins in the system; and (4) emerging distributed power generation. The key problems associated with the power quality are: damage to sensitive equipment, interference, malfunction, extra losses, personnel safety issues, poor utilization, and poor power factor.

The quality of electrical power may be described as a set of values of parameters such as: continuity of service, variation in voltage magnitude, transient voltages and currents, and harmonic content in the waveforms for AC power. Among these, harmonics play a vital role in all the segments of the electrical system. Harmonics are sinusoidal voltages or currents having frequencies that are integer multiples of the frequency at which the supply system is designed to operate at the fundamental frequency. Harmonic distortion originates in the nonlinear characteristics of devices and loads on the power system. The problems caused by harmonics are: transformer feeder overheating, circuit breaker inadvertent trips, fuse blowing especially on distribution feeder laterals, equipment malfunction (sensitive equipment), increased kVA demand and need to oversize, total power factor and its reduction, waste of electric energy, and light flickering.

Total harmonic distortion (THD) is a measure of the effective value of the harmonic components of a distorted waveform as given in the Equation (4.1). This index can be calculated for either voltage or current. It is the measure that quantifies "how close the waveform is to pure sine."

$$\text{THD} = \frac{\sqrt{\sum_{h>1}^{h_{\max}} X_h^2}}{X_1} \tag{4.1}$$

where X_h is the root mean square (RMS) value of harmonic component h of the quantity X and X_I is the fundamental component. The THD is calculated up to the harmonic $h_{\max}$, which is typically 20. The THD may be calculated with either the RMS values or the peak values. There exists a connection between the THD and true RMS value X_{RMS} of the waveforms as shown in Equation (4.2).

$$X_{\text{RMS}} = X_1 + \sqrt{1+\text{THD}^2} \tag{4.2}$$

provided that if no harmonics exist above $h_{\max}$ and the waveform is periodic with a fundamental wave period, then the THD can be represented as in the Equation (4.3).

$$\text{THD} = \sqrt{\left(\frac{X_{\text{RMS}}}{X_1}\right)^2 - 1} \tag{4.3}$$

4.3 Solutions to Power Quality Problems

There are two approaches to the mitigation of power quality problem that can be done from customer side or from utility side. The first approach is load conditioning, which ensures that the equipment is less sensitive to power disturbances, allowing the operation even under significant voltage distortion. The second solution is to install line conditioning systems that suppress or counteract the power system disturbances.

A versatile and flexible solution to voltage quality problems is offered by active power filters. Shunt active power filters operate as a controllable current source and series active power filters operates as a controllable voltage source. Both schemes are implemented preferable with voltage source pulse-width modulation (PWM) inverters, with a direct current (DC) bus having a reactive element such as a capacitor. Active power filters can perform one or more of the functions required to compensate power systems and to improve power quality. Their performance also depends on the power rating and the speed of response. However, with the restructuring of power sector and with shifting trend towards distributed and dispersed generation, the line conditioning systems or utility side solutions will play a major role in improving the inherent supply quality. From the utility perspective, power quality has been defined as the parameter of the voltage that affects the customer's supersensitive equipment.

4.4 Multilevel Inverters and Their Structures

A multilevel power converter structure has been introduced as an alternative in high power medium voltage applications. Even though the conventional PWM inverters are widely used in industrial applications, they possess several drawbacks:

- The carrier frequency must be very high. If the output waveform has frequency of 50 Hz, then the carrier frequency should be greater than 1 kHz.
- The pulse height is very high. In a normal PWM waveform (not multistage PWM), all the pulse height is equal to DC link voltage and its output voltage has a large jumping span. This results in large *dv/dt* and strong electromagnetic interference (EMI).
- The pulse width would be very narrow when the output voltage has a low value.
- PWM causes plenty of harmonics to produce poor THD. In a very rigorous switching condition, the switching devices experience large switching power losses.
- Control circuitry for the switching pulses generation and devices used is complex and costly.
- A multilevel power converter accumulates the output voltage in horizontal levels and overcomes the drawbacks of conventional inverters as given below:
- The switching frequencies of the switching devices are low, which are equal to or only a small multiple of the output signal frequency.
- The pulse heights are quite low. For an "m"-level inverter with output amplitude V_m the pulse heights are V_m/m or only a small multiple of it. Usually, it causes low *dv/dt* and ignorable EMI.
- The pulse widths of all pulses have reasonable values that are comparable to the output signal.
- Multilevel converters cannot cause high levels of harmonics and produce less THD.
- Offers smooth switching condition and the switching devices have small switching power losses.
- Control circuitry is simple and the devices are not costly.

The term multilevel comes from the three-level converters. The commutation of the power switches aggregates these multiple DC sources in order to achieve high voltage at the output. The advantages of this multilevel approach include good power quality, better electromagnetic compatibility, low switching losses, and high voltage capability. The three structures of MLI are neutral point-clamped (NPC) or diode-clamped multilevel inverter (DCMLI), flying capacitor multilevel inverter (FCMLI), and cascaded multilevel inverter (CMLI).

The first multilevel converter can be attributed to Baker and Bannister who patented the cascaded H bridge (CHB) in 1975. This cascaded inverter was first defined with a format that connects separately DC-sourced full-bridge cells in series to synthesize a staircase AC output voltage. Nabae, Takahashi, and Akagi presented a diode-clamped topology in 1981 that utilizes a bank of series capacitors to split the DC bus voltage. The diode-clamped inverter was also called the NPC inverter when it was first used in a three-level inverter; the mid voltage-level was defined as the neutral point. The NPC inverter effectively doubles the device voltage level without requiring precise voltage matching. Meynard and Foch in 1992 patented the flying capacitor (FC) architecture, which uses floating capacitors to clamp the voltage levels.

In 1992, Osagawara presented a standard current source inverter (CSI), but increased the number of current levels instead of voltage levels (Babaei, 2010). Vázquez et al. (2010) pointed the drawback of the current source topologies, which lies in the limited dynamic performance due to the use of large DC chokes as DC link whereas the voltage source inverter (VSI) has high dynamic performance. Although the cascade inverter was invented earlier, its applications did not prevail until the mid-1990s. Rodríguez et al. (2002) have surveyed that due to the great demand for medium voltage high power inverters, the cascade inverter has drawn tremendous interest.

The two terminologies that constitute the working of MLIs are "stages" and "levels." The individual H bridge inverter is termed as a single "stage." During its conduction it produces the output voltage with square shape in both positive and negative half cycles, which are termed as two "levels." The common zero potential is also included as the additional level, which results in a "single stage three level" inverter. Any number of inverter output voltage levels can be achieved by increasing the number of inverter stages. While increasing the levels, consideration should be focused on the number of switching devices used and their corresponding control circuits. Figure 4.1a and b represents the power circuit of a fifteen-level diode-clamped and FC-type MLI with solar PV array at its input. These two MLI configurations are well-suited for the applications with single input supply. The design of solar PV array in these cases requires a high power rating or a boost converter circuit to step up its voltage in order to reach the higher levels.

4.4.1 Diode-Clamped Multilevel Inverter

Figure 4.1a shows the power circuit of a fifteen-level diode-clamped multilevel inverter (DCMLI). It is also termed a neutral point-clamped multilevel inverter (NPCMLI) where the switching devices are connected in series to make up the desired voltage rating and output levels. The inner voltage points are clamped by either two extra diodes or one high frequency capacitor. The switching devices of an m-level inverter are required to block a voltage level of $V_{dc}/(m–1)$. The clamping diodes need different voltage ratings for different inner voltage levels.

In the power circuit, the DC bus voltage (PV array) is split into fifteen levels by seven series-connected bulk capacitors. The middle point of each two capacitors can be defined as the neutral point. The key components that distinguish this circuit from a conventional inverter are the diodes. These diodes clamp the switch voltage to half the level of the DC bus voltage.

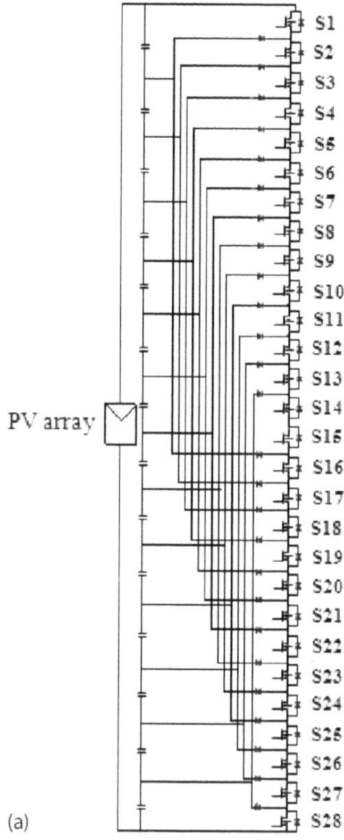

FIGURE 4.1 (a) Power circuit of fifteen-level DCMLI. *(Continued)*

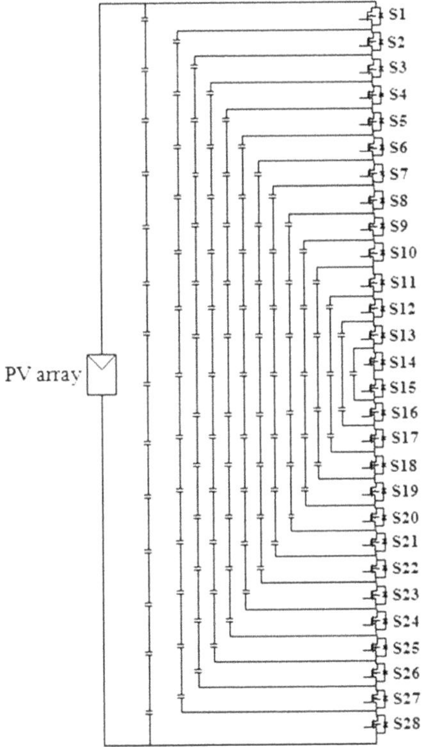

FIGURE 4.1 (Continued) (b) power circuit of fifteen-level FCMLI.

The disadvantages of DCMLI are: (a) Real power flow is difficult for a single inverter because the intermediate DC levels will tend to overcharge or discharge without precise monitoring and control and (b) the number of clamping diodes required is quadratically related to the number of levels, which can be cumbersome for units with a high number of levels.

4.4.2 Flying Capacitor Multilevel Inverter

The circuit has been called the capacitor-clamped inverter with dependent capacitors clamping the device voltage to one capacitor voltage level. This topology has a ladder structure of DC side capacitors, where the voltage on each capacitor differs from that of the next capacitor. The voltage increment between two adjacent capacitor legs gives the size of the voltage steps in the output waveform. The FCML inverter power circuit illustrated in Figure 4.1b provides a fifteen-level output.

The voltage synthesis in a fifteen-level capacitor-clamped inverter has more flexibility than a diode-clamped converter. Unlike the diode-clamped inverter, the FC inverter does not require all of the switches that are ON (conducting) in a consecutive series. Moreover, the FC inverter has phase redundancies, whereas the diode-clamped inverter has only line–line redundancies. These redundancies allow a choice of charging/discharging specific capacitors and can be incorporated in the control system for balancing the voltages across the various levels.

The disadvantages of FCMLI are: (a) Control is complicated to track the voltage levels of all the capacitors. Also, pre-charging all of the capacitors to the same voltage level and start up are complex. (b) Switching utilization and efficiency are poor for real power transmission. (c) The large numbers of capacitors are both more expensive and bulkier than clamping diodes in multilevel diode-clamped converters. Packaging is also more difficult in inverters with a high number of levels.

In order to overcome these drawbacks of both NPCMLI and FCMLI, the structure suitable for solar PV with the aid of power quality improvement is considered.

4.4.3 Cascaded H Bridge Multilevel Inverter (CHBMLI)

A cascaded multilevel inverter (CMLI) consists of series H bridge (single-phase full-bridge) inverter units. The general function of this MLI is to synthesize a desired voltage from separate DC sources (SDCs), which may be obtained from batteries, fuel cells, or solar cells. The resulting phase voltage is synthesized by the addition of the voltages generated by different H bridges. If the DC link voltages of the HBs are identical, then the CMLI is termed as symmetrical. However, it is possible to have different values among the DC link voltages of HBs, and the circuit can be called asymmetrical. As solar PV voltages are variable with respect to environmental factors, asymmetrical inverters are highly recommended. Figure 4.2 shows the power circuit of a solar-fed cascaded fifteen-level inverter along with its output voltage waveform. The AC terminal voltages of different level inverters are connected in series. Unlike the NPCMLI or FCMLI, the CMLI does not require any voltage clamping diodes or voltage balancing capacitors.

Among the various MLI topologies, cascade configuration has been utilized for medium voltage and high voltage renewable energy systems such as solar PV due to its modular and simple structure. Cascaded inverters are ideal for connecting renewable energy sources with an AC grid, because of the need for separate DC sources. This is the case in regard to applications such as PVs or fuel cells. The higher number of voltage levels can effectively decrease harmonics content of staircase output, thus significantly simplifying the output filter design.

Figure 4.2 shows the phase voltage waveform of a fifteen-level cascaded inverter with seven PV array inputs. The phase voltage is synthesized by the sum of seven inverter outputs given by the relation as

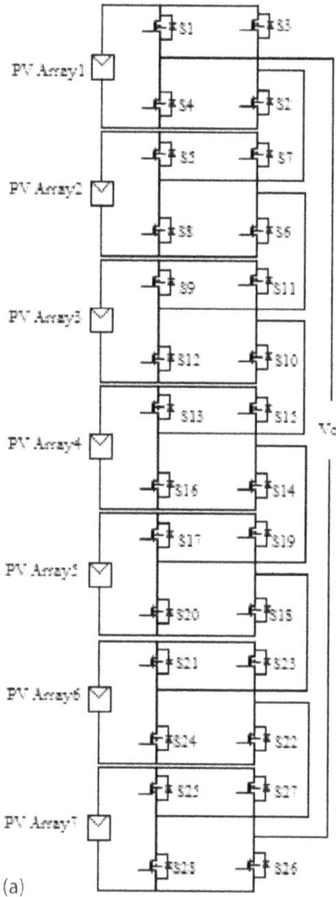

FIGURE 4.2 Solar-fed cascaded fifteen-level inverter (a) and its output voltage waveform (b). *(Continued)*

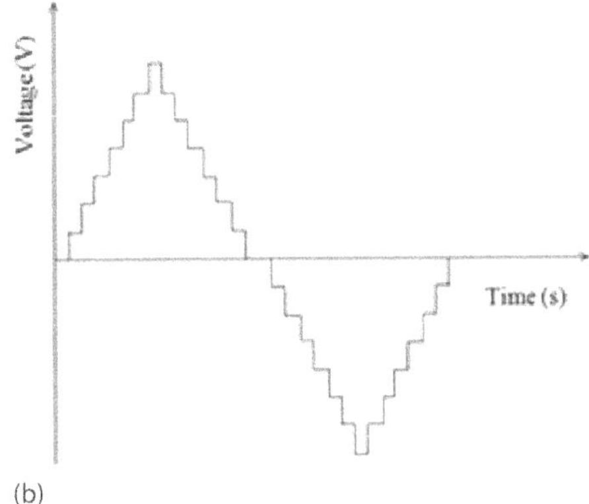

(b)

FIGURE 4.2 (Continued) Solar-fed cascaded fifteen-level inverter (a) and its output voltage waveform (b).

$v_{an} = v_{a1}+v_{a2}+v_{a3}+v_{a4}+v_{a5}+v_{a6}+v_{a7}$. Each inverter level can generate three different voltage outputs, $+V_{dc}$, 0 and $-V_{dc}$ by connecting the PV array source to the AC output side by different combinations of the four switches in the individual inverter stage. As a case, in the first stage of the inverter, turning S1 and S4 ON yields the output $+V_{dc}$ and turning S2 and S3 ON yields the output $-V_{dc}$. Turning OFF all the switches yields the output 0. Similarly, the AC output at each level can be obtained in the same manner. If N_s is the number of input PV sources, the output phase voltage level is $m = N_s+1$. Thus, a fifteen-level cascaded inverter needs seven SDCs and seven full bridges. Controlling the switching angles at different inverter stages can minimize the harmonic distortion of the output voltage, which in turn improves the power quality.

Compared with NPCMLI and FCMLI, CMLI requires the fewest number of components to achieve the same number of voltage levels. Optimized circuit layout and packaging are possible because each level has the same structure. The only disadvantage of the CMLI is that it needs separate DC sources for real power conversions. However, this disadvantage can be compensated by utilizing solar PV at its input.

4.4.4 Reduced Order Multilevel Inverter

Besides the three basic MLI structures, new topologies based on the existing multilevel concepts have been classified as hybrid MLI. The hybrid multilevel topologies are constituted by using the combination of two basic topologies such as NPCMLI or FCMLI to replace the H-bridge as the basic module of the CMLI, in order to reduce the number of the separated DC sources. The asymmetric hybrid MLIs synthesize the output voltage waveforms with reduced harmonic content. This advantage is achieved by using distinct voltage levels in different modules, which can generate more levels in output voltage waveform. It can also reduces the THD value, while preventing the increase in number of switching devices and sources.

Each power module of a hybrid MLI can be operated at distinctive DC voltage and switching frequency, improving the efficiency and THD compensation characteristics of the inverter. Nevertheless, conventional PWM strategies, which generate switching frequency at fundamental frequency (50 Hz), are not appropriate for the hybrid inverters due to switching devices of the higher voltage modules. Besides, they would have to operate at high frequencies only during some inverting instants. To achieve this control strategy, hybrid modulation methods can be used in such a way that higher power cells are switched at low frequency and low power cells switched with high frequency.

Hybrid MLIs promise significant improvements for medium voltage and high-power industrial drives. Asymmetrical multilevel inverters help in minimizing the harmonic contents of output voltage without increasing the number of power devices. The use of various DC voltages in supply leads the hybrid MLI topologies in an effort to optimize the power processing of the entire system.

4.4.5 Comparison of Multilevel Inverters

NPCMLI offers the advantage of having common DC input capacitors for the three phases. Its main disadvantage is that too many diodes are used for clamping, which also make the implementation of the physical layout difficult. NPCMLI structure can be extended to a higher number of levels but these are less attractive because of additional losses and uneven distribution of these losses in the outer and inner devices. The clamping diodes, which have to be connected in series to block the higher voltages, introduce more conduction losses, and produce reverse recovery currents during commutation that affects the switching losses of the other devices even more.

In a FCMLI, the clamping capacitor voltages can be balanced within a few cycles by using voltage synthesis redundancies. Its main disadvantage is that many capacitors are used for clamping, though these capacitors are cheaper compared to the clamping diodes of the DCMLI. Furthermore, for low switching frequency, the clamping capacitors become large in size, thus decreasing the power density of the inverter.

In a CMLI, no clamping capacitors or diodes are necessary. Furthermore, a low switching frequency can be used. Modularized circuit layout and packaging is possible because each cell has the same structure and there are no extra FCs or clamping diodes. Its main disadvantage is the use of an independent DC source for every cell. For this reason, the CMLI can be used in the applications pertaining to solar PV and fuel cell-based power generation systems.

Examining the power circuits of the three MLIs it is observed that they use the same number of semiconductor switches per phase. However, the NPCMLI uses $(m-1)$ $(m-2)$ clamping diodes, thus increasing the cost compared to the other two structures. The FCMLI and CMLI need 2 $(m-1)$ heat sinks whereas the NPCMLI needs $2(m-1) + (m-1)$ $(m-2)$. In order to generate an m-level inverter output, the CMLI uses the least amount of semiconductor devices and consequently requires the lowest implementation cost. One additional advantage of the CHB converter is that if any device fails in the H-bridges, the inverter can still be operated at reduced power level.

4.4.6 Applications of Multilevel Inverters

Multilevel converters are finding increased attention in industry and commercial sectors as the preferred choices of electrical power conversion for medium and high-power applications. There are many applications for MLIs, such as flexible AC transmission system (FACTS) equipment, high voltage direct current (HVDC) lines, and electrical drives. MLIs are currently commercialized in standard and customized products that power a wide range of applications, such as compressors, extruders, pumps, fans, grinding mills, rolling mills, conveyors, crushers, blast furnace blowers, gas turbine starters, mixers, mine hoists, reactive power compensation, marine propulsion, hydro pumped storage, wind energy conversion, and railway traction.

4.4.7 Integration of MLI with Solar PV Systems

The ever-rising demand for electrical energy and depleting fossil fuel reserves are compelling reasons to use existing resources more efficiently. New highly efficient power electronic technologies and proper control strategies are therefore needed to reduce energy waste and to improve power quality. The increasing demand for energy has stimulated the development of alternative power sources such as PV modules, fuel cells, and wind turbines. The PV modules are particularly attractive as renewable sources due to their relatively small size, noiseless operation, simple installation, and the possibility of installing them closer to the user PV power generators convert the energy of solar radiation directly to electrical energy without any moving parts. PV power generators can be classified into standalone and grid-connected systems. In standalone system, the energy storage has a big influence on the design of the systems. In grid-connected system, the grid acts as an energy storage into which the PV power generator can inject power whenever power is available.

In PV modules, the output voltage has low DC amplitude. In order to be connected to the grid, the PV modules output voltage should be boosted and converted into an AC voltage. This task can be performed using one or more conversion stages (multistage). Many topologies for PV systems are multistage, having a DC to DC converter with a high frequency transformer that adjusts the inverter DC voltage and isolates the PV modules from the grid. However, the conversion stages decrease the efficiency and make the system more complex.

In PV systems where series modules are connected to a conventional two-level inverter, the occurrence of partial shades and the mismatching of the modules lead to a reduction of the generated power. In addition, the conventional two-level voltage source converters will not be able to deliver the performance parameters such as improved power quality, maximum allowed switching frequency, higher voltage operation and reduction in filter size. To overcome these problems, the connection of the modules can be made using a multilevel converter. Multilevel converter maximizes the power obtained from the arrays, reduces the device voltage stress, and generates output voltages with lower THD.

Grid connection of PV systems has been traditionally performed by three different types of configurations: centralized conversion topology (large three-phase system), string topology (medium single-phase system), and the AC module topology (small single-phase system). More recently, a hybrid between the centralized and the string configuration, called multi-string topology (for medium to large, single- or three-phase system) has gained more attention. The centralized topology uses a single three-phase inverter to connect to the grid. The advantages are its simple structure and control, which come at the expense of reduced power generation due to module mismatch and partial shading. This topology is considered nowadays obsolete.

The string topology uses one inverter per string, improving the total generated power. It also increases modularity, since additional strings can be added to the system without the need of changing the inverter dimensions. Depending on the size of the string, a boost DC to DC converter or a step-up transformer is necessary to reach the grid voltage. The string topology is the most widely installed solution for PV grid-connected systems today.

The AC module topology, or converter integrated module, is the most modular and has the best maximum power tracking capability, since one converter is dedicated per module. This is intended for smaller systems and more domestic use. The main disadvantage is that a DC–DC boost stage or step-up transformer is a must, which increases the cost if an AC module system of the same power of a string system is compared.

The CMLI has attracted attention for the PV integration as each H Bridge (or power cell) needs isolated DC sources, which can be easily given by PV modules or strings. Furthermore, it adds interesting benefits such as higher voltage operation by interconnecting enough modules or strings in series to reach grid voltage, eliminating the need of step-up transformer or boost DC–DC converters. In addition, the inherent improved power quality of multilevel converters reduces filter size and switching frequency, improving the system efficiency.

CMLI can be used as a series connection of string inverters or as a series connection of AC module inverters. In the first case, less PV modules are necessary per string, as voltage will be elevated by series connection. This improves maximum power point tracking (MPPT), since lesser modules are concentrated to a single converter. The second case, i.e., if used as the series connection of several AC modules, the internal DC–DC boost stage is no longer necessary, simplifying each converter (fewer semiconductors, no boost inductor, lower switching frequency, improved efficiency, etc.). In addition, since they are connected in series at the AC side, DC cables are no longer necessary as with the multi-string topology. Finally, compared to any of the traditional topologies, the CHB multilevel approach has by far the best power quality, reducing filter needs while improving efficiency and overall performance.

The same configurations can also be extended to the standalone PV systems. The major advantage of this system is its potential to supply abundant electricity in areas not provided by the general power grid. In many standalone PV inverters, alternating current is required to operate at 230 V, 50 Hz for home or office appliances. Generally, standalone inverters operate at voltages of 12, 24, 48, 96, 120, or 230 relying on the power level. In the case where the amplitude of voltage produced by solar array is low, there is the need for an additional boost converter or a step-up transformer to obtain high output voltage. It converts power from DC to AC and is commonly connected in series with a PWM inverter. However, a somewhat high switching frequency of PWM inverter and its stress result in low efficiency and occasionally EMI problems. In addition, an output filter is required to reduce high switch frequency components and to produce sinusoidal output from the inverter. The advantages of transformerless PV inverters are higher efficiency, lower cost, less complexity, and smaller volume compared to their counterparts with transformer galvanic isolation. Hence the CMLI is well-suited for both standalone and grid-connected systems with respect to output voltage regulation and harmonic reduction.

4.5 Power Quality Improvement Techniques for a Solar-Fed CMLI

In the recent years the demand for solar electric energy has grown consistently, which is mainly due to the decreasing costs and prices. This decline has been driven by the following factors: (1) increasing efficiency of solar cells, (2) improvements in manufacturing technology, and (3) economies of scale. PV inverter, which is the heart of a PV system, is used to convert DC power obtained from PV modules into AC power before it fed into the grid. Improving the output waveform of the inverter reduces its respective harmonic content and, hence, the size of the filter used. It also reduces the level of EMI generated by switching operation of the inverter. Multilevel converters are particularly suitable for renewable PV energy, where a great concern is efficiency and power quality. The performance of the MLI can be improved by: (a) harmonic elimination, (b) control strategies, and (c) new MLC topologies. Attenuating the lower order harmonics using a larger output filter inductance is not a good option as it increases losses in the system along with a larger fundamental voltage drop and with a higher cost.

4.5.1 Intelligent Techniques

In spite of the fact that selective harmonic elimination (SHE) was introduced in 1973 considerable work needs to be addressed with aid of intelligent techniques. In recent years, the control techniques are modified into intelligent based in order to achieve high precision output in spite of the situations of imprecision, uncertainty, and partial truth. In order to improve the performance parameters of solar-fed CMLI, the intelligent techniques such as artificial neural networks (ANN), fuzzy logic controllers (FLC) and optimization techniques such as genetic algorithm (GA), particle swarm optimization (PSO), and bees optimization (BO) were developed. In fact, each and every technique has its own control algorithm and based on the natural phenomenon. Neural networks are simplified models of the biological nervous system and therefore have drawn their motivation from the kind of computing performed by a human brain. In general, it is a highly interconnected network of a large number of processing elements called neurons in an architecture inspired by the brain. It can massively parallel and intend to exhibit parallel distributed processing. Neural networks learn from examples. They can be trained with known examples of problem to acquire knowledge about it. Once appropriately trained, the network can be put to effective use in solving unknown or untrained instances of the problem.

Fuzzy logic is a generalization of classical set theory. Fuzzy logic representations founded on fuzzy set theory try to capture the way humans represent and reason with real world knowledge in the face of uncertainty. A fuzzy set can be defined mathematically by assigning to each possible individual in the universe of discourse, a value representing its grade of membership in the fuzzy set. This grade corresponds to the degree to which the individual is similar or compactable with the concept represented by the fuzzy set.

Optimization techniques provide the best solution in the given search space based on the condition that the solution satisfies the constraints. This can be classified as derivative-free and derivative-based methods. GA, PSO, and BO fall under the category of derivative-free optimization methods. The GA repeatedly modifies a population of individual solutions. At each step, the GA selects individuals at random from the current population to be parents and uses them in generating the children for the next generation. Over successive generations, the population "evolves" toward an optimal solution.

PSO was inspired by the sociological behavior associated with swarms such as flocks of birds and schools of fish. The individuals in the population are called particles. Each particle is a potential solution for the optimization problem and tries to search the best position through flying in a multidimensional space. BO is an optimization algorithm based on the natural foraging behavior of honeybees to find the optimal solution. This algorithm proceeds with the three phases representing the three types of bees available in the colony.

4.5.2 Problem Statement

In solar PV systems enhancing the power quality is one of the challenging problems. The major problem in the solar PV conversion system lies in providing a pure form of power to the consumers. In addition to the power quality, availability, and reliability of the power also need to be accounted. On considering these problems, the power quality improvement techniques are proposed for a solar-fed fifteen-level inverter to enhance both power quality and power quantity in the system.

The output voltage waveform of the MLIs is affected by the following factors: (1) the topology used, (2) application, (3) control algorithm, (4) size of the filter, and (5) choice of switching frequency. This chapter provides the solutions for each factor such that (1) the control algorithm involving intelligent techniques such as ANN, FLC, and optimization techniques to reduce the harmonic distortions, (2) the usage of filter is completely eliminated in the system in meeting the objective of improving power quality without using the output filter circuits, and (3) a minimum switching frequency (1000 Hz) is used for the modulation techniques and fundamental switching frequency (50 Hz) for SHE techniques. In this connection low switching frequency methods are preferred as they reduce the switching losses.

4.6 Literature Review

Fei et al. (2009) have formulated the quarter wave symmetry SHE PWM for a five-level CMLI. The number of switching angles for half wave symmetry SHE PWM is twice that of quarter wave symmetry SHE PWM. The methods for obtaining initial values are based on the rule of equality of area and superposition of center of gravity of the PWM section with the sine reference signal. A comprehensive method is required to assess the multiple valid solutions of SHE PWM if a global optimized solution is required for the specific applications.

Liu et al. (2009a) have introduced the real-time algorithm for calculating switching angles in the context of step modulation that minimizes THD in an MLI, but the approach was not extended for unequal DC sources. The data of switching angles stored in the lookup table increases if the required resolution of the fundamental component of the voltage increases. Hence the proposed method aims for real-time calculation for a seven-level CMLI. For the variation of modulation indexes of 0.7, 0.8 and 0.9, the obtained THD values are 17.12%, 12.89% and 14.98% respectively.

Liu et al. (2009b) have focused on real-time algorithm for calculating switching angles in the context of stair case modulation to minimize voltage THD in a multilevel inverter with unequal voltage steps. Non real-time algorithms for the staircase modulation cannot be applied practically in MLIs with varying voltage steps, since the sizes of lookup tables would be huge. The proposed algorithm is implemented for a seven-level CMLI where the input DC voltages are in the range of (0–30) V. For the varying input DC voltages, the THD obtained is 12.20% to 16.69%.

Taghizadeh and Hagh (2010) have expounded the elimination of harmonics in a CMLI by considering the non-equality of separated DC sources using PSO. An eleven-level inverter is considered and the optimal switching angles required to eliminate four numbers of harmonic order is determined using PSO. With modulation index $m_a = 0.47$, the line to line voltage THD obtained is 9.822%.

Wang and Ahmadi (2010) have proffered the SHE PWM for an MLI using equal area criteria and harmonic injection, which in turn provides only four equations to achieve the optimal switching angles irrespective of the number of voltage levels. In spite of equal area criteria providing the possible solution for the initial guess in Newton–Raphson (NR) method, the junction point of the modulation waveform in the first equation has to be found out using NR based iterations. Multiple values for junction point of the modulation waveform also results in a major drawback of the four equations-based method.

Fei et al. (2010) have formulated the half wave symmetry SHE PWM for a five-level CMLI. In this topology, 20 switching angles are considered in a half cycle incorporating its rising and falling edges, so

20 equations can be formulated. The equations and initial values are dependent on the initial phase of the output fundamental voltage. Hence an appropriate method is required to obtain the initial values. The methods such as multi-carrier PWM and equal area criteria are suggested. If these methods are invalid and require more solutions, the trial and error method needs to be adopted.

Ahmadi et al. (2011) have applied SHE-based optimal switching angle calculation for high power inverters. They developed harmonics injection and equal area criteria-based four equation method to realize optimal pulse-width modulation (OPWM) for two-level inverters and weight-oriented distribution function for a five-level MLI with unbalanced DC sources. The method is iterative based and requires the initial starting point for switching angles in equal area criteria. In addition, the fundamental voltage compensation is needed as the resulting fundamental component is usually different from the desired fundamental component.

Filho et al. (2011) have approximated the selective harmonic elimination (SHE) problem using ANNs to generate the switching angles in an eleven-level full-bridge cascade inverter powered by five varying DC input sources. A nondeterministic method is used to solve the system for the angles and to obtain the data set for the ANN training. The switching angles generated by the ANN may not provide a satisfactory result, or harmonic elimination at some points will generalize. However, a fast result can be obtained, and more angles can be easily added to provide a better output waveform. The THD is greater than the expected range of 7%–14% as a consequence of those components added up by the voltage transient. Harmonics are not completely eliminated in this approach, but they are minimized.

Farokhnia et al. (2011) have dealt with an analytical algebraic method for formulating the line voltage THD of MLIs with unequal DC sources. For a three-phase structure, in delta connection the line voltage is equal to the phase voltage whereas in wye connection, the line voltage is obtained from subtracting the phase b voltage from phase a voltage. A five-level inverter is taken into consideration for which the methods are proposed to calculate the line voltage THD. With the input voltages of 9 V and 18 V, a five-level inverter is implemented and the line voltage THD obtained is 14.03%. The advantages of the extracted formulas of line-voltage THD are simplicity and rapidity in calculations, considering all harmonics and the possibility of finding optimal switching angles analytically.

Kavousi et al. (2012) have devised the BO method for harmonic elimination in a cascaded multilevel inverter. Bee algorithm (BA) is applied to a seven-level inverter for solving the harmonic elimination equations. The algorithm is based on the food foraging behavior of a swarm of honeybees and it performs a neighborhood search combined with a random search. Switching angles are found such that low-order harmonics (fifth and seventh) are eliminated and the magnitude of the fundamental harmonic reaches its desirable value. The methodology is implemented in hardware with symmetrical DC voltage of 12 V. With modulation index = 0.8, the line to line voltage THD obtained is 8.99%.

Yousefpoor et al. (2012) have formulated the THD minimization applied to the seven-level CMLI phase voltage waveform due to its simplicity. They emphasized the difference between the phase voltage and line voltage waveforms. In MLIs' output voltage, attention is mostly paid to the phase voltage, which has simpler waveform and easy formulation. The objective function required for the GA is derived for phase voltage waveform. On execution, three optimal switching angles are obtained for a seven-level CMLI. On determination of optimal switching angles, by phase voltage minimization approach, the resultant THD is 17.06% and by line voltage minimization approach the resultant THD is 26.72%.

Filho et al. (2013) have enunciated the approach for modulation of eleven-level CMLI using SHE using ANN. The data set required for ANN training is obtained from GA. The coefficients of the objective function need to be obtained are made by trial and error method. The THD obtained for eleven-level inverter is 9.7%.

4.7 Modeling of Solar Panel

The PV panel that serves as the input source for the individual MLI stages is modeled according to the commercial PV used in the hardware implementation. One of the problems in the PV generation systems is that the amount of electric power generated by the solar arrays is always changing with weather conditions, i.e., the intensity of solar radiation. The modeled PV panel considers the variation in both temperature and irradiance by observing the real-time data of these parameters for the geographic region. In order to achieve the panel of required rating, the mere consideration is focused on modeling of both solar cell and solar modules to develop a solar PV array.

Solar cell is basically a PN junction fabricated in a thin wafer or layer of semiconductor. The main advantage of solar energy when compared to other renewable energy sources is that the electromagnetic radiation of sunlight can be directly converted to electricity through PV effect. Being exposed to sunlight, photons with greater energy are absorbed and create electron hole pairs, which are proportional to the incident irradiation. Under the influence of the internal electric fields of the PN junction, a photocurrent that is directly proportional to the solar insolation is created. The PV system will naturally exhibit the Voltage–Current and Power–Voltage characteristics, which will vary with the radiant intensity and cell temperature.

The equivalent circuit of the solar cell, which is considered as general single-diode model, is given in Figure 4.3. Based on the equivalent circuit, the solar cell can be modeled as a current source in parallel with a diode that generates the current density J_{sc}. The dark current flows in the opposite direction and is caused by a potential between the positive and negative terminals.

In addition, two resistances – one in series (R_s) and one in parallel (R_p) are connected. The series resistance is caused by the fact that solar cell is not a perfect conductor and describes an internal resistance to the current flow. The parallel resistance is caused by leakage of current from one terminal to the other, due to poor insulation.

For an ideal solar cell, $R_s = 0$ and R_p = infinity. The voltage current characteristic equation of a solar cell is given in Equation (4.4).

$$I = I_{\text{ph}} - I_S\left[\exp\left(q\frac{(V + IR_S)}{kT_cA}\right) - 1\right] - \frac{(V + IR_S)}{R_p} \tag{4.4}$$

where I_{ph} is a light generated current or photocurrent, I_s is the cell saturation of dark current, q (=1.6 × 10^{-19} C) is an electron charge, k (=1.38 × 10^{-23} J/K) is a Boltzmann's constant, T_c is the cell's working temperature, and A is an ideal factor. The ideality factor A of 1.3 is used as a typical value as 4 for silicon-based PV modules. The photocurrent mainly depends on the solar insolation and cell's working temperature, which is described in Equation (4.5).

$$I_{\text{ph}} = \left[I_{SC} + T_1\left(T_C - T_{\text{ref}}\right)\right]\lambda \tag{4.5}$$

where I_{sc} is the short circuit current of solar cell at 25°C and 1000 W/m^2. These temperature and irradiance levels are denoted as standard test conditions (STC) for any solar cells. T_1 is the short circuit current temperature coefficient. T_{ref} is the reference temperature of solar cell and λ is the solar insolation in W/m^2.

The solar cell saturation current varies with the cell temperature described in Equation (4.6).

$$I_s = I_{RS}\left(\frac{T_c}{T_{\text{ref}}}\right)^3 \exp\left[qE_G\left(\frac{1}{T_{\text{ref}}} - \frac{1}{T_C}\right)/kA\right] \tag{4.6}$$

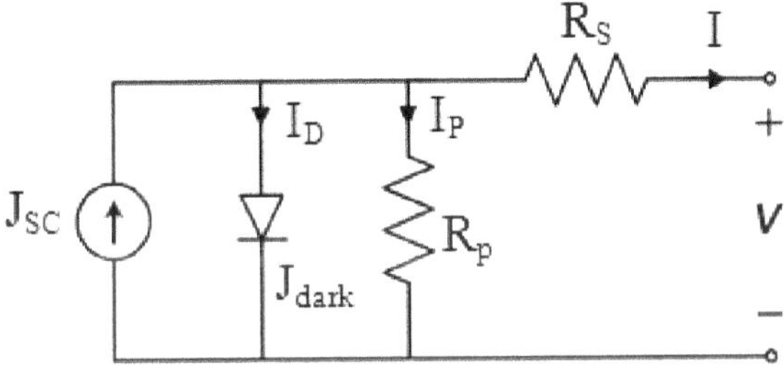

FIGURE 4.3 Equivalent circuit of solar cell (general model).

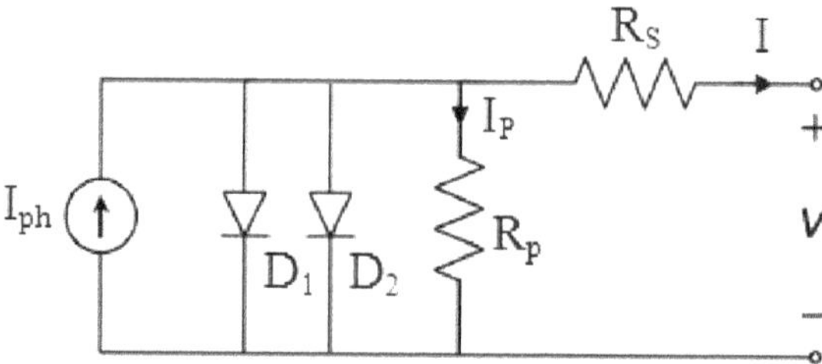

FIGURE 4.4 Equivalent circuit of solar cell (double exponential model).

where I_{RS} is the reverse saturation current of solar cell at reference temperature and solar radiation. E_G is the band gap energy of the semiconductor material used in the cell. The ideal factor A is dependent on PV technology used such as monocrystalline, polycrystalline, etc. Despite the simplicity in the single-diode model, it exhibits serious deficiencies when subjected to temperature variations. In addition, the single-diode model assumes that the recombination losses in the depletion region are absent. Consideration of these drawbacks leads to the design of the two-diode model.

As the polycrystalline silicon-based solar panel is used for the implementation, a double exponential model as shown in Figure 4.4 is modeled. It is derived from the physical behavior of solar cell constructed from polycrystalline silicon. This model is composed of light-generated current source, two diodes, a series resistance, and a parallel resistance. The shunt resistance R_p is inversely related to shunt leakage current to the ground. The efficiency of PV is insensitive to the variation in R_p and the shunt leakage resistance can be assumed to approach infinity without leakage current to ground. A small variation in R_s will significantly affect the PV output power.

The other models, which are termed as "appropriate" and "simplified," were obtained by reducing the above equation into the following form given in Equations (4.7) and (4.8) by considering the ideal property of the solar cell.

$$I = I_{\text{ph}} - I_s\left[\exp\left(q\frac{(V + IR_s)}{kT_cA}\right) - 1\right] \tag{4.7}$$

$$I = I_{\text{ph}} - I_s\left[\exp\left(\frac{qV}{kT_cA}\right) - 1\right] \tag{4.8}$$

Most of the models based on equivalent circuit are mainly used for the MPPT applications. Hence, the model required for the CMLI applications by considering the physical property of the solar panel used in the experimental setup is desired. Therefore, the double exponential model with two diodes is chosen and modeled along with the possibility of inserting the variations due to environmental conditions. The output current equation of this model is given from Equations (4.9) through (4.11).

$$I = I_{\text{ph}} - K_1 - K_2 - \frac{(V + 1 \times R_S)}{R_p} \tag{4.9}$$

where:

$$K_1 = I_{S1} \times \left(\exp\frac{(V + 1 \times R_S)}{(N_1 \times V_t)} - 1\right) \tag{4.10}$$

$$K_2 = I_{S2} \times \left(\exp\frac{(V + 1 \times R_S)}{(N_2 \times V_t)} - 1\right) \tag{4.11}$$

I_{s1} and I_{s2} are the diode saturation currents and V_t is the thermal voltage. N_1 and N_2 are the quality factors (diode emission coefficients), which are in the range of 1 to 2. The solar generated current I_{ph} is given in Equation (4.12), where I_{ph0} is the measured solar-generated current for irradiance I_{r0}.

$$I_{ph} = I_r \times \left(\frac{I_{ph0}}{I_{r0}} \right) \tag{4.12}$$

As the PV cell produces less than 2 W at 0.5 V approximately, the cells must be connected in series and parallel configuration on a module to produce high power. The series connection increases the voltage whereas in parallel connection the value of current is increased. A PV array is a group of several PV modules that are electrically connected in series to generate the required voltage and current. Here it should be noted that the series connection of PV cells is prone to mismatch power losses if the electrical characteristics of the PV cells are not similar or the cells do not operate under uniform conditions.

Let N_s and N_p be the number of cells connected in series and parallel for a PV module. Then the terminal equation for the current and voltage of the array is given by the Equations (4.13) and (4.14).

$$I = I_{ph}N_p - I_s N_s \left[\exp\left(q\left(\frac{V}{N_s} + \frac{IR_s}{N_p} \right) kT_c A \right) - 1 \right] - W \tag{4.13}$$

$$W = \frac{\left(N_p V / N_s + IR_s \right)}{R_p} \tag{4.14}$$

The PV efficiency is sensitive to small change in R_s but less sensitive to the variation in R_p. For a PV array, the resistance is apparently important and the parallel resistance approaches infinity, which is assumed to be open. In most of the commercial PV products, PV cells are generally connected in series configuration to form a PV module to obtain adequate working voltage.

The most important parameters widely used for describing the performance of the solar cell are open circuit voltage V_{oc} and short circuit current I_{sc}. V_{oc} is the maximum voltage generated across the terminals of the solar cell when they are kept open (infinite load resistance). I_{sc} is the maximum current flowing in a solar cell when the terminals are shorted with each other, which can vary due to various technical and environmental reasons. Technical reasons can be minimized during the manufacturing process and during system design phase whereas environmental reasons are harder to avoid.

It is essential to note that higher band gap materials will have lower I_{sc} and higher V_{oc}. The PV cell with the lowest I_{sc} limits the current of the whole series connection. The uneven I_{sc} is due to partial shading of the PV array resultant from clouds, trees, buildings, etc.

Under partial shading conditions, if one PV cell of the generator composed of series-connected cells is shaded, the short circuit currents of the non-shaded cells will be higher than the short circuit current of the shaded cell. If then the current of the PV power generator is higher than the short circuit current of the shaded cell, the shaded cell will be reverse biased due to the other cells in the series connection. In this case, the reverse-biased cell acts as a load in the series connection dissipating part of the power generated by the other cells leading to power losses. This can also lead to hot spots in the shaded cell and the cell can be damaged.

The worst situation is when the series connection is short circuited. Then, the shaded cell dissipates all of the power generated by the other cells in the series connection. In order to prevent PV cells from damaging due to hot spots, manufacturers of PV modules have connected bypass diodes in antiparallel with PV cells. When some of the PV cells of the PV module become shaded, they become reverse biased and the bypass diode connected in antiparallel starts to bypass the current exceeding the short circuit current of the shaded cells and limits the power dissipated in the shaded cells. Hence it is required to include the effects of nonuniform conditions on different PV power generator configurations.

Normally I_{ph} is much greater than I_{sc}, whereas ignoring the small diode and ground leakage currents under zero terminal voltage, I_{sc} is approximately equal to I_{ph}. On the other hand, V_{oc} is obtained by assuming the output current equal to zero. For the given V_{oc} at reference temperature and ignoring the shunt leakage current, the reverse saturation current is approximately obtained as given in Equation (4.15). The maximum power can be expressed as per the Equation (4.16).

$$I_{RS} = \frac{I_{sc}}{\left[\exp(qV_{oc}/N_s kAT_c) - 1\right]} \tag{4.15}$$

$$P_{max} = I_{max}V_{max} = \gamma V_{oc} I_{sc} \tag{4.16}$$

where V_{max} and I_{max} are the maximum terminal voltage and output current of PV module and γ is the fill factor, which is a measure of cell quality.

4.8 Design Specifications

In most of the countries, the consumer loads operate at the input supply of 230 V_{RMS} (root mean square) with frequency 50 Hz. Hence for the design considered in both the simulation and experimental setup, this specification holds at predominant level. To achieve this, the design procedure is started from the solar cell modeling. The system considered is single-phase standalone PV system with battery storage, which enables it to operate even during weak weather conditions.

Solar PV systems can be utilized in any of these three forms: (a) standalone, which operates independent of the utility grid, (b) grid connected, which is utility interactive system and requires a bidirectional interface, and (c) bimodal systems, which operate in either utility interactive or standalone, but not concurrently. The proposed system is application specific especially intended towards rural areas where there is less concentration of utility grid.

The solar cell of rating $V_{oc} = 0.5$ V and $I_{sc} = 7$ A is chosen based on the data sheet of the commercial solar PV specifications. Nearly 24 solar cells are connected in series at standard test conditions (STCs) to develop a 12 V, 7 A, which constitutes a single solar module. Four such solar modules are connected in series to achieve 48 V, 7 A solar PV array or panel. The series connection of the module is the same as that of the cell. This 48 V, 7 A solar PV serves as the input source for the single inverter stage.

Seven such input sources are modeled to power the seven inverter stages, thereby a fifteen-level output waveform is obtained. The following relations given in Equations (4.17) and (4.18) hold good for the desired design requirement:

$$V_{peak} = 48\text{V} \times 7 = 336\text{V} \tag{4.17}$$

$$V_{RMS} = \frac{336}{\sqrt{2}} = 237.59\text{V} \tag{4.18}$$

In order to extend the system for making it suitable for grid-connected system, the condition given in Equation (4.19) has to meet.

$$V_{dc} > \sqrt{2}V_{grid} \tag{4.19}$$

In the proposed design, the total $V_{dc} = 336$ V is greater than the square root of the grid voltage (230 V_{RMS}) as per the condition given in Equation (4.19).

As the design made for STC, it produces the fixed DC output from solar panel without any variations. Most of the models in various literature deal with the fixed output supply panels. Hence the PV panel model, which exhibits the variations thus occurring due to temperature and irradiance, is required to adhere to the real-time specifications. In order to achieve this, a detailed and analytical study on temperature and irradiance variations is undertaken throughout the year by solar PV observatory for the geographic location where the experiment is conducted. The radiation measurements used are beam and diffuse horizontal surface radiation gathered with a PV pyranometer.

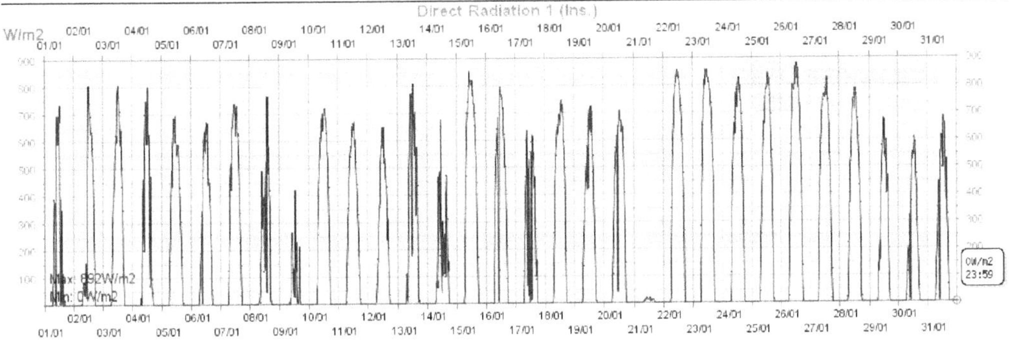

FIGURE 4.5 Solar data for the month of January.

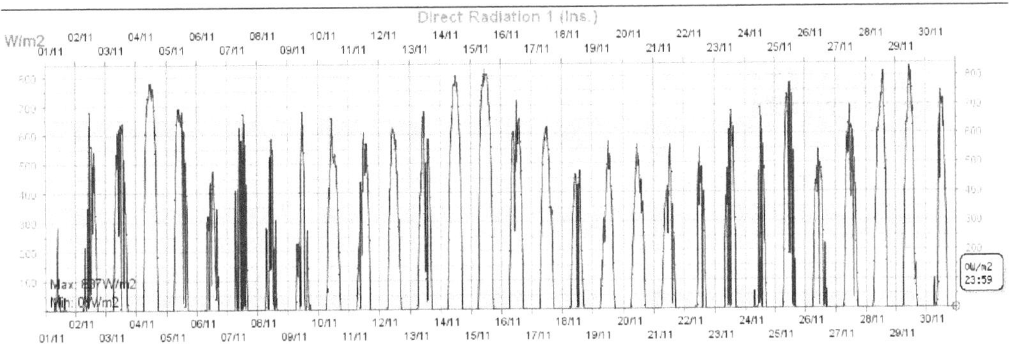

FIGURE 4.6 Solar data for the month of November.

Figures 4.5 and 4.6 show the analysis that depicts the irradiation and temperature levels measured for the months of January and November. Based on the analysis it is found that the irradiance varies from 0 W/m^2 to nearly 1000 W/m^2.

The Solectric 9000 model is taken into consideration for modeling, which provides 115 W of nominal maximum power and has 24 series-connected polycrystalline silicon cells for a single module. It consists of two bypass diodes, each of which is connected in antiparallel with 12 series-connected PV cells to protect them against hot spots. For providing a PV array input, 96 series-connected cells are used for modeling.

In order to verify the model, the voltage current and power voltage characteristics are plotted and shown in Figures 4.7 and 4.8. These graphs are plotted for the various irradiance levels and temperature limits. This graph clearly shows the V_{oc}, I_{sc}, V_{mp}, I_{mp}, and P_m which are the important criteria for maximum power tracking.

Based on the characteristics, it is found that the nonlinear nature of the PV cell is apparent. The output current and power of the PV cell depend on the cell's terminal operating voltage, temperature, and solar insolation. With the increase in working temperature, the short circuit current of the PV cell increases whereas the maximum power output decreases. If the increase in the output current is much less than the decrease in the voltage, the net power decreases at high temperatures. The increase in solar irradiance increases the short circuit current of the PV module and the maximum power output increases. This is due to the fact that the V_{oc} is logarithmically dependent on the solar irradiance and I_{sc} is directly proportional to the radiant intensity.

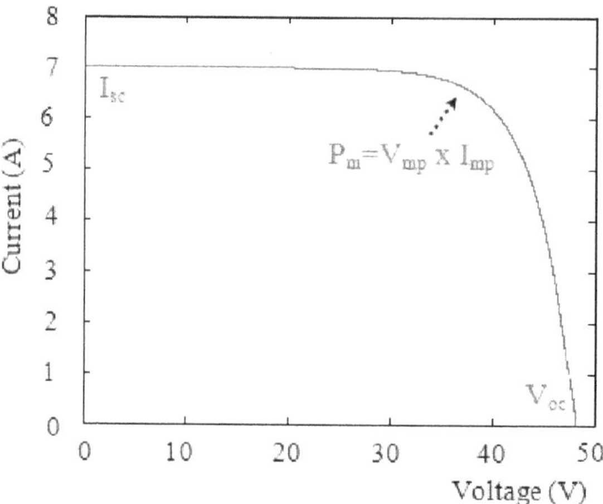

FIGURE 4.7 *V–I* characteristics of solar PV array.

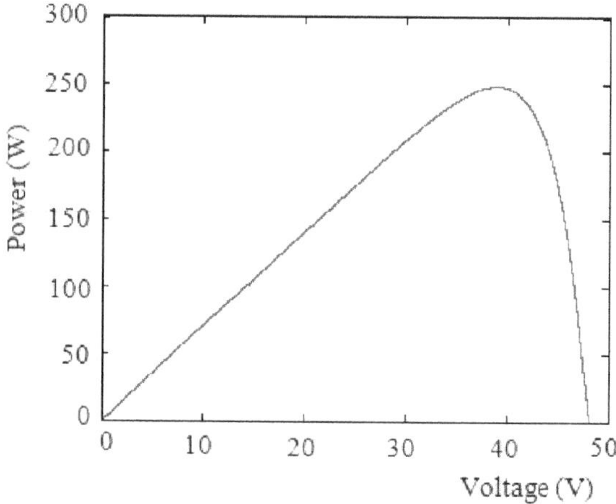

FIGURE 4.8 *V–P* characteristics of solar PV array.

4.9 Experimental Setup

A 3 kW_p solar PV power supply unit is designed and implemented for the seven-stage fifteen-level solar-fed CMLI with multiple carrier PWM generation in a single chip. In most of the countries, much of the domestic loads operate from single-phase 230 V, 50 Hz AC system. Hence the output voltage of the system is designed for 230 V AC supply. In order to achieve this, the input supply used in the setup is 48 V for the individual inverter stages.

Table 4.1 shows the rating of the individual solar PV module. Each module has a rated power of 115 W with voltage variation of 16 to 21 V (nominal 12 V) depending on the operating conditions such as light intensity and so on. Four PV panels are connected in series to get a nominal voltage of 48 V, which constitutes the single individual source for CMLI.

TABLE 4.1

Solar PV Panel Specifications

Parameter	Value
Model	Solectric 9000
P_{mpp}	115 Wp
V_{oc}	21.2 V
I_{sc}	7.4 A
V_{pm}	16.5 V
I_{pm}	6.95 A
Max system voltage	540 V
Tolerance at peak power	±5%
Number of panels	28
Total power	3220 Wp

Solar panels are connected to the loads in any of three divisions: direct, standalone, and grid connected. For application oriented, the first one is not applicable. In the proposed set up, standalone type is used and can also be extended to grid-connected systems. Four 12 V, 100 Ah battery packs are connected in series to get a nominal DC bus of 48 V. These batteries are charged from the PV unit through a controlled charging circuit. There are seven sets of individual PV supply source used for the hardware implementation as they produce the required $48 \times 7 = 336\ V_p$ ($230\ V_{RMS}$) in accordance with the simulation model rating.

Table 4.2 reveals the specifications of the entire hardware setup. Figure 4.9 shows the 3 kW_p solar PV plant and Figures 4.10 and 4.11 shows the complete hardware setup and the corresponding output waveform of the proposed CMLI. Here the notation INV specifies the individual inverter stage.

TABLE 4.2

Specifications of Experimental Set Up

Parameter	Rating
Charge Controllers (CC)	
Make	Sukaam
V–I rating	48 V, 10 A
Number of CC	7
Battery Bank	
Make & model	EXIDE 6LMS100L
Rating	12 V, 100 Ah
Number of batteries	28
Power Circuit	
Semiconductor devices	MOSFET IRF 840
Number of devices	28
Control Circuit	
Controller	DSP – TMS320F2812
Measuring Instruments	
PQ analyzer	Yokogawa WT3000
Oscilloscope	Tektronix

FIGURE 4.9 3 kWp solar PV plant with 28 modules.

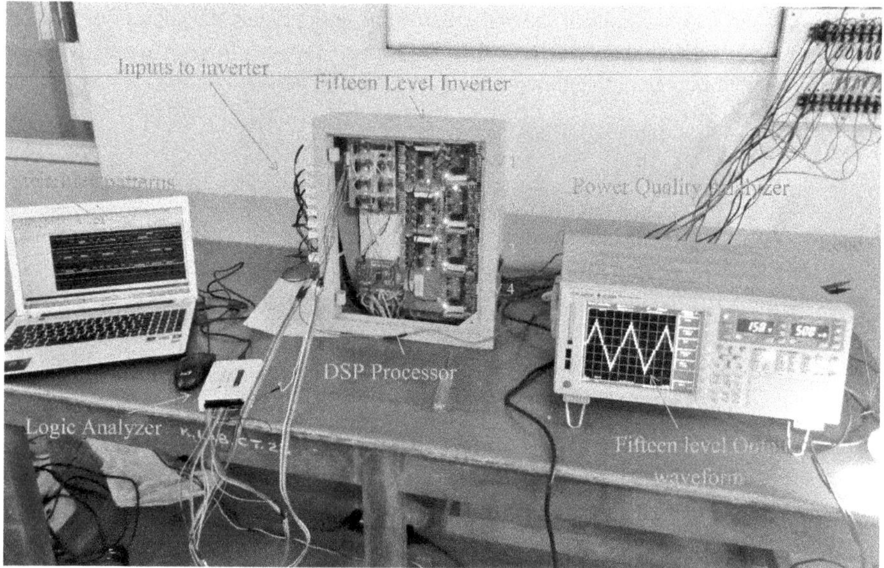

FIGURE 4.10 Experimental setup of the proposed fifteen-level inverter. (*Note*: See color eBook for improved color differentiation in figures and labeling.)

The key issue for a high efficiency and reliability transformerless PV inverter is that, in order to achieve high efficiency over a wide load range, it is necessary to utilize MOSFETs for all switching devices. MOSFETs are utilized in MLIs to reduce the inverter cost or to provide a high bandwidth sinusoidal output voltage at high efficiency. This means that it cannot be achieved with linear amplifiers.

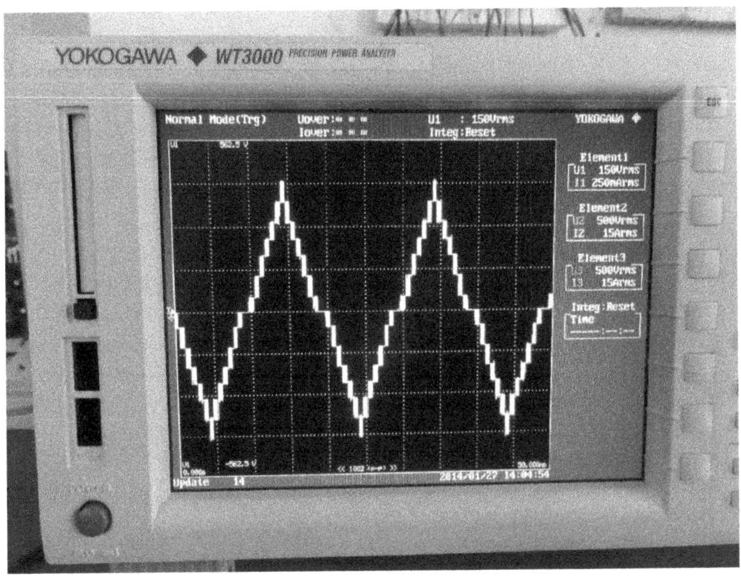

FIGURE 4.11 Fifteen-level inverter output voltage waveform.

4.10 Selective Harmonic Elimination

MLIs have many advantages and many structures. Unfortunately, most of the existing inverters are unable to produce good output waveforms because of their poor THD. This is because each level switching angle is not properly arranged. In order to gain good power quality, it is necessary to arrange the switching angles carefully to achieve lowest THD. Harmonics is defined as the integral multiples of the fundamental frequency and it appears as the waveform distortion of the voltage or current.

The presence of harmonics in inverter output will lead to malfunctioning of microprocessor-based equipment, overheating in neutral conductors, transformers or induction motors, deterioration or failure of power factor correction capacitors, erratic operation of breakers and relays, and pronounced magnetic fields near transformers and switchgear. To eliminate undesired harmonics in MLI with equal DC sources, various modulation methods are presented by Kouro et al. (2007). However, PWM-based methods are not able to eliminate the lower order harmonic completely. In addition, for the methods pertaining to unequal DC sources such as from solar PV a more efficient method is required to deteriorate the harmonics.

SHE is an approach that suppresses the specific order harmonics in the output voltage of the inverter by choosing appropriate switching angles. A fundamental issue associated with such method is to obtain the arithmetic solution of nonlinear transcendental equations, which contain trigonometric terms and naturally present multiple solutions. Dahidah and Agelidis (2008) have put forth a generalized approach for SHE in MLI.

The set of nonlinear equations can be solved by various iterative methods. Fei et al. (2010) have remarked on the disadvantage of iterative methods like Newton–Raphson as their dependence on an initial guess and divergence problems for a large number of inverter levels. In the resultant theory, the transcendental equations that describe the SHE problem are converted into an equivalent set of polynomial equations and then the resultant mathematical theory is utilized to find all possible sets of solutions for this equivalent problem. The resultant theory is limited to finding up to six switching angles for equal DC voltages and up to three switching angles for unequal DC voltage. Moreover, it is complicated and time-consuming, which requires new expression when voltage level or input DC voltage is changed. Step modulation introduced by Liu et al. (2009b) calculates the switching angles in real time but this approach is not extended for unequal DC sources, which are most required for solar-fed CMLI.

4.10.1 Problem Statement

Applying the inequality of DC sources results in the asymmetry of the transcendental equation set to be solved and requires the solution of a set of high degree equations, which are beyond the capability of contemporary computer algebra software tools. This limitation is rectified in the proposed method by adopting the various intelligent techniques such as the ANN and optimization methods (GA, PSO, and BO) in addition to the optimal harmonic stepped waveform (OHSW) strategy for a solar-fed fifteen-level inverter.

In the ANN-based approach the harmonic equations are solved and switching angles are obtained such that the fundamental is kept constant and lower order harmonics are minimized or eliminated. The data sets with varying input voltages and switching angles are trained with ANN. The trained network is integrated with the solar PV system to reduce the harmonic distortions. Similarly, the objective function for the proposed methodology is formulated and corresponding operations are performed with respect to GA, PSO, and BO to obtain the optimal switching angles required for the inverter switches. In OHSW, by considering the symmetrical nature of the waveform the optimal switching angles are obtained. Experimental setup for the method that provides the least THD is implemented for a 3 kWp solar PV plant to show the effectiveness of the proposed system.

By using Fourier series, the staircase output of MLI with non-equal input sources is described in Equation (4.20).

$$V(\omega t) = \sum_{n=1,3,5}^{\infty} \frac{4V_{PV}}{n\pi}\left[k_1\cos(n\alpha_1) + k_2\cos(n\alpha_2) + \cdots + k_m\cos(n\alpha_m)\right]\sin(n\omega t) \tag{4.20}$$

where $k_i V_{PV}$ is the ith DC voltage, n is the order of harmonics, m is the inverter output levels, V_{PV} is the nominal DC voltage from solar PV, and the switching angles α_1 to α_m must satisfy the condition given below.

$$\alpha_1 \le \alpha_2 \le \alpha_3 \le \cdots \le \alpha_m \le \frac{\pi}{2}$$

The number of harmonics that can be eliminated by the MLI is (m–1). For a seven-stage inverter, six harmonic orders can be minimized or eliminated. The expression governing the SHE for a fifteen-level inverter is given in Equation (4.21). Here the objective is to determine the switching angles α_1 to α_7 as depicted in Figure 4.12.

$$V(\omega t) = \sum_{n=1,3,5}^{\infty} \frac{4V_{PV}}{n\pi}\begin{bmatrix} V_{PV1}\cos(n\alpha_1) + V_{PV2}\cos(n\alpha_2) + V_{PV3}\cos(n\alpha_3) + \\ V_{PV4}\cos(n\alpha_4) + V_{PV5}\cos(n\alpha_5) + V_{PV6}\cos(n\alpha_6) + \\ V_{PV1}\cos(n\alpha_7) \end{bmatrix}\sin(n\omega t) \tag{4.21}$$

In CMLI-based PV systems, the operating DC voltages of standard PV cells range from 15 to 35 V and from the CMLI-based energy storage system with batteries, the voltages of batteries also change due to their states of charge (SoC). Hence, the different voltages that arrive at the inverter stages will cause unregulated fundamental with higher magnitude of low order harmonics that also need to be considered. The elimination of triplen harmonics for the three-phase power system applications is not necessary, because these harmonics are automatically eliminated from the line to line voltage.

4.10.2 Optimal Harmonic Stepped Waveform

The harmonic stepped waveform technique employing the quarter wave symmetry concept is much suitable for the MLI to determine the optimal switching angles in order to provide the output voltage waveform with reduced THD. Switching losses can also be minimized by turning the switch ON and OFF only once per cycle.

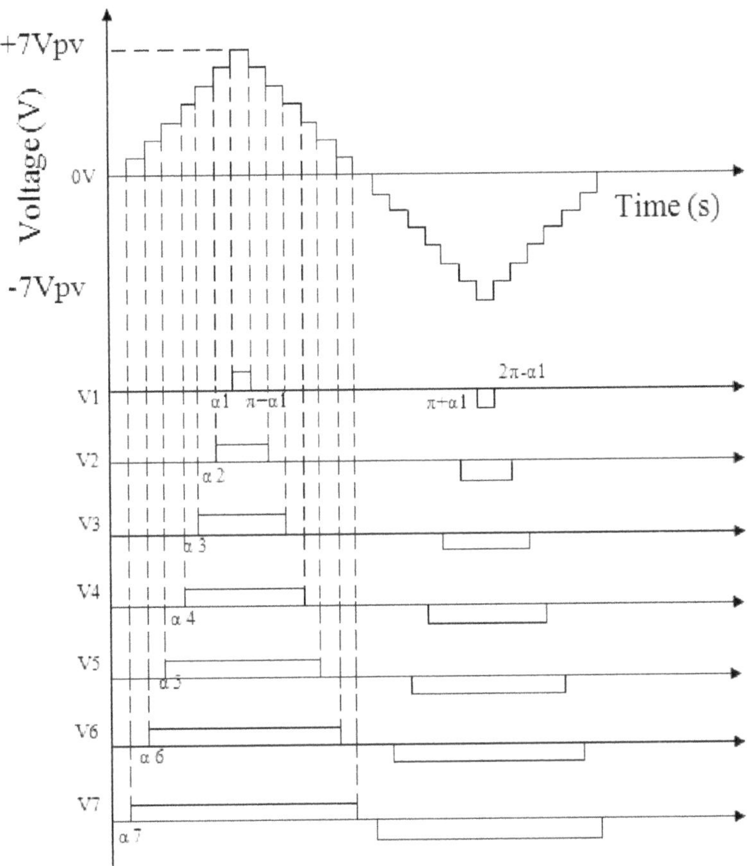

FIGURE 4.12 Output voltage waveform of seven-stage fifteen-level inverter.

There are three possible optimization techniques for reducing the low order harmonics. They are: (1) step heights are optimized with equally spaced steps, (2) step spaces are optimized with the steps of equal height, and (3) optimizing both heights and spaces. In the variable input scheme, the third optimization technique is more feasible than the other two techniques. The Fourier series expression for this method is given in Equation (4.21). OHSW technique with equal input sources is investigated by (Sirisukprasert, 1999).

The OSHW is depicted as quarter wave symmetry where the first half cycle of the output waveform is given in Figure 4.13. Consider the switching angles for the complete cycle as α_1 to α_{28}, in which each set of α can be obtained by sampling the given waveform into four sets or quarters.

In the first quarter, from the interval zero to α_1 the output voltage is zero and at α_1, the output voltage approaches to V_{PV1}. At α_2, the output voltage changes from V_{PV1} to $(V_{PV1}+V_{PV2})$ and the process continues up to $\pi/2$ where the voltage becomes $(V_{PV1}+V_{PV2} + V_{PV3}+V_{PV4}+ V_{PV5}+V_{PV6}+ V_{PV7})$.

In the second quarter the level of output voltage decreases to $(V_{PV1}+V_{PV2} + V_{PV3}+V_{PV4}+ V_{PV5}+V_{PV6})$ at $\pi-\alpha_7$. The process will be repeated up to $\pi-\alpha_1$ where the output voltage again approaches to zero to complete a half cycle. In the next half cycle, the same process will be continued in a similar manner other than the amplitude of the input voltage changing from positive to negative.

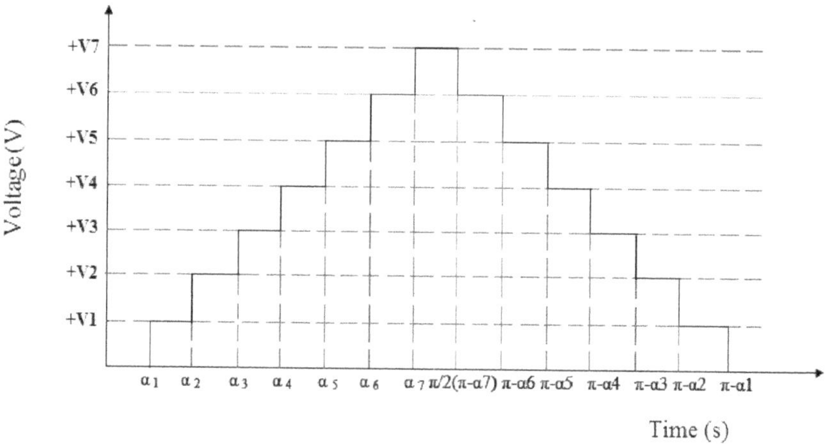

FIGURE 4.13 Quarter wave symmetry waveform for fifteen-level inverter.

First quarter:

Consider the switching angles α_1 to α_7 during the interval 0 to $\pi/2$.

Second quarter:

$$\alpha_8 = \pi - \alpha_7$$

$$\vdots$$

$$\alpha_{13} = \pi - \alpha_2$$

$$\alpha_{14} = \pi - \alpha_1$$

Third quarter:

$$\alpha_{15} = \pi + \alpha_1$$

$$\vdots$$

$$\alpha_{20} = \pi + \alpha_6$$

$$\alpha_{21} = \pi + \alpha_7$$

Fourth quarter:

$$\alpha_{21} = 2\pi - \alpha_7$$

$$\vdots$$

$$\alpha_{27} = 2\pi - \alpha_7$$

$$\alpha_{28} = 2\pi + \alpha_1$$

While considering the harmonic contents in the output waveform, the amplitude of DC components and all even order harmonics approach near zero. Hence the odd harmonic in the quarter wave symmetric CMLI is to be estimated. The set of nonlinear equations to obtain optimal switching angles is given in Equations (4.22) to (4.28).

$$V_{PV1}\cos(\alpha_1)+V_{PV_2}\cos(\alpha_2)+\cdots+V_{PV7}\cos(\alpha_7)=\frac{M\pi}{4} \tag{4.22}$$

$$V_{PV1}\cos(3\alpha_1)+V_{PV2}\cos(3\alpha_2)+\cdots+V_{PV7}\cos(3\alpha_7)=0 \tag{4.23}$$

$$V_{PV1}\cos(5\alpha_1)+V_{PV2}\cos(5\alpha_2)+\cdots+V_{PV7}\cos(5\alpha_7)=0 \tag{4.24}$$

$$V_{PV1}\cos(7\alpha_1)+V_{PV2}\cos(7\alpha_2)+\cdots+V_{PV7}\cos(7\alpha_7)=0 \tag{4.25}$$

$$V_{PV1}\cos(7\alpha_1)+V_{PV2}\cos(7\alpha_2)+\cdots+V_{PV7}\cos(7\alpha_7)=0 \tag{4.26}$$

$$V_{PV1}\cos(11\alpha_1)+V_{PV2}\cos(11\alpha_2)+\cdots+V_{PV7}\cos(11\alpha_7)=0 \tag{4.27}$$

$$V_{PV1}\cos(13\alpha_1)+V_{PV2}\cos(13\alpha_2)+\cdots+V_{PV7}\cos(13\alpha_7)=0 \tag{4.28}$$

where $M = \frac{h_1}{V_{PV(1-7)}}$

h_1 is the amplitude of the fundamental component. As the proposed method involves the variable input sources, the V_{PV} is taken at the left-hand side in the above expressions. From Equation (5.3), varying the modulation index (M) it can control the amplitude of the fundamental component. The nonlinear equations listed are the undesirable harmonic components, which can be eliminated and are set to be zero.

The Newton–Raphson (NR) method is used to solve the nonlinear equation systems using successive approximation procedure, which is suitable for implementing in a computer program. The algorithm to calculate optimal switching angles involves Equations (4.29) to (4.39).

Step 1:

Formation of switching angle matrix is given in Equation (4.29) where the unknown switching angles α_1 to α_7 are to be determined.

$$\alpha^j=\left[\alpha_1^j,\alpha_2^j,\alpha_3^j,\alpha_4^j,\alpha_5^j,\alpha_6^j,\alpha_7^j\right]^T \tag{4.29}$$

Step 2:

Formation of nonlinear system matrix as given in Equations (4.30) and (4.31).

$$F^j=\begin{bmatrix}\cos(\alpha_1^j)+\cos(\alpha_2^j)+\cdots+\cos(\alpha_7^j)\\ \cos(3\alpha_1^j)+\cos(3\alpha_2^j)+\cdots+\cos(3\alpha_7^j)\\ \cos(5\alpha_1^j)+\cos(5\alpha_2^j)+\cdots+\cos(5\alpha_7^j)\\ \cos(7\alpha_1^j)+\cos(7\alpha_2^j)+\cdots+\cos(7\alpha_7^j)\\ \cos(9\alpha_1^j)+\cos(9\alpha_2^j)+\cdots+\cos(9\alpha_7^j)\\ \cos(11\alpha_1^j)+\cos(11\alpha_2^j)+\cdots+\cos(11\alpha_7^j)\\ \cos(13\alpha_1^j)+\cos(13\alpha_2^j)+\cdots+\cos(13\alpha_7^j)\end{bmatrix} \tag{4.30}$$

$$\left[\frac{\partial F}{\partial \alpha}\right]^j = \begin{bmatrix} -\sin(\alpha_1^j),-\sin(\alpha_2^j),\ldots,-\sin(\alpha_7^j), \\ -\sin(3\alpha_1^j),-\sin(3\alpha_2^j),\ldots,-\sin(3\alpha_7^j) \\ -\sin(5\alpha_1^j),-\sin(5\alpha_2^j),\ldots,-\sin(5\alpha_7^j) \\ -\sin(7\alpha_1^j),-\sin(7\alpha_2^j),\ldots,-\sin(7\alpha_7^j) \\ -\sin(9\alpha_1^j),-\sin(9\alpha_2^j),\ldots,-\sin(9\alpha_7^j) \\ -\sin(11\alpha_1^j),-\sin(11\alpha_2^j),\ldots,-\sin(11\alpha_7^j) \\ -\sin(13\alpha_1^j),-\sin(13\alpha_2^j),\ldots,-\sin(13\alpha_7^j) \end{bmatrix} \tag{4.31}$$

Step 3:

Formation of harmonic amplitude matrix as per the expression given in Equation (4.32).

$$T = \left[\frac{M \times 3 \times \pi}{4} 0 \cdot 0 \cdot 0 \cdot 0 \cdot 0 \cdot 0\right]^T \tag{4.32}$$

Equation (4.3) and (4.13) can be rewritten as per Equation (4.33).

$$F(\alpha) = T \, (4.33)$$

Using the matrices given in Equations (4.29)–(4.33) and NR method, the statement of algorithms is as follows to incorporate it in the program using MATLAB software.

Step 1:

Guess for initial values of α as per Equation (4.34). The process of finding the initial guess through equal area criterion is given by Wang and Ahmadi (2010).

$$\alpha^0 = \left[\alpha_1^0, \alpha_2^0, \alpha_3^0, \alpha_4^0, \alpha_5^0, \alpha_6^0, \alpha_7^0\right] \tag{4.34}$$

Step 2:

This step includes Equations (4.35) to (4.39).

$$F(\alpha_0) = F^0 \, \mathrm{W} \tag{4.35}$$

Linearizing the Equation (4.33) at α_0 we get,

$$F^0 + \left[\frac{\partial f}{\partial \alpha}\right]^0 d\alpha^0 = T \tag{4.36}$$

and

$$d\alpha^0 = \left[d\alpha_1^0, d\alpha_2^0, d\alpha_3^0, d\alpha_4^0, d\alpha_5^0, d\alpha_6^0, d\alpha_7^0\right] \tag{4.37}$$

Solving Equation (4.37) using the inverse of the matrix, which is given in Equation (4.38).

$$d\alpha^0 = \text{Invmatrix}\left[\frac{\partial f}{\partial x}\right]^0 \left[T - F^0\right] \tag{4.38}$$

Step 3:

Updating of initial values is made as per Equation (4.39).

$$\alpha^{j+1} = \alpha^j + d\alpha^j \tag{4.39}$$

The steps (2)–(3) are repeated until $d\alpha^j$ is fulfilled to the degree of accuracy and the condition to be satisfied is $\alpha_1, \alpha_2, \alpha_3, \ldots, \alpha_7 < \pi/2$.

4.10.3 Artificial Neural Network

An ANN is composed of simple information processing elements known as "neurons" operating in parallel that tend to emulate the human brain. They are trained to perform a particular function, which is done by adjusting the values of connection "weights" between the elements. The advantage of the ANN approach is its capability to auto tune the application with no requirement of explicit function required for the control. Neural nets are trained by providing sample classification data and making adjustments in their weight vectors so that they become "experienced" enough to classify unknown patterns successfully. Learning can be either supervised or unsupervised depending on the availability of training data, which is the set of pairs of input output vectors. In the presence of such data, the ANN can measure the deviation from desired output values when the net is presented with an input pattern.

The characteristics of the ANN can be summarized into the following three entities.

- Architecture: pattern of connection between the neurons, which may be either feed forward or recurrent networks
- Activation function: a function of the net input to the processing unit, which maps the net input value to the output signal value, i.e., the activation
- Learning methods: method of determining the appropriate set of weights on the interconnections

Neural networks can be trained online or offline. In online learning, the weight changes are applied to the network after each training pattern, i.e., after each forward and backward pass. In offline learning, the weight changes are accumulated for all patterns in the training file and the sum of all changes is applied after one full cycle (epoch) through the training pattern file. Prior to the implementation of the ANN-based switching angle determination method for CMLI, the initial step includes the choice of input–output parameters and their data collection.

4.10.4 Data Set Collection

The dataset required for the ANN training is obtained by solving the harmonic elimination equations given from Equations (4.4) through (4.9) along with the basic expression given in Equation (4.2) by considering the variation of input voltages at all the solar panels. For the given variation in the input voltages it results in seven switching angles. Harmonic elimination equations are solved in MathCAD Prime 2.0 because it solves the linear and nonlinear system of equations within a fraction of a second. It provides the results in milliseconds after giving the guess values as input that has to be converted into equivalent angles in radians before being delivered to the respective inverter switches.

For a fifteen-level inverter, the harmonic elimination equations are solved. The guess values provided for solving the equations in MathCAD are $\alpha_1 = 11.25°$, $\alpha_2 = 22.5°$, $\alpha_3 = 33.75°$, $\alpha_4 = 45°$, $\alpha_5 = 56.25°$, $\alpha_6 = 67.5°$, $\alpha_7 = 78.75°$. In the equations, it is clear that by providing the fundamental voltage with desired value and other harmonic components equal to zero, the lower order harmonic contents such as 3rd, 5th, 7th, 9th, 11th, and 13th are eliminated such that the desired fundamental is kept constant. Some of the datasets obtained are listed in Table 4.3. Nearly 1500 datasets are collected in such a way that for the corresponding input voltages (V_{PV}) from the solar panels the corresponding set of switching angle (α in radians) values are to be obtained.

In order to reduce the size of the data set, Filho et al. (2011) considered this problem as a combination instead of permutation. If not for a seven-stage converter with ten points equally spaced between 48 and 58 V, it would generate 10^5 different combinations. Hence, the data set can be reduced by considering the voltage vector $v_1 = [40.4, 52.6, 52.7, 41.6, 53.7, 42.8, 42.5]$ as being the same input as $v_2 = [52.6, 40.4, 42.5, 52.7, 42.8, 53.7, 41.6]$.

TABLE 4.3
Sample Data Set Required for ANN Training

Input Voltage ($V_{PV(1-7)}$)	Switching Angles (α_1 to α_7 In Radians)
[40.4, 52.6, 52.7, 41.6, 53.7, 42.8, 42.5]	[0.112, 0.198, 0.432, 0.587, 0.812, 1.104, 1.53]
[42.4, 52.6, 52.7, 43.6, 53.7, 45.8, 42.5]	[0.117, 0.196, 0.427, 0.584, 0.802, 1.093, 1.526]
[44.4, 52.6, 52.7, 45.6, 53.7, 40.8, 44.5]	[0.136, 0.19, 0.444, 0.594, 0.834, 1.117, 1.533]
[44.4, 40.6, 41.7, 42.6, 52.7, 40.8, 44.5]	[0.111, 0.211, 0.413, 0.559, 0.798, 1.1, 1.535]
[51.4, 43.6, 49.9, 50.8, 42.8, 51.5, 48.5]	[0.162, 0.166, 0.417, 0.598, 0.782, 1.062, 1.523]

4.10.5 ANN Architecture

The ANN architecture considered is a multilayered feed-forward network as shown in Figure 4.14. There are two hidden layers with 20 neurons in the first hidden layer and 10 neurons in the second hidden layer with activation function "tansig" in both and "purelin" activation function with seven neurons at the output layer. The input PV voltage and switching angle data sets obtained from the MathCAD software are used to train the ANN by Levenberg Marquart back-propagation (LMBPN) algorithm. Table 4.4 shows the specifications of the ANN.

Levenberg Marquart (LM) algorithm is the combination of the steepest descent (error back propagation [BPN]) and Gauss–Newton algorithm and switches between these two methods during training process. LM is the fastest back-propagation algorithm for training moderate sized feed-forward neural networks and they are highly recommended for supervised learning applications. The standard BPN algorithm comprises three steps such as feed-forward of input data, error-back propagation, and weight updating. In addition to these steps, LM algorithm was designed to approach second order training speed

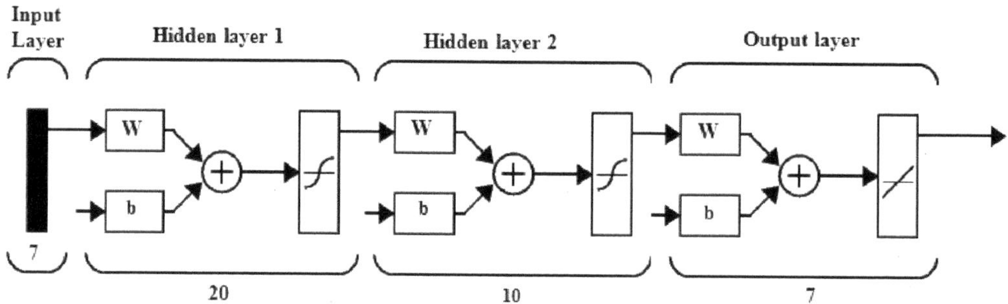

FIGURE 4.14 ANN architecture.

TABLE 4.4
ANN Specifications

Parameters	Values
Input layer	7 (Input PV voltages)
Hidden layer 1	20
Hidden layer 2	10
Output layer	7 (switching angles)
Activation function (input–hidden1)	Tansig
Activation function (hidden1–hidden2)	Tansig
Activation function (hidden2–output)	Purelin
Training algorithm	Levenberg Marquart
Performance parameter	Mean square error

without computing the Hessian matrix. The advantage of LM algorithm includes faster, stable convergence and it does not require more memory when compared to other algorithms.

The trained network is then validated for its performance and then Simulink model of the trained network is generated. For the generated model, when provided with input voltage values, the network provides the corresponding switching angles. The switching angles generated by the neural network are converted to the corresponding gating pulses by triggering circuit and then given to the gate terminal of the MOSFET switches in the cascaded H bridge inverter. The triggering circuit consists of a triangular waveform, which is compared with the modulating signal whose variations are determined by the angle sets obtained from the neural network. The pulses are produced that are then fed to the gates of the MLI switches. Based on the variation in the input to the inverter, the corresponding switching angles are generated. This results in the minimum THD and also the elimination of selective order harmonics thereby improving the quality of power in solar PV.

4.11 Optimization Techniques

Optimization is the process of obtaining the best result under any given circumstances. It is extensively used in science and technical fields to minimize the effort required and maximize the desired benefit. The formulation of optimization problem begins with identifying the underlying design variables, which are primarily used during the optimization process. The next task is to identify the constraints associated with the optimization problem, which represents some functional relationship among the design variables and other design parameters satisfying certain physical phenomena and resource limitations.

Constraints are of two types: equality and inequality. Equality constraints are usually more difficult to handle and need to be avoided. The third task in the formulation procedure is to find the objective function in terms of the design variables and other problem parameters. The objective function can be of two types, either maximization or minimization. Based on the duality principle, the algorithm developed for the minimization problem can be used to solve the maximization problem by simply multiplying the objective function by –1 and vice versa. The final task is to set the minimum and maximum bounds on each design variable, but certain optimization algorithms do not require this information.

4.11.1 Problem Formulation

For a solar-fed fifteen-level inverter, the objective is to find the optimal switching angles (α_1, α_2, α_3, α_4, α_5, α_6 and α_7) that will minimize the THD. From Figure 4.2, the expression for the RMS value of fundamental and harmonic components of the phase voltage is obtained through Fourier analysis as given in Equation (4.40).

$$V_n = \begin{cases} \dfrac{2\sqrt{2}}{n\pi} V_{PV(1-7)}\left(\mathrm{Cos}(n\alpha_1) + \mathrm{Cos}(n\alpha_2) + \mathrm{Cos}(n\alpha_3) + \ldots.. + \mathrm{Cos}(n\alpha_7)\right), \text{odd} \\ 0, \text{even} \end{cases} \quad (4.40)$$

Considering the waveform of quarter wave symmetry, the maximum possible value of the fundamental component is given in Equations (4.41) and (4.42) by equating all the switching angles to zero (i.e., α_1 to $\alpha_7 = 0$).

$$V_{1\max} = \frac{2\sqrt{2}}{\pi}.\left(V_{PV(1-7)} \times 7\right) \quad (4.41)$$

$$V_{1\max} = \frac{14\sqrt{2}}{\pi} V_{PV(1-7)} \quad (4.42)$$

Equation (4.43) shows the normalization of the fundamental component based on its maximum value and Equation (4.44) describes the RMS value of the phase voltage. Equation (4.45) shows the expression for the resultant V_{RMS}.

$$V_1(p.u.) = \frac{V_1}{V_{1(\max)}} = \frac{2\sqrt{2}V_{PV(1-7)}}{n\pi} \times \frac{\pi}{14\sqrt{2}V_{PV(1-7)}}$$

$$V_1(p.u.) = \frac{V_1}{V_{1(\max)}} = \frac{1}{7}\{(\cos\alpha_1) + (\cos\alpha_2) + (\cos\alpha_3) + \cdots + (\cos\alpha_7)\} \tag{4.43}$$

$$V_{\text{RMS}} = \sqrt{\frac{2}{\pi}\int_0^{\pi/2} V^2 d\omega t} \tag{4.44}$$

$$V_{\text{RMS}} = V_{PV(1-7)}\sqrt{\frac{2}{\pi}(\alpha_2-\alpha_1)+4(\alpha_3-\alpha_2)+9(\alpha_4-\alpha_3)+16(\alpha_5-\alpha_4) + 25(\alpha_6-\alpha_5)+36\left(\frac{\pi}{2}-\alpha_7\right)} \tag{4.45}$$

In per unit, by dividing Equation (4.45) by (4.42), the resultant RMS voltage is given in Equation (4.46).

$$V_{\text{rms}}^{p.u.} = \frac{V_{\text{RMS}}}{V_{1(\max)}}$$

$$V_{\text{rms}}^{p.u.} = V_{PV(1-7)}\sqrt{\frac{\pi}{196}\begin{bmatrix}(\alpha_2-\alpha_1)+4(\alpha_3-\alpha_2)+9(\alpha_4-\alpha_3)+ \\ 16(\alpha_5-\alpha_4)+25(\alpha_6-\alpha_5)+36\left(\frac{\pi}{2}-\alpha_7\right)\end{bmatrix}} \tag{4.46}$$

THD is defined as the ratio of all harmonic components of RMS value given in Equation (4.44) to the fundamental component of RMS value given in Equation (4.46). It is expressed in the Equation (4.47).

$$\text{THD} = \frac{\sqrt{\sum_{n=2}^{\infty} V_n^2}}{V_1}$$

$$\text{THD} = \sqrt{\frac{\pi}{4}\frac{\begin{bmatrix}(\alpha_2-\alpha_1)+4(\alpha_3-\alpha_2)+9(\alpha_4-\alpha_3)+ \\ 16(\alpha_5-\alpha_4)+25(\alpha_6-\alpha_5)+36(\pi/2-\alpha_7)\end{bmatrix}}{(\cos\alpha_1)+(\cos\alpha_2)+(\cos\alpha_3)+(\cos\alpha_4)+ (\cos\alpha_5)+(\cos\alpha_6)+(\cos\alpha_7)} - 1} \tag{4.47}$$

For a fifteen-level inverter, the objective function and constraints are given in Equations (4.48) and (4.49), respectively.

$$\text{Min } Z = \{W(k) \times (\text{per unit equivalent} - \text{Equation (4.43)}) + \text{Equation (4.47)}\} \tag{4.48}$$

Constraints

$$0 \le \alpha 1, \ldots, \alpha 7 \le \frac{\pi}{2} \tag{4.49}$$

where, α_1, α_2, α_3, α_4, α_5, α_6, and α_7 are the switching angles and $W(k)$ is the weighting factor, which is used for making the error small enough and per unit equivalent of the output peak voltage whose value is between 0 and 1. In the objective function, the term (per unit equivalent – Equation [4.43]) is the absolute value of error in adjusting the fundamental component.

4.11.2 Genetic Algorithm

GA is a derivative-free optimization stochastic method used for solving both constrained and unconstrained optimization problems that are based on natural selection, the process that drives biological evolution. It handles both maximization and minimization problems. The GA repeatedly modifies a population of individual solutions. At each step, GA selects individuals at random from the current population to be parents and uses them to produce the children for the next generation. Over successive generations, the population "evolves" toward an optimal solution.

GA was initially developed by John Holland, University of Michigan in 1975. GAs encode each point in a solution space into a binary bit string called chromosomes, and each point is associated with a fitness value that, for maximization, is usually equal to the objective function evaluated at the point. GAs usually keep a set of points as a population (or gene pool), which is then evolved repeatedly toward a better overall fitness value. In each generation, the GA constructs new population using genetic operators such as cross over and mutation. The members with higher fitness value are more likely to survive and participate in mating operations. After a number of generations, the population contains members with better fitness values.

4.11.3 Computation of Switching Angles

GA is a search heuristic that mimics the process of natural evolution. In the proposed method, GA is used for determining the optimal switching angles required for a fifteen-level inverter. GA consists of 6 major stages: (1) generation of initial chromosomes, (2) population, (3) fitness evaluation, (4) crossover, (5) mutation, and (6) termination.

4.11.3.1 Generation of Initial Chromosomes

The initial process that takes place in GA is the generation of initial chromosomes. The initial chromosomes are generated between the minimum and maximum values of the entire chromosome. A fifteen-level inverter requires seven H bridges; thus, each chromosome for this application will have seven switching angles, i.e., α_1 to α_7. The number of variables specific to the problem is the number of genes in the chromosome, which is the number of H bridges in a CMLI. After generating the initial chromosomes, the next step is to compute the fitness function.

4.11.3.2 Population

The GA starts with a group of chromosomes known as population. The size of the population is an important parameter that needs to be tuned. The complexity of the problem, which is reflected by the size of the search space, is a factor to be considered while fixing the size of the population. The initial population is normally randomly generated. Higher population might increase the rate of convergence but it also increases the execution time. The population considered has 20 chromosomes, each containing seven switching angles. The population is initialized with random angles between 0° and 90° taking into consideration the quarter-wave symmetry of the output voltage waveform.

4.11.3.3 Fitness Function

Fitness function is one of the most important processes in GA applied to identify the best chromosome. Fitness functions are objective functions that are used to evaluate a particular solution represented as chromosomes in a population. There are two types of fitness functions. The first one, the fixed type, does not allow the fitness function to change. In the second case, the fitness function is mutable. The most important factor in deciding a fitness function is that it should correlate to the problem closely and be simple enough to be computed quickly.

The most important item for the GA to evaluate the fitness of each chromosome is the cost function. The objective of the proposed method is to minimize specified harmonics; therefore, the cost function has to be related to these harmonics. In this work, the 3rd, 5th, 7th, 9th, 11th, and 13th harmonic orders at the output of a fifteen-level inverter are to be minimized. Then the cost function (*f*) can be selected as the sum of these two harmonics normalized to the fundamental given in Equation (4.50).

$$f(\alpha_1,\alpha_2,\alpha_3,\alpha_4,\alpha_5,\alpha_6,\alpha_7)=100.\frac{|V_3|+|V_5|+|V_7|+|V_9|+|V_{11}|+|V_{13}|}{|V_1|} \tag{4.50}$$

For each chromosome a multilevel output voltage waveform is created using the switching angles in the chromosome and the required harmonic magnitudes are calculated using FFT techniques. The switching angle set producing the maximum fitness value (minimum THD) is the best solution of the first iteration.

4.11.3.4 Crossover Operation

The crossover operation is performed between two chromosomes to obtain a new chromosome. The crossover operation is done based on the crossover rate. Based on this crossover, the genes are selected and a new child chromosome is generated. After generating a new chromosome, the fitness function is applied to this new child chromosome.

4.11.3.5 Mutation Operation

The mutation operator imparts a small change at a random position within a chromosome. The mutation operation is performed based on the mutation rate, which is usually kept very low. The mutation is done by mutating the genes randomly based on the given mutation rate.

4.11.3.6 Termination

In the termination stage, the best chromosome is selected based on the fitness function. The above process is repeated until it reaches the maximum number of iterations. After the completion of termination process, the best switching angles are obtained.

4.11.4 Particle Swarm Optimization

PSO is a population-based, stochastic, optimization technique developed by Dr. James Kennedy and Dr. Russell Eberhart in 1995. PSO is inspired by the social behavior of bird flocking or fish schooling. The PSO algorithm maintains a swarm of individuals (called particles), where each particle represents a candidate solution. In PSO, the potential solutions, called particles, fly through the problem space by following the current optimum particles. Particles follow a simple behavior: emulate the success of neighboring particles and their own achieved successes.

The modeling of PSO includes the following functional parameters. A particle j in the swarm has three attributes, such as:

1. Position in search space X_j
2. Velocity Y_j
3. Personal best Pbest_j: encountered by the particle so far

Another important attribute of the swarm is the global best position Gbest, which is the best position of the particles in the entire population. The position and velocity are updated by Equations (4.51) and (4.52), respectively.

$$y_{ij}(k+1)=w(k)y_{ij}(k)+C_1r_1(P\text{best}_{ij}(k)-X_{ij}(k))+C_2r_2\big(G\text{best}_i(k)-X_{ij}(k)\big) \tag{4.51}$$

$$X_{ij}(k+1)=X_{ij}(k)+Y_{ij}(k+1) \tag{4.52}$$

where k is the number of iteration, $w(k)$ is the inertia constant or weighting function, C_1 and C_2 are the weighting factors or acceleration coefficients, and r_1 and r_2 are the uniformly distributed random numbers within [0 1].

Equation (4.51) is composed of three attributes such as inertial component, cognitive component, and social component. The inertial behavior of the bird to fly in the previous direction is called inertial component. Cognitive component includes the memory of the bird about its previous position whereas social component denotes the memory of the bird about the best position among the particles and collaborative behavior of the particles to find the global optimal solution. Equation (4.52) specifies the calculation of new velocity by the summation of previous position and new velocity.

The algorithmic steps involved in solving the SHE problem for a solar-fed fifteen-level inverter are as follows:

Step 1:

Start with random position (X_i) and velocity (V_i) vectors. Each particle in the population is randomly initialized between 0° and 90°. Similarly, the velocity vector of each particle has to be generated randomly between $-V_{max}$ and V_{max}.

Step 2:

Evaluate the particles using the fitness function of the harmonic minimization problem. The switching angles α_1 to α_7 can be calculated such that the selected harmonic orders can be eliminated from the output phase voltage of the CMLI to minimize the cost function. The fitness function given in Equation (4.53) is derived based on the expression used for an eleven-level inverter by Taghizadeh and Hagh (2010).

$$f(\alpha_1,\ldots,\alpha_7)=100\times\left[\left|M-\frac{|V_1|}{V_{PV(1-7)}}\right|+\left(\frac{|V_3|+|V_5|+|V_7|+|V_9|+|V_{11}|+|V_{13}|}{|V_{PV(1-7)}|}\right)\right] \tag{4.53}$$

where M is the modulation index. On evaluation of fitness function, *P*best and *G*best can be obtained.

Step 3:

Update the Pbest (P_i) position of the particles. If the current position of the ith particle is better than its previous personal best position, replace P_i with the current position X_i. In addition, if the best position of the personal bests of the particles is better than the position of the global best, replace Gbest (P_g) with the best position of the personal bests.

Step 4:

Update the velocity and position vectors. All the particles in the population are updated by velocity and position update rules given in Equations (4.51) and (4.52).

Step 5:

The maximum number of iteration counts is set as termination criteria. If the iteration counter reaches maximum, stop; else, increase the iteration counter and go back to step 2.

4.11.5 Bees Optimization

BO is a derivative-free optimization method developed by Pham in 2007. This algorithm is inspired by honey bees' foraging behavior. In nature, bees are well-known as social insects with well-organized colonies. Their behaviors such as foraging, mating, and nest site location have been used by researchers to solve many difficult combinatorial optimization and functional optimization problems.

The BO has proved to give a more robust performance than other intelligent optimization methods for a range of complex problems.

4.11.6 Natural World of Bees

A colony of honey bees can disperse in multiple directions simultaneously to exploit a large number of food sources. In principle, flower patches with plentiful amounts of nectar or pollen that can be collected with less effort should be visited by more bees, whereas patches with less nectar or pollen should receive fewer bees.

A bee colony consists of three kinds of bees: employed bees, on-looker bees, and scout bees. Employed bees carry information about place and amount of nectar in a particular food source. They transfer the information to on-looker bees with a dance in the hive. The time of dance determines the amount of nectar in a food source. This dance contains three pieces of information regarding a flower patch: its distance from the hive, the direction in which it will be found and its quality rating (or fitness). This dance is necessary for colony communication and the information helps the colony to send its bees to flower patches precisely, without using guides or maps.

The information provided from the dance enables the colony to evaluate the relative merit of different patches according to both the quality of the food they provide and the amount of energy needed to harvest it. An on-looker chooses a food source based on the amount of nectar in a food source. A good food source attracts more on-looker bees to itself.

Scout bees seek in search space and find new food sources. Scout bees control the exploring process, while employed and on-looker bees play an exploiting role. In this algorithm, food sources are considered as possible solutions to a problem. The food source is a D-dimensional vector, where D is the number of optimization variables. The amount of nectar in a food source determines the value of fitness as expressed by Kavousi et al. (2012).

4.11.7 Computation of Switching Angles

A colony prospers by deploying its foragers to good fields. In the proposed method, BO algorithm is used to determine the optimal switching angles required for a solar-fed fifteen-level inverter. The bees algorithm consists of the following steps:

Step 1:

Initialization of initial food sources randomly. The number of initial food sources is half of the bee colony.

Step 2:

Employed bees are sent to the food sources to determine the amount of nectar and calculate its fitness. For each food source, there is only one employed bee. The number of food sources is equal to the number of employed bees. The employed bees modify the solutions, saved in memory, by searching in the neighborhood of its food source. The employed bees save the new solution if its fitness is better than the older one. Employed bees go back to the hive and share the solutions with the onlooker bees.

Step 3:

On-looker bees, which are another half of the colony, select the best food sources using a probability-based selection process. Food sources with more nectar attract more on-looker bees. On-looker bees are sent to the selected food sources.

The on-looker bees improve the chosen solutions and calculate their fitness. Similar to employed bees, the on-looker bees save a new solution if its fitness is better than an older solution.

Step 4:

The food sources that are not improved for a number of iterations are abandoned. So, the employed bee is sent to find new food sources as a scout bee. The abandoned food source is replaced by the new food source.

Step 5:

The best food source is memorized. The maximum number of iterations is set as a termination criterion, which is checked at the end of the iteration. If it is not met, the algorithm returns to step 2 for the next iteration.

4.12 Simulation Results

Simulation is carried out in MATLAB/Simulink 2013a for all three topologies by using the solar PV panel as the input. The seven-stage inverter, which provides 15 levels, is taken into consideration, which has the capability to minimize six numbers of harmonic orders in the output voltage and also reduces the THD.

4.12.1 Optimal Harmonic Stepped Waveform

The program depicting Equations (4.29) through (4.39) is written in MATLAB-M file editor. Based on the initial guess value, the execution of the program results in the optimal switching angles. Initially the program results in radiant, which is then converted to degree, followed by time equivalent conversion for providing the signals to the inverter switches.

A programmable pulse generator (PPG) is proposed that will directly feed the optimal switching signals to the switches without any delay after the successive solutions have been arrived based on this algorithm. Figure 4.15 shows the PPG where the resultant switching angles stored in the array are directed towards inverter gates after the conversion segments.

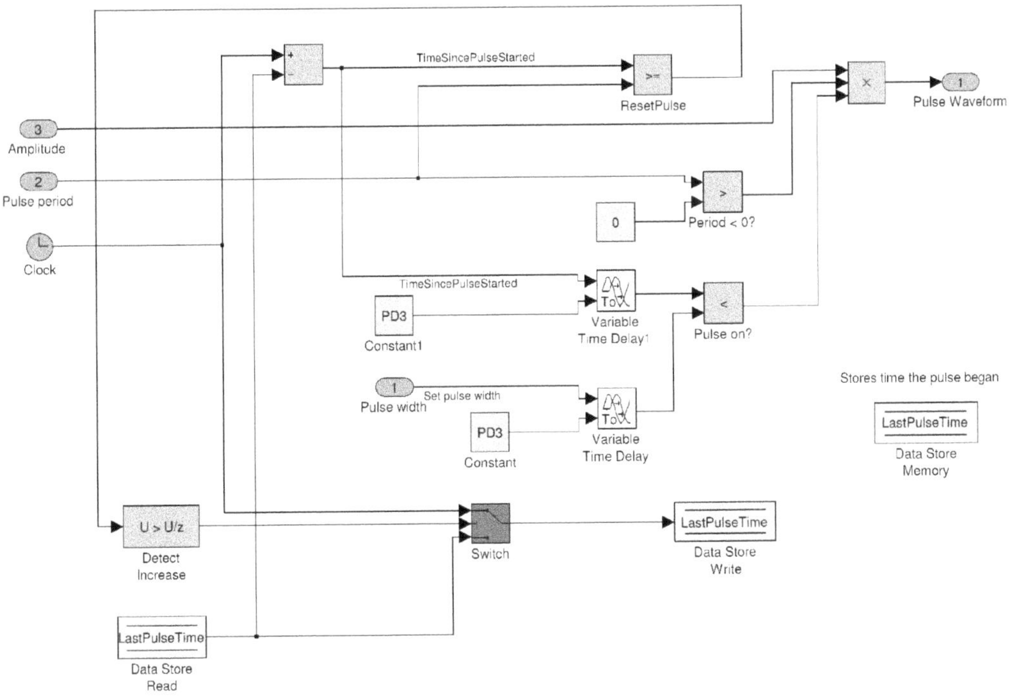

FIGURE 4.15 Programmable pulse generator. (*Note*: See color eBook for improved color differentiation in figures and labeling.)

TABLE 4.5
Switching Angles Determination

		M = 0.36		
α	1st Quarter	2nd Quarter	3rd Quarter	4th Quarter
α_1	8.857	171.1143	188.8857	351.1143
α_2	14.8027	165.1973	194.8027	345.1973
α_3	27.9257	152.0743	207.9257	332.0743
α_4	47.2589	132.7411	227.2589	312.7411
α_5	57.27771	122.7229	237.2771	302.7229
α_6	72.8232	107.1768	252.8232	287.1768
α_7	87.8539	92.1461	267.8539	272.1461
		M = 1		
α_1	5.0126	174.9874	185.0126	354.9874
α_2	14.2451	165.7549	194.2451	345.7549
α_3	25.9953	154.0047	205.9953	334.0047
α_4	37.6040	142.3960	217.6040	322.3960
α_5	52.0406	127.9594	232.0406	307.9594
α_6	65.0413	114.9587	245.0413	294.9587
α_7	86.2279	93.7721	266.2279	273.7721

Table 4.5 displays the various sets of switching angles obtained for the modulation index $M = 0.36$ and $M = 1$. The advantage of the proposed method needs to satisfy the condition $\alpha_1, \alpha_2, \alpha_3 \ldots \alpha_7 < \pi/2$ rather than the condition $\alpha_1 \leq \alpha_2 \leq \alpha_3 \leq \ldots \leq \alpha_m \leq \pi/2$ for the other SHE problems.

Figure 4.16 shows the output voltage waveform of a fifteen-level inverter and its FFT analysis is given in Figure 4.17, which shows the THD value of 8.55%. Adopting OHSW technique by (Sirisukprasert, 1999) the THD obtained is about 32.78% for a seven-level inverter whereas for a solar-fed fifteen-level inverter the THD obtained is 8.55%.

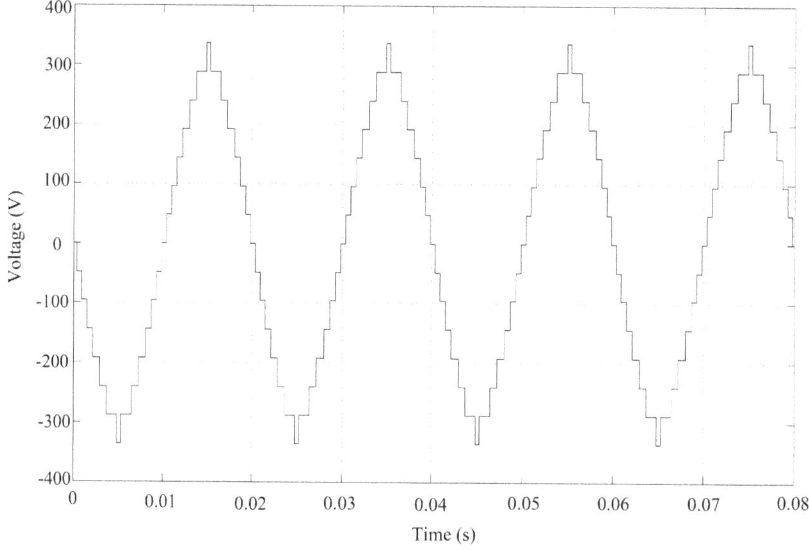

FIGURE 4.16 Fifteen-level output voltage waveform with OHSW technique.

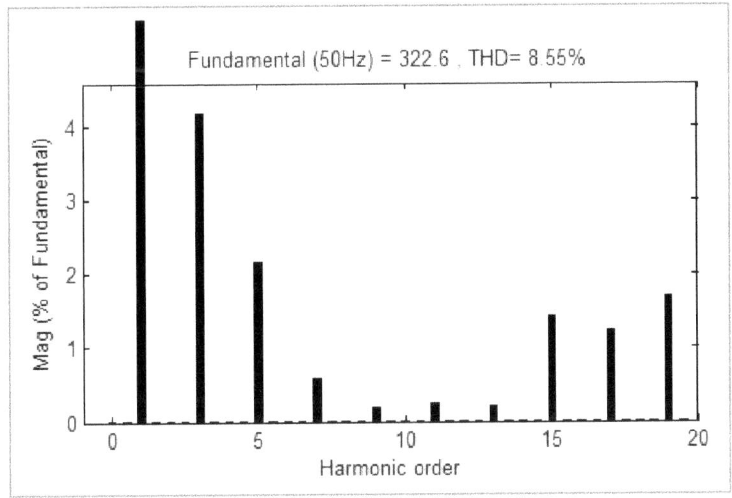

FIGURE 4.17 FFT analysis of the output waveform with OHSW.

4.12.2 Artificial Neural Networks

The simulation and analysis of ANN-based solar-fed CMLI are developed for both eleven levels and fifteen levels, respectively. In an eleven-level inverter, by solving four harmonic equations under various input PV voltage conditions the data set is obtained. Neural network is trained with these datasets and the trained network is then implemented with CMLI. The sets of harmonic orders considered are 5, 7, 11, and 13. The output voltage waveform of an eleven-level CMLI is shown in Figure 4.18 and its corresponding FFT analysis is given in Figure 4.19. Figure 4.20 shows the FFT analysis of CMLI without adopting the ANN technique. The input PV voltages considered are {43 V, 50 V, 48 V, 43 V, 40 V}. The results show

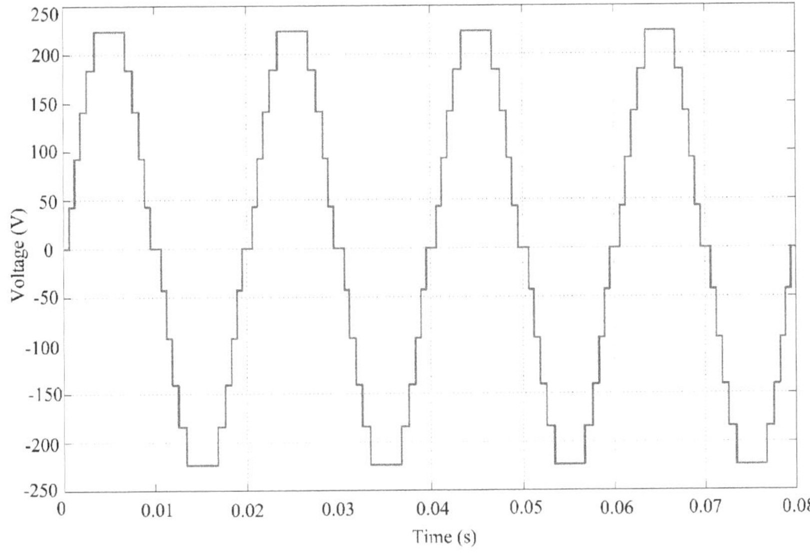

FIGURE 4.18 Output voltage waveform of eleven-level inverter with ANN.

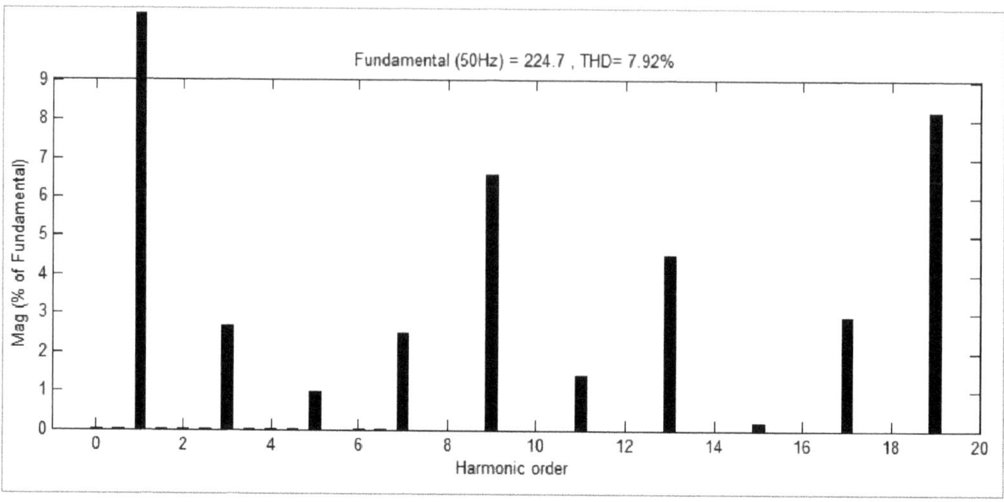

FIGURE 4.19 FFT analysis of eleven-level inverter with ANN.

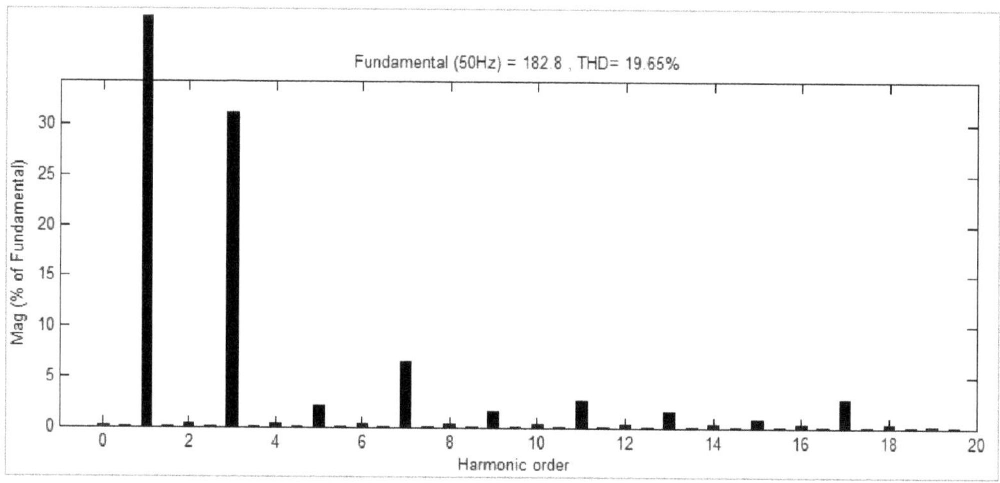

FIGURE 4.20 FFT analysis of eleven-level inverter without ANN.

that there is a considerable reduction in the THD value when the ANN approach is used. In addition, the magnitudes of the harmonic orders are also considerably reduced.

Similarly, the fifteen-level inverter is simulated and the results are obtained. The neural network is trained with the datasets obtained by solving six SHE equations. The trained network is then implemented with the solar-fed CMLI. The analysis is performed with the input PV voltage values of {49.9 V, 50.7 V, 52.6 V, 54.9 V, 44.7 V, 51.4 V, 43.5 V} and the output voltage waveform and current waveform for the *RL* load of R=100 Ω and L=10 mH are shown in Figures 4.21 and 4.22, respectively. Here the floating-point data of voltage values is also considered, which normally occurs in practical implementations. FFT analysis of output voltage and current obtained from solar-fed CMLI with ANN is given in Figures 4.23 and 4.24. The comparison of harmonic order for the solar-fed eleven-level and fifteen-level inverters with ANN is given in Figure 4.25.

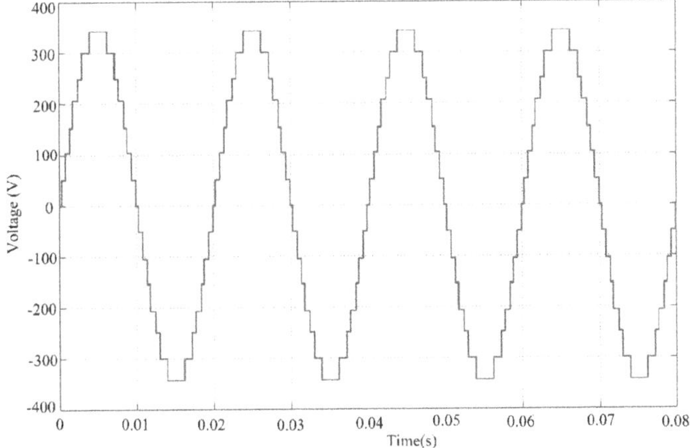

FIGURE 4.21 Output voltage waveform of fifteen-level inverter with ANN.

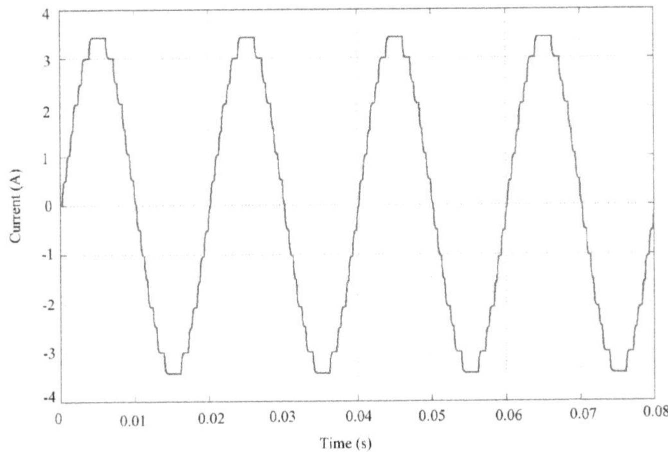

FIGURE 4.22 Output current waveform of fifteen-level inverter with ANN.

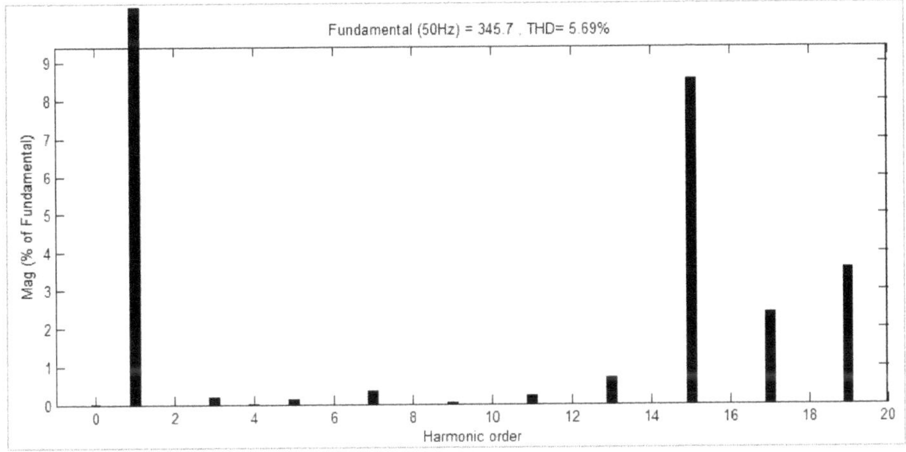

FIGURE 4.23 FFT analysis of fifteen-level inverter output voltage with ANN.

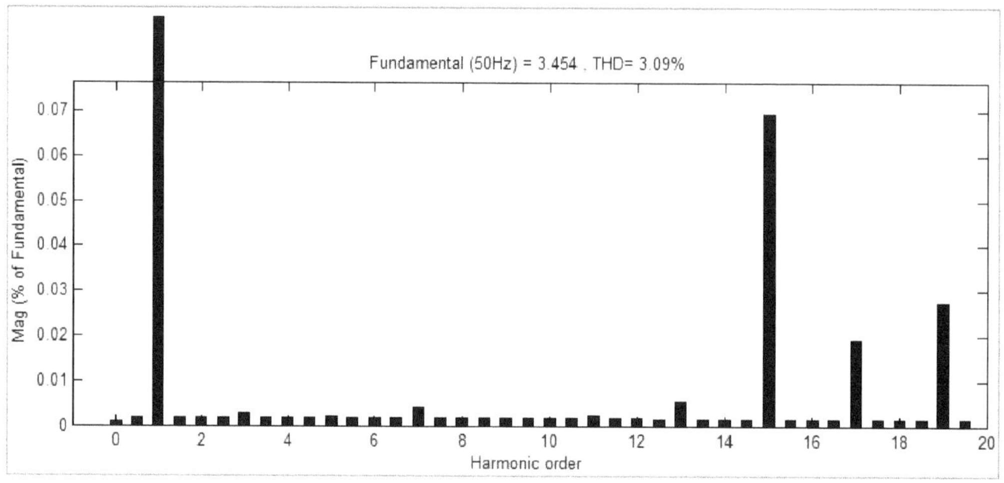

FIGURE 4.24 FFT analysis of fifteen-level inverter output current with ANN.

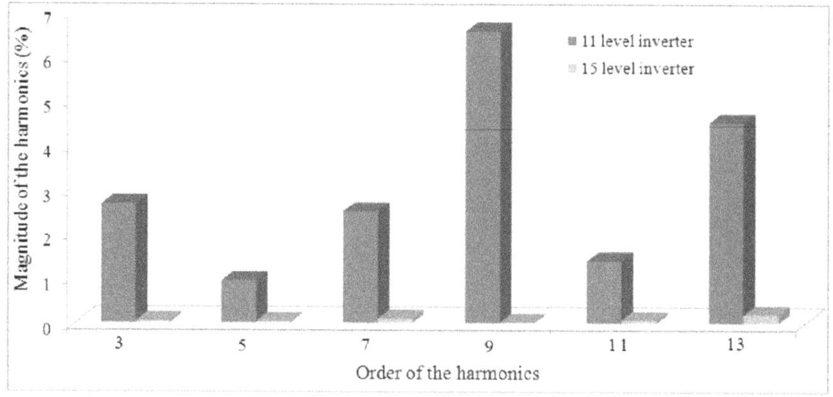

FIGURE 4.25 Comparison of harmonic orders for eleven- and fifteen-level inverters.

4.12.3 Optimization Techniques

Table 4.6 shows the GA parameters used for the execution of the program. Figure 4.26 shows the output voltage waveform of a fifteen-level inverter and the corresponding FFT analysis is given in Figure 4.27.

Table 4.7 shows the PSO parameters used for the execution of the program. Figure 4.28 shows the output voltage waveform of a fifteen-level inverter and the corresponding FFT analysis is given in Figure 4.29.

Table 4.8 shows the BO parameters used for the execution of the program. Figure 4.30 shows the output voltage waveform of a fifteen-level inverter and the corresponding FFT analysis is given in Figure 4.31.

TABLE 4.6
GA Specifications

Parameter	Value
Population size	24
Cross over rate	0.5
Mutation rate	0.234
Stopping criteria	100 iterations

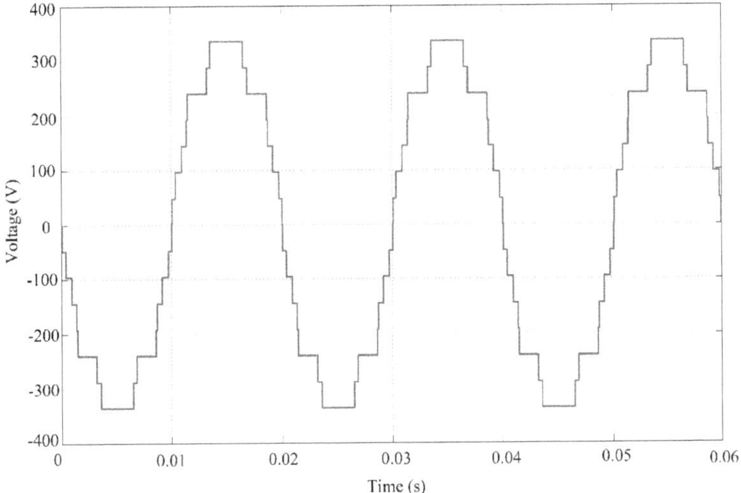

FIGURE 4.26 Output voltage waveform of fifteen-level inverter with GA.

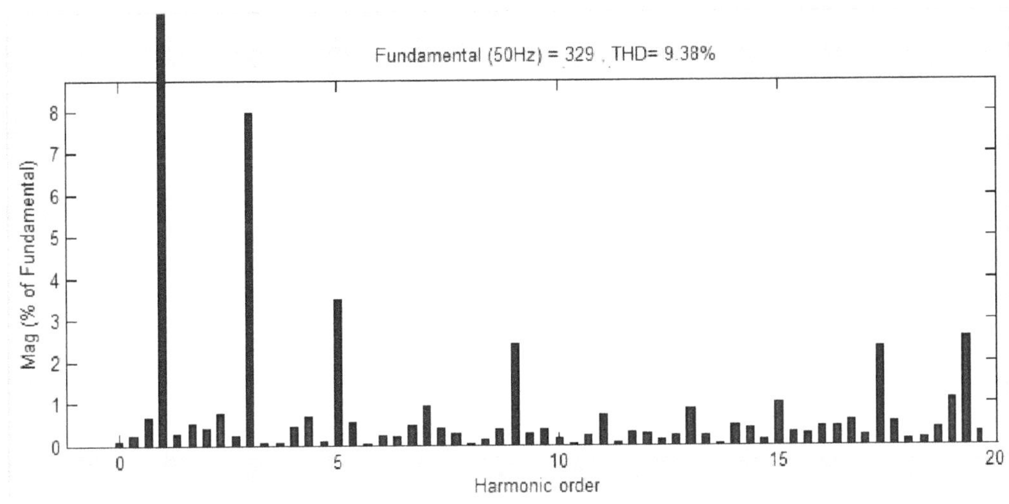

FIGURE 4.27 FFT analysis of fifteen-level inverter output voltage with GA.

TABLE 4.7
PSO Specifications

Parameter	Value
C_1, C_2	0.3
r_1, r_2	0.9 [0 to 1]
$w(k)$	0.9
Stopping criteria	100 iterations

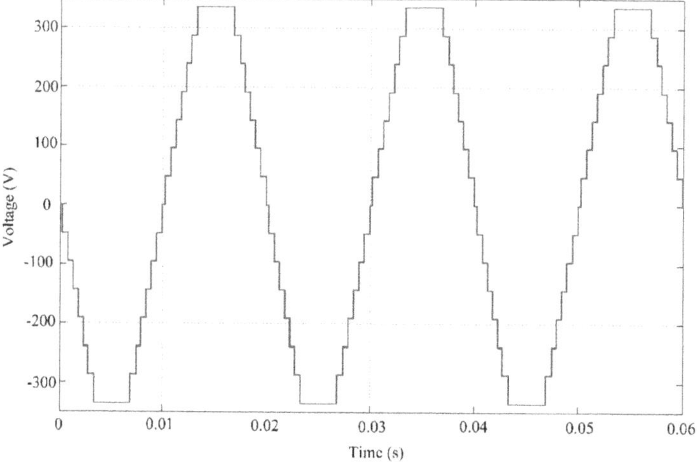

FIGURE 4.28 Output voltage waveform of fifteen-level inverter with PSO.

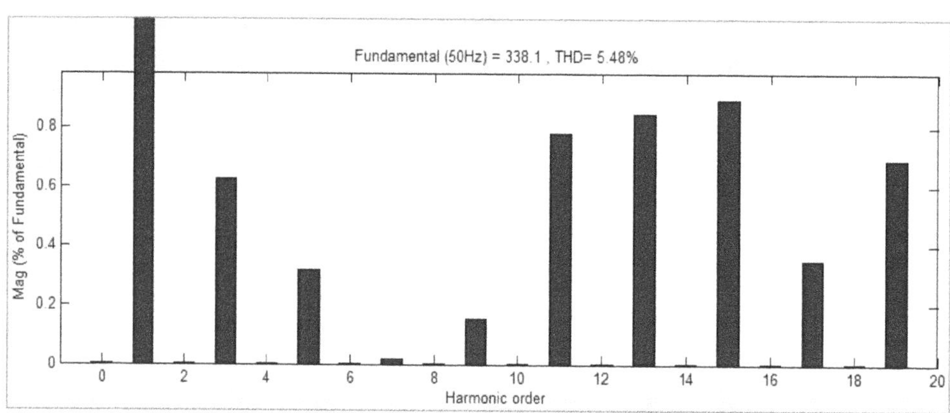

FIGURE 4.29 FFT analysis of fifteen-level inverter output voltage with PSO.

TABLE 4.8
BO Specifications

Parameter	Value
Number of employed bees	30
Number of scout bees	20
Number of sites selected for neighborhood search	15
Number of elite sites	7
Stopping criteria	100 iterations

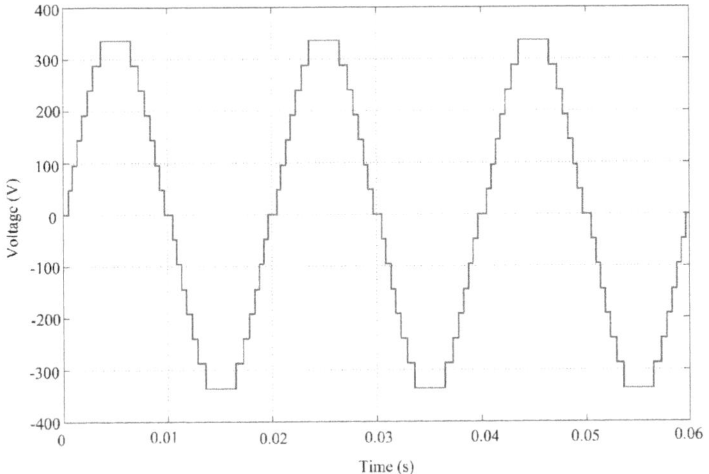

FIGURE 4.30 Output voltage waveform of fifteen-level inverter with BO.

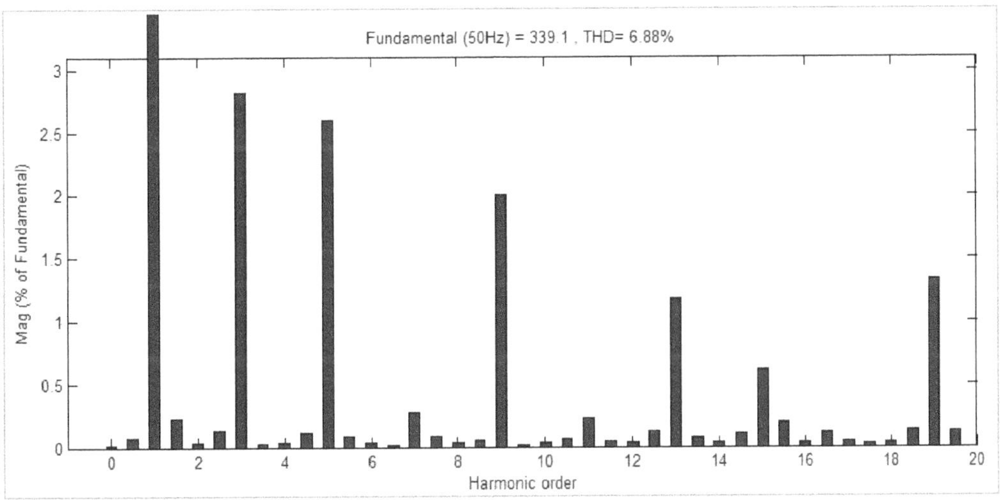

FIGURE 4.31 FFT analysis of fifteen-level inverter output voltage with BO.

4.13 Experimental Results

There are seven sets of individual PV supply source used for the hardware implementation as they produce $48 \times 7 = 336$ Vp (230 V_{RMS}). The methodology used in the ANN-based system is extended for hardware implementations as it provides less THD. Figure 4.32 shows the hardware setup including the seven inverter stages with the FPGA SPARTAN 3E processor. In the figure, INV depicts the term inverter. Each input source to the inverter stages provides the variation up to ±10% and these inputs are given to the FPGA processor after being scaled down to the appropriate range. In the processor, offline procedure is utilized where the switching angles are precalculated and then programmed. Due to the variations in the solar PV, the switching angles are altered at every interval of time.

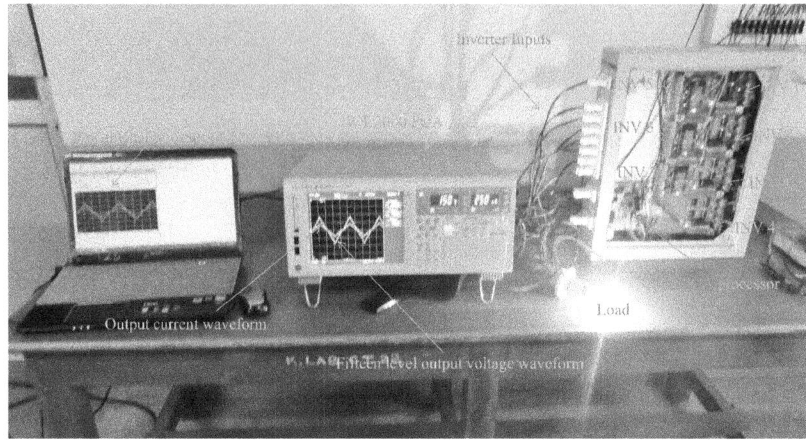

FIGURE 4.32 Hardware set up for fifteen-level inverter with SHE. (*Note*: See color eBook for improved color differentiation in figures and labeling.)

Figures 4.33 and 4.34 show the variations of the switching angles monitored for the change in input PV voltage marked as Case I and II, respectively.

For each set of source voltages, the processor finds the relevant switching angles and transfers it to the driver circuit, which is connected to MOSFET for turning it ON. The voltage sensing is made through seven sets of signal processing elements. The desirable parameters of Xilinx SPARTAN 3E processor are: (i) very low cost, high-performance logic solution for high volume, consumer oriented applications, (ii) proven advanced 90-nanometer process technology, (iii) multi-voltage, multi-standard selector interface pins, (iv) enhanced double data rate (DDR) support, (v) efficient wide multiplexers with wide logic functions, (vi) fast look-ahead carry logic, and vii) eight digital clock managers that make the processor suitable for SHE-based applications. Figure 4.35 shows the interconnection of developed hardware to the solar PV and Figure 4.36 shows the fifteen-level output voltage waveform obtained.

Figure 4.37 shows the fast Fourier transform (FFT) analysis using the power quality analyzer (PQA) WT3000 where the THD obtained in this method is about 5% with the possible minimization of low order harmonics considered in this work. The variation of input PV voltage is measured at every instant, which is used for solving SHE equations. Based on the results, it satisfies the IEEE standard for voltage harmonics 519-1992, such that for a system whose bus voltage is less than 69 kV, the individual magnitude of the harmonic order should be 3% and the voltage THD should be 5%. For the given system on considering the six harmonic orders, the magnitude is reduced substantially to less than 3%. It makes the developed hardware suitable for reducing the harmonic orders, thereby reducing the overall harmonic distortions.

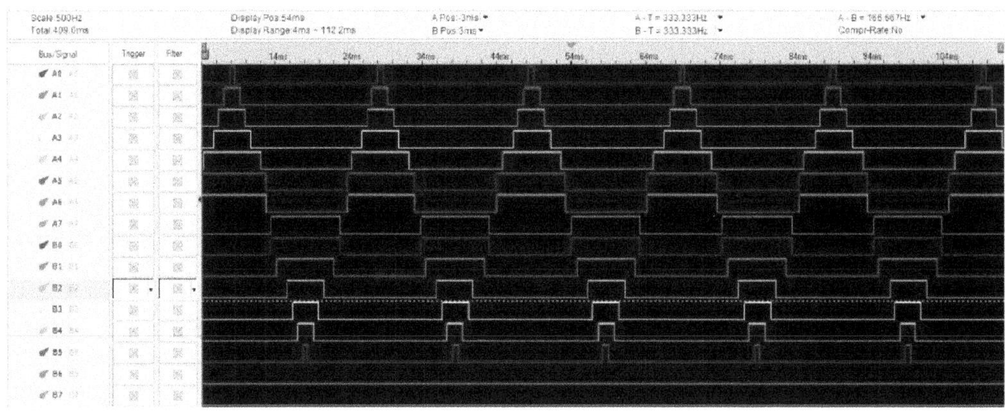

FIGURE 4.33 Variation of switching angles (case-I).

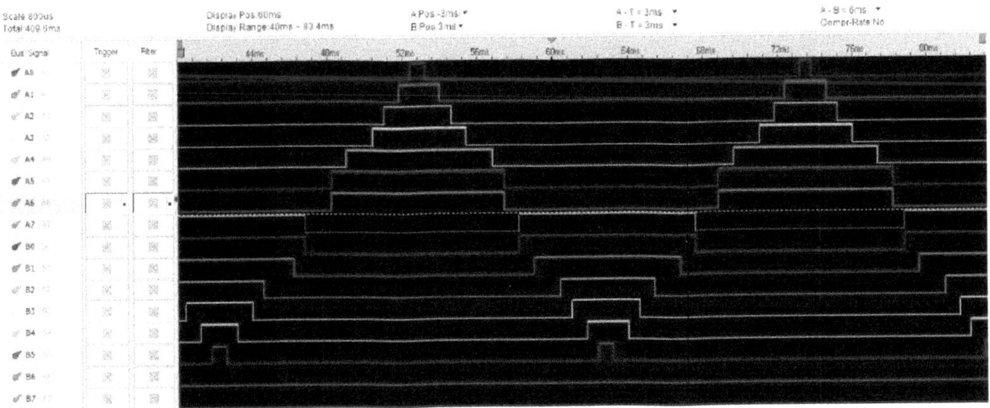

FIGURE 4.34 Variation of switching angles (case-II).

FIGURE 4.35 Hardware set up with solar PV integration. (*Note*: See color eBook for improved color differentiation in figures and labeling.)

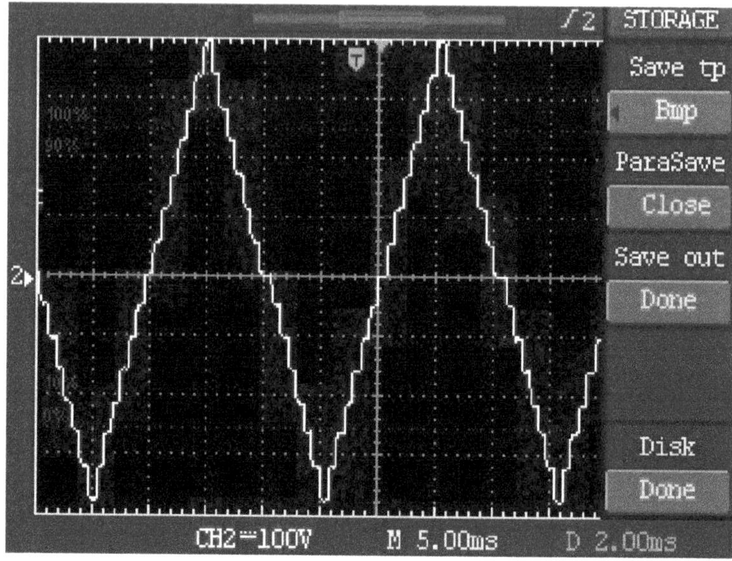

FIGURE 4.36 Fifteen-level output voltage waveform.

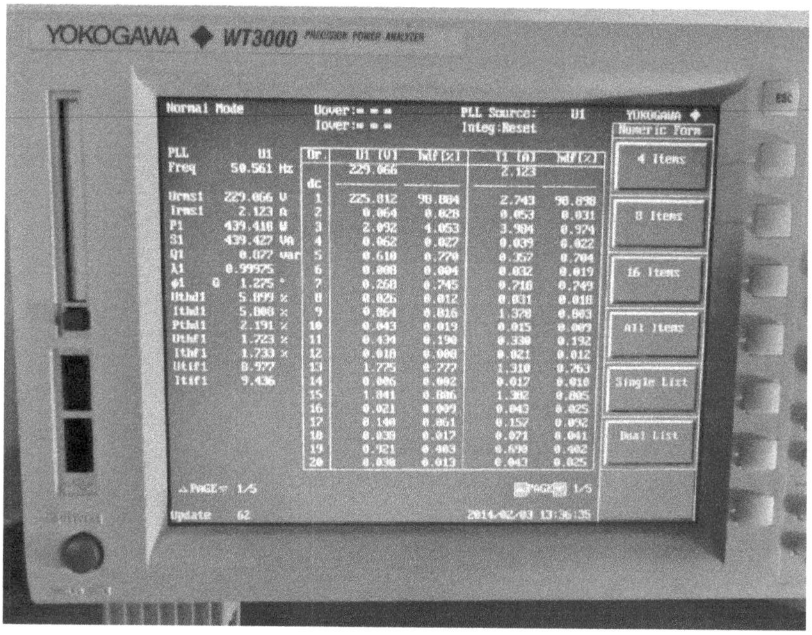

FIGURE 4.37 Harmonic analysis using WT3000.

The comparison of the methods proposed for SHE is given in Table 4.9, which shows that the ANN-based method provides better results and is widely suitable for solar PV systems. The OHSW technique requires the initial guess values for switching angles on formulating the problem. The optimization technique is less preferable for solar PV systems as the voltage value given in the objective function is altered often to achieve a higher degree of freedom in finding the optimal switching angles. While comparing the three different optimization techniques (GA, PSO, and BO), PSO provides the better result with respect to reduction in THD. Table 4.10 shows the comparison of the SHE problem with various algorithms.

TABLE 4.9

Comparison of Results for the Proposed Methodologies

S.No.	Method	Level	THD (%)
1	OHSW	15	8.55
2	ANN	11	7.92
4	ANN	15	**5.73**
5	GA	15	9.38
6	PSO	15	**5.48**
7	BO	15	6.88

TABLE 4.10

Comparison with Other Methods

S.No.	Authors	Levels	Methodology	THD
1.	Liu et al. (2009a)	7	Step modulation	14.98%
2.	Liu et al. (2009b)	7	Staircase modulation	16.69%
3.	Du et al. (2009)	7	Fundamental frequency	11th = 2% 13th = 3.8%
4.	Farkhnia et al. (2011)	7	Analytical	14.03%
5.	Fei et al. (2010)	7	Half wave symmetry	–
6.	Filho et al. (2011)	11	ANN	7%–14%
7.	Filho et al. (2013)	11	ANN	8.7%
8.	Yousefpoor et al. (2012)	7	GA	17.06%
9.	Taghizadeh and Hagh (2010)	11	PSO	9.822%
10.	Kavousi et al. (2012)	7	BO	12.52%
11.	Proposed	11	ANN	7.92%
12.	Proposed	15	ANN	5.89% 11th = 0.434% 13th = 1.775%

A fifteen-level solar-fed CMLI for the elimination of certain harmonic orders is developed for the power quality improvement. The method is proposed with the intelligent techniques such as ANN and optimization techniques (GA, PSO, and BO). In addition, the OHSW-based harmonic elimination method has also been considered. The experimental investigation is carried out for a 3 kWp system with FPGA processor in arriving at the reduction of six possible harmonic orders. The advantages of this method include simple computational algorithm and no requirement of filters, detailed look-up tables and output transformers.

4.14 Lower Order Harmonics Mitigation in a PV Inverter

The fundamental requirements to interface PV systems with the grid are as follows:

- The voltage magnitude and phase must equal that required for the desired magnitude and direction of the power flow. The voltage is controlled by the transformer turn ratio or the power electronic converter firing angle in a closed-loop control system.
- The frequency must be exactly equal to that of the grid or else the system will not work. To meet the exacting frequency requirement, the only effective means is to use the utility frequency as the inverter switching frequency reference.

The PV power systems have made a successful transition from small standalone sites to large grid-connected systems. The utility interconnection brings a new dimension to the renewable power economy by pooling the temporal excess or the shortfall in the renewable power with the connecting grid that generates base-load power using conventional fuels. This improves the overall economy and load availability of the renewable plant site. The array by itself does not constitute the PV power system. The PV power system needs a sun tracker to point the array to the sun, various sensors to monitor system performance, and power electronic components that accept the DC power produced by the array, charge the battery, and condition the remaining power in a form that is usable by the load. If the load is AC, the system needs an inverter to convert the DC power into AC at 50 or 60 Hz.

The peak-power tracker senses the voltage and current outputs of the array and continuously adjusts the operating point to extract the maximum power under varying climatic conditions. The output of the array goes to the inverter, which converts the DC into AC. The array output in excess of the load requirement is used to charge the battery. The battery charger is usually a DC–DC buck converter. If excess power is still available after fully charging the battery, it is shunted in dump heaters, which may be a room or water heater in a standalone system. When the sun is not available, the battery discharges to the inverter to power the load. The battery discharge diode is used to prevent the battery from being charged when the charger is opened after a full charge or for other reasons. The array diode is used to isolate the array from the battery, thus keeping the array from acting as the load on the battery at night. The controller collects system signals, such as the array and the battery currents and voltages, and keeps track of the battery state of charge by bookkeeping the charge/discharge ampere-hours. It uses this information to turn on or off the battery charger, discharge converter, and dump loads as needed. Thus, the controller is the central controller of the entire system. In the grid-connected system, dump heaters are not required, as all excess power is always fed to the grid lines. The battery is also eliminated, except for a few small critical loads, such as the start-up controller and the computer. DC power is first converted into AC by the inverter, ripples are filtered, and only then is the filtered power fed into the grid lines. In the PV system, the inverter is a critical component, which converts the array DC power into AC for supplying the loads or interfacing with the grid.

4.14.1 Methodology

The power topology used is shown in Figure 4.38. The advantages of this topology are the switches are all rated for low voltage that reduces the cost and increases the reliability of the system. This is the good choice for low-rated PV inverters having the rating less than kilowatt. Ideally this topology will not have lower order harmonics, but will be injected by distorted magnetizing current drawn by transformer, dead time between switching of devices in the same leg, on-state voltage loss, and grid voltage distortion.

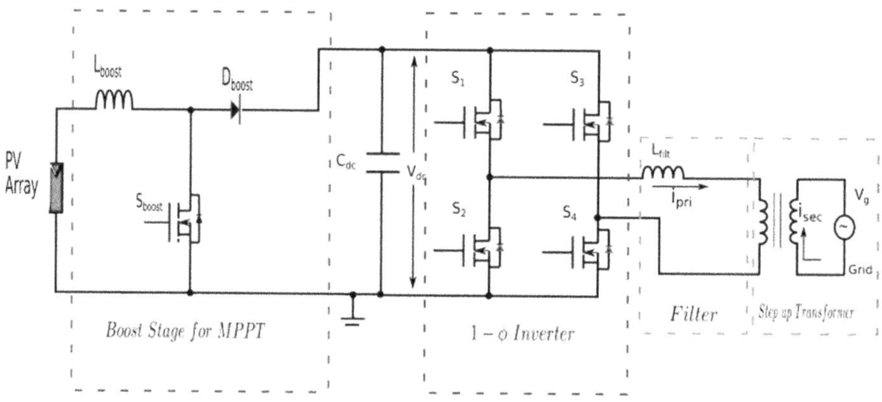

FIGURE 4.38 Power circuit topology of 1-Ø PV system for a low voltage inverter with 40 V DC bus connected to 230 V grid using step-up transformer.

The varying power reference from MPPT block from which AC current reference is generated, offsets in sensors, and A/D converter in digital controller (which will result in the even order harmonics drawn from grid) contribute to lower power quality. To meet IEEE 519-1992 and IEEE 1547-2003, these harmonics must be attenuated. A large output filter for attenuation will increase the system losses and large voltage drop with higher cost.

Even when there are shifts in grid frequency, the adaptive filter-based method is used for the compensation of harmonics. The implementation is simple. This may estimate the particular harmonic in the grid current using least mean square (LMS) adaptive filter. Using a proportional controller, it generates the harmonic voltage reference, which is added to fundamental voltage reference to attenuate the harmonic. This paper also includes the design of value of gain of the proportional controller to achieve harmonic compensation.

The DC in the terminal voltage of inverter results in DC current flow into the transformer primary. If a PR controller is used, DC offset in control loop will propagate into the system and the terminal voltage of inverter is non-zero. Hence a modification is proposed, which includes the integral block along with a PR controller to ensure that there is no DC output current in the inverter.

4.14.2 Origin of Lower Order Harmonics and Fundamental Current Control

The section discusses the origin of lower order harmonics. The current control using a proportional resonant integral (PRI) controller is also discussed.

4.14.3 Origin of Lower Order Harmonics

4.14.3.1 Odd Harmonics

The predominant cause for lower order odd harmonics is the distorted magnetizing current drawn by transformer, inverter switches dead time, voltage drops in semiconductor devices, distortion in grid voltage, and DC bus voltage ripple. Due to nonlinear characteristics of *B–H* curve of transformer core, the transformer contains the lower order harmonics. The amplitude of harmonics can be obtained by *B–H* curve of transformer and the phase angle depends upon on the power factor of the system. The current will be in phase with the voltage due to UPF, but the magnetizing current lags grid voltage by 90°. Harmonic current will have +90° or −90° phase displacement that depends on harmonic order.

The dead time effect will introduce the lower order harmonics, which are proportional to dead time, DC bus voltage, and switching frequency. The dead time effect can be modeled as square wave error voltage out of phase with current at pole of leg. The amount of distortion is smaller due to the drop in devices. The amplitude of harmonic voltage for *h*th harmonic is expressed as

$$V_{\text{error}} = \left(\frac{4}{h}\Pi\right)\left(2\frac{V_{\text{dc}}t_d}{T_s}\right) \tag{4.54}$$

where, t_d – delay time, T_s – switching frequency, V_{dc} – dc bus voltage.

The harmonic current amplitude can be obtained from filter inductance, transformer leakage inductance, and series resistance. The phase angle will be 180° for UPF. Depending on the relative magnitude of distortion caused due to dead time and magnetizing current, the harmonic content will have phase angle with respect to fundamental current.

4.14.4 Even Harmonics

The circuit is very sensitive to the presence of DC offset in the terminal voltage of the inverter. Varying power reference is given by the MPPT block, A/D converter offset, and sensors that cause the DC offset. Assume certain amount of DC exits in current control loop, this will apply voltage with a DC offset

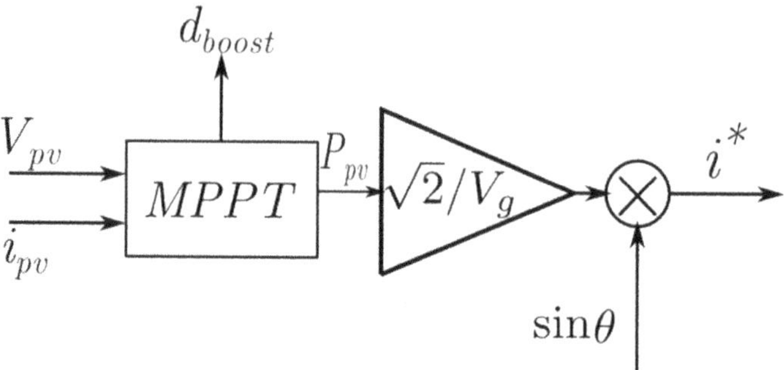

FIGURE 4.39 Generation of inverter AC current reference from MPPT block.

across the L-filter and transformer primary. The net current will be determined by net resistance present in loop. The average current will cause a DC shift in *B–H* curve of the transformer, which will mean asymmetric nonlinear saturation characteristics that may cause the magnetizing current of transformer to lose its half symmetry. The result of this is the occurrence of even harmonics. The DC in the system can be eliminated by PRI controller.

Figure 4.39, shows d_{boost} is the duty ratio given to the boost converter switch, V_{pv} and i_{pv} are the panel voltage and current, P_{pv} is the panel power output, V_g is the grid rms voltage, and i^* is the reference to current controller loop from the MPPT block.

4.15 Fundamental Current Control

Conventional reference frame control consists of a PR controller to generate voltage reference. A modification is proposed, by adding the integral block, G_I, which is termed as PRI controller. Figure 4.40 shows the block diagram of the fundamental current control with the PRI controller

$$G_I = \frac{K_i}{s} \tag{4.55}$$

$$G_{\text{PR}(s)} = K_p + \left(\frac{K_r s}{s^2} + \omega_0{}^2 \right) \tag{4.56}$$

The plant transfer function is given by

$$G_{\text{plant}(s)} = \frac{V_{\text{dc}}}{(R_s + sL_s)} \tag{4.57}$$

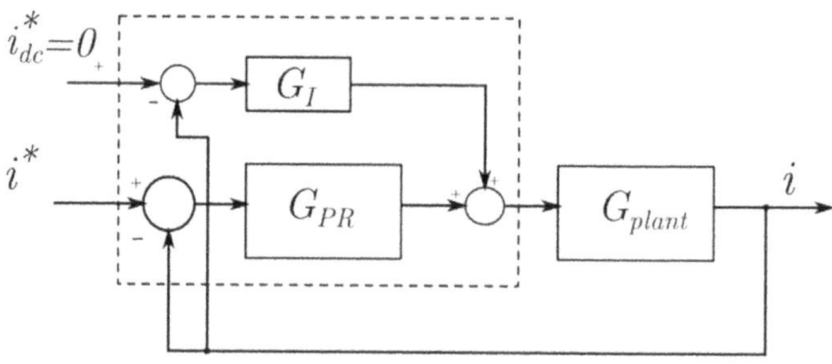

FIGURE 4.40 Block diagram of the fundamental current control with the PRI controller.

The inverter will have a gain of V_{dc} to voltage reference generated by the controller and impedance offered is $R_s + sL_s$ in the s domain. R_s and L_s are the net resistance and impedance referred to the primary side of the transformer. L_s include the filter inductance and leakage inductance or transformer. R_s is the net series resistance due to transformer and filter inductance. The PRI controller is proposed to ensure that the output current does not contain a DC offset, which introduces zero at $s = 0$. Hence the output current will not contain any DC offset.

4.16 Design of PRI Controller Parameters

The fundamental current corresponds to power injected into the grid and the control objective is to achieve unity power factor (UPF) operation. A PR controller is designed, assuming the integral block is absent, $K_i = 0$. Design of PR controller is done by considering PI controller. The PI parameters are chosen based on plant transfer function and current controller bandwidth. The PI controller parameters are used for the PR controller parameters.

Let

$$G_{\text{PI}(s)} = K_{p1} * \frac{(1+sT)}{sT} \tag{4.58}$$

With a PI controller as the compensator block as in Figure 4.40 and without an integral block, the forward transfer function is

$$G_{\text{for}}w(s) = \left(K_{p1} * \frac{(1+sT)}{sT}\right) \frac{V_{\text{dc}}}{Rs + sLs} \tag{4.59}$$

The pole is cancelled by zero given by the PI controller where ω_{bw} is the required bandwidth, then following relations are obtained:

$$T = \frac{L_s}{R_s} \tag{4.60}$$

$$K_{p1} = \frac{(\omega_{\text{bw}} R_s T)}{V_{\text{dc}}} \tag{4.61}$$

$$K_{i1} = \frac{(\omega_{\text{bw}} R_s)}{V_{\text{dc}}} \tag{4.62}$$

For the PI controller, expressions obtained in (4.61) and (4.62) are used for proportional and resonant gain, thus

$$K_p = \frac{(\omega_{\text{bw}} R_s T)}{V_{\text{dc}}} \tag{4.63}$$

$$K_r = \frac{(\omega_{\text{bw}} R_s)}{V_{\text{dc}}} \tag{4.64}$$

4.17 Adaptive Harmonic Compensation

The adaptive harmonic compensation is based on least mean square (LMS) adaptive filter to estimate the particular harmonic in output current, which is used to generate a counter voltage reference using proportional controller to mitigate particular harmonic. Input vector and filter output is given by

$$\bar{x}(n) = \left[x(n)\ x(n-1)\ \ldots\ x(n-N+1)\right]^T \tag{4.65}$$

$$\bar{y}(n) = \bar{w}^T x(n) \tag{4.66}$$

The error signal is

$$e(n) = d(n) - y(n) \tag{4.67}$$

Here *d*(*n*) is primary input. The frequency component of *d*(*n*) is adaptively estimated by *y*(*n*). The performance function of LMS adaptive filter is defined as

$$\tau = e^2(n) \tag{4.68}$$

In adaptive filter, weight vector w is updated with the performance function moves towards its minimum.

$$\bar{w}(n+1) = \bar{w}(n) - \mu\Delta(e(n)^2) \tag{4.69}$$

The μ is the step size and convergence of the adaptive filter depends on step size μ. A smaller value will make the adaptation process very slow where large value will make the system oscillatory. Δ is defined as a gradient performance function with respect to weights of filter. The equation for weights of the LMS adaptive filter

$$\bar{w}(n+1) = \bar{w}(n) + 2\mu e(n)\bar{x}(n) \tag{4.70}$$

To reduce the particular lower order harmonics of grid current, the i_k is estimated from samples of grid current and phase-locked loop (PLL) unit vectors at that particular frequency. A voltage reference is generated from i_k, and generated voltage is subtracted from main control voltage reference.

The block diagram shown in Figure 4.41 depicts the adaptive filter that estimates the particular *k*th harmonic i_k of grid current *i*. The adaptive block is given by two inputs $\sin(k\omega_0 t)$ and $\cos(k\omega_0 t)$ from PLL. This samples are multiplied by $W_{\sin}$ and $W_{\cos}$. The output is subtracted from sensed grid current sample, which is taken as error for LMS algorithm. The weights are updated and output will estimate

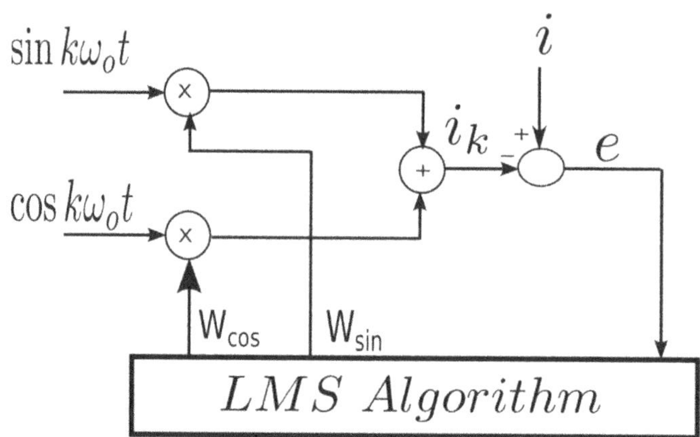

FIGURE 4.41 Block diagram of adaptive estimation of a particular harmonic grid current.

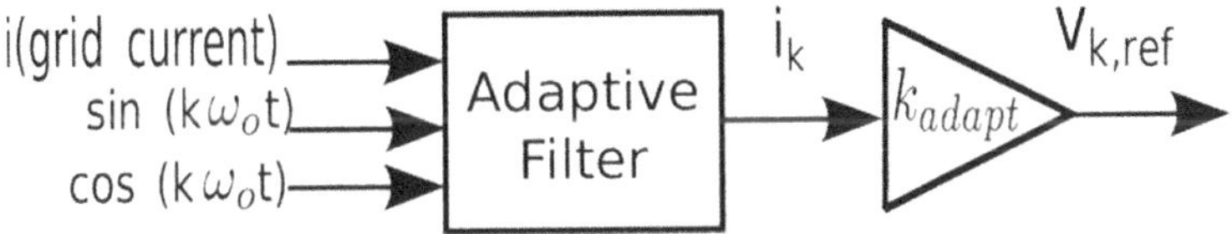

FIGURE 4.42 Generation of voltage reference from estimated *k*th harmonic compensation current using LMS adaptive filter.

*k*th harmonic of grid current *i*. The following figure shows the generation of voltage reference from estimated *k*th harmonic compensation of current using LMS adaptive filter. Figure 4.42 shows the generation of voltage reference from estimated *k*th harmonic compensation of current using LMS adaptive filter

$$W_{\cos}(n+1) = W_{\cos}(n) + 2\mu e(n)\cos\left(k\omega_0 nT_s\right) \tag{4.71}$$

$$W_{\sin}(n+1) = W_{\sin}(n) + 2\mu e(n)\sin\left(k\omega_0 nT_s\right) \tag{4.72}$$

The reference voltage has to be generated from estimated current. The proportional gain is used for design and implementation. Figure 4.43 shows the scheme used for harmonic voltage reference generation from estimated harmonic current. If third, fifth, and seventh need to mitigated, then three adaptive filters and three gain terms k_{adapt}.

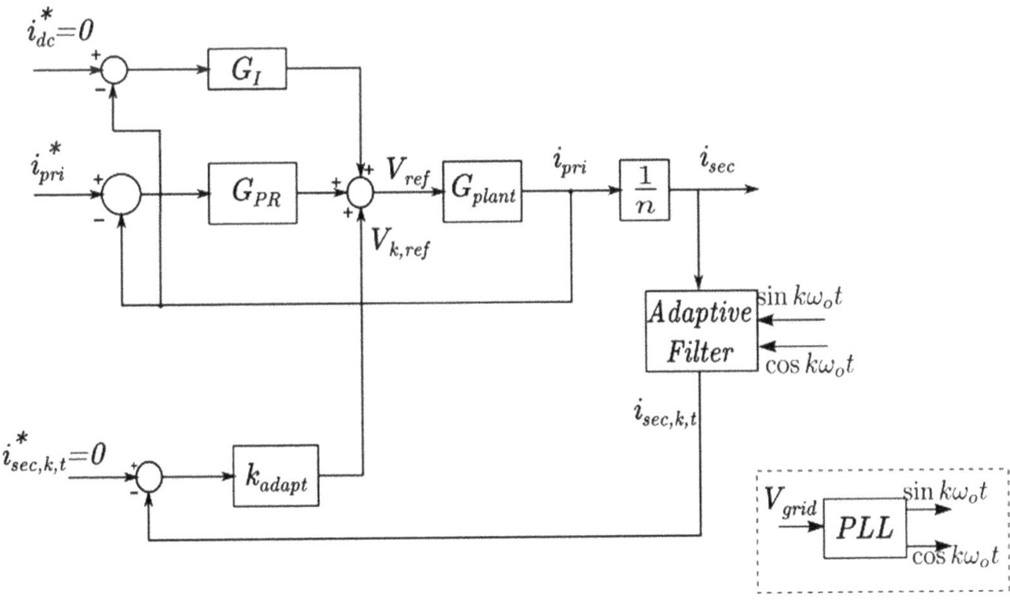

FIGURE 4.43 Complete AC current control structure of the inverter.

4.18 Simulink Model

To design the MATLAB/Simulink model, choose the following blocks from the Library Simulink Browser:

AC voltage source

Path: Simulink Library Browser>>Simscape>>SimPowerSystems>>Electrical Sources>>AC voltage source

Function: Ideal sinusoidal AC voltage source

Path: Simulink Library Browser>>Simscape>>SimPowerSystems>PowerElectronics>> Universal Bridge

Function: This block implements a bridge of selected power electronic devices

Voltage and current measurement

Path: Simulink Library Browser>>Simscape>>SimPowerSystems>Measurements>>Voltage measurements, Current measurements

Function: Ideal voltage/current measurement

R, L, C

Path: Simulink Library Browser>>Simscape>>SimPowerSystems>Elements> RLC branch

Powergui

Path: Simulink Library Browser>>Simscape>>SimPowerSystems>>Powergui

Function: The Powergui block is necessary for simulation of any Simulink model containing SimPowerSystems blocks. It is used to store the equivalent Simulink circuit that represents the state-space equations of the model.

Scope

Simulink>>Sinks>>Scope

Figure 4.44 shows the Simulink model without filter and its corresponding FFT analysis in Figure 4.45. The above figure shows the THD level for the Simulink model without using an LMS adaptive filter. The THD is found to be 34.70%. The THD can be reduced by using an LMS filter. Figure 4.46 shows the Simulink model build by using an LMS adaptive filter. Figure 4.46 shows the Simulink model with filter and its corresponding FFT analysis in Figure 4.47. From the above FFT analysis it can be seen that the THD has been reduced to 8.94% by using an LMS adaptive filter and PRI controller.

Modification to the inverter current control for a grid-connected single-phase PV inverter has been proposed in this paper, for ensuring high quality of the current injected into the grid. The proposed method uses an LMS adaptive filter to estimate a particular harmonic in the grid current that needs to be attenuated. The estimated current is converted into an equivalent voltage reference using a proportional controller and added to the inverter voltage reference. The design of the gain of a proportional controller to have an adequate harmonic compensation has been explained. The PRI controller and the adaptive compensation scheme together improve the quality of the current injected into the grid. The complete current control scheme consisting of the adaptive harmonic compensation and the PRI controller has been verified experimentally and the results show good improvement in the grid current THD.

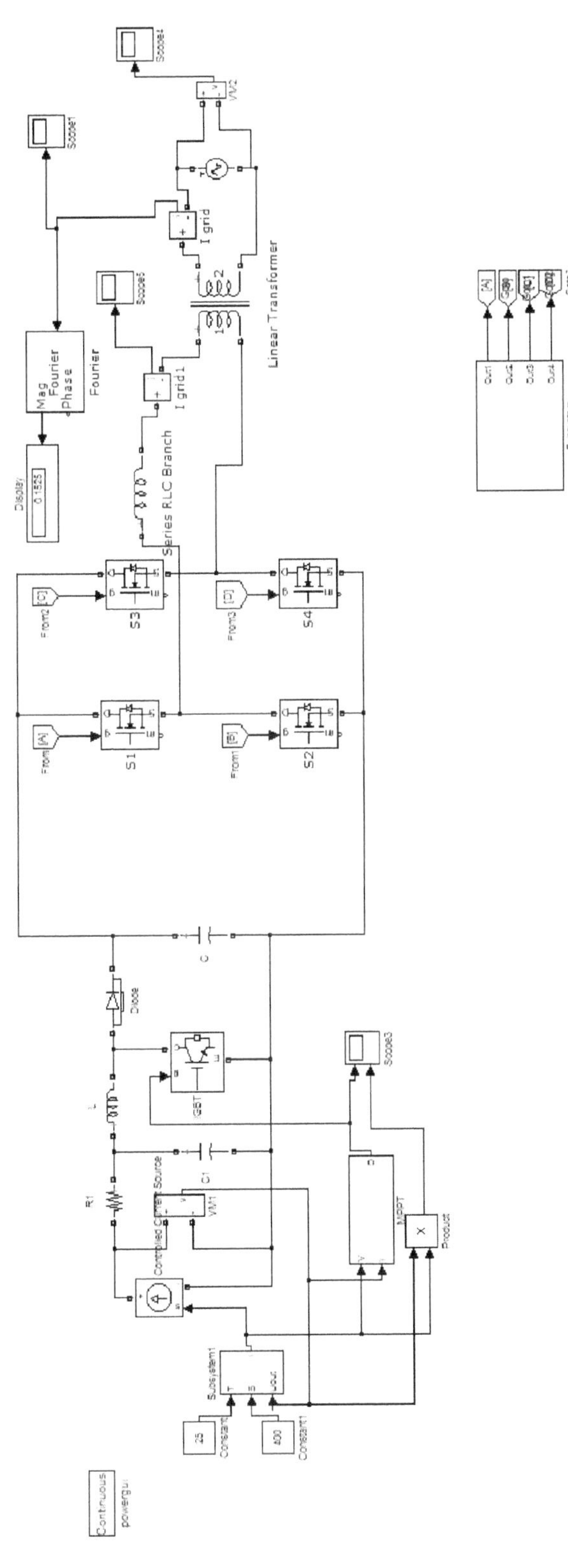

FIGURE 4.44 Simulink model without filter.

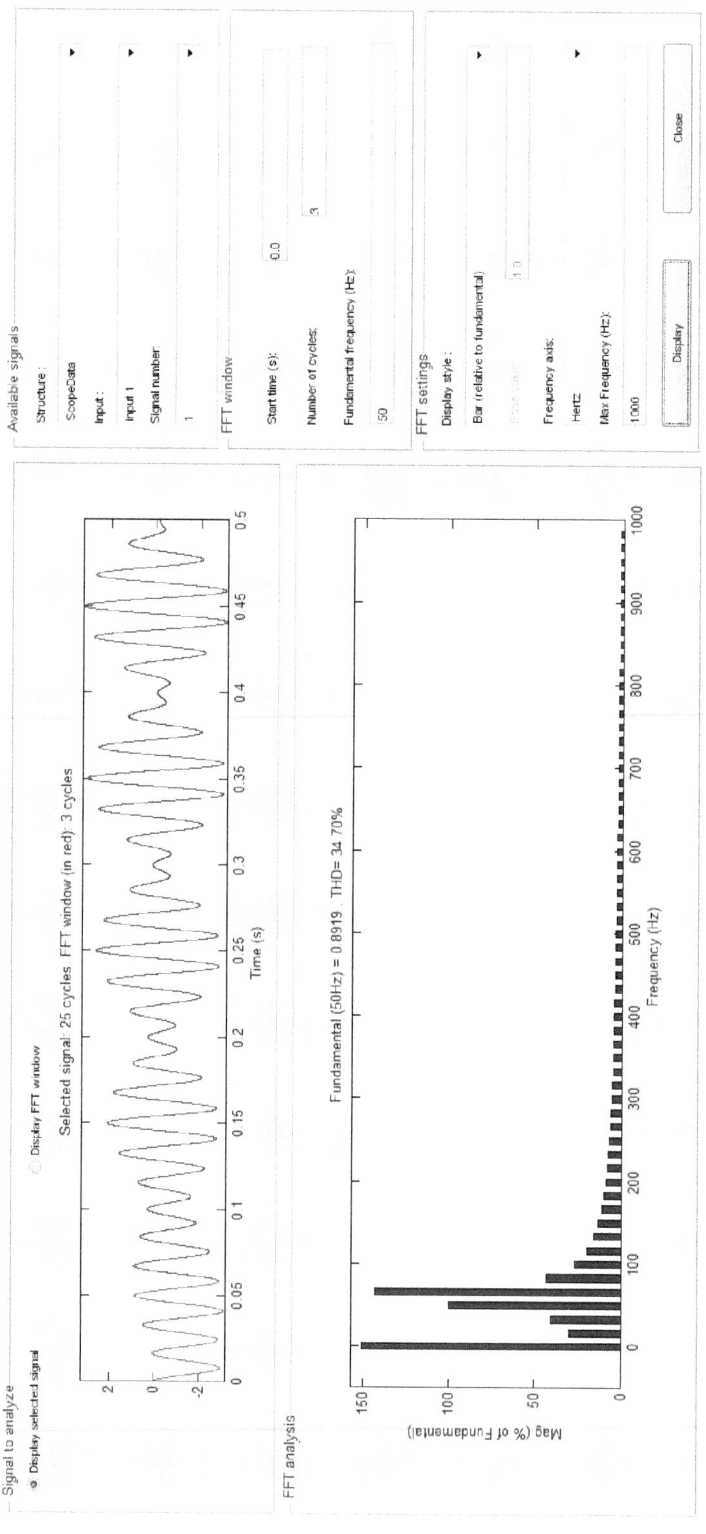

FIGURE 4.45 FFT analysis. (*Note:* See color eBook for improved color differentiation in figures and labeling.)

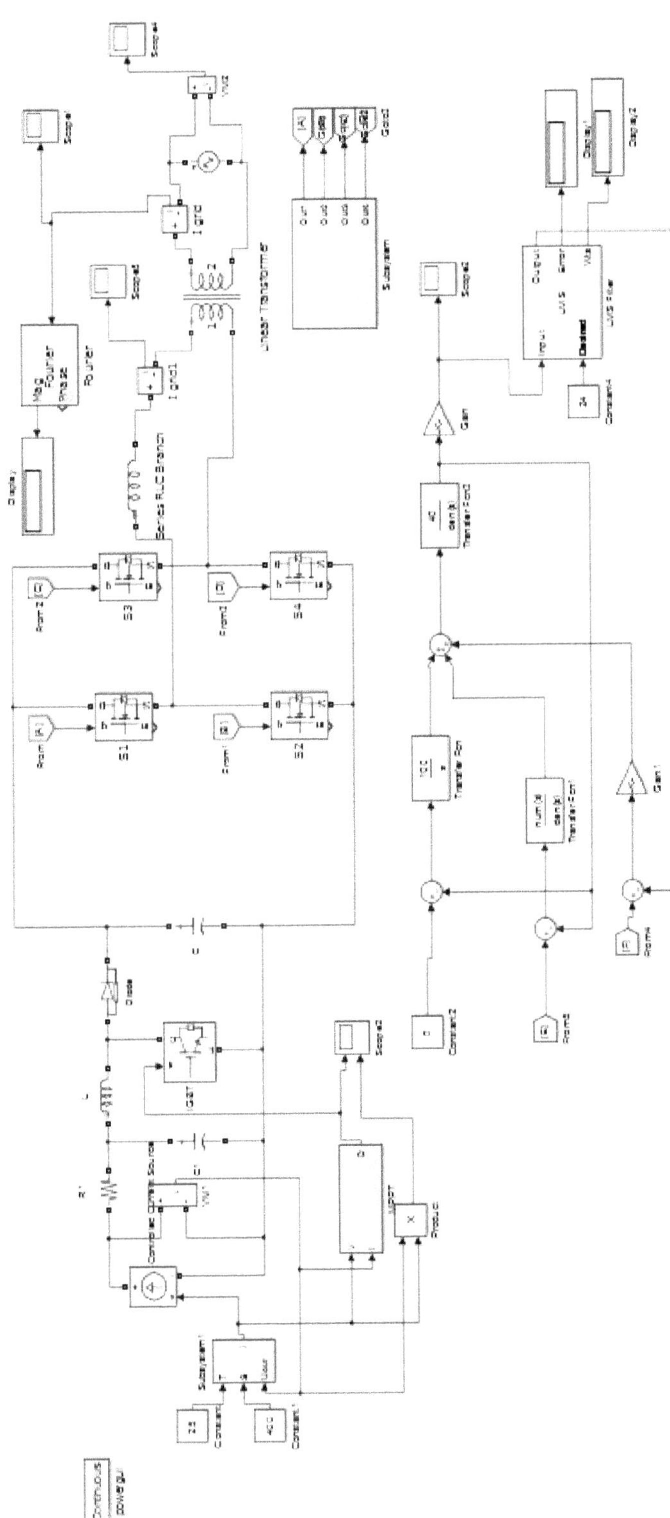

FIGURE 4.46 Simulink model with filter.

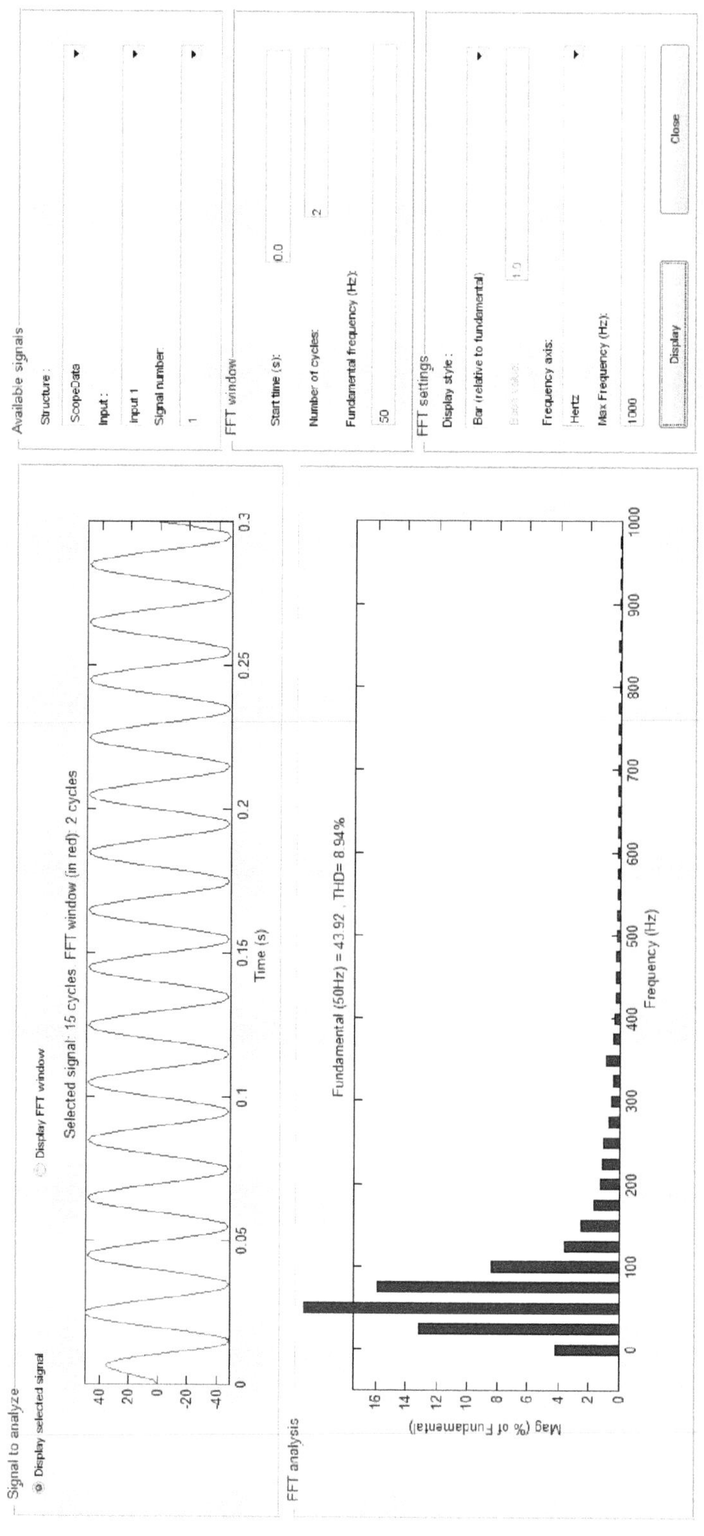

FIGURE 4.47 FFT analysis. (*Note:* See color eBook for improved color differentiation in figures and labeling.)

REFERENCES

1. Ahmadi, D, Zou, K, Li, C, Huang, Y and Wang, J. 2011. A universal selective harmonic elimination method for high-power inverters. *IEEE Transactions on Power Electronics*, 26(10): 2743–2752.
2. Babaei, E. 2010. Optimal topologies for cascaded sub-multilevel converters. *Journal of Power Electronics*, 10(3): 251–261.
3. Dahidah, SA and Agelidis, VG. 2008. Selective harmonics elimination PWM control for cascaded multilevel voltage source converters: A generalized formula. *IEEE Transactions on Power Electronics*, 23(4): 1620–1630.
4. Du, Z, Tolbert, LM, Ozpineci, B and Chiasson, JN. 2009. Fundamental frequency switching strategies of a seven level hybrid cascaded H bridge multilevel inverter. *IEEE Transactions on Power Electronics*, 24(1): 25–33.
5. Farokhnia, N, Vadizadeh, H, Fathi, SH and Anvariasl, F. 2011. Calculating the formula of line-voltage THD in multilevel inverter with unequal DC sources. *IEEE Transactions on Industrial Electronics*, 58(8): 3359–3372.
6. Fei, W, Du, X and Wu, B. 2010. A generalized half wave symmetry SHE PWM formulation for multilevel voltage inverters. *IEEE Transactions on Industrial Electronics*, 57(9): 3030–3038.
7. Fei, W, Ruan, X and Wu, B. 2009. A generalized formulation of quarter-wave symmetry SHE-PWM problems for multilevel inverters. *IEEE Transactions on Power Electronics*, 24(7): 1758–1766.
8. Filho F, Tolbert, LM, Cao, Y and Ozpineci, B. 2011. Real time selective harmonic minimization for multilevel inverters connected to solar panels using artificial neural network angle generation. *IEEE Transactions on Industry Applications*, 47(5): 2117–2124.
9. Filho, F, Maia, HZ, Mateus, THA, Ozpineci, B, Tolbert, LM and Pinto, JOP. 2013. Adaptive selective harmonic minimization based on ANNs for cascade multilevel inverters with varying DC sources. *IEEE Transactions on Industrial Electronics*, 60(5): 1955–1962.
10. Kavousi, A, Vahidi, B, Salehi, R, Bakhshizadeh, MK, Farokhnia, N and Fathi, SH. 2012. Application of the bee algorithm for selective harmonic elimination strategy in multilevel inverters. *IEEE Transactions on Power Electronics*, 27(4): 1689–1696.
11. Kouro, S, Rebolledo, J and Rodríguez, J. 2007. Reduced switching-frequency-modulation algorithm for high power multilevel inverters. *IEEE Transactions on Industrial Electronics*, 54(5): 2894–2901.
12. Liu, Y, Hong, H and Huang, AQ. 2009a. Real time calculation of switching angles minimizing THD for multilevel inverters with step modulation. *IEEE Transactions on Industrial Electronics*, 56(2): 285–293.
13. Liu, Y, Hong, H and Huang, AQ. 2009b. Real time algorithm for minimizing THD in multilevel inverters with unequal or varying voltage steps under staircase modulation. *IEEE Transactions on Industrial Electronics*, 56(6): 2249–2258.
14. Rodríguez, J, Lai, JS and Peng, FZ. 2002. Multilevel inverters: A survey of topologies, controls, and applications. *IEEE Transactions on Industrial Electronics*, 49(4): 724–738.
15. Sirisukprasert, S. 1999. Optimized harmonic stepped waveform for multilevel inverter. M.Sc. thesis, Virginia Polytechnic Institute and State University, Blacksburg, VA.
16. Taghizadeh, H and Hagh, MT. 2010. Harmonic elimination of cascade multilevel inverters with non equal DC Sources using particle swarm optimization. *IEEE Transactions on Industrial Electronics*, 57(11): 3678–3684.
17. Vázquez, N, López, H, Hernández, C, Vázquez, E, Osorio, R and Arau, J. 2010. A different multilevel current source inverter. *IEEE Transactions on Industrial Electronics*, 57(8): 2623–2632.
18. Wang, J and Ahmadi, D. 2010. A precise and practical harmonic elimination method for multilevel inverters. *IEEE Transactions on Industry Applications*, 46(2): 857–865.
19. Yousefpoor, N, Fathi, SH, Farokhnia, N and Abyaneh, HA. 2012. THD minimization applied directly on the line-to-line voltage of multilevel inverters. *IEEE Transactions on Industrial Electronics*, 59(1): 373–380.

5

Review of Control Topologies for Shunt Active Filters

5.1 Background

In general, an electrical distribution system is highly unbalanced and nonlinear, resulting in power quality issues. The power quality (PQ) is mainly exaggerated due to current harmonics injected by the nonlinear loads into the distribution system. Nonlinear loads are characteristically electronic switching power supplies, like computers, printers, copiers, uninterruptible power supplies, adjustable speed drives, electronic lighting ballasts, medical apparatus, telecommunications utensils, entertainment gadgets, etc. Poor PQ leads to undesirable operation of equipment, interference with communication lines, increased power losses, etc. It is very significant to maintain the electric PQ of the standard limits. Among many different active filters, shunt active power filter (ShAPF) is most extensively accepted and is a final choice for many industries. This chapter gives a detailed picture of topologies of ShAPF for the system containing three-phase three-wire and three-phase four-wire system. A comparison is made summarizing various topologies available in the literature with PQ detection methods, their merits, and demerits. In addition, technical factors and recent improvements on ShAPF are discussed, which is followed by summary and future scope sections.

The PQ detection and classification methods are based on the actual details extracted from measured and stored waveforms are detailed in the literature [1–18]. The passive filters [19–23] and active filters [24–28] are commonly used to improve the PQ. The ShAPF topologies are classified based on the type of power sources (AC or DC, renewable), number of phases (1 or 3 ph), inverter topologies (VSI or CSI), use of isolation transformers, use of neutral current compensation transformers, number of switching devices, advanced strategies (multi-level inverters) control strategy, etc. These ShAPFs are developed to meet the necessities of different topologies like three-phase three-wire and three-phase four-wire distribution power systems. The active power filters (APFs) consist of mainly power electronic switching devices and passive energy storage elements such as inductors and capacitors, also called as active power quality conditioners (APQC), active power line conditioners (APLC), and instantaneous reactive power (IRP) compensator (IRPC) [24–28]. These active filters aim to supply compensation for voltage harmonics, voltage unbalance to the firms over and above current harmonics, and current imbalance to the clients. The active filters offer mitigation of voltage and current harmonic disturbance, sudden voltage distortions, voltage notches, transient instabilities, and power factor (PF) enhancement efficiently in low and medium power appliances [29–34].

Most commercial and residential three-phase systems are star with neutral ground connected and have a high magnitude of neutral current. This is often observed due to the substantial applications of nonlinear loads and, usually, to the degree of fundamental load unbalance. Therefore, it is quite usual to detect heavy neutral currents containing zero-sequence fundamental current superimposed with a large amount of zero-sequence harmonics. Some of the undesirable effects arising from this are: the increase of neutral to ground voltage, the cable heating, and the transformer overheating. Most of these problems are expected to increase in number in the coming years since they are associated with the operation of power electronic devices. The examples of nonlinear-phase-to-neutral loads are rectification circuits, switched and linear power supplies, compact fluorescent lighting, variable speed drives, etc. Some of these are very efficient at producing zero-sequence harmonics. The third harmonic is the most important component. This harmonic current can easily reach levels as high as 100% of the fundamental component [1] and, in these circumstances, high neutral-harmonic-current circulation may occur. With the intention of reducing the circulation of offset zero-sequence harmonic currents all over the mains (phase and neutral cables, transformers, etc.) and to avoid undesirable consequences, classical solutions have been pointed out, such as transformer derating procedures, harmonic filters, and an independent neutral to feed nonlinear loads, etc. Filters can be divided as:

1. Passive filters: resistive, inductive, and capacitive combinations that provide low or high impedance paths to harmonic frequency currents [2]
2. Active filters: their purpose is the cancellation of harmonic components through the generation of opposing frequencies that cancel out those produced by loads [2–4]

Another approach to passive filtering emerges from the sole use of electromagnetic arrangements in order to achieve a high (blockade) or low (passing) impedance performance. This approach was inspired by the technology used to produce the saturated reactor described in [5,6], which from here on will be referred to as the electromagnetic filter. In comparison to the conventional passive filter, the electromagnetic filter is a simpler, less expensive device, especially for low-voltage application.

The working principle of SHAF is production of a suitable reference current signal equal to the compensating reference current signal. For getting a compensating reference current signal, it is necessary to get the harmonic content in the load by estimating each harmonic component or ultimately subtracting the load current with the fundamental component. Depending on the load type connected to the power system, APFs are basically divided into three configurations: two-wire system normally single-phase, three-wire system with three-phase without neutral, and four-wire having three phases with neutral [35]. Based on the connections to supply, the APF topologies are classified as series, shunt, and hybrid (combinations of both) [36]. The series and shunt APF hybrid combination is called a unified power quality conditioner (UPQC). There are two types of power-fed filters. They are current-fed and voltage-fed. The current-fed shunt current source active filter type with inductor as storage element is generally used to meet the harmonic current condition of nonlinear loads. This hybrid APF has the drawbacks of more losses and requires high rated AC power capacitors, which adds cost to the filters as shown in Figure 5.1a. The voltage-controlled converter type shunt active filter is generally located at the load end, as shown in Figure 5.1b, and is used for compensation of current harmonics, compensation of reactive power, and current balancing.

Power system distortion is generally expressed in terms of harmonic components. AC voltages are generally sinusoidal and have a fundamental frequency of 60 Hz. Harmonic currents and/or voltages are typically present in an electrical system and exist at some odd multiple of the fundamental frequency. Typical values are the third harmonic component (180 Hz), the fifth harmonic component (300 Hz), and the seventh harmonic component (420 Hz). One of the problems caused by harmonic currents is voltage waveform distortion. Other problems due to harmonic currents are interference with communication signals, excessive losses and heating in motors and transformers, excessive distribution neutral current, errors in power measurements, malfunction of protective relays, and resonance conditions at a bus that contains a harmonic source and where PF correction capacitors are connected.

In an office building, the equipment that employs switch-mode power supplies is a major source of harmonic currents. In a switch-mode power supply the 60 Hz AC voltage is converted into DC through a single-phase diode bridge rectifier. At the output of the rectifier is a large capacitor that "smoothes out" the output voltage waveform. The capacitor charges and discharges during the peak of the voltage waveform. This causes the load current to flow in sharp pulses rather than in a sinusoidal manner. Due to the deviation of the current waveform from a sinusoidal waveform, harmonic currents are created. Because of their efficiency, low cost, and light weight, switch-mode power supplies are used extensively by manufacturers of computer and other electronic equipment.

Of particular importance is the effect of harmonic distortion on the three-phase, four-wire distribution systems that exist in most commercial and industrial buildings. This type of distribution system is susceptible to excessive neutral current and transformer heating, which can lead to the eventual failure of the neutral conductor or the transformer.

In three-phase, four-wire systems, electrical loads are connected from the line to the neutral of the three phases. In an ideal balanced sinusoidal three-phase power system, the neutral current, which is

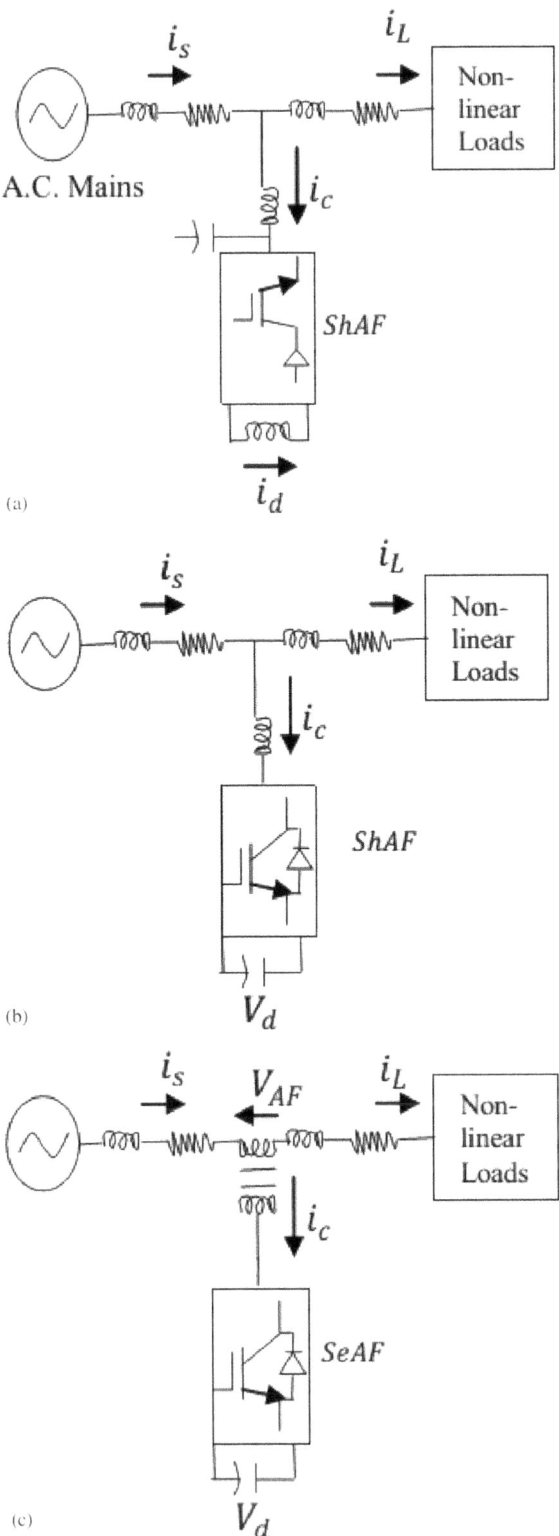

FIGURE 5.1 Single line diagram of (a) current fed type shunt active filter, (b) voltage fed type shunt active filter, (c) voltage fed type series active power filter.

equal to the vector sum of the three phase currents, is equal to zero. Under normal operating conditions some phase unbalance occurs resulting in a small neutral current.

Non-sinusoidal loads, such as office equipment, draw highly distorted phase currents. When the load is nonlinear, significant neutral current flow into the system that can be as high as 1.7 times the phase current. The triplen harmonic currents (third, sixth, ninth, etc.) add in phase in, the neutral conductor of a three-phase system. The main problem associated with excessive neutral currents is the overheating of the neutral wire. This can be hazardous in an office building that has computers and other nonlinear loads and where the neutral conductor is designed to hand only allow levels of neutral currents arising mainly from small phase imbalances. In building wiring systems, the first indication of high neutral current is often a fire or a failed transformer.

The series APF shown in Figure 5.1c can be used anywhere in the power system to compensate nonlinear load, reactive power, and voltage balance. The circuit in Figure 5.1c is connected close to load end using an identical transformer to reduce voltage harmonics. It helps in balancing the load as well as source terminal voltage and also improves voltage regulation. Large number of works is reported in the literature on various configurations and control methodologies of single-phase APFs for PQ betterment with single-phase nonlinear loads [37–46]. Single-phase APFs are classified into two topologies, namely, current-source inverter (CSI) and voltage-source inverter (VSI). CSI-based APFs are inductor-based energy storage devices whereas VSI-based APFs are fed from the energy stored in the capacitor. The performance of three-phases three-wire VSI [47–49] and three-phase four-wire VSI [50,51] APFs has been detailed in the literature. These APFs are used to mitigate PQ instability in the distribution network. For PQ improvement, the series APF [44,52], shunt active filter [37–43,53–63] is reported. Aruchamy et al. [64] proposed the ant colony optimization (ACO) control method for SAPF to improve dynamic performance and PQ. The authors have demonstrated with the hardware execution of the proposed ShAPF to authenticate the outcome of the anticipated methodology. Regarding hybrid filters [65–71], UPQC [72,73], static var compensator [74], and distribution static compensator [75,76] are extensively used terminologies in the specialization of PQ betterment. These proposed topologies have made it likely to mitigate PQ disturbances in the perspective of both utilities and consumers.

The PQ is largely affected owing to current harmonics introduced by the nonlinear loads into the distribution system. These harmonics produce the harmonics in the system voltage resulting to hamper the performance of equipment in the distribution system [60]. These disturbances spread to the consumer equipment connected at the point of common coupling (PCC). The circulation of disturbances leads in overheating of transformers, motors, and maloperation of protection devices and compensation capacitors, cables, etc. [56]. It results in rising power losses in transmission and distribution systems. The communication systems also suffer because of EMI generated by these PQ disturbances [59]. Hence, there is a requirement to explore different PQ mitigation techniques and topologies to improve the quality of power supplied.

ShAPF is also described as DSTATCOM in the literature on current control mode, induce harmonic components and reactive components of load current tackling PQ issues [77]. Voltage control mode regulates load voltage to a constant value protecting sensitive loads from voltage disturbances [78,79]. The ShAPF performance depends on control algorithm adopted for removal of reference current components [80] like IRP theory, synchronous reference frame (SRF) theory, symmetrical component (SC) theory, sliding mode control average unit power factor (AUPF) theory, and adaline-based neural network [81,82]. The optimal placing and sizing of the ShAPF play a vital function in PQ improvement. It is also proved that employing the firefly algorithm [83] and particle swarm optimization technique [84] will also improve the performance of the filter.

Commercial buildings commonly have a 208/120 V transformer in a delta–wye configuration. These transformers commonly feed receptacles in a commercial building. Single-phase nonlinear loads (such as personal computers) produce harmonic currents that cause extra losses in the transformer feeding the loads. The additional losses are the results of increased eddy currents. Excessive eddy current losses occur due to the high frequency harmonic currents because these losses are proportional to the squares of the frequencies. A transformer feeding a nonlinear load is subject to overheating and consequently should be derated or operated below rated specifications. Single-phase, nonlinear, loads connected to the receptacles feeding off a transformer produce triplen harmonic currents that algebraically add up in the neutral conductor on the secondary side of the delta–wye connected transformer. When this neutral current reaches the transformer, it is reflected into the primary winding where it circulates and causes overheating and transformer failures.

According to vast literature on the subject, the traditional approach for harmonics compensation in industrial applications assumes that the nonlinear loads can be modeled as pure harmonic current sources. Depending on the nonlinear load characteristics, the usual representation of nonlinear loads as a simple combination of harmonic current sources may greatly simplify the analyses of their effects on the overall system under similar conditions where distortion has been derived. However, additional conclusions may not be reliable if any modification is introduced in the circuit, such as the connection of a filter or a change in the loading condition. This certainly limits the usefulness of linear models to study nonlinear processes.

A more general model should include not only the harmonic current or voltage sources but also their corresponding admittances or impedances. The main problem of this approach is that the equivalent admittance varies with the circuit operating condition, due to the nonlinear nature of the real load. Obviously, according to the Norton–Thevenin equivalence, it is possible to represent the harmonic source either as current or voltage sources. Nevertheless, this equivalence is not valid for pure current or voltage sources.

Another important aspect is that some nonlinear loads should be modeled by harmonic voltage sources rather than by harmonic current sources. Among such loads are diode rectifiers with capacitive output (DC) filter, which constitute the usual interface between electronic loads and the AC feeder. Typical applications of this case are TV sets, computers, electronic ballasts for fluorescent lamps, and even the input rectifier stage of inverters, if line reactors are not used. The classification as harmonic voltage or current source depends on what is being imposed by the nonlinear load. In the case of motor-type loads, the motor inductances limit the current changing rate, which confers the behavior of harmonic current source to this kind of loads. In the case of rectifiers with DC output capacitors and without significant series impedance (such as electromagnetic interference [EMI] filters), it is the DC capacitor that limits the voltage changing rate and confers the behavior of harmonic voltage source to these types of loads. This paper uses data from laboratory measurements and simulations of typical residential loads (refrigerator, TV set, and fluorescent compact lamps) in order to verify whether they behave as current or voltage-source-type loads.

To decide which type of source model best fits each case, a harmonic shunt filter is added to the circuit and the load response is then analyzed. For pure current source behavior, it is expected that the load imposes fixed harmonic currents and the shunt filter absorbs part of this current according to the admittance, relative to the main feeder. For a pure voltage source type, it is expected that a shunt filter will increase the harmonic current emission of the nonlinear load, since the filter represents a new low-impedance path for the imposed harmonic voltages. It should be stressed that in the simulation studies of this paper, neither harmonic current nor voltage source models were used, but the electronic circuit models representing the respective loads were. The knowledge of the characteristic of nonlinear loads is important in order to understand the behavior of these kinds of loads under different operating conditions, and is especially useful to select the most effective compensation method to be used.

The share between linear and nonlinear loads is evaluated considering field measurements on a residential low-voltage distribution feeder. Finally, the prevailing behavior of the combined nonlinear loads is evaluated and tested by installing a capacitive bank at the transformer secondary.

5.1.1 Nonlinear Load Types: Current Source or Voltage Source

5.1.1.1 Current Source Load Type

Typical home appliances, which impose current distortion, are the electromagnetic devices, such as motors and transformers. Refrigerators, freezers, washing machines, and air-conditioning devices can also be included in this group. In this case, the current distortion depends on the motor's design and varies with the voltage level. If this load effectively behaves as a harmonic current source, then a shunt harmonic filter should be capable of absorbing the load's harmonic currents, according to the relative admittances of the filter ($1/Z_f$) and the feeder ($1/Z_i$). To verify this, a shunt filter, tuned 5% below the 5th harmonic (285 Hz), was designed and connected at the refrigerator (load) input terminals. The filter capacitance (20 F) was selected to fully compensate the displacement factor (fundamental PF), thus resulting in a filter inductance of 15.6 mH.

With the filter connected, the 5th current harmonic at the main supply was reduced 10.8 dB (3.4 times). This result agrees with the relation obtained from impedance measurements. At 300 Hz, the measured filter impedance is 4.1, the source impedance is 9.4, and the expected 5th harmonic reduction is 3.3 times. Thus, one can conclude that the refrigerator's nonlinearity behaves similar to a harmonic current source and, for this kind of load (refrigerator), the shunt filter was effective.

5.1.1.2 Voltage-Source-Type Load

The circuit of a diode rectifier with a capacitive output filters is an example of nonlinear load with harmonic voltage source behavior. As mentioned before, this kind of circuit is present in almost all residential and commercial nonlinear loads, such as computers, video monitors, TV sets, electronic lamp ballasts, etc. For the second test, a 20-in TV set was supplied by the same supply source used in the previous test. Such load usually absorbs constant power, and the current varies in order to compensate the input voltage changes. The voltage distortion observed at the PCC is THDV%, while the current THDI reached 108%, and the resulting PF was only 0.66. However, the fundamental PF (displacement factor) was 0.97, capacitive. The capacitive behavior at the fundamental frequency of this kind of rectifier is due to the capacitor loading process, which occurs just before the peak value of the mains voltage, so the fundamental component of the current results slightly in advance of the mains voltage. Since any LC harmonic filter also presents a capacitive behavior at the fundamental frequency, it is not possible to design a passive shunt tuned filter that simultaneously reduces the current distortion and improves the fundamental PF.

The equivalent circuit and voltage and current waveforms of respective different types of nonlinear loads are shown in Figure 5.2. The single-phase diode uncontrolled rectifier equivalent circuit [12] and its DC output voltage and current are shown in Figure 5.2a. The θ_1 to θ_4 are different commutation angles where changes in voltage and current are observed for a particular cycle of AC input waveform. The current is nearly constant, but voltage is highly distorting in nature.

In Figure 5.2b, a three-phase uncontrolled diode rectifier with its voltage and current waveforms is presented. Two modes of operation, continuous conduction mode (CCM) and discontinuous conduction mode (DCM) type rectifier behaviors are studied. The θ_1 to θ_{12} are different commutation angles describes the operation and behavior of DC voltage and AC current. The AC input voltage is continuous in general, output DC voltage waveforms become continuous or discontinuous as described in [13]. These types of loads are also very distorting particularly voltage source waveforms. In Figure 5.2c, a three-phase-controlled thyristor-based rectifier with voltage and current waveforms at different commutation angles θ_1 to θ_8 is shown. It is observed that output DC current I_D is not DC, but is in square wave shape, which will distort output current waveform [14]. The discharge lamp based equivalent circuit with its

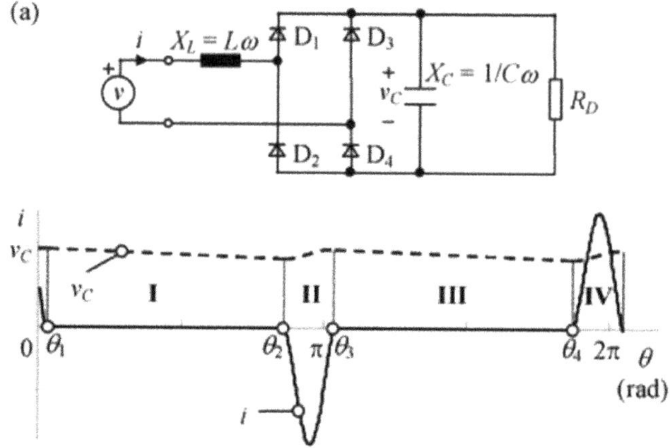

FIGURE 5.2 Equivalent circuits and current and voltage waveforms of the nonlinear loads: (a) single-phase uncontrolled rectifier.

(Continued)

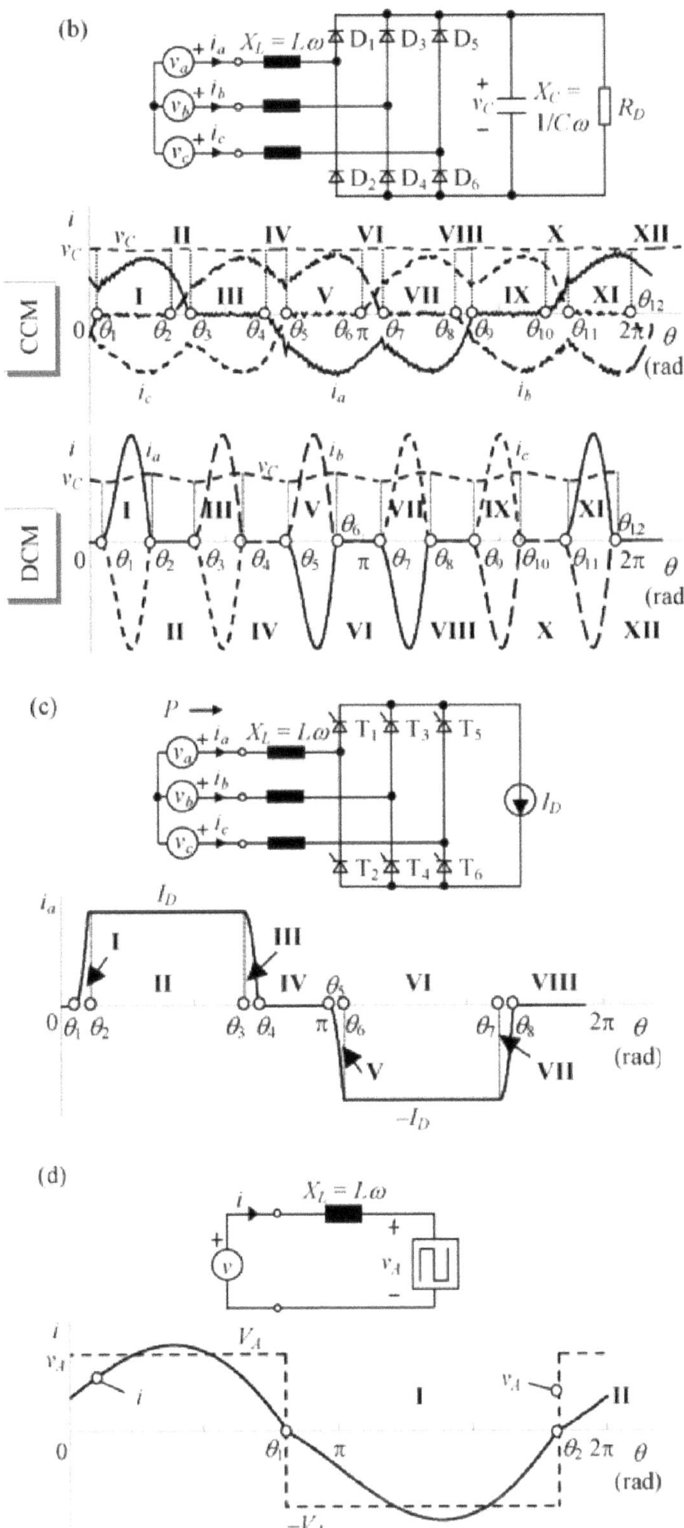

FIGURE 5.2 (Continued) Equivalent circuits and current and voltage waveforms of the nonlinear loads: (b) three-phase uncontrolled rectifier, (c) three-phase six-pulse rectifier and (d) discharge lamp.

waveforms is shown in Figure 5.2d. The arc voltage V_A, source current, which is the same as load current and commutation angles θ_1 to θ_2, is also shown here. Due to this type of load, source voltage, which is initially sinusoidal in nature, is becoming a square-shaped waveform. If such type of load is dominating, the source voltage will become very much distorted [15]. Hence the above four types of loads are said to be more distorting and compensation in source voltage and current distortions need to be considered.

5.2 Three-Phase Three-Wire Systems

Three-phase three-wire SAPFs are used for the PQ improvement in the three-phase three-wire without neutral in distribution network where the loads are typically nonlinear, unbalanced and need quick support of reactive power for PF adjustments. The topologies for the three-phase three-wire ShAPFs are classified based on isolations as (a) non-isolated and (b) isolated VSC-based ShAPFs.

1. Non-isolated VSC-based ShAPFs: Three-leg VSC based topology is shown in Figure 5.3a and is widely demonstrated in the literature [41–46]. The two-leg VSC based topology with split capacitors [9,50] is helpful as it has fewer semiconductor switching devices as shown in Figure 5.3b. However, control and regulation of equal DC voltages of DC capacitors and requirement of quite high DC bus voltages are major problems.
2. Isolated VSC-Based ShAPFs: Three 1Φ VSCs as 3Φ three-wire ShAPF is reported [9,54], but has more switching devices. This makes this topology less attractive with more switches as in Figure 5.3c. Topology with star/star (Y–Y) is shown in Figure 5.3d and two other isolated topologies of ShAPF using star/delta (Y–Δ) transformer are shown in Figure 5.3e and f. The Y–Δ transformer is required with kV A rating equal to the desired reactive power injection. Here, the transformer arrangement provides isolation from the network and also offers the flexibility to use an "off-the-shelf" VSC for the desired application. However, many configurations using different transformers may be used in this type of ShAPF.

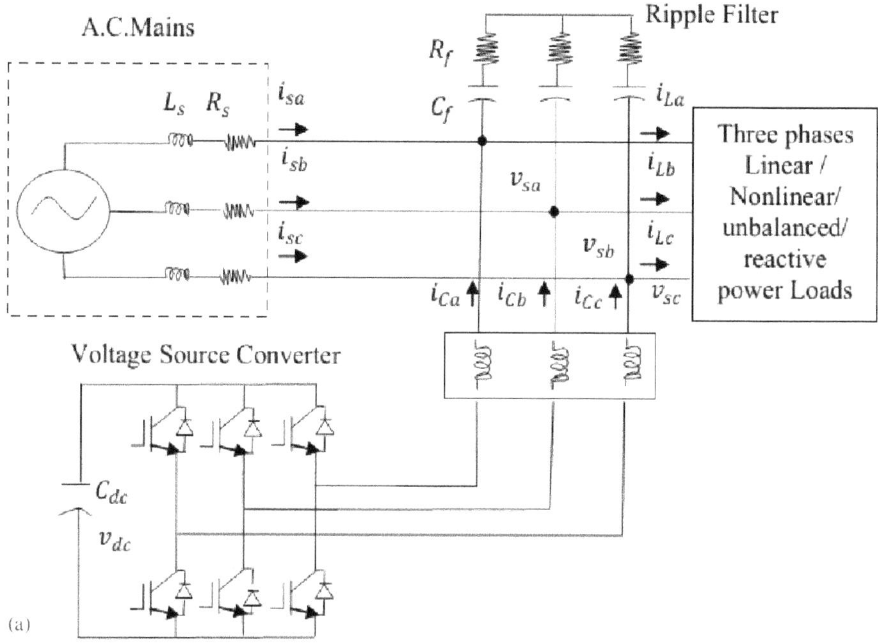

FIGURE 5.3 (a) Three-leg topology of the three-phase three-wire ShAPF. *(Continued)*

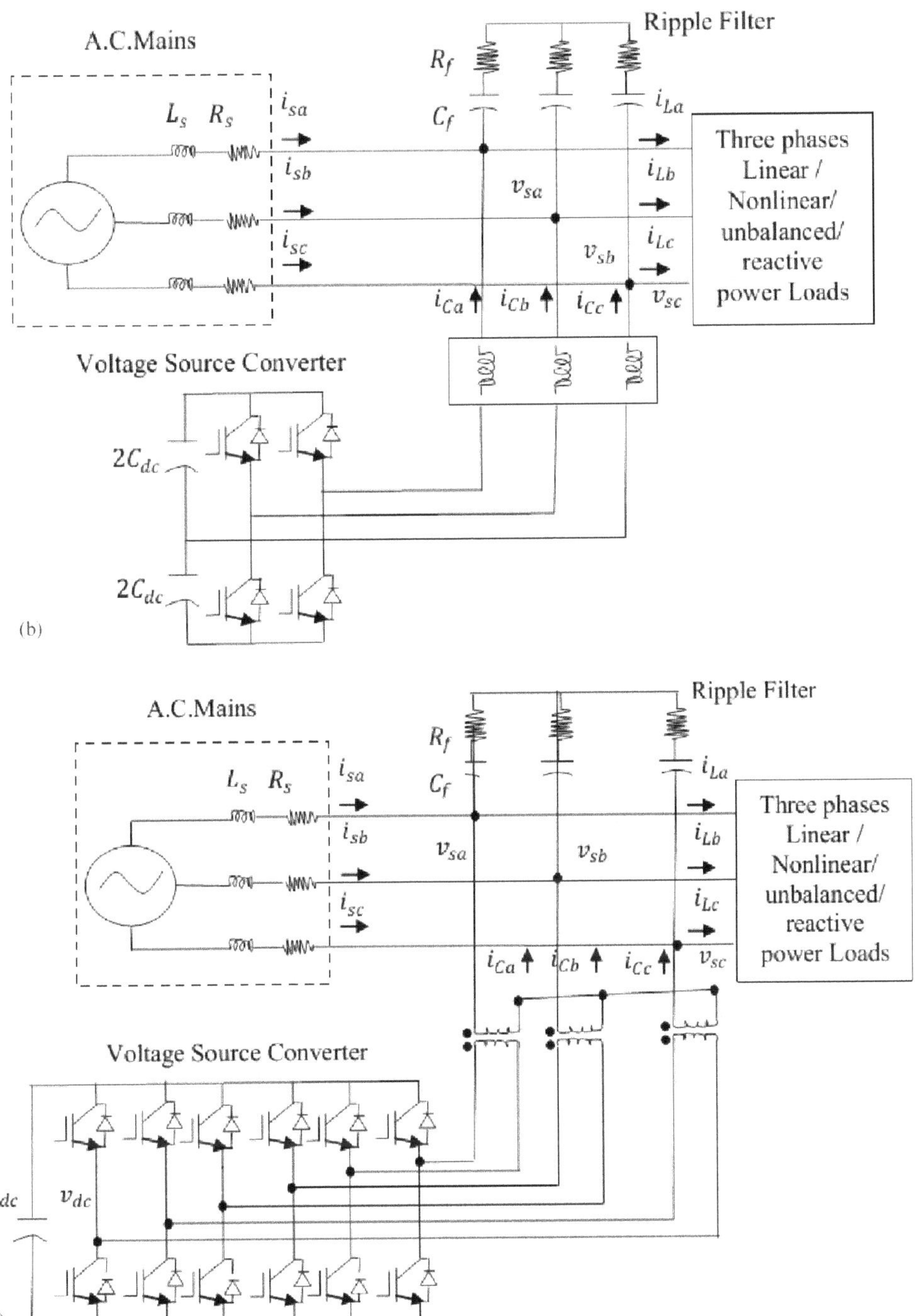

FIGURE 5.3 (Continued) (b) two-leg with split capacitor, (c) three single-phase topology. *(Continued)*

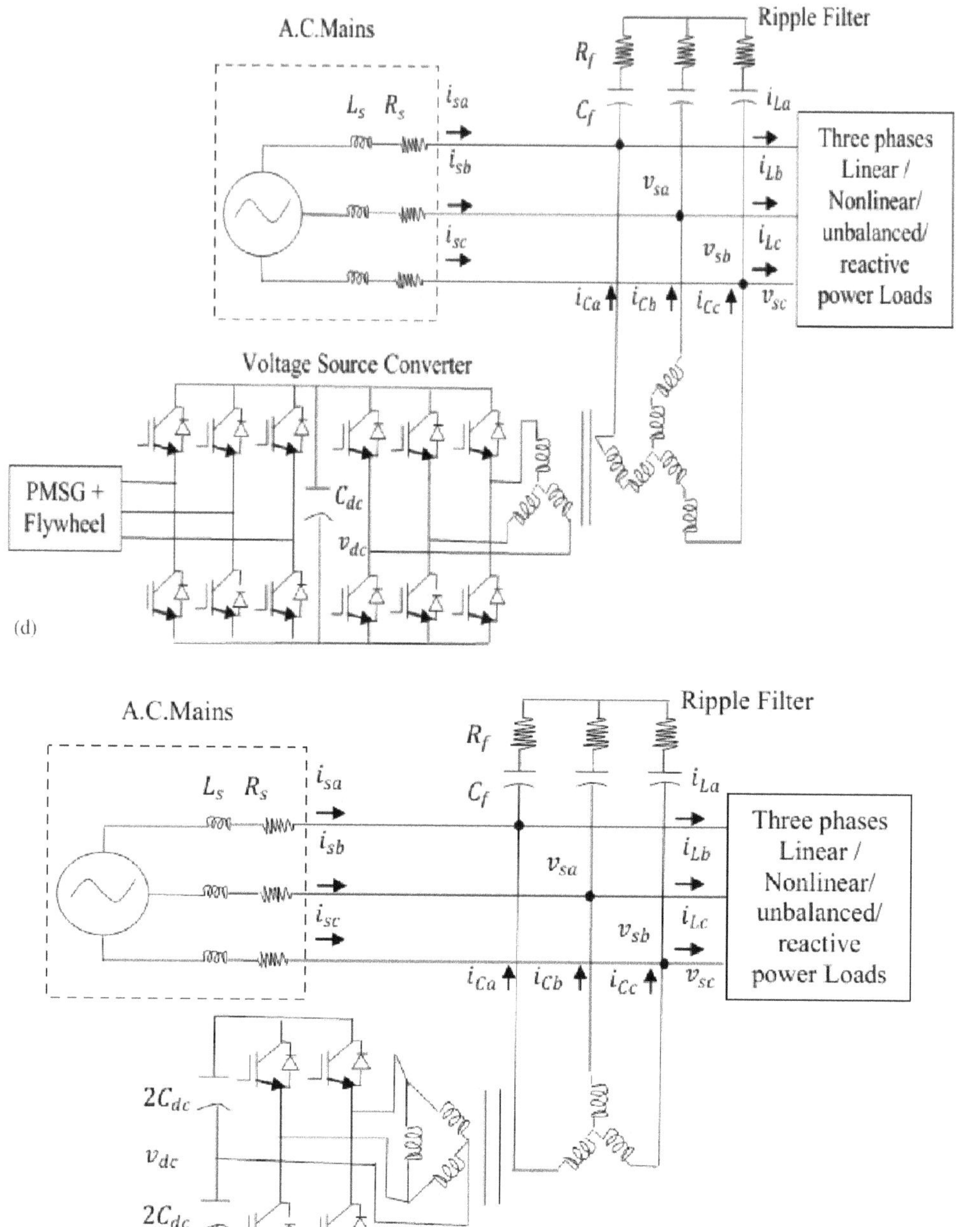

FIGURE 5.3 (Continued) (d) isolated three-leg VSC with PMSG and flywheel arrangement, (e) isolated two-leg VSC.

(*Continued*)

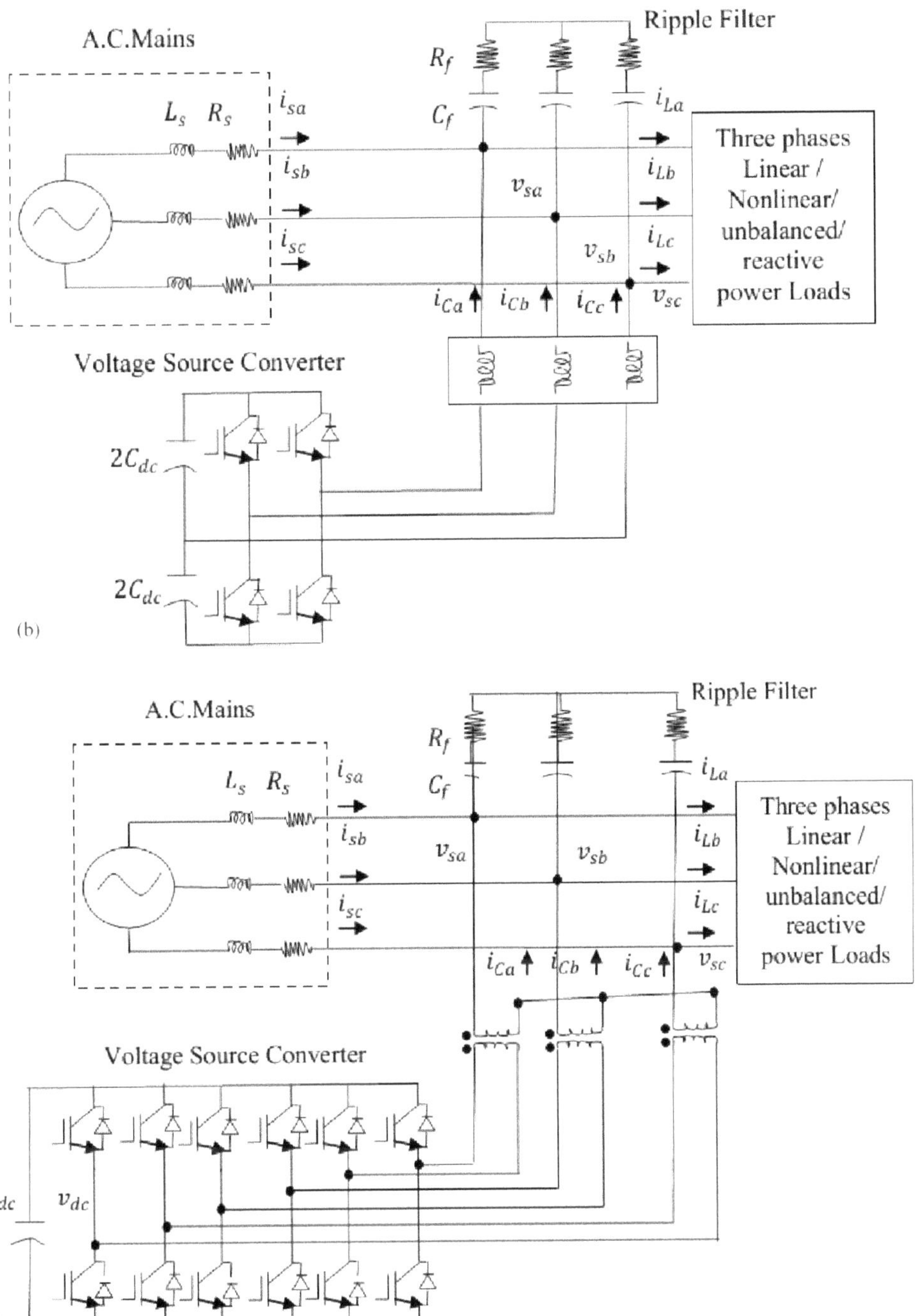

FIGURE 5.3 (Continued) (b) two-leg with split capacitor, (c) three single-phase topology. *(Continued)*

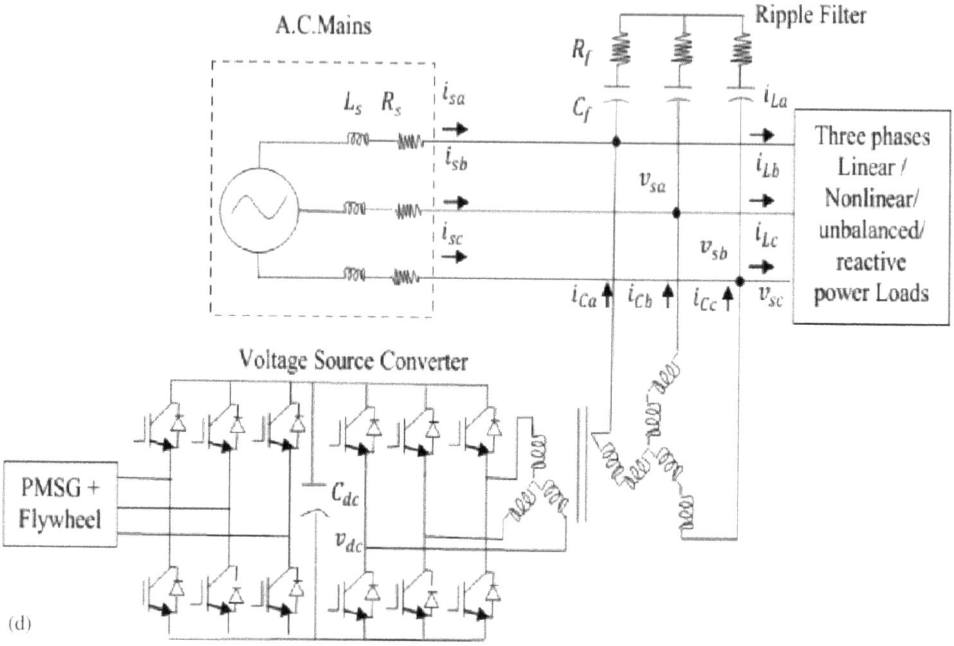

(d)

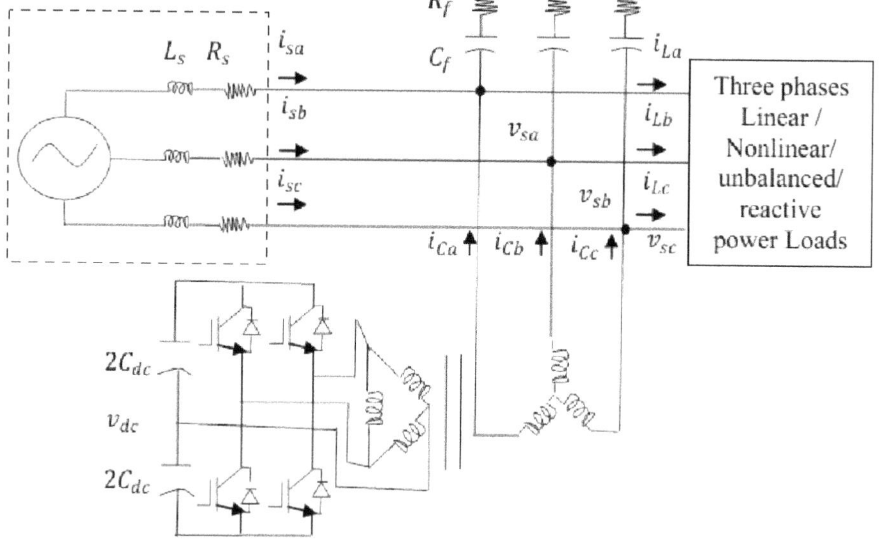

(e) Voltage Source Converter

FIGURE 5.3 (Continued) (d) isolated three-leg VSC with PMSG and flywheel arrangement, (e) isolated two-leg VSC.

(Continued)

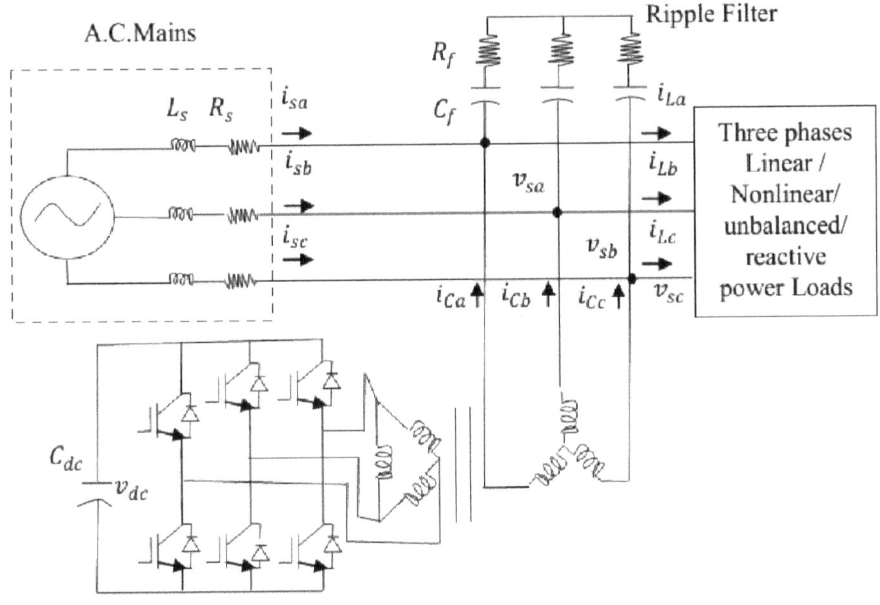

FIGURE 5.3 (Continued) (f) isolated three-leg VSC.

5.3 Design of Transformer, Passive Filters, IGBT

The main components in ShAPF are fast switching semiconductor devices such as IGBTs and MOSFETs controlled as ON and OFF switches and are used for PWM-based switching. IGBTs are used for high power rating, while MOSFETs are used for small rating and are mostly bi-directional devices with anti-parallel diode arrangement. Although ShAPF major component is the VSC, some other components are also used like DC bus capacitor, interfacing inductors, and coupling transformers. Based on the ripple current and switching frequency, the interfacing inductor is chosen and the design of the DC bus capacitor is based on the energy storage ability desired during transient conditions [56]. The switches' rating depends on compensation required for harmonics current, unbalance loading, and reactive power. The switches' voltage rating depends on the DC bus voltage across the capacitor. The tolerance in the sign is measured sometimes, which allows the ShAPF power circuit to bear events of over-current situations. The switching frequency is preferred based on the elimination of the highest order harmonics. Further, the processor speed also affects the switching frequency. To avoid violation of current deformation and to limit the current control, the DC bus voltage is desired to be maintained at a minimum voltage [12].

5.3.1 Design of Transformers

It is be seen that the harmonic current source is provided by a three-phase saturated-core reactor, which has a common star terminal connected to the three-phase AC supply neutral. Besides the fundamental frequency current, this load produces a large number of odd harmonic components, of which the third has the highest magnitude. For such purposes, a saturable reactor or a transformer is commonly used.

Neutral current compensation using zigzag transformer has been proposed in 90s. The windings are connected in a zigzag fashion as described in the literature. The windings are connected in such a way that the magneto motive force (MMF) should cancel each other. This transformer has been proposed to attenuate the neutral current of the three-phase four-wire distribution system. Scott-connected transformers were widely used as a means of interconnecting three-phase and two-phase systems, and a Scott-connected transformer without secondary winding has been proposed for neutral current compensation. This was done by installing Scott-connected transformer units near the loads that produce unbalance current. A star delta transformer has been proposed for the neutral current mitigation. This transformer is shunt connected with the load and prevents the neutral current from flowing towards the supply transformer side. The use of a star delta transformer along with a voltage source converter (VSC) as a SHAF for load balancing has been proposed. With the help of two single-phase transformers, a T-connected transformer has been proposed for the neutral current compensation and a T-connected transformer is also proposed along with a VSC as a SHAF (Distribution Static Compensator). Generally, a star hexagon transformer is used for producing six output phases, a neutral current compensation with star–hexagon transformer has been proposed in the literature. A star polygon transformer similar to a star hexagon transformer has been proposed for neutral current minimization. The previously mentioned transformer topologies are concentrated on the neural current compensation. Neutral current compensation using zero sequence blocking reactor (ZSBR) and zigzag transformer has been successfully implemented in China. The ZSBR connected in series with the utility provides high impedance to the zero-sequence current and the zigzag transformer connected in shunt with the load provides a low impedance path for the zero-sequence harmonic current so the zero-sequence harmonic current will flow through the zigzag transformer instead of the source.

The transformers for PQ applications are used in two ways: (1) non-isolated and (2) for providing isolation to the VSC. The former is used in the condition for compensating the neutral current only and the latter is used to provide isolation along with neutral current compensation. Transformers are to be designed for MMF balance to attain the neutral current compensation. The designs of the PQ transformers are as follows.

5.3.1.1 Zigzag Transformer [85]

The connection of the zigzag transformer and its phasor diagram is shown in Figure 5.4a. If V_a, V_b, and V_c are the winding voltages and V_{za} is the resultant voltage, then

$$V_{za} = K_1 V_a - K_2 V_c \tag{5.1}$$

where K_1 and K_2 are the fractions of winding in the phases.

Considering $V_a = V\angle 0°$ and $V_{za} = \sqrt{3}V\angle 30°$

Then,

$$\sqrt{3}\,V\angle 30° = K_1 V\angle 0° - K_2 V\angle 120°$$

We get $K_1 = 1$, $K_2 = 1$.

The voltage per phase is $V_{sa} = 415/\sqrt{3} = 239.60$ V, then

$$V_a = V_b = V_c = 239.60/\sqrt{3} = 138.33 \text{ V}.$$

Three single-phase transformers each of rating 1.4 kV A, 140/140 V is selected.

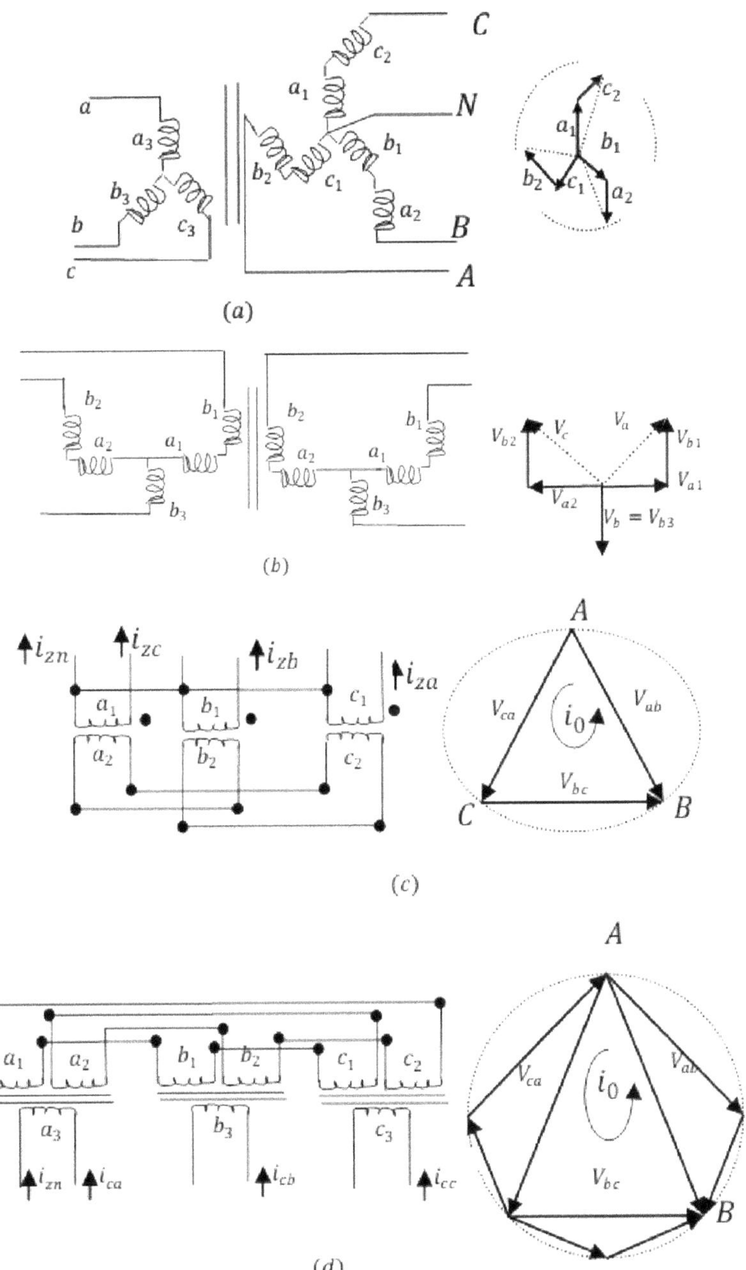

FIGURE 5.4 Isolated transformer and phasor diagram: (a) zigzag, (b) isolated T-connected, (c) star/delta, (d) isolated star/hexagon.

5.3.1.2 T-Connected Transformer [85]

Generally, a Scott transformer is used for interfacing three-phase and two-phase supply system. It uses a main transformer and a teaser transformer. Here the methodology focused only on the neutral current so secondary winding is excluded and the primary winding of Scott-connected transformer is connected in shunt with the load and ZSBR connected in series with the source transformer. The windings of the Scott transformer will attenuate the neutral current and ZSBR will block the zero-sequence components.

Two single-phase transformers, with 240/120 V, 2.4 kV A and 208/208 V, 2.1 kV A are used as shown in Figure 5.4b. The suffixes 1 and 2 represents the primary winding and secondary windings, respectively. The schematic diagram for neutral current compensation and harmonic mitigation using Scott transformer is shown in Figure 5.4b. The transformer terminals a, b, c, and n are connected to phases a, b, c, and neutral, respectively.

In this method, a T-connected transformer is connected in shunt with the load and ZSBR is connected in series with the source transformer. The windings of the transformer will attenuate the neutral current and ZSBR will block the zero-sequence components. For this proposal, three single-phase transformers are used; the first transformer has two windings and the second transformer has three windings. The connection diagram of a T-connected transformer is shown in Figure 5.4b.

The connection of two single-phase transformers for interfacing with a 3Φ four-wire system is shown in Figure 5.4b. The T-connected transformer windings gives a path for zero-sequence fundamental current and harmonic currents. Hence it offers a path for the neutral current when connected in parallel at PCC. For unbalanced loads, zero-sequence load neutral current is generally divided equally into three currents and acquires a path from the T-connected transformer windings. The windings current rating is determined based on the neutral current compensation requirement. The winding voltage is designed as shown in Figure 5.4. The phasor diagram in Figure 5.4b provides the following relations to obtain the windings turn's ratio. If V_{a1} and V_{b1} are each winding voltages and V_a is the resultant voltage, then

$$V_{a1} = V_a \cos 30°$$

$$V_{b1} = V_b \sin 30°$$

Considering $|V_a| = |V_b| = |V_c$, the line voltage is $V_{ab} = 415$ V, then

$$V_a = V_b = V_c = 415/\sqrt{3} = 239.60 \text{ V}.$$

$$V_{a1} = 207.49 \text{ V}, V_{b1} = 119.80 \text{ V},$$

Hence, two single-phase transformers of rating 2.4 kV A, 240/120/120 V and 2.1 kV A, 208/208 V are selected.

The T–T connected transformer is a combination of two single-phase transformers to produce the same phasor characteristic relation similar to a standard three-phase transformer. But, due to irregular nature of the T–T transformer connected, adverse zero-sequence and negative sequence voltages will be generated into the electric network. Fortunately, these sequence voltages are fairly small for an accurately designed T–T transformer connected under normal load conditions. The T–T transformer, consists of two single-phase transformers, is being used gradually more as a substitute or the costly standard three-phase distribution transformer. Since the T–T transformer uses single-phase transformers, definite merits, like lower manufacturing costs than for an equivalent three-phase transformer. Different from the open-delta transformer, which also has similar design with two single-phase transformers, the T–T transformer has a substantial neutral that grounds the transformer. When connecting the two single-phase transformers properly, the T–T transformer can signify delta–wye, delta–delta, or wye–wye with or without grounding (floating type) wye neutral design available in a generalized three-phase transformer.

Ideally, the T–T transformer will defer the voltage phasor diagram at no load as an equivalent generalized three-phase transformer. But, when the T–T transformer is loaded, the voltage phasor diagram turns into distorted. This distortion in voltage waveform is an outcome of the asymmetrical impedances existing in the T–T transformer to the three-phase system, and so cannot be removed completely.

As the T–T transformer is asymmetrical, the task of result of transformer impedance uniqueness by application of the transformer voltage equations will be really difficult. The technique [1] for estimating asymmetrical three-phase transformers, though, has really eased this so hard, if not unfeasible for analysis. The method produces transformer sequence impedance, which explains the transformer suitably in provisions of its winding-to-winding impedances. Once sequence impedance of transformer circuit is found for the particular investigated transformer; the remaining analysis of getting transformer distortion voltage possesses using the circuit to get the positive-, negative-, and zero-sequence voltages at the terminals of the transformer. These sequence voltages, combined by method of symmetrical voltage components, gives the line-to-ground voltages at the terminals of the transformer. Thus, the distortion in the voltage produced by the T–T transformer is derived.

The transformer has two T-connected single-phase transformers given in such a way as to get the phasor relation of a delta–wye neutral grounded transformer. Its coupled sequence impedance network, shown in Figure 5.4b, is applicable only for this transformer type design. The voltages EH_I, EH_2, and EH_O, and EL_I, EL_2, and EL_O are the voltage sequences referred to high-voltage (HV) and the low voltage (LV) transformer terminals. The HV and LV sequence terms may be interchanged on sequence impedance obtained at HV and LV terminals are also interchanged on the winding diagram of the transformer. To make simpler the sequence impedance diagram, short-circuit winding impedances are united into groups of self and mutual-impedance, as, Z_A, Z_B, Z_C, and Z_D. The result of the mutual coupling among these positive (+ve), negative (−ve), and zero (o) sequence impedances is obtained by 1-to-1 ideal transformers. The phase shift produced by the Y–Δ connection is signified by the 90-degree phase shifting transformers in the +ve and −ve sequence components. Despite the transformer connection, the 0 sequence is not phase shifted. The voltage, current, and impedance values are in per unit (p.u.) notation.

The impedances obtained in the sequence impedance network are in terms of the transformer short-circuit impedances. The information necessary to successfully determine the transformer sequence impedances is as follows:

1. A diagram of the transformer winding configuration
2. The short-circuit impedance values specified in the sequence impedance circuit
3. The rated no-load voltage across the excited transformer terminals used in finding the short-circuit impedance
4. The transformer kV A rating

Arrangement information of the transformer winding is essential as equivalent sequence impedance network will differ for each and every type of connection for the transformer. The short-circuit impedance tests must be performed on the actual transformer. The required short-circuit winding impedances, given in the sequence impedance diagram, are subscripted in terms of the transformer winding designations, so any confusion between the diagram windings and the actual transformer terminals will produce incorrect results. The short-circuit impedances for the T–T connected transformer are determined by the standard procedure of applying an exciting voltage to one winding and short circuiting any one of the remaining windings on the same magnetic circuit. The excitation voltage should be sufficient to force rates current through the energized winding. Each impedance will require a short-circuit impedance test of the actual transformer, with the subscripts of the impedances present in the sequence impedance circuit indicating which windings are to be tested. Since the short-circuit impedance between any two windings is a reciprocal impedance, the excitation source may be applied to either of the two windings while short circuiting the non-excited one. However, in the development of the equivalent sequence impedance circuit, certain nonreciprocal sequence mutual impedances may be introduced. An assumption made to prevent the occurrence of nonreciprocal sequence mutual impedances are that certain per unit short-circuit impedances is equal. For example, by reason of transformer symmetry, the impedances (per unit) Z1_3 and Zl-4, Figure 5.3b, should be nearly equal. This fact, however, is dependent upon the transformer winding arrangement and location, which may vary considerably with transformer design. If this condition of equal per unit short-circuit impedances as specified in the sequence impedance diagram is not supported from the actual test data, the presented method of analysis will be voided.

In addition to the applied voltage, current, and real power (watts) measured at the excited terminals during the short-circuit test, the no-load voltage across the excited winding is also needed to convert the short-circuit impedance to a per unit value. A base kV A of one third of the transformer total kV A is used in all per unit computations. To determine the amount of voltage distortion introduced by a T–T connected transformer, the sequence impedance circuit (representing a particular T–T transformer connection) must be utilized in a SC network. The sequence impedance network should contain the equivalent transformer circuit plus the circuits of any adjacent element in the electric distribution system. The adjacent circuits might include the entire system of which the transformer is a component; however, as in most cases, this not necessary. Usually the surrounding network may be reduced to a reasonable size by including only those elements that will have predominant effects upon the operation of the transformer. Following the determination of the adjacent network, the problem becomes a relatively straightforward process of solving the SC network for the desired voltages and currents. The sequence impedance circuit representing the T–T connected transformer is handled as any other SC circuit by placing it in the proper network location. The solution for a particular current or voltage may be accomplished by almost any method of circuital analysis and will not be discussed in detail.

In a trial solution for the positive-, negative-, and zero-sequence currents introduced by a T–T connected transformer, an equivalent delta–wye grounded transformer was chosen as shown in Figure 5.4b. Certain simplifying assumptions concerning the external elements connected to the transformer have been made, as is apparent from the diagram. First, the excitation voltage for the transformer was assumed to be supplied by a balanced three-phase infinite bus with no intervening voltage drops, and second, the configuration of the load impedance was chosen as a static grounded wye with no mutual couplings between phases. These assumptions, although somewhat ideal, permit a thorough analysis of the sequence currents introduced by the T–T connected transformer alone, since neither the excitation voltage nor the load impedance in the example induce any negative or zero-sequence voltages. It should be emphasized that these assumptions may be modified or abandoned if the complexity of the adjacent circuits demands a more acute and detailed analysis. The solution of the sequence impedance network in Figure 5.4b for the sequence currents was obtained by using the concept of phasor power. Instead of assuming a one per unit positive sequence exciting voltage and calculating the mesh currents IL1, IL2, and ILO, a one per unit current was assumed to be flowing in the positive-sequence circuit thereby eliminating the need to solve three simultaneous equations for the currents IL2 and ILC. The magnitudes of the resulting negative- and zero-sequence currents as calculated are expressed as a percentage of the positive-sequence current.

5.3.1.3 Star/Delta (Y–Δ) Transformer [86]

Here, a star–delta transformer is connected in shunt with the load and ZSBR is connected in series with the source transformer for neutral current and harmonic mitigation. It uses three single-phase transformers, each having ratings of 240/240 V, 2.4 kV A. The connection diagram of a star–delta transformer is shown in Figure 5.4c.

The star/delta (Y–Δ) transformer connection and its phasor diagram are shown in Figure 5.3c. The transformer winding current rating is based on the circulating current because of the zero-sequence components in the load current. The voltage of the primary winding is

$$V_a = V_{LL}/(\sqrt{3}) = 415/\sqrt{3} = 239.60 \text{ V}.$$

Therefore, 240-V winding is selected in Y–Δ transformer. The secondary line voltage is selected for the same current to flow in the windings. The transformer voltage ratio is 1:1. Hence, three numbers of 1Φ transformers each of rating 2.4 kV A, 240/240 V are preferred.

5.3.1.4 Star/Hexagon Transformer [87]

The primary winding of the star–hexagon transformer is connected in shunt with the load and ZSBR is connected in series with the source transformer. The windings of the transformer will attenuate the neutral current and ZSBR will block the zero-sequence components. For this proposal, three single-phase transformers, each having ratings of 240/140/140 V, 2.4 kV A, are used. The connection diagram of star hexagon transformer is shown in Figure 5.4d.

The hexagon-connected transformer secondary winding provides a path for the harmonic currents and zero-sequence fundamental current. Therefore, it provides a neutral current path when connected in shunt at PCC. In 1Φ loads, the zero-sequence load neutral current circulates in the hexagon windings of the transformer. Each primary winding voltage is the phase voltage. The star–hexagon transformer windings voltage rating is designed as in Figure 5.3. This star/hexagon transformer with the phasor diagram shown in Figure 5.3d explains the subsequent relations to get the turns ratio of windings. If V_a, V_b and V_c are the per-phase voltages across each transformer winding and V_{ca} is the resultant voltage, then

$$V_{ca} = K_1V_a - K_2V_c,$$

where K_1 and K_2 are fractions of windings in the phases. Considering $V_a = V \angle 0°$ and $V_{za} = \sqrt{3}V \angle -120°$, we get $K_1 = 1$, $K_2 = 1$. The line voltage is $V_{ca} = 200$ V, then

$$V_a = V_b = V_c = 200/\sqrt{3} = 115.50 \text{ V}.$$

Hence, three single-phase transformers, each of rating 2.6 kV A, 240/120/120 V, are selected.

A three-leg VSC is a combination with a star/hexagon transformer for the reactive power mitigation for PF correction or voltage regulation along with elimination of harmonics currents, load balancing, and neutral current mitigation. The transformer-connected star or hexagon offers a path to zero-sequence current in the three-phase four-wire distribution system. To optimize the VSC voltage rating, the transformer with star/hexagon is intended to have an appropriate voltage rating for the secondary windings for the three-leg VSC integration. This connection for the transformer presents the choice of "off-the-shelf" VSC for this purpose and it also gives isolation for the VSC circuit. Star/hexagon transformer is accounted in the literature for various purposes. For optimal voltage rating of the three-leg VSC, the secondary of the transformer is calculated for low voltage. The dynamic behavior is studied for PF correction and voltage regulation modes of the SHAF. The load harmonic current's mitigation and load balancing are also obtained along with PF correction or voltage regulation. The SHAF features the following characteristics.

a. Use of readily available three-phase three-leg VSC as SHAF for three-phase four-wire system
b. Neutral current compensation with linear and nonlinear loads using the star/hexagon-connected transformer
c. Isolated operation of three-leg VSC as it is integrated with the star/hexagon-connected transformer
d. Harmonic current compensation and load balancing
e. Reactive current compensation for unity power factor (UPF) or the ZVR at the PCC
f. Capacitor-supported operation of SHAF

The SHAF is connected in shunt with the load at the PCC. The inductor (L_s) corresponds to line inductance and the resistor (R_s) corresponds to the effective resistance of the line. The current injected by the SHAF is controlled based on the required compensation. Figure 5.1b depicts the compensator operation in the PF correction mode using the phasor diagram. The SHAF reactive current (I_c) is injected to revoke the load current reactive power component so that the source current is reduced to the active power current component only (I_s). Figure 5.3d shows the phasor diagram for ZVR operation. In this mode, SHAF injects a current I_c, such that the voltage at PCC (V_S) and source voltage (V_M) are in the locus of the same circle. The SHAF currents are adjusted dynamically under varying load conditions.

Ripple Filter: To filter the high-frequency noise from the voltage at the PCC, a low-pass first-order filter tuned at half the switching frequency is used. To provide low impedance for the harmonic voltage at a 5 kHz frequency, the ripple filter capacitor is calculated as $C_f = 5$ μF [88]. A series resistance (R_f) of 5 Ω is integrated in series with the capacitor (C_f).

5.4 Design of Capacitors for VSC

Design of VSC: Some topologies design of ShAPF is reported in the literature. However, selection of the component for a three-leg VSC in three-phase three-wire ShAPF is considered here for a sample calculation. The ShAPF is designed for a rating of 12 kV A for the reactive power compensation of the load. The DC bus capacitor value in the DC bus capacitor voltage and interfacing inductor are obtained as follows.

1. DC Bus Capacitor: The DC capacitor (C_{dc}) value depends on the instantaneous energy available to the ShAPF during transients [89]. The energy conservation principle is applied as

$$\frac{1}{2}C_{dc}\left[(V_{dc}^2)-(V_{dc1}^2)\right]=3V(aI)t$$

where V_{dc} is the reference DC voltage and V_{dc1} is the minimum voltage level of DC bus, "a" is the overloading factor, V is the phase voltage, I is the phase current, and t is the time by which the DC bus voltage is to be recovered.

Considering the minimum voltage level of DC bus $V_{dc1} = 690$ V and $V_{dc} = 700$ V, $V = 239.60$ V, $I = 27.82$ A, $t = 350$ μs, $a = 1.2$, the calculated value of C_{dc} is approximately 3000 μF.

2. DC bus capacitor voltage: The minimum DC bus voltage should be greater than twice the peak of the phase voltage of the system [89]. The DC bus voltage is calculated as

$$V_{dc} = \sqrt{2}V_{LL}/\sqrt{3m}$$

where "m" is the modulation index and is considered here as 1 and in general lies between 0 and 1. It is selected as 700 V for a of 415 V power distribution system.

3. AC Inductor: The selection of the AC inductance (L_f) depends on the current ripple [89], $i_{cr,\,p\text{-}p}$ switching frequency f_s, and DC bus voltage (V_{dc}), and L_f is given as

$$L_f = (\sqrt{3m}V_{dc})/(12af_s i_{cr(p-p)})$$

where m is the modulation index and "a" is the overload factor. Considering $i_{cr,\,p\text{-}p} = 5\%$, $m = 1$, $V_{dc} = 700$ V, $f_s = 10$ kHz, and $a = 1.2$, the L_f value is approximately 2.5 mH.

5.5 Topologies-Design Consideration

The topologies for ShAPFs are based on either three-phase three-wire or four-wire network and also based on selection of isolation. Isolation is provided from the power system to the VSC using different types of converter transformers as described below.

The zigzag transformer: It provides rugged, passive compensation and fewer complexes over the active compensation techniques. It has the advantage of reduction in load unbalance and reducing the neutral current on the source side. It has the lowest kVA rating. However, performance of a zigzag transformer is dependent on the location close to the load. Performance of reducing the neutral current on the source side is affected during the conditions of distorted and unbalanced voltages.

Star–delta: Easily available on the market, simple design, less costly. Star-connected primary winding offers a low impedance path for zero-sequence currents. The delta-connected secondary winding provides a path for the induced zero-sequence currents to circulate. But, its compensation characteristics depend on the impedance of the transformer, location, and source voltage. It will not completely compensate for the neutral current and high kVA rating.

T-connected Transformer: Is low in height, small size in floor space, and lower weight than any other types of transformers. Two single-phase transformers are used, which make the core more economical to build and easy to assemble. It is observed as open-circuit for the positive as well as negative sequence components of currents, so current flowing from the transformer is only zero-sequence component. However, it has the disadvantage that the transformer rating depends on the magnitude of the load harmonic content and the imbalance. Their compensation behaviors depend on the location, source voltage, and imbalance of the transformer. Impedance offered for these zero-sequence component of current depends on the zero-sequence utility system impedance.

Star–hexagon: Primary winding of a star-connected transformer facilitates a low impedance path for the zero-sequence harmonic currents. The secondary winding in hexagon shape offers a path for the zero-sequence induced currents. It will decrease zero-sequence harmonic current to a major extent. The main drawback is, it will not entirely mitigate zero-sequence currents. Its mitigation behavior depends on the transformer impedance, source voltage, and location. It has high cost and complex design and is not that easily available in the market. It has highest kVA rating.

The selection of particular topology is chosen mainly with the availability of a neutral wire called the fourth wire, availability of transformer, choice of passive filters, inductor and capacitor rating, spacing, sizing, and economic values.

5.6 Three Phase Four Wire Systems

Three-phase four-wire ShAPFs for the PQ improvement are useful where three-phase four-wire electric power distribution system is available. The classification of the three-phase four-wire ShAPF topologies is shown in Figure 5.5 and is classified with and without transformer-based ShAPF topologies.

1. Non-isolated VSC without transformers [90–106]: The ShAPF topologies without transformers are classified as four-leg VSC topology and three-leg as shown in Figure 5.5. The four-leg topology of ShAPF is shown in Figure 5.5a and is extensively described in the literature [107–112], and the VSC's fourth leg is connected to the neutral conductor for neutral current compensation in the main power distribution network. Similarly, other topologies of ShAPF for three-phase four-wire system for the mitigation of neutral current along with PQ improvement in the supply current are three-leg with split capacitors (Figure 5.5b) [9], three-leg with neutral terminal at the positive or negative terminals of DC bus (Figure 5.5c) [105], and a hybrid ShAPF (Figure 5.5d) [106].
2. Non-isolated three-leg VSC with transformers [85–89,113–128]: The ShAPF topologies with transformers are further classified as non-isolated and isolated VSC-based ShAPFs. A topology of ShAPF based on non-isolated three-leg VSC with a zigzag transformer is shown in Figure 5.6a. The zigzag transformer here reduces the current in the neutral wire. This topology has advantages like ruggedness, passive compensation, and lower complexity over active compensation methods in [129,130] and [85,88,113]. Other topology based on star/delta transformer with three-leg VSC is shown in Figure 5.6b [86,114]. The other transformers such as T-connected [89] and star/hexagon transformer [115] are also used along with three-leg VSC as three-phase four-wire ShAPF as shown in Figure 5.6c and d, respectively. The merits of zigzag transformer are also applicable to other types of transformer configurations.
3. Non-isolated two-leg VSC with transformers [116–119]: A two-leg VSC with split capacitors and a zigzag transformer for three-phase four-wire ShAPF is shown in Figure 5.6e [116]. A zigzag transformer and other transformers like star/delta transformer, T-connected transformer, and star/hexagon transformer with two-leg VSC as three-phase four-wire ShAPF are shown in Figure 5.6f [117], g [118], and h [119], respectively. The power electronic switches are less in four-leg than three-leg VSC topologies.

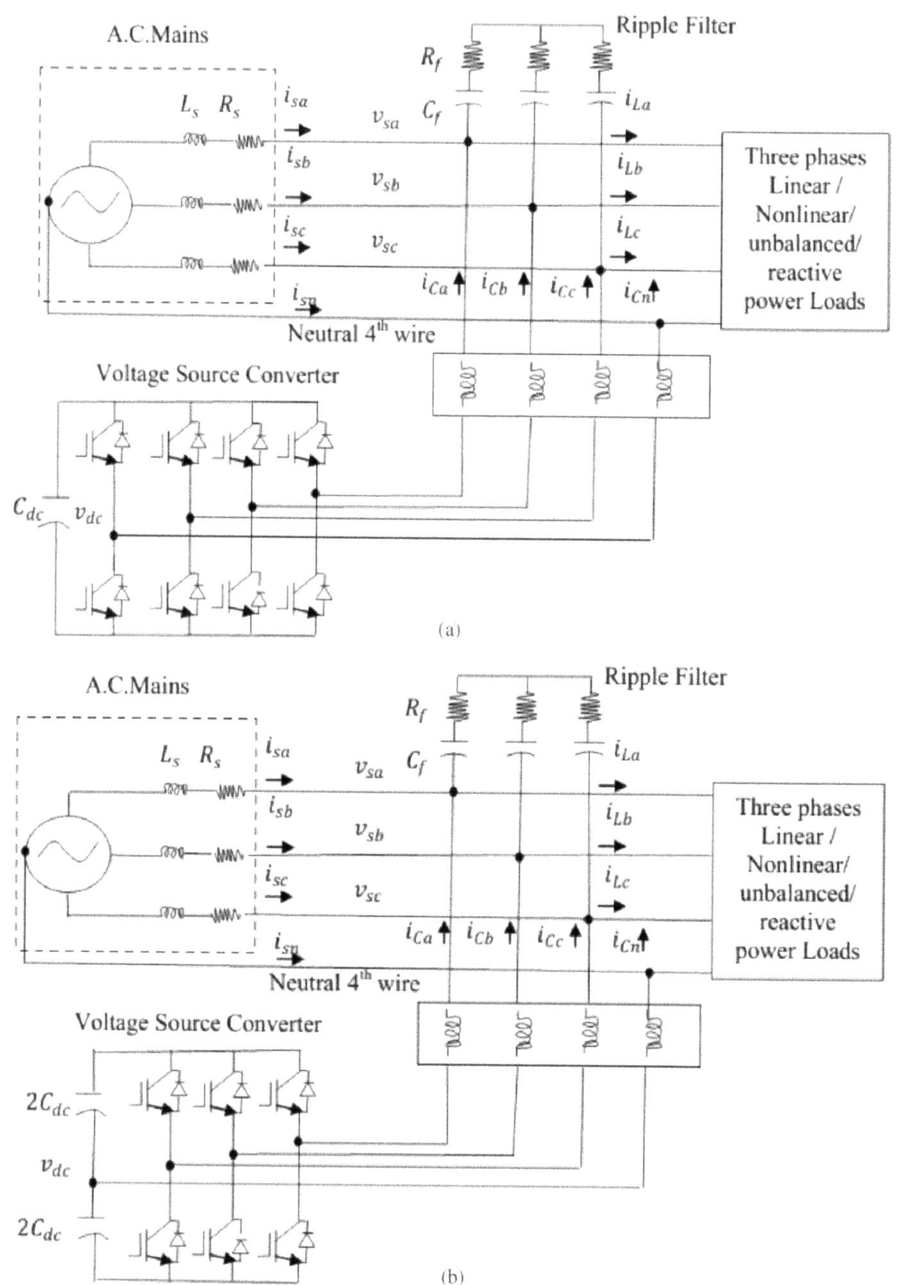

FIGURE 5.5 Topologies for three-phase four-wire ShAPF (a) four-leg topology of the three-phase four-wire VSC based, (b) three-leg with split capacitor.

(Continued)

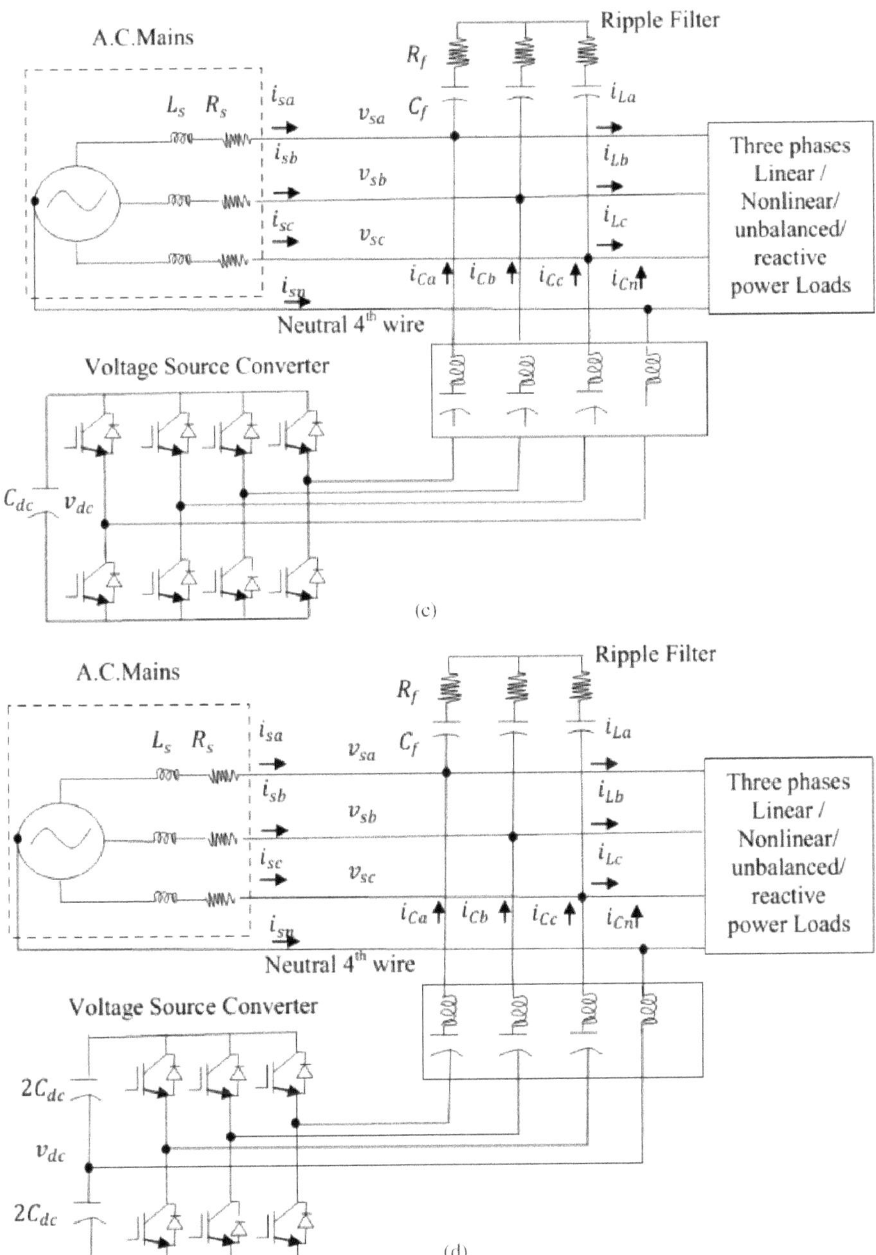

FIGURE 5.5 (Continued) Topologies for three-phase four-wire ShAPF (c) three-leg with neutral terminal at DC bus, (d) three-leg VSC with three DC capacitors.

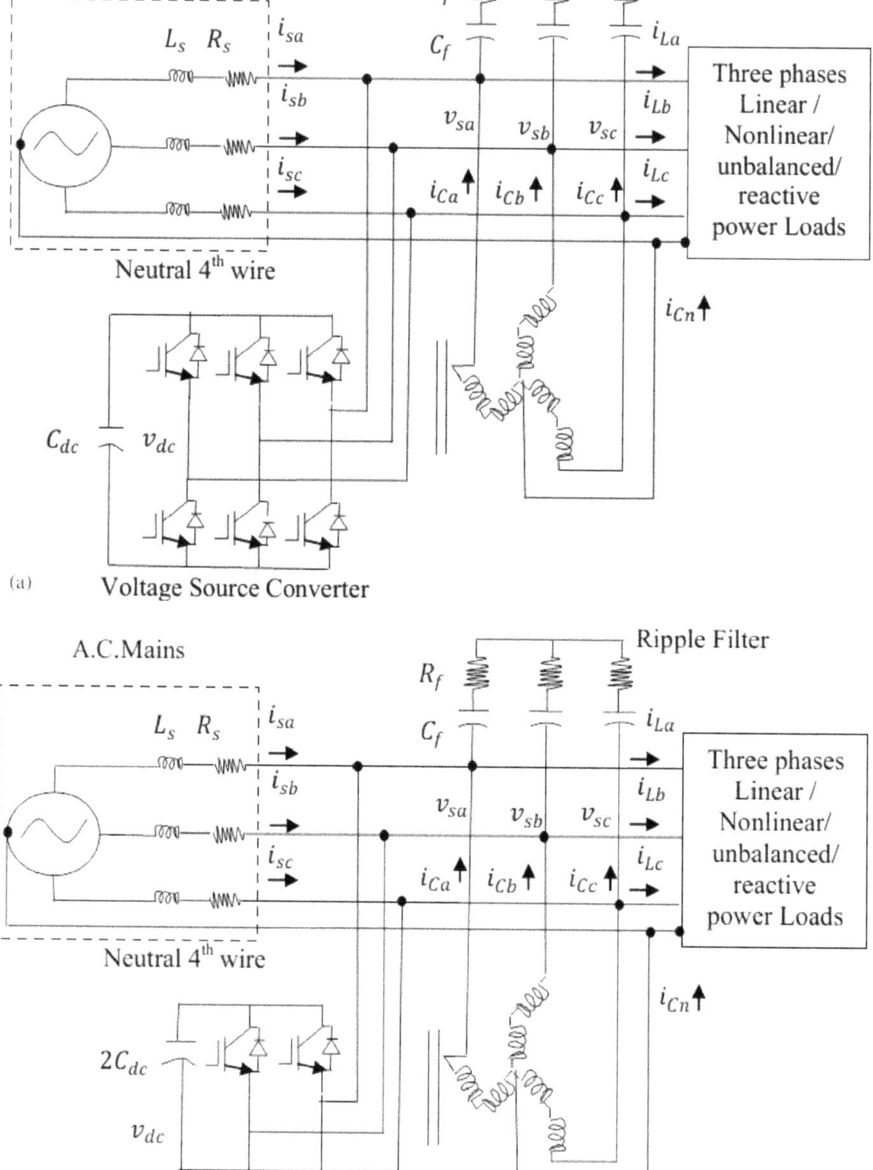

FIGURE 5.6 Topologies for three-phase four-wire ShAPF with non-isolated VSC using transformers (a) three-leg VSC based with zigzag, (b) three-leg split capacitor with star–delta. (*Continued*)

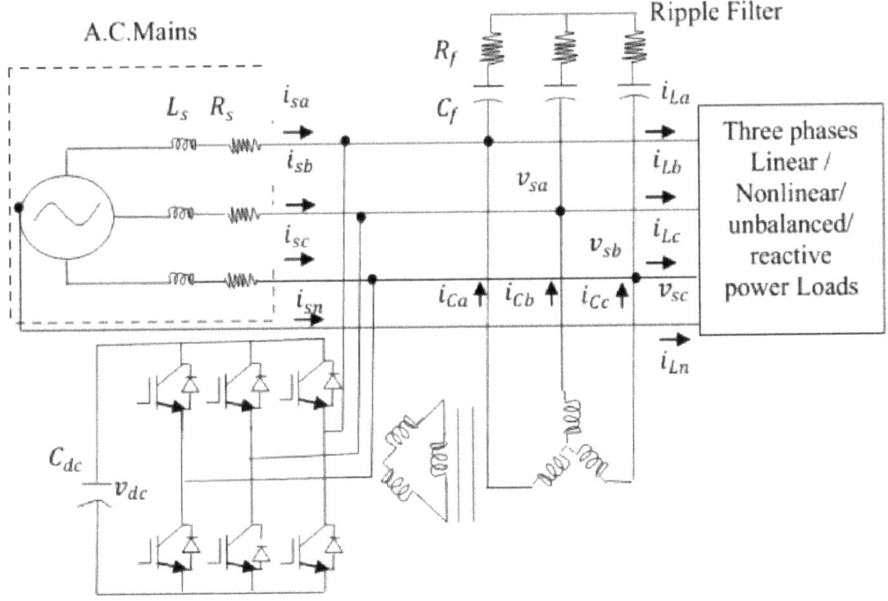

(c) Voltage Source Converter

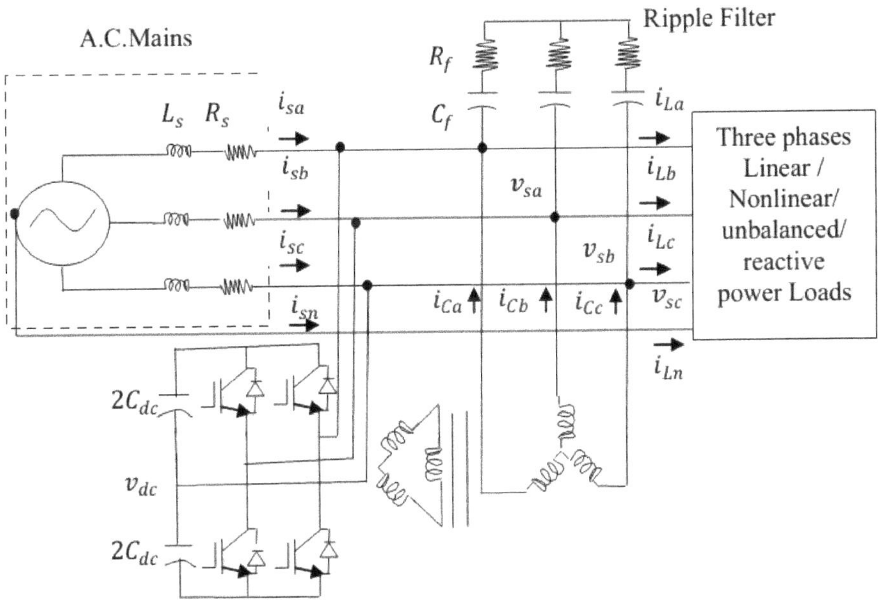

(d) Voltage Source Converter

FIGURE 5.6 (Continued) Topologies for three-phase four-wire ShAPF with non-isolated VSC using transformers (c) three-leg with T-connected, (d) three-leg VSC with star–hexagon. *(Continued)*

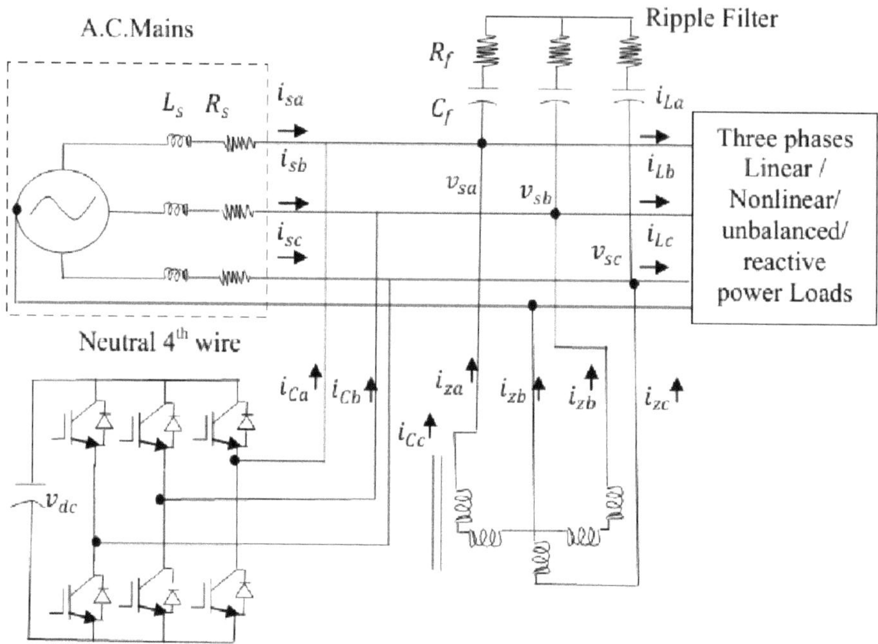

(e) Voltage Source Converter

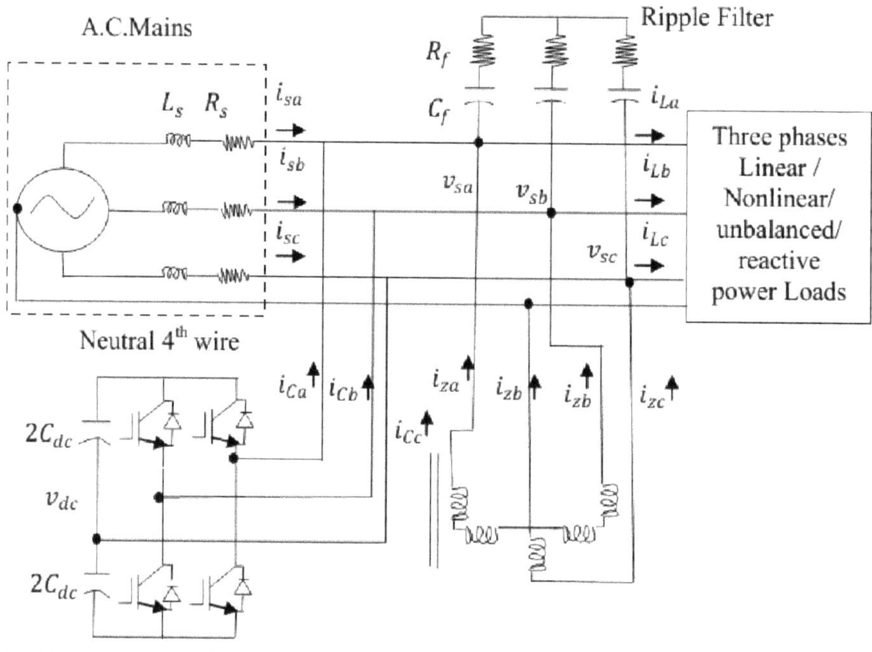

(f) Voltage Source Converter

FIGURE 5.6 (Continued) Topologies for three-phase four-wire ShAPF with non-isolated VSC using transformers (e) two-leg VSC with zigzag, (f) two-leg split capacitor with star–delta. *(Continued)*

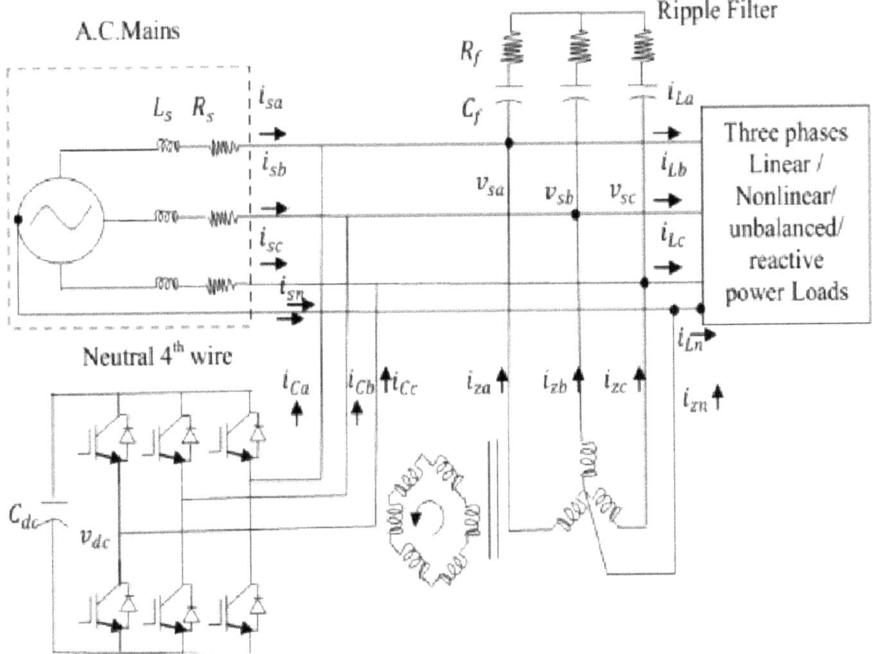

(g) Voltage Source Converter

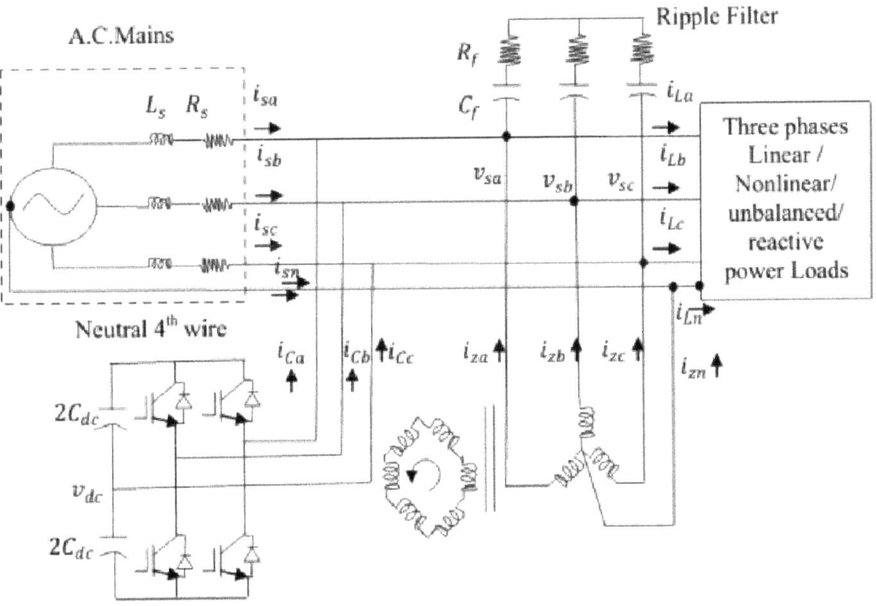

(h) Voltage Source Converter

FIGURE 5.6 (Continued) Topologies for three-phase four-wire ShAPF with non-isolated VSC using transformers (g) two-leg with T-connected, (h) two-leg VSC with star–hexagon.

4. Isolated three single-phase VSCs: Three single-phase VSCs as three-phase four-wire ShAPF are given in [110,120] and shown in Figure 5.7a. The ShAPF consists of three H-bridge VSCs that are maintained by a common DC link storage capacitor. Three single-phase transformers are connected to the outputs of these VSCs to provide isolation between source and converter and also to provide inductance between the VSCs and PCC.
5. Isolated three-leg VSC with transformers [87,121–125]: The three-phase four-wire ShAPF based on three-leg VSC-connected to the secondary of a zigzag transformer is shown in Figure 5.7b [121,122]. The other transformers such as star/delta transformer, T-connected transformer, and star/hexagon transformer are also used along with isolated three-leg VSC as three-phase four-wire ShAPF as shown in Figure 5.6c [123], d [124], and e [87,125], respectively.
6. Isolated two-leg VSC with transformers [126–128]: Isolated H-bridge VSC with split capacitors used along with a transformer as 3Φ four-wire ShAPF with a reduced number of semiconductor devices connected to the secondary of a zigzag transformer is shown in Figure 5.7c [126]. The other transformers such as star/delta transformer [127], T-connected transformer, and star/hexagon transformer [128] are also used along with two-leg VSC as three-phase four-wire ShAPF as shown in Figure 5.7d–i, respectively.

Figures 5.6d and 5.7h, i show the star/hexagon transformer-based three-phase four-wire SHAF for PQ improvement. A three-leg VSC is connected to the secondary of the transformer, which is hexagon-connected winding. The neutral current compensation is also achieved because this hexagon winding provides a circulating path for the zero-sequence components of the load currents. The star-connected primary winding of the transformer is connected to the PCC. The SHAF consists of a three-phase pulse-width modulated (PWM) VSC using six insulated-gate bipolar transistors (IGBTs), three interface inductors, and one DC capacitor. This star/hexagon-connected transformer provides isolation to the SHAF as well as the suitability of selecting an off-the-shelf three-leg VSC.

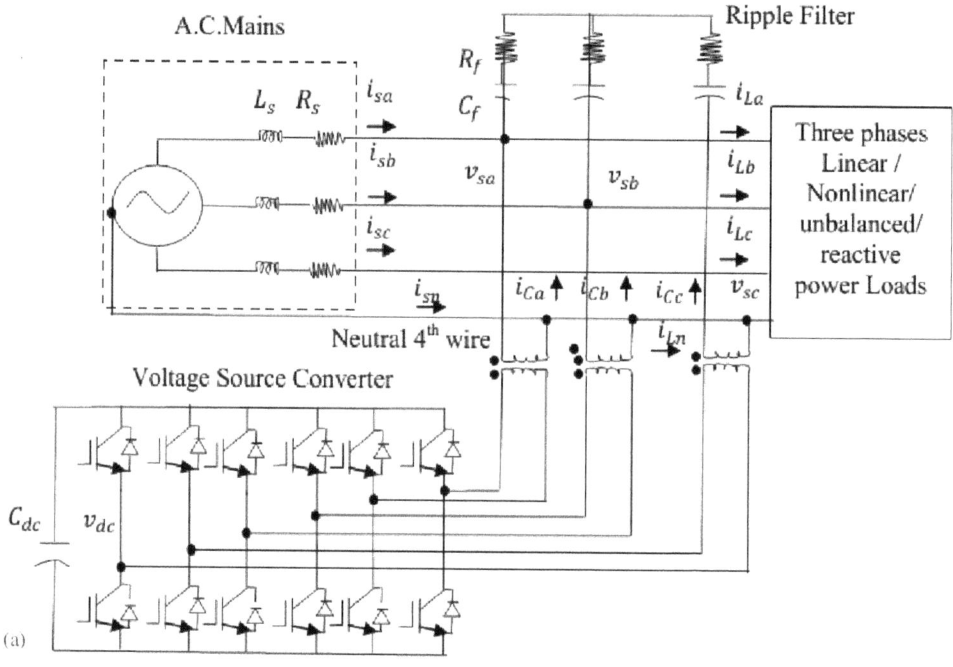

FIGURE 5.7 Topologies for three-phase four-wire ShAPF with isolated VSC using transformers (a) four single-phase topology.

(Continued)

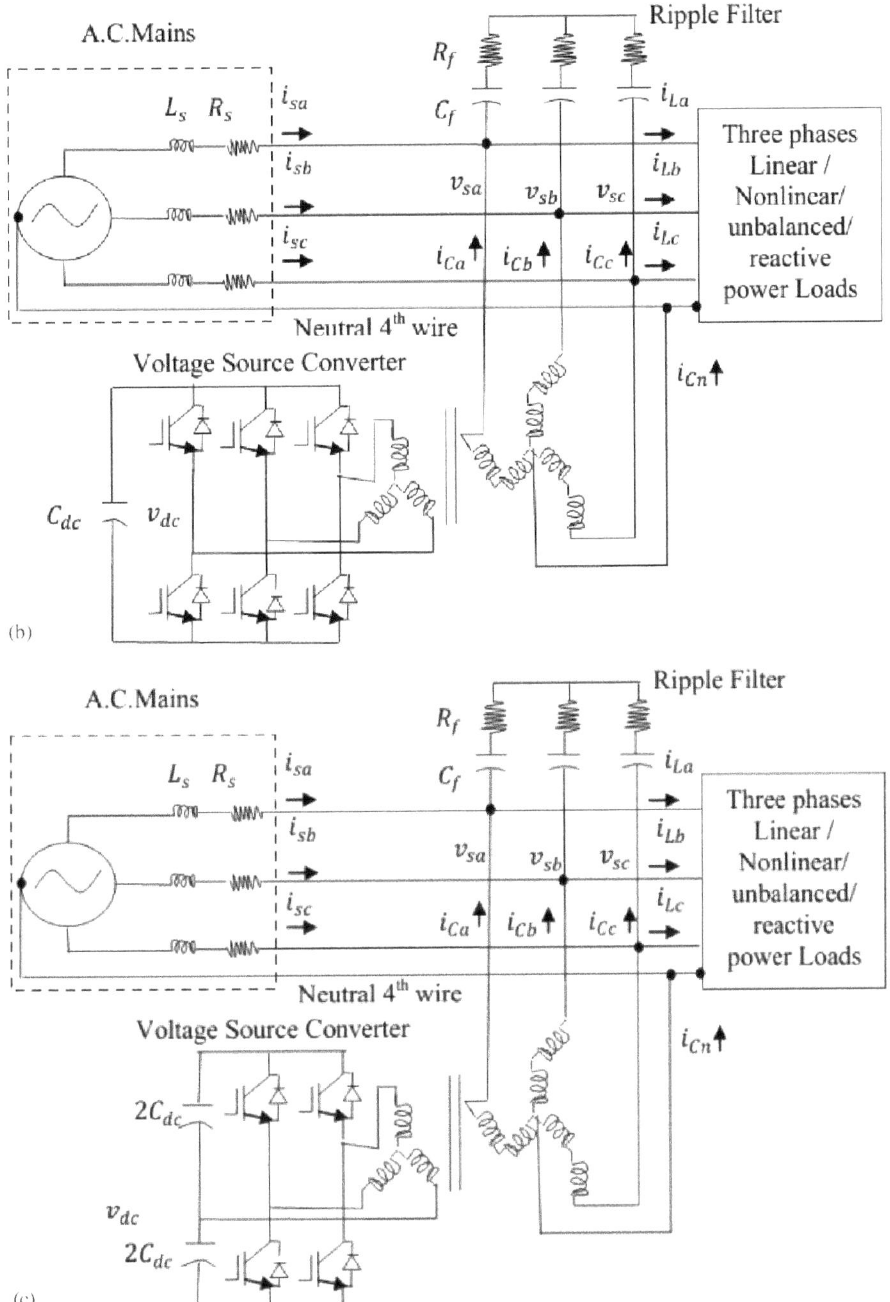

FIGURE 5.7 (Continued) Topologies for three-phase four-wire ShAPF with isolated VSC using transformers (b) three-leg VSC based with zigzag, (c) two-leg VSC based with zigzag. *(Continued)*

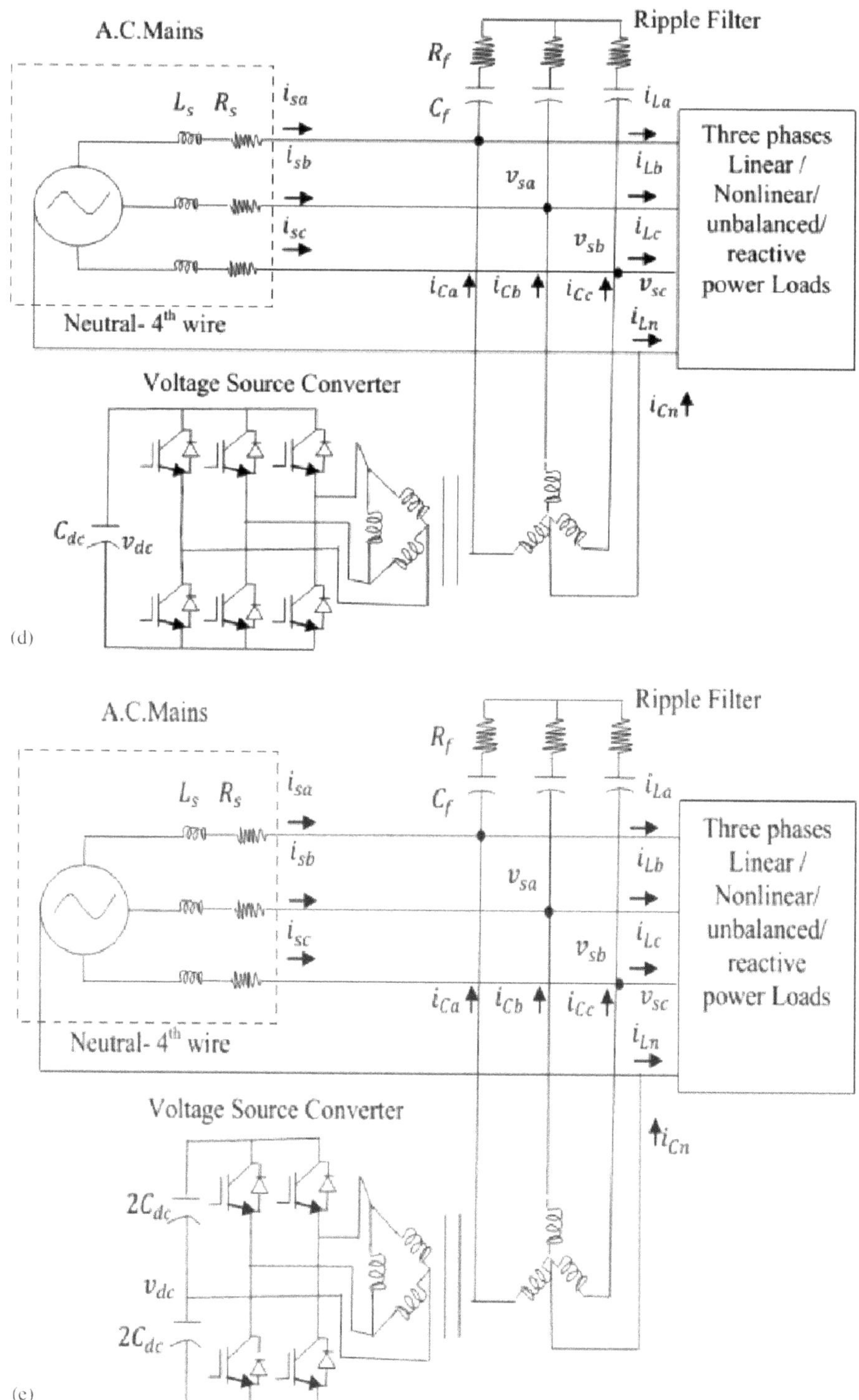

FIGURE 5.7 (Continued) Topologies for three-phase four-wire ShAPF with isolated VSC using transformers (d) three-leg with star–delta, (e) two-leg split capacitor with star–delta. *(Continued)*

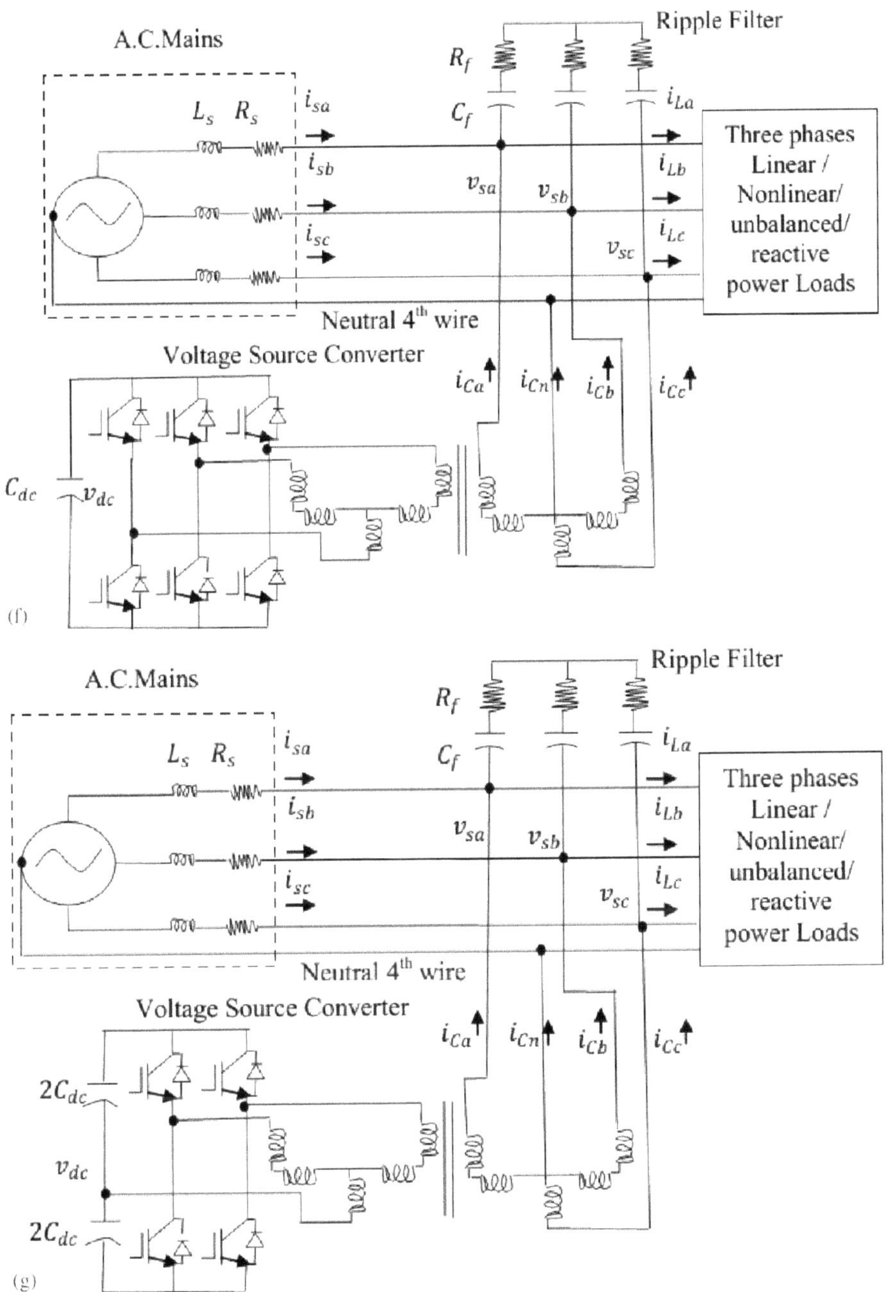

FIGURE 5.7 (Continued) Topologies for three-phase four-wire ShAPF with isolated VSC using transformers (f) three-leg with T-connected, (g) two-leg with T-connected.

(Continued)

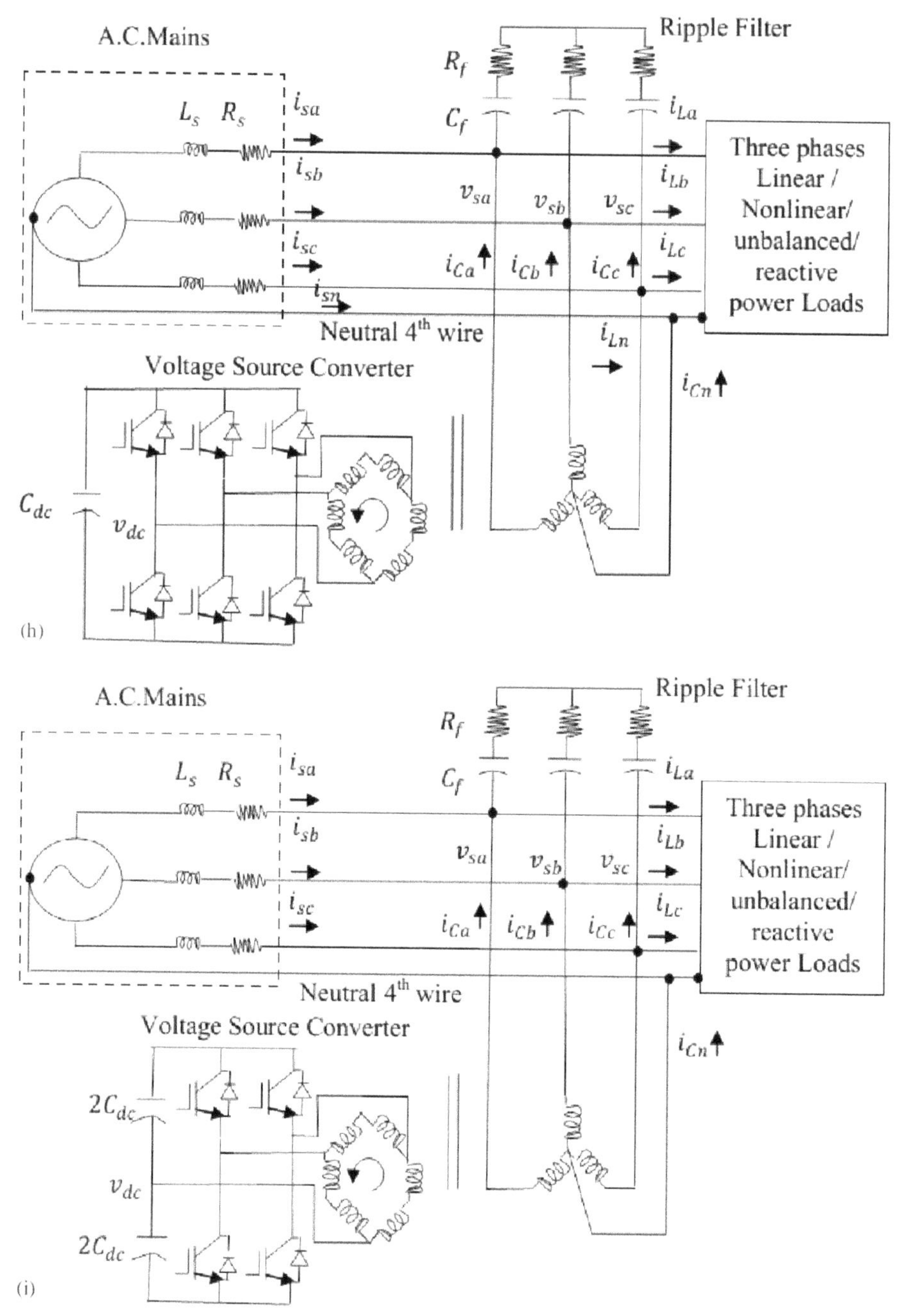

FIGURE 5.7 (Continued) Topologies for three-phase four-wire ShAPF with isolated VSC using transformers (h) three-leg VSC with star–hexagon, (i) two-leg VSC with star–hexagon.

The SHAF provides neutral current compensation, harmonic elimination, and load balancing along with PF correction or line voltage regulation. The detailed design of the SHAF, star/hexagon transformer, and control of the SHAF is given in the following sections. The hexagon-connected transformer secondary winding provides a path for the harmonic currents zero-sequence fundamental current. Therefore, it provides a path for the neutral current when connected in shunt at PCC. Under single-phase load, the zero-sequence load neutral current circulates in the hexagon windings of the star/hexagon transformer.

The voltage across each primary winding is the phase voltage. The three-leg VSC is connected to this transformer, as shown in Figure 5.3d. The voltage rating of the star/hexagon transformer windings is designed as shown below. The performance of SHAF system has been demonstrated for neutral current compensation along with reactive power compensation, harmonic elimination, and load balancing for both linear and nonlinear loads. The voltage regulation and PF correction modes of operation of the SHAF have been observed as expected. The star/hexagon transformer has been found effective for compensating the zero-sequence fundamental and harmonics currents. The DC bus voltage of the SHAF has been regulated to the reference DC bus voltage under all varying loads. The SHAF has been found to meet IEEE 519-1992 standard recommendations of harmonic levels.

A three-leg, PWM-controlled VSC is used as a SHAF and it has six insulated-gate IGBTs, three interface inductors, and one DC capacitor. The voltage rating of the secondary of the transformer is used for using an off-the-shelf three-leg VSC. The line voltage of the VSC is considered as 200 V and the rating required for meeting the reactive power compensation of the load considered is found to be 12 kV A.

The approach is passive and has an interconnection of a star–delta transformer between the AC and DC sides of the diode rectifier topology (see Figure 5.6d). This interconnection has 120-degree conduction angles of each diode and will generate a third harmonic circulating current among the AC and DC sides of the converter bridge. The third harmonic current is circulating in nature is mostly improves the performance of the diode rectifier. The input resultant current is almost having sinusoidal shape with main decrease in the line current. A Y/Δ transformer is interconnected between the AC input and DC output sides as shown. The capacitors on the DC side offer the midpoint for output voltage V_o. The transformer secondary is connected in Δ and is unloaded.

The transformer is chosen to be a standard three-limb core type. The voltage drop between the transformer neutral "N" and the capacitor midpoint "e" due to the 120-degree conduction intervals of each diode is essentially third harmonic. Therefore, the interconnection of points "N" and "e" via inductor L results in a circulating third harmonic current I, The inductor L is a saturable core type, whose value can be altered to regulate the 1, magnitude. The current I, is then equally divided in three limbs of the transformer. The circulating third harmonic current is shown to drastically improve the input performance of the rectifier interface. The primary winding of the transformer therefore carries zero-sequence third harmonic currents. The zero-sequence flux ($\varphi 0$) in each limb is co-phase and the sum $3\varphi o$, must seek a path through the air or through the transformer tank, either of which presents a high reluctance. This result is a low zero-sequence excitation impedance. The zero-sequence leakage impedance, however, is nearly the same as the positive sequence impedance, which is also low in a typical transformer. Therefore, the circulating third harmonic current I, has a low impedance path via the transformer. It should be noted that the transformer draws negligible fundamental current from the input source (equal to the excitation current) since the delta secondary is unloaded. The third harmonic currents in the primary are balanced by currents circulating in the delta winding. The transformer is commonly known as a grounding transformer in power distribution systems.

This interconnection of Y–Δ in arrangement with conduction intervals of 120 degree of each diode is revealed to produce a circulating third current harmonic between the AC and DC side of the rectifier bridge. The third harmonic circulating current is therefore produced by design and is shown to considerably improve input PF and reduce current harmonics. The input transformer required is a basic Y/Δ type usually known as a grounding transformer in power distribution systems [7]. The merits of the system are as follows:

- The scheme is passive and does not interfere with the AC to DC rectification process of the diode rectifier topology.
- The resulting input current is near sinusoidal in shape and is in phase with the voltage.
- The circulating third harmonic current is automatically generated by the interconnection. The required VA rating of the passive components is a fraction of the diode rectifier VA.
- The transformer draws negligible fundamental current (equal to the no-load current) from the input supply since the secondary winding (delta) is unloaded. The impedance of the transformer (three limb core) for zero-sequence third harmonic currents is nearly equal to the fundamental leakage impedance value, which is small.

- The delta-connected secondary permits a path for zero-sequence third harmonic currents to circulate, thereby automatically balancing the flux in the transformer core.
- The scheme can be viewed as a cost-effective retrofit to the existing wide range of diode rectifier type utility interface applications.

Recent surveys of 208/120 V three-phase four-wire electric systems, industrial plants, and buildings with computers and nonlinear loads lead to excessive currents in the neutral. These neutral currents are generally third harmonic in nature and their existence is tied to wiring failures, transformer overheating, elevating of neutral potentials, etc. The star/delta transformer topology along with a two-switch PWM-controller-based active filter with the closed loop control of the APF promise abolition of neutral current harmonics under unstable load conditions.

The Y-connected primary, as shown in Figure 5.7d, provides a relative neutral point "N" between the APF and distribution system neutral "n." The delta-connected secondary provides a path for zero-sequence currents to circulate, thereby maintaining star point "N" at the same potential as the supply neutral "n." Moreover, the secondary of the transformer is connected to a diode-based three-phase bridge rectifier that gives the small amount of power necessary to sustain the DC voltage across the capacitors. The operation of the APF is as follows; the neutral current I_n is sensed via a current sensor and is processed through a 60 Hz notch filter in order to remove any fundamental current component in I_n. The filtered current signal is now compared with I_{ref}, which is intentionally set to zero. The resultant error signal is fed to the PWM control logic in order to introduce an equal and opposite current I_N, so attaining abolition in closed loop. This current insertion method neutralizes any harmonic current flowing in the neutral conductor, and thus protects the upstream distribution system and transformer. Hence, if the filter is cancelling 100% of the neutral current, then IN = In and In′ = 0. It should be noted that the injected harmonic current IN, from the APF, is essentially zero-sequence (3rd, 9th, etc.).

The zero-sequence current IN is injected and equally divided among the three primary windings of the transformer such that $I_p = I_N/3$. In this system, the three-limb core construction for the transformer is preferred for the following reasons. In the core type construction, the zero-sequence flux in the three legs does not add to zero as in the positive sequence case. Instead, the sum 340 must seek a path through the air or through the transformer tank, either of which presents a high reluctance. The result is a low zero-sequence excitation impedance, so low that it can be neglected. Therefore, the zero-sequence currents encounter only small leakage impedance (nearly the same as the zero-sequence impedance) in the path through the three-legged core type transformer. However, the delta-connected secondary carries circulating third harmonic current to balance the primary limb $I_N/3$.

In this topology, the two capacitors will maintain the DC bus voltage, which is chosen to supply a ripple-free DC voltage and is duly rated to carry the harmonics in the neutral current. The inductor L_f is chosen to filter the switching harmonics generated by the PWM operation of the two switches.

The advantages of this approach are as follows.

1. Active power filters offer continuous measurement and cancellation of neutral current harmonics.
2. Active power filters do not consume any real power other than that required to account for internal losses. The APF is expected to be over 90% efficient.
3. The system can adapt to changing load conditions.
4. The APF has fast response characteristics and sufficient bandwidth to cancel several zero-sequence harmonics appearing in the neutral. The proposed APF operates at a high frequency (= 20 kHz), and an input filter stage can be designed to bypass the switching harmonics, thereby avoiding interference with the line.
5. The APF in Figure 5.1 employs state-of-the art power semiconductor devices, and therefore is compact, light in weight, and occupies less space.

The active filter requires two magnetic components, a delta/star transformer and an inductor filter L_n. The transformer kV A rating is decided by the product of the supply mains frequency voltage and the currents circulated inside the windings. These currents contain harmonics generated by the inverter in the star windings, $3n$ harmonic currents circulating in the low impedance delta windings, and the fundamental currents needed to feed the inverter losses. Assuming reasonable inverter efficiencies, the burden of the inverter can be ignored. Since the transformer is excited at utility frequencies, core material can be conventional grain-oriented steel.

5.7 Effect of Neutral and Grounding Practices for Power Quality Improvement

Shared or downsized neutral conductors are mainly vulnerable to overheating from triple-n ($3n$) harmonics. These harmonics are odd integer multiples of the third harmonic of the fundamental frequency like 50 or 60 Hz, including the third harmonic itself (150 Hz), the ninth harmonic (450 Hz), the fifteenth harmonic (750 Hz), etc. In star-connected three-phase systems, $3n$ harmonic waveforms are in phase, do not revoke in the neutral conductor, and are arithmetically additive. Properly grounded shielded cables, metal-clad cables, or metal conduits offer electromagnetic shielding and diminish intrusion for circuits that serve sensitive electronic equipment. Many design techniques are employed to reduce EMI. A recent survey of PQ experts shows that 50% of all PQ problems are related to grounding, ground loops, ground bonds ground current and neutral-to-ground voltages, or other ground-associated issues.

Neutral current in three-phase power systems is often thought to be only the result of the imbalance of the phase currents. With computer systems, very high neutral currents have been observed even when the phase currents are balanced. In three-phase star-connected power systems, the neutral current is the vector sum of three-phase line-to-neutral currents. Underbalanced, three-phase, linear currents, of sine waves phase angle of 120 electrical degrees apart, the sum at any moment in time is zero, and hence there will be no flow in the neutral current. In the majority of three-phase power systems delivering power to single-phase loads, some phase current imbalance and partial neutral current will be produced. These neutral currents of small magnitude resulting from somewhat light unbalanced loads do not create issues for usual large power distribution systems. There are conditions where even perfectly balanced single-phase loads can result in significant neutral currents. Nonlinear loads like rectifiers and switched mode power supplies will have phase currents that are not sinusoidal in shape. The vector addition of balanced, three-phase non-sinusoidal currents is not essentially equal to zero in magnitude. For instance, balanced square-wave shaped currents will lead to significant change in the current in the neutral wire. In three-phase networks, the triplen harmonics with reference to neutral currents (e.g., third, sixth, ninth, etc.) are cancelled. These triplen ($3n$) harmonics have three times fundamental power frequency and are spaced 120 electrical degrees in time base on the fundamental power frequency. The $3n$ harmonic currents are in phase to each other and are additive in neutral circuit.

Nonlinear loads in power system affect the other devices and distribution system. Harmonic distortion of current affects the interruption capability of circuit breakers. PF correction capacitors tune the system to resonate near a harmonic frequency, present in large load current and large system voltage. Harmonic current causes overheating of conductors due to skin effect, proximity effect and increasing of total rms value of current because of harmonics. Transformer is overloaded with harmonic currents and losses due to eddy current and stray losses are increased with harmonics. Neutral current problem in three-phase four-wire systems is important since most of the nonlinear loads are single-phased. For balanced systems triple harmonic components of current are added in the neutral conductor, and other components are cancelled. In unbalanced condition, triple harmonics and other components of current flow through the neutral conductor. This excessive neutral current increases the neutral-to-ground voltage, affecting power system and sensitive electronic devices. To study the effects of nonlinear loads to neutral ground voltage a model is used. In the model, balanced and unbalanced condition of fundamental, 3rd and 5th harmonic components of current, neutral current, neutral-to-ground voltages, and load voltages are obtained for standard and full size of neutral conductor.

Computer loads are generally nonlinear loads. The typical current waveform and harmonic content of the two popular computer power supply connections. The current of typical balanced line-to-line-connected power supplies contains no triplen harmonics. However, the typical current of balanced line-to-neutral-connected power supplies is very rich in triplens. For balanced line-to-neutral power supplies the neutral current would be 1.61 times the phase current. Under worst case conditions with rectifier conduction angles of 60°, the neutral current will be 1.73 times the phase current. Hence, when a line-to-neutral-connected system is used on three-phase power systems, major change in the flow of neutral current can be predictable.

Modern trends in computer-based systems have improved chances of momentous neutral currents. There is a shift from three-phase power to single-phase power supplies. The advance in switched-mode power supplies (SMPS), with the merits of better efficiency and lower cost, lead to increase in the $3n$ harmonic current waveform content over the prior linear power supplies. SMPS are-connected directly to the line-to-neutral supply without using a step-down transformer. If a step-down transformer is used with linear power supplies, the $3n$ harmonic currents are lessened due to its series inductance and hence it lets connection to the phase-to-phase voltage.

Another trend in computer-based systems is the change from AC motors with linear input currents to DC motors with nonlinear input currents in such devices as tape drives, fans, and disk drives. The DC motors need power with similar distinctiveness as other computer-based power supplies. The improved application of DC motors leads to increased content in the harmonic current distortion of the whole computer system.

High neutral currents in computer power systems can cause overloaded power feeders, overloaded transformers, voltage distortion, and common mode noise. Three-phase, four-wire building power feeders are often sized based on three current-carrying conductors in a conduit in accordance with the National Electrical Code (NEC). When the neutral conductor carries harmonic currents, additional heat is generated and reduces the power feeder effectiveness. With four-phase-based current carrying conductors, the effectiveness of the power feeder must be decreased to about 80% of the three-phase current carrying conductor rating in accord with NEC tables. Neutral conductors, which are typically sized the same as the phase conductors, can still be overloaded since the neutral current can exceed the rated phase current. The balanced $3n$ harmonic currents, which add in the neutral conductor, are cancelled in a delta–wye transformer. The balanced $3n$ currents flow as circulating currents in the transformer's delta primary. As a result, more current flows in the transformer's windings (causing additional heating) than is detected by the transformer's primary circuit over-current protection device. Overloading of the transformer can result. Since the power supplies that generate the harmonic neutral currents have high peak-to-rms current waveforms, the voltage waveform will be distorted. "Flat-topping" of the voltage waveform due to change in the impedance of the power system is observed at the harmonic current frequencies. Since the power supplies use the peak voltage of the sine wave to keep the capacitors at full charge, reductions in the peak voltage rms of the voltage may be normal.

The waveform distortion can also lead to extra heating in motors and other magnetic devices that are driven from the distorted same voltage source. The common mode noise in one form in three-phase power systems is the difference in the voltage between neutral and ground. With more harmonic neutral current flow, the impedance of the neutral wire at the harmonic frequencies can produce large neutral conductor voltage dips. These neutral conductor voltage dips appear as common-mode noise to the computer-based system. The low-frequency common-mode noise voltages fairly effect the computer system is rather questionable, yet computer dealer provisions usually call for less than 0.5–3 V RMS, neutral-to-ground voltage, despite frequency.

The characteristics of neutral current for computer system depend on the switch-mode power supplies type, drive loading, distribution transformer, and feeder supplying system. The harmonic distortion in neutral current cannot vary over a wide range. However, it is possible to identify two basic waveform types that can be used for analysis purposes.

A. Type 1-Balanced Condition Waveform

This is mainly harmonic component consisting of $3n$ harmonics only. The computer power supplies that produce the neutral currents under balanced current condition have high peak current waveforms. Consequently, the voltage waveform can become distorted. The total harmonic distortion in current for the selected waveform is 24,853%. The most predominant harmonics are 3rd, 15th, 9th, 21st, and 5th harmonic and other orders that are considerably small.

B. Type 2-Unbalanced Condition Waveform

The neutral currents of this type are associated with linear load. The $3n$ harmonics are responsible for nonlinear loads and fundamental current responsible for load unbalances. The selected waveform has a distortion level of 933%. The most predominant harmonics are 3rd, 15th, 19th, 21st, and 5th harmonic and others order that are considerably small. The $3n$ harmonic currents due to balanced and unbalanced loads, which add in the neutral conductor, flow as circuiting currents in a transformer's delta primary causing additional heating in the transformer. In order to remedy these effects, various harmonic filter approaches have been installed for reduction of harmonics in neutral conductors. The effect of unbalanced load leads to fundamental current, which may flow through filter. Excessive fundamental current levels in this case may lead to overload/burned filter and additional losses in systems.

For computer system, the $3n$ harmonic currents in the three phases are all added in the neutral, and the neutral current is higher than the phase current. The waveforms have high %THD. For an unbalanced system, both fundamental and triplen currents are contained. Adding fundamental current of neutral current is a common way for this to occur. This event is due to unbalanced load condition and unbalanced voltage. Because of effects of unbalanced load on neutral filter for harmonic reduction, it can encounter fundamental current in neutral conductors. Filter overheating problems may arise. This could lead to filter failure. The unbalanced voltage caused by voltage sags may lead to draw phase current that contains all zero, positive, and negative sequence affecting neutral current. For the computer system including inductive loads (i.e., single-phase motor and fluorescent lamps), very high neutral current is possible.

The suitable grounding of the electric network is serious to achieve its efficient and sure action. When considering electric systems, besides the fundamental frequency (50 or 60 Hz), certain harmonics are included that will distort the telephone and telecommunication systems. To minimize these effects, grounding of the fourth wire called the neutral wire plays a vital role. The electric grounding helps in:

- Protecting people and animals from the electric shock risk and protecting the equipment from damages for over-voltages and short circuits
- Supplying a zero-sequence reference point
- Controlling of noises produced due to nonlinear loads and other devices
- Offering a path to earth for the lightning or surge currents

This active filter must be able to act as a neutral current electronic compensator. It is realized with a four-wire three-phase PWM VSC connected to the AC mains with available neutral. The neutral of AC mains is connected to the midpoint of the converter on the DC side capacitors. The converter AC currents are controlled with a conventional space vector-based hysteresis controller in the $\alpha\beta0$ space coordinate system. The AC mains neutral currents are determined and injected by the PWM voltage converter. A consumer's unbalances will produce first harmonic neutral current and it has to be compensated. Zero-sequence harmonic currents caused by single-phase rectifiers and other electronic appliances can also be compensated.

In low voltage three-phase electricity supply networks the neutral wire is usually provided. The existence of neutral wire currents results in several undesirable phenomena in the supply network. Unbalances in the mains voltages are originated. These unbalances can be minimized by constructing the power transformers with a small zero-sequence impedance. Also, as the neutral wire is usually smaller than phase wires, some precautions must be taken.

Single-phase consumers are connected to the AC network in a way that they can constitute a three-phase balanced load. However, if this is not the case, a first harmonic neutral current will result. The neutral current can also be produced by a different reason. Due to the widespread use of single-phase rectifiers and other single-phase electronic appliances a wide current harmonic spectrum is produced in the mains AC currents. Assuming that the single-phase nonlinear loads are connected to the three-phase mains in a balanced way, the mains currents will be generally determined by several positive sequence systems of harmonic orders 1, 7, 13, …, negative sequence systems of harmonic orders 5, 11, …, and zero-sequence systems of harmonic orders 3, 9, 15 …. All the positive and negative sequence harmonics systems will lead to a zero neutral wire current. However, the zero-sequence harmonic systems will add up in the neutral wire. Therefore, it is desirable that the neutral wire current can be compensated.

Power electronic compensators for the three-phase electricity supply network are being studied and developed. One of these systems is the APFs, which are used to compensate mains current or voltage harmonics. These types of electronic filters have filtering characteristics that are independent of the network topology and present inherent over-current protection.

The AC mains neutral current compensator can be implemented with a four-wire PWM voltage converter or a four-wire PWM current converter. The four-wire PWM voltage converter has three inductors (or current sources) in series with the AC side and capacitors (or voltage sources) on the DC side. The AC mains neutral conductor is connected to the middle point of the converter DC side capacitors. Figures 5.6 and 5.7 present the general control circuit of the neutral current power electronic compensator based on the PWM voltage converter.

The AC currents i_a, i_b, and i_c must be controlled by the PWM voltage converter. This can be easily accomplished using three independent two-level hysteresis current controllers for each phase. However, with this current controller the neutral current error will be three times greater than the corresponding phase current errors. In order to have equal current errors, a control system based on the αβ0 coordinate system is used. In this coordinate system, the phase currents information is much more interesting. The zero-sequence component is decoupled from the α and β components. If the converter is connected to a three-wire network, without accessible neutral, the converter currents zero-sequence component is always null. Thus, it is only necessary to control the two other converter current components i, and i_p. This can be realized with two hysteresis current controllers placed in the αβ reference frame. The converter currents i, and i_p are controlled choosing the appropriate converter voltage vector in the αβ plane. If the converter is connected to a four-wire network, with accessible neutral, the three components i_o, i_α, and i_β, must be controlled. The converter AC currents in αβ0 coordinate system can be controlled choosing the appropriate AC voltage vector through a switching table. The AC mains neutral current to be compensated is sensed. The zero-sequence current component is thus obtained, $i_{R0} = i_{\mathrm{Rn}}/\sqrt{3}$. So, the converter zero-sequence reference current i_o* is equal to the load zero-sequence current to be compensated i_{R0} (with a minus sign).

The system injects AC currents on the mains so that a null mains neutral wire current is obtained. Thus, consumer's unbalances that may result in a first harmonic neutral current can be compensated. Zero-sequence currents harmonics caused by electronic appliances can also be neutralized. The system is implemented with a four-wire PWM voltage converter. The PWM voltage converter AC currents are controlled with a generalized space vector-based hysteresis controller in the αβ0 coordinate system. It is designed so that a low semiconductor switching frequency is achieved. Thus, high power semiconductor devices can be used and so a high power-rating converter can be implemented.

Causes for excessive neutral currents amounting to 75% and 157% of phase current values in two installations supplying predominantly fluorescent lighting circuits have been identified. These high neutral currents are accounted for by three main components, viz., fundamental phase current unbalance, third harmonic currents caused by saturation of ballasts, and more third harmonic current caused by third harmonic voltage waveform distortion in the supply. Solutions are also provided in the literature to overcome the problem, which cancels third harmonic currents along a common neutral conductor, in the form of a six-phase system supplied by two transformers connected in anti-phase.

In a three-phase four-wire system, under normal operating conditions the loads are reasonably balanced. The current in the neutral is expected to be small, typically not exceeding 20% of the normal load current in the phases. However, the problem of excessive neutral current in three-phase systems occasionally arises, particularly in circuits that supply predominantly fluorescent lighting loads. The neutral current magnitudes experienced on such systems can be quite high at about, typically, 60%–80% of the phase current values. In one installation, the neutral current exceeded the phase current values at 150%, creating great concern. The first and immediately suspected reason for this excessive neutral current of magnitude and phase angle unbalance between phases can usually be quickly dismissed by a simple measurement of the circuit's phase currents and PFs. Examination of phase and neutral current oscillograms will, however, show significant amounts of third harmonic components.

Examination of the corresponding neutral current waveforms shows that the neutral current consists of nearly all third harmonic currents. Again, this is not unexpected as in a balanced three-phase four-wire system it can be shown that, while the currents of fundamental frequency (50 Hz) sum to zero at the neutral, the third harmonic components and all triple n components remain in phase, resulting in the summation of all third harmonic and its multiple component currents of each phase in the neutral wire. Thus, the neutral wire of any three-phase system with third harmonic currents in the lines will always carry the sum of all the third harmonic currents and its multiples of each of the three phases.

1. Capacitors provide PF correction and reduction of 50 Hz phase currents.
2. Neutral current, which is mainly of third harmonics, is contributed by saturation in the iron-cored inductive choke.
3. With PF correction capacitors included in a balanced configuration, the neutral current is unchanged.
4. The increase in ratio of I_N/I_R is due to a decrease in I_R and not to an increase in I_N.
5. Unbalanced lamps create unbalance in phase current and neutral current. Such unbalance is of the 50-Hz variety. Unbalanced capacitors also create unbalance in phase current and neutral current, which is also of the 50-Hz variety.
6. It is common in practical installations to remove the occasional lamp (by simply unscrewing the lamp tube) to save energy in places where lower light levels can be tolerated. Where these light fittings are connected in pairs sharing a PF correction capacitor (say, 8 μF for two lamps), removal of one lamp accentuates the unbalance effects cited under (4) and (5).

The principal causes of excessive neutral current in the installations have now been identified. It can be stated that the presence of the main component, viz., the third harmonic currents, is inherent in the system with an abundance of fluorescent lamp circuits and PF correction capacitors. Electrically, there is "nothing wrong" with the system since it is balanced and healthy. In practical installations, however, such excessive currents in the neutral may create concern. This excessive and "unexpected" neutral current was blamed for noise and vibration experienced in the bus bar riser system. In installation with feeder lights, there is a genuine cause for concern as the neutral current exceeded the phase currents. This is aggravated by the fact that, in most installations, panel metering only shows phase currents and not the neutral current as it is normally assumed that neutral currents are small. This effect can, therefore, be allowed to persist undetected. In fact, the problem in installation was only detected when overheating occurred on the neutral conductor, resulting in the burning of its insulation near a joint to the miniature circuit breaker (MCB), leading to a fire at the switchboard. Depending on how much of a problem this effect of excessive neutral current is causing an installation, one or more of the following possible solutions may be considered.

5.7.1 Reduction of Neutral Current Carried by Existing Problem Conductor

1. The excessive neutral current carried by a problem running an additional parallel neutral conductor to share the present neutral current (this is effectively the same as increasing the size of the neutral conductor to carry the large current); or diverting the phase and neutral current to another distribution board to relieve loading on the problem circuit.
2. Reduction of neutral current in the circuit can be achieved to a limited extent by ensuring balance in lighting loads as well as PF correction capacitors, i.e., taking care not to remove lamp tubes indiscriminately, which also leaves capacitors in unbalanced states, particularly when twin-shared PF correction capacitors are used in light fittings; or reviewing whether the values of PF correction capacitors used are too high, particularly in installations with significant harmonic voltage distortion.

3. As excessive neutral current is caused principally by an abundance of fluorescent lamp circuits on the same distribution board, the effect will be reduced by sharing the distribution board to include some other loads that do not generate so many third harmonic components.

5.7.2 Scheme to Cancel Neutral Current along Parts of Bus Bars

In the case of installation where excessive neutral current was present in a long bus bar riser, the neutral current here can be significantly reduced by using a two-transformer scheme supplying six phases. Two three-phase transformers (1 and 2) are connected with their phase windings in ant phase to produce an effectively six-phase system of line voltages. Each floor of lighting circuits, which is assumed to be equally loaded (a reasonable practical assumption), is supplied alternately by transformers 1 and 2, respectively. As the neutrals $n1$ and $n2$ at each alternate floor are connected to a common neutral bus bar riser N, the neutral current Z in the riser will effectively carry the sum of currents I_{N1}, and I_{N2} of the transformers 1 and 2. Being in ant phase, the excessive third harmonic currents generated by fluorescent lamp circuits on each floor will cancel at this common neutral bus bar riser. The current I_N in this neutral riser will be significantly reduced. The above scheme was tested in the laboratory and successfully confirms the workability of the proposed solution. It is realized, of course, that the small sections that carry the return current to the transformers IN1 and IN2 still contain the usual high values of neutral current of its respective transformer. Nevertheless, the main neutral bus bar riser is now relieved of its problem.

5.8 Advantages with Three Phase Four Wire System

Three-phase four-wire ShAPF is used in three-phase four-wire distribution system to filter out load current to meet the terms for the utility [55,56]. This is used to minimize the effect of poor load PF so that source current is nearly at UPF. It provides harmonic compensation in loads so that the source current becomes almost sinusoidal. It also provides compensation for unbalanced loads to balance source currents, to cancel out DC offset in loads and for PQ betterments. The advantage of a four-wire system is when a fault occurs, there is more flexibility to keep the system running since it may only affect one of the phases.

5.9 Topologies-Design Consideration

The three-phase four-wire distribution systems using Δ–Y transformer or Y–Y transformer are beneficial if the three phases have about the same amount of current in them; then, the neutral will have very little current or we can say that the system is balanced. Single-phase loads are connected between the different phase conductors and the neutral conductor. Therefore, the neutral conductor serves as the common return for the entire single-phase load currents. This may create unbalance in the distribution system and devices like computers, photocopiers, and other power electronic devices producing harmonics may damage neutral conductors and destroy the distribution transformer too.

During balanced condition, the neutral currents cancel each other and there will be almost zero current through the neutral conductor, so the designer will consider a negligible current through the neutral wire and the neutral wire is sized smaller than the phase wires. Sometimes due to unbalanced loads or due to harmonics the neutral current may increase up to 1.7 times larger than the rated current. Thus, the neutral current exceeds the phase current, possibly causing damage to the neutral conductor as well the transformer. For solving this problem, many compensating topologies are proposed from time to time. But there is an absence of reliable passive compensation methods even till now. The passive neutral current compensation method involves the use of transformers, inductors, and capacitors. Passive compensation using transformer has been used in recent years due to the advantages of low cost, high reliability, and simplified circuit connection. Zigzag transformers have been used for the past few decades for neutral current compensation.

The neutral current mitigation techniques include a Scott-transformer, T-connected transformer, star–hexagon transformer, and star–polygon transformer designed for MMF balance. These transformer configurations along with the conventional transformer configurations such as a star–delta transformer and a zigzag transformer are analyzed, designed, simulated, and verified by developing prototypes by many authors. The transformers present a source to the zero-sequence fundamental current component as well as harmonics currents. Two single-phase transformers are necessary for the generalized Scott-connection and T-connection technique whereas three single-phase transformers are essential in other methods.

Three-phase four-wire distribution systems are facing the severe problem of excessive neutral current and harmonic neutral current along with other PQ problems such as high reactive power demand, harmonic currents, and unbalanced loads. The typical loads in a three-phase four-wire distribution system may be computer loads, lighting ballasts, and small rating adjustable speeds drives (ASD) in air conditioners, fans, refrigerators, and other domestic and commercial appliances, etc. It has a significant portion of the third harmonic current component. The iron-cored inductive ballasts as well as electronic ballasts in fluorescent lighting also contribute to third harmonic currents. Observations on computer power systems have indicated harmonic neutral currents from zero to 1.73 times the phase current in the United States in a survey. It also exposed that 22.6% of the sites have neutral currents exceeding the full load phase currents and this scenario has become worse in recent years due to the proliferation of such nonlinear single-phase loads. The major reason for such excessive neutral current is the proliferation of nonlinear loads as well as unbalanced loads in the three-phase four-wire distribution system.

Residential and commercial supply networks show dominant nonlinear loads and they are modeled as harmonic voltage sources. The excessive neutral currents in the neutral conductor result in wiring failure due to improper sizing of the neutral conductor. Moreover, excessive neutral-to-ground voltage may appear and this results in the malfunction of sensitive equipment. The transformer is also overheated due to harmonic currents and these result in insulation damage and failure. Some corrective measures suggested are given below.

1. Derating of transformers
2. Oversizing of neutral conductor
3. Use of four-leg active filters
4. Use of zigzag transformers
5. Use of star/delta transformer

The derating of the transformer leads to low efficiency operation and increased heat due to losses. However, the application of the zigzag transformer or a star/delta transformer for reduction of the neutral current has an advantage of passive compensation, ruggedness, and less complexity.

The first one is based on a Scott-connected transformer arrangement and the Scott-connected windings of the transformer provide a path to the zero-sequence fundamental as well as harmonics currents. There are only two single-phase transformers required for the proposed Scott-connection. Some other new configurations investigated are based on T-connected transformer, star/hexagon transformer, and star–polygon transformer.

The magnitude of the neutral current in three-phase four-wire distribution system depends on the harmonic content and phase unbalance of the load currents. The passive technique for neutral current mitigation involves LC filters, zigzag transformers, and star/delta transformers [30–38]. But the LC filters are not effective for voltage source type loads. A phase shifting and zero-sequence current filter is patented and a zigzag transformer for harmonic neutral current reduction is also patented.

Figure 5.4a shows a zigzag transformer connected in parallel to the load for filtering the zero-sequence components of the load. The zigzag transformer consists of three single-phase transformers with turn ratio 1:1. It acts as an open circuit for positive sequence and negative sequence component of load currents and the current flowing through the zigzag transformer is only the zero-sequence components. The phasor diagram of the transformer shows that the resultants voltages are 120° phase shifted. The zero-sequence component of load currents can be locally circulated using a star/delta transformer as shown in Figure 5.4c. The secondary delta-connected winding provides a path for the zero-sequence currents and hence the supply neutral current is reduced to zero. The relative value of a series inductance (Z_{sn}) with transformer impedance plays a significant role in the effectiveness of the neutral current compensation. Therefore, the transformers must be installed near the load or it is required to connect a high value of series inductance. In this paper, some new transformer configurations are proposed for the neutral current compensation in a three-phase four-wire distribution system. The Scott-transformer and T-connected transformer are used in the power system for different applications. The proposed configuration of T-connected transformer for neutral current compensation is shown in Figure 5.4b and in this two single-phase transformers are used for neutral current compensation. These single-phase transformers are connected in T-configuration and it shows that the resultant voltages are 120° phase shifted. The transformer is simple and the windings are suitably designed for MMF balance. A Scott-transformer for neutral current compensation with fourth-wire topology in Figure 5.7 (e and f) and as compared to earlier T-connected transformer, the required numbers of windings is reduced. A star–hexagon transformer is reported along with VSCs as SHAF, which also helps in neutral current compensation as shown in Figure 5.4d. A star–polygon transformer is reported for some other application, but it is proposed for neutral current compensation. Similarly, a zigzag transformer is analyzed in detail for zero-sequence harmonic compensation in the literature.

The neutral current mitigation techniques for a three-phase four-wire distribution system using magnetic have been demonstrated in the literature and some new cost-effective method using transformers have been proposed for neutral current compensation. The neutral current mitigation techniques include a Scott-transformer, T-connected transformer, star–hexagon transformer, and star–polygon transformer designed for MMF balance. These transformer configurations along with the conventional transformer configurations such as a star–delta transformer and a zigzag transformer have been applied practically for distribution systems nowadays.

5.9.1 Topology with Three-Leg VSC-Based with Zigzag

The zigzag transformer provides a low impedance path for the zero-sequence fundamental current and harmonic currents and hence offers a path for the neutral current when connected in shunt at PCC. Under single-phase load, nearly half of the load current flows through the windings of the zigzag transformer. Similarly, the voltage across each winding is one third of the line voltage and hence all six windings (two windings each of three phases) are selected as 150 V. In order to connect the DSTATCOM to this transformer, a set of third windings in each phase is introduced at its secondary as shown in Figure 5.4a. The secondary line voltage is selected as 200 V and hence the voltage across each star-connected winding is selected as 120 V.

The performance of a three-phase VSC-based DSTATCOM and the zigzag transformer for PCC voltage regulation, along with neutral current compensation and load balancing of a three-phase four-wire load, is shown in Figure 5.6a. The PCC voltage (v_s), balanced source current (i_s), load current (i_l), compensator current (i_C), load neutral current (i_{Ln}), zigzag neutral current (i_{Zn}), source neutral current (i_{sn}), amplitude of PCC voltage (V_s) and DC bus voltage (v_{dc}) are demonstrated under change of load conditions. It is observed that the amplitude of PCC voltage is regulated to the reference amplitude by injecting the required reactive power compensation. The zero-sequence fundamental current of the load neutral current resulted from the unbalanced load currents is circulated in the zigzag transformer and hence the source neutral current is maintained at nearly zero. The DC bus voltage of the capacitor of the VSC of DSTATCOM is regulated by the controller and the DC voltage is maintained near the reference DC voltage under all load disturbances.

5.9.2 Topology with Three-Leg Split Capacitor with Star–Delta

The conventional three-leg VSI split capacitor-based DSATCOM topology is shown in Figure 5.6b. Here, v_{sa}, v_{sb}, and v_{sc} are source voltages in a, b, and c phases, respectively. The three-phase source currents are represented by i_{sa}, i_{sb}, and i_{sc}, load currents by i_{la}, i_{lb}, and i_{lc}, respectively. The load draws reactive, unbalanced, and nonlinear currents. The terms, i_{ln}, i_{sn}, and i_{fn} represent the load neutral current, source neutral current, and filter neutral current, respectively. In this paper, the distribution feeder is considered as stiff and hence feeder impedance is neglected. The inductance and resistance of interfacing filter are represented by L_f and R_f, respectively. This topology has two DC-link capacitors and a total DC-link voltage of 2 Vdc is maintained across them using DC-link voltage controller. Here, the voltage Vdc is selected as 1.6 times the peak value of the source phase voltage.

The distinguishing feature of this topology is the splitting of DC bus capacitor into two capacitors. The midpoint of the two capacitors is utilized for the neutral return path, and thus, the entire neutral current flows through the DC bus capacitors. In such topology, it is important to maintain equal voltages across two capacitors. Therefore, two voltage control loops are used to maintain a reference DC bus voltage across the inverter switches. Typically, two voltage sensors are required either to sense voltage across two capacitors or voltage across one capacitor and total DC bus voltage. The performance of 2C topology, thus, is significantly determined by the voltages across split capacitors. A difference between these two voltages may cause a DC circulating current to flow, affecting the overall compensation. Additionally, in 2C topology, there is no direct control over the neutral current as it gets compensated due to the algebraic difference of currents injected by the remaining phases. The 2C topology is suitable for low-to-medium voltage and low power rating applications.

5.9.3 Topology with Three Leg with T-Connected

Figure 5.4b shows the wind power generation system (WPGS) configuration of the three-leg VSC is used with a capacitor or battery energy storage system (BESS) to store and dissipate power. The VSC is connected to the PCC through an interface reactor in each leg. An excitation capacitor is used across generator terminals to build the rated voltage under no-load conditions. The VSC supplies the reactive power to the IAG system to regulate the voltage during application of loads. With the battery at its DC bus, a frequency regulation takes place by exchanging the active power under varying wind speeds and loads. The battery absorbs the surplus power when the supply frequency increases above reference value and delivers the deficit power to common bus when the supply frequency decreases below the reference value. The supply frequency depends directly on the instantaneous load demand. The VSC consists of IGBT-based three half-bridges. The three-legs consist of two IGBTs and the common point of each leg is connected with an individual phase at PCC through an interfacing inductor. The load neutral terminal is provided by the VFC using the neutral terminal of a T-connected transformer. The non-isolated T-connected transformer provides the path for the neutral current. This topology with instantaneous p-q theory has the capabilities to work as a load leveler, a load balancer, a harmonic eliminator, and a voltage regulation.

For controlling the voltage and frequency of induction generator (IG) systems, the electronic load control consists of a three-leg VSC with a chopper switch, an auxiliary load, and a DC bus capacitor on the DC link for controlling active and reactive powers. The T-connected transformer is used to mitigate the triplen harmonics and zero-sequence currents, thus reducing the rating of VSC devices. A chopper with an auxiliary load at its DC link is used to absorb surplus active power is not used by consumer loads.

A simple connection of suitably designed windings of Scott-transformer for MMF balance has been found to be a better topology for neutral current compensation for highly unbalanced loads. The Scott-transformer requires only two single-phase transformers and hence this topology is found to be simple and cost effective compared to other methods of neutral current compensation. The application of magnetic mitigates the source neutral current to zero and reduces the magnitude of the source current, The THD of the source current is also reduced using the Scott-transformer connection.

5.9.4 Topology with Three Leg with Star–Hexagon Connected

Each VSC leg consists of a pair of IGBTs, and the common point of each leg is connected with an individual phase of the generator bus through an intermediate inductor. A three-phase four-wire system is developed for nonlinear and unbalanced load in Figures 5.4d, 5.6g and h, and 5.7i. The load neutral terminal is provided by the star terminal of the primary windings of the star/hexagon transformer. The secondary windings of a star–hexagon transformer are connected in the hexagon configuration and there is an external connection to this side. The use of a star–hexagon transformer serves the purpose of neutral current compensation and triplen harmonic elimination and, sequentially, it reduces the rating of VSC due to these features. For realizing the VFC, it requires an appropriate selection of VSC, a BESS, and an intermediate inductor, a scaling circuit for sensing the input currents and voltages signals and the gate driver circuit for output signals. Figure 5.4d shows a star/hexagon transformer. The transformer is designed for MMF balance to provide a low impedance path for the zero-sequence current. The phasor diagram of secondary windings is also shown in the figure. The star–hexagon transformer is used at PCC to realize the three-phase four-wire system.

Figure 5.4d shows the power circuit of a proposed H-bridge VSC-based SHAF along with a star/hexagon transformer connected in the three-phase four-wire distribution system. The linear and nonlinear, balanced, and unbalanced loads are connected at the PCC. The SHAF consists of a two-leg PWM VSC using four insulated-gate IGBTs, three interface inductors, and two DC capacitors. The star/hexagon transformer connected at the load terminal provides a circulating path for zero-sequence harmonic and fundamental currents. The SHAF provides neutral current compensation, harmonics elimination, and load balancing along with PF correction or line voltage regulation. The compensator current is equal to the reactive power component of the load current. Figure 5.4d shows the phasor diagram for ZVR operation. The SHAF injects a current I_c, such that the load voltage, VS, and source voltage, VM are in the locus of the same circle.

The star/hexagon transformer has been effective for compensating the zero-sequence fundamental and harmonic currents. The DC bus voltages of these split capacitors of DSTATCOM have been regulated to equal magnitude and the total DC voltage is also regulated under all varying loads. The use of four-switch, H-bridge VSC for three-phase four-wire system reduces the complexity and hence reduces the cost of DSTATCOM. The star/hexagon transformer requires three single-phase transformers similar to the star/delta transformer and zigzag transformer.

5.10 Comparison of Topologies

The three-phase three-wire and four-wire system is compared and is tabulated as shown in Table 5.1. The comparison is made between these wiring arrangements and is also made without and with isolation topology. When three-phase three-wire without neutral arrangement, three-phase three-system is chosen. When neutral is available as fourth wire, three-phase four-wire topology can be chosen. Many advantages are described for four-wire distribution system compared to a three-wire. Also, fewer switches with split capacitor can be chosen; zigzag, T-connection, and hexagon type of special features transformers are also helpful to limit the unbalanced current passing through the neutral. Many control strategies are available in the literature proposed for four-wire systems to compensate unbalanced neutral current and fast acting strategies with quick response and economic feasibility. The main advantages of a three-phase four-wire strategy with transformer is the low weight of the transformer with isolation from the main power system network.

A four-pole topology for the SHAF system exhibits the facility of operating it as a three-pole device. In general, single-phase linear and nonlinear loads are powered with a three-phase four-wire supply system. Often it is found that a single-phase load on a three-phase supply system creates a problem of load unbalancing. Moreover, three-phase and single-phase nonlinear loads are the main cause of current harmonics in the supply system. A harmonic current passing through the supply impedance creates problems of voltage harmonics and voltage unbalance at the PCC. Since inception of the novel concept, which is harmonic current cancellation by an active filter, there have been numerous advances in the field of harmonic elimination using various types of multi-pulse converters. The SHAF system and the load arrangement are shown in Figure 5.3. With the presence of switch SW, the SHAF system has the facility of being operated as a three-pole device or a four-pole device. In Figure 5.2, nonlinear load is a three-phase, three-wire system. Figures 5.5 and 5.6 show a typical load arrangement for a four-pole SHAF system. Although there may be several other arrangements of the loads, the arrangement shown

TABLE 5.1

Comparison of Three Phase Three Wire and Four Wire System

Source Type	Advantages	Disadvantages
Three-phase three-wire system	Least number of semiconductor devices, less cost. Better applications are with PV, wind type systems. Better performance is observed with isolated load than with grid-connected system Two-leg VSC with split capacitor. Fewer switching devices Three one-phase VSCs—independent phases control-efficient in unbalanced system YΔ transformer kVA rating = Q injection. It provides flexibility to use an "off-the-shelf" VSC	Requires bulky transformer Control and regulation of equal DC voltage of DC capacitor and requirements of quite high DC bus voltages are major problems More switching devices, hence less attractive The transformer provides isolation from the system. Have better performance with transformer than without transformer.
Three-phase four-wire system without transformer with transformer	Less cost, size, and weight Four-leg VSC is connected to neutral conductor, hence better neutral current compensation, and elimination of harmonics, load balancing, and voltage regulation. Better power rating, lower filter size, compensation better performance, and less power loss compared to system with isolation (transformer). Zigzag, YΔ, Y-hex, T-conn transformer: reduces neutral current flow. It provides advantages due to passive compensation, ruggedness, and less complexity. Besides these, advantages with trf are: better stability, operation under various power sources, flexibility in operation either in voltage or current control mode, fast acting DC link voltage, quicker transient response	Excessive neutral current along with other PQ problems such as poor voltage regulation, high reactive power burden, harmonic current injection, load balancing Control of DC bus voltage and higher voltage rating of switching devices are factors to be considered

in Figure 5.4 covers a wide variety of possible loads. It can be considered that a three-phase four-wire system can feed three single-phase loads. These single-phase loads may or may not be of the same types that cause load unbalance, load current harmonics, and excessive current in the neutral conductor. The four-pole SHAF system shown in Figure 5.5 is capable of compensating for harmonics and reactive power, and balancing the three single-phase loads.

The four-pole SHAF system is also capable of compensating for neutral current, thereby relieving the need for a neutral conductor with an excessive rating. The SHAF systems shown in Figures 5.3, 5.5, and 5.6 are voltage-controlled VSIs connected in parallel with the nonlinear loads. As shown in Figures 5.3, 5.5, and 5.6, the nonlinear load may be three-phase/single-phase, balanced/unbalanced load, requiring harmonic elimination, reactive power compensation, power-factor correction, and load balancing. The basic function of the SHAF system is to eliminate harmonics and to meet reactive power requirements for the load locally so that the AC supply system feeds only sinusoidal UPF currents. For the supply system, the AF and the load combined together behave as a balanced three-phase resistive load. The SHAF system connected in shunt with the load also enhances the system efficiency, as the supply system does not process harmonic and reactive power.

In Figures 5.5 and 5.6, a three-phase load is connected at the PCC in parallel with the SHAF system. The fourth pole (center point of two equal halves of the DC bus capacitor) of the SHAF system is connected to the neutral point of the three-phase four-wire supply system through the inductor (R_n and L_n). Therefore, a return path is made between the center-tapped DC capacitor of the AF system and the neutral point of the supply system. In this configuration the SHAF system behaves as a four-pole device. It is observed with this topology that, as soon as SHAF system is switched into the system, the stepped-wave-shaped supply currents turn into a sinusoidal shape with UPF and the SHAF system compensates for load harmonics and reactive power. The voltages across both halves of the DC bus capacitors are found to be the same. The SHAF system exhibits the desired self-supporting DC bus voltage. The current in the neutral conductor consists only of switching noise generated due to the PWM switching of the SHAF devices. With the SHAF system in action, the system behaves as a three-phase UPF-balanced network on AC supply. However, owing to the presence of a fourth wire, the supply current consists of a third harmonic. The supply current results for three-pole SHAF system do not consist of a third harmonic owing to the absence of a return path for co-phase harmonics. This return path is provided by a physical connection between the center-tapped DC bus capacitor of the SHAF system and the neutral point of the three-phase four-wire supply system. Therefore, it can be observed that the four-pole AF system operates as a return path for co-phasal harmonics and that is the reason for the appearance of a third harmonic in the frequency spectrum of the supply current. Also, from the study by previous authors, it is clear that the neutral wire of the supply has co-phasal harmonics. However, the AF system compensates for all these harmonics and makes the supply neutral conductor free from co-phasal harmonics. It is clear from the literature that, during load change from a lower value to a higher value, the AF system experiences a sag in its DC bus voltage resulting in loss of current control. Therefore, during load change the supply currents are distorted because of insufficient DC bus voltage, and the neutral conductor also has co-phasal harmonics. However, as soon as transient operating conditions due to load perturbation are over, the AF system recovers the above-mentioned voltage sag in its DC bus and current in the supply neutral conductor disappears. On the other hand, if it is desired to have zero current in the supply neutral conductor even during transient conditions of load increase, the SHAF DC bus voltage reference (V_{d_c}) should be set at a higher value leading to higher losses and voltage stresses in the AF power circuit.

The SHAF with three-phase four-wire system has been found to be capable of compensating for current in the supply neutral conductor as well as shaping the three-phase supply currents to be sinusoidal and in phase with the supply voltages. The harmonic spectrum of the supply as well as the neutral current is very low compared to systems without SHAF. It is found that the THD in the load current varies from one phase to another, based on the loading on the phases, respectively. The THD in the load neutral conductor current is 147%, which includes the dominant fundamental component along with co-phasal and odd harmonics, whereas the THD in the supply neutral conductor current is very low, demonstrating that the SHAF system compensates for current in the neutral conductor. The THD of the currents in all three phases of the supply system is well below the 5% mark of the IEEE 519 standard when using three-phase four-wire topology. It is observed that with unequal loading on the three phases, unlike phases a and b, during load perturbation the supply current in phase c experiences higher distortion. This is because the DC bus voltage requirement of the SHAF system is different for all three phases, with the highest for phase c, as this is the maximum distorted load. However, to prevent the reappearance of the supply neutral current and to prevent distortion in any of the supply currents, the DC bus voltage reference (Vdc) of the SHAF system must be set based on the load current that has the maximum value of THD. The AF system is found to be capable of balancing three single-phase loads of different characteristics from the literature.

Therefore, for a load that generates third harmonics, the four-pole AF system has been found to be effective compared to a three-pole AF system. It has also been found that three single-phase balanced loads on a three-phase four-wire supply system generate a third harmonic in the neutral conductor. On the other hand, three single-phase unbalanced loads on a three-phase four-wire supply system generate third harmonics, and all odd harmonics, including the fundamental current component in the neutral conductor. From the simulation study it has been found that the four-pole AF system is an ideal candidate for eliminating all harmonics in the supply neutral conductor and providing a sinusoidal shape to the three-phase supply current in phase with the supply voltage. The harmonics spectrum of the load and supply currents has been given to demonstrate the effectiveness of the AF system.

To evaluate the extent of the computer system neutral current problem, the following information was deemed to be of interest.

1. The ratio of average load to rated system capacity (percent capacity)
2. The ratio of neutral current to average phase current
3. The ratio of neutral current to rated full load current
4. The ratio of neutral triplen harmonic current to average phase current
5. The ratio of neutral triplen harmonic current to rated full load current

Items (1), (2), and (3) are straightforward calculations directly from the survey data (rms voltages, rms phase and neutral currents, and rated system capacity). Items (4) and (5) are of interest because they indicate the extent of the harmonic neutral current problem by excluding neutral current, which is the result of simple phase current imbalance, a problem that can usually be readily corrected by better balancing of the loads. A rigorous mathematical analysis indicates that if only the rms phase and neutral currents are known, the exact harmonic content of the neutral current cannot be determined.

However, a practical approximation of the harmonic content of the neutral current can be made by assuming that the fundamental and all non-triplen harmonics exist in the neutral according to the rms phase current imbalance, and that all neutral current in excess of the phase current imbalance is due to triplen harmonics. Implicit in this approximation is the assumption that the phase angles of harmonic currents of the same frequency are approximately the same.

The magnitude of the neutral current in three-phase computer power systems depends on the harmonic content and phase balance of the load currents. While very high neutral currents are possible due to the additive nature of triplen harmonic currents, a low percentage of data processing sites in the United States are actually experiencing neutral currents in excess of the rated phase current. However, recent trends in computer systems make high harmonic neutral currents more likely. Power system problems associated with high harmonic neutral currents include overloaded four-wire power feeders and branch circuits, overloaded transformers, voltage distortion, and common mode noise. Whenever three-phase, four-wire power systems are used to supply power to computer systems or other similar electronic loads, the power system design should allow for the possibility of high harmonic neutral current to avoid potential problems.

5.11 Summary

This chapter presents a thorough review and planning on the ShAPF to enhance the PQ at utility grid and industrial amenities. Different aspects of ShAPFs and new progress in this field of research have been discussed in detail to highlight their advantages and disadvantages. The review and classification of research papers published summarize that ShAPFs are beneficial to alleviate both voltage and current related PQ disturbances. This chapter also gives vital knowledge to research beginners in the field of PQ. A wide categorization of PQ improvement topologies into four categories with supplementary sub-classification of various topologies is expected to give an easy selection of suitable technique for a particular purpose. The selection norm of PQ upgrading devices for precise applications with techno-economic considerations is also summarized. The passive filters are the simplest and most economical despite their demerits of huge size and regulation issues. To overcome these restrictions with passive filters QPF is used. ShAPFs are more successful with nonlinear loads and unbalanced loads to sustain PQ. Hybrid filters offer optimal results for PQ compensation with simple design, reduced cost, and control with elevated reliability for PQ improvement. However, hybrid filter technology is an expensive solution. ShAPF is extremely effective for improving PQ at the distribution voltage level and has merits of making voltage stable. A complete state of the art for ShAPF for PQ improvement in three-phase three-wire and four-wire distribution system is explained and is tabulated as shown in Figure 5.8. Voltage regulation or the mitigation of reactive power

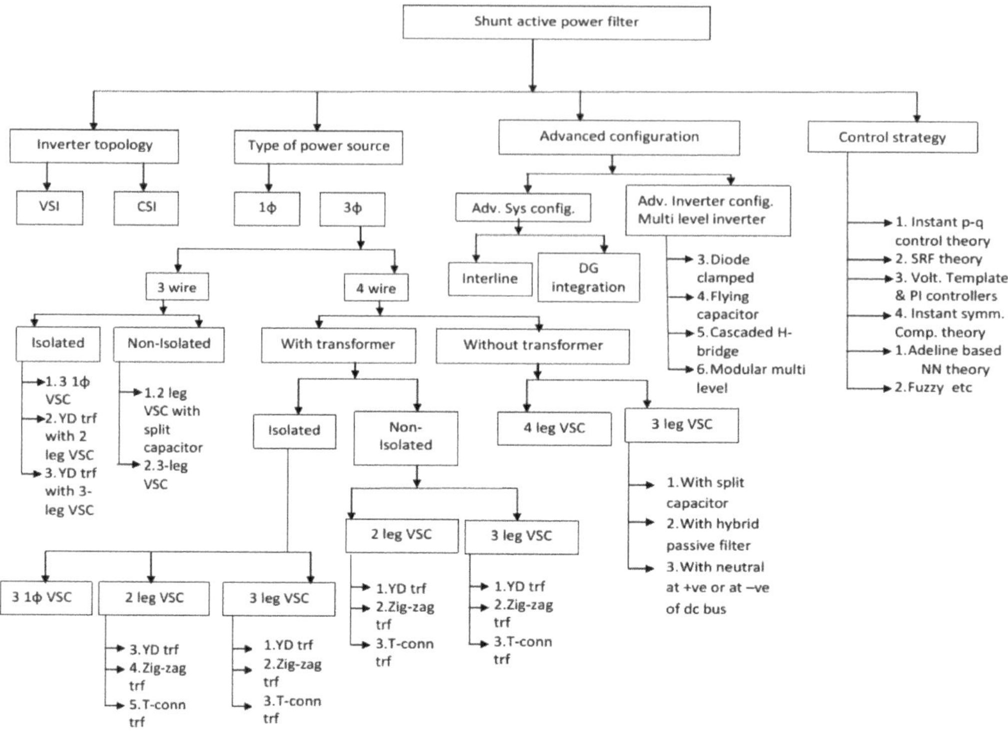

FIGURE 5.8 Summary of topologies and classification of shunt active power filter (ShAPF).

for PF correction, load balancing, harmonics elimination, and neutral current compensation has been explained briefly for three-phase three-wire and three-phase four-wire ShAPF for distribution medium voltage power systems.

An APF has to cancel harmonic currents in the neutral of a three-phase four-wire system. The APF displays the ability to effectively cancel undesirable excessive harmonics flowing in the neutral line of a three-phase four-wire system, and is highly efficient and low in cost. The star/delta topology drastically improves the system performance, contributes to efficient use of electric energy, and virtually eliminates excessive heating of distribution transformers due to harmonic currents.

While considering the kV A rating, T-connected and Scott transformers are lower compared to other schemes and, on the basis of comparing the economic aspect, Scott transformer with ZSBR is preferred since it requires only two single-phase transformers. The T–T-connected transformer investigated introduced values of negative- and zero-sequence currents of less than 1.5%. The level of voltage unbalance (dependent upon the amount of negative- and zero-sequence current flowing in the circuit) introduced by the T–T-connected transformer is quite acceptable in a distribution system, since the winding-to-winding impedances of the T–T-connected transformer are a small percentage of the total load impedance(s). The T–T-connected transformer, however, cannot be used as a high-voltage power transmission device since the higher per unit winding-to-winding impedances normally associated with this type of transformer would cause a prohibitive increase in the voltage unbalance. It can be concluded that

1. The negative- and zero-sequence currents introduced by the T–T-connected transformers may be found by utilizing the interconnected positive-, negative-, and zero-sequence impedance circuits of the transformer.
2. The T–T-connected transformer, when used as a distribution transformer, introduces negligible negative- and zero-sequence voltages and therefore maintains an acceptable three-phase voltage balance under full load conditions. This is due primarily to the low per unit winding-to-winding impedances inherent in the constituent single-phase transformers.

Unlike components of fundamental frequency, which sum vectorially at the neutral to give the unbalance value, third harmonic currents remain in phase allowing themselves to sum algebraically in the neutral conductor. Investigations on two installations show that these third harmonic currents in the phases can amount to about 25% and 64% of the fundamental current magnitude giving neutral current values of 75% and 157% of the phase values. These high neutral current values are accounted for by three main components, viz., 50-Hz unbalance, third harmonic currents caused by saturation of ballasts and nonlinear lamp arc characteristics, and more third harmonic currents caused by the presence of third harmonic voltage distortion in the phase voltage waveform. Solutions for the reduction and diversion of this excessive current from problem neutral conductors are given. A scheme using two transformers connected in ant phase providing a six-phase system, which cancels third harmonic currents along a common neutral conductor, is also tested in the laboratory and found to be workable.

Voltage sags and short interruption can affect neutral current for the three-phase four-wire system for computers without a UPS system. High neutral currents increase due to voltage sags and short interruption causing overloaded power feeder, overload transformer, voltage distortion, and common mode noise. The inrush currents of neutral conductor may lead to wiring failure. It has been shown that, when adding the RL loads (i.e., computer loads connected to lines with other loads like motor, etc.), neutral currents will increase during voltage sags leading to damage of neutral conductors. Also increased neutral currents may destroy neutral third harmonic blocking filter mostly installed in modem buildings. Future work will begin next year for further monitor of computer systems in industries in terms of rms current caused by various voltage sag conditions. This will be used for calculating proper conductor sizes.

Measurement results show that harmonics are a significant problem for building electrical distribution systems. To improve the PQ in distribution systems, filters can be used. But this seems to be not applicable for practical and economic reasons. Instead, buildings' central compensation can be designed as a filter. For neutral-to-ground problems, the neutral conductor size can be increased.

REFERENCES

1. Dalai S, Chatterjee B, Dey D, Chakravorti S, Bhattacharya K. Rough-set-based feature selection and classification for power quality sensing device employing correlation techniques. *IEEE Sens J* 2013;13 (2):563–573.
2. Afroni MJ, Sutanto D, Stirling D. Analysis of nonstationary power quality waveforms using iterative Hilbert Huang transform and SAX algorithm. *IEEE Trans Power Deliv* 2013;28 (4):2134–2144.
3. Valtierra-Rodriguez M, de Jesus Romero-Troncoso R, Alfredo Osornio Rios R, Garcia-Perez A. Detection and classification of single and combined power quality disturbances using neural networks. *IEEE Trans Ind Electron* 2014;61 (5):2473–2482.
4. Biswal B, Biswal M, Mishra S, Jalaja R. Automatic classification of power quality events using balanced neural tree. *IEEE Trans Ind Electron* 2014;61(1):521–530.
5. Ghosh AK, Lubkeman DL. The classification of power system disturbance waveforms using a neural network approach. *IEEE Trans Power Deliv* 1995;10 (1):109–115.
6. Dash PK, Mishra S, Salama MMA, Liew AC. Classification of power system disturbances using a fuzzy expert system and Fourier linear combiner. *IEEE Trans Power Deliv* 2000;15 (2):472–477.
7. Dash PK, Panigrahi BK, Sahoo DK, Panda G. Power quality disturbance data compression, detection, and classification using integrated spline wavelet and S-transform. *IEEE Trans Power Deliv* 2003;18 (2):595–605.
8. Zhu TX, Tso SK, Lo KL. Wavelet based fuzzy reasoning approach to power quality disturbance recognition. *IEEE Trans Power Deliv* 2004;19 (4):1928–1935.
9. Youssef AM, Abdel-Galil TK, El-Saadany EF, Salama MMA. Disturbances classification utilizing dynamic time warping classifier. *IEEE Trans Power Deliv* 2004;19 (1):272–278.
10. Chilukuri MV, Dash PK. Multi-resolution S-transform based fuzzy recognition system for power quality events. *IEEE Trans Power Deliv* 2004;19(1):323–330.
11. Abdel-Galil TK, Kamel M, Youssef AM, El-Saadany EF, Salama MMA. Power quality disturbance classification using the inductive inference approach. *IEEE Trans Power Deliv* 2004;19 (4):1812–1818.

12. Kelley AW, Yadusky WF. Rectifier design for minimum line-current harmonics and maximum power factor. *IEEE Trans Power Electron* 1992;7(7):332–341.
13. Grotzbach M, Bauta M, Redmann R. Line side behaviour of six-pulse diode bridge rectifiers with AC-side reactance and capacitive load. In: *Proceedings of the Third European Power Quality Conference*, Bremen, Germany; 1995. pp. 525–534.
14. Grotzbach M, Redmann R. Analytical predetermination of complex line-current harmonics in controlled AC/DC converters. *IEEE Trans Ind Appl* 1997;33(3):601–612.
15. Mayordomo JG, Hernandez A, Asensi R, Beites LF, Izzedine M. A unified theory of uncontrolled rectifiers, discharge lamps and arc furnaces. Part I. An analytical approach for normalised harmonic emission calculations. In: *Proceedings of the Eighth IEEE International Conference on Harmonics Quality of Power (ICHQP)*, Athens, Greece; 1998. pp. 740–748.
16. Mishra S, Bhende CN, Panigrahi BK. Detection and classification of power quality disturbances using S-transform and probabilistic neural network. *IEEE Trans Power Deliv* 2008;23 (1):280–287.
17. Tse NCF, Chan JYC, Lau W-H, Lei Lai L. Hybrid wavelet and Hilbert transform with frequency shifting decomposition for power quality analysis. *IEEE Trans Instrum Meas* 2012;61 (12):3225–3233.
18. Prakash Mahela O, Gafoor Shaik A, Gupta N. A critical review of detection and classification of power quality events. *Renew Sustain Energy Rev* 2015;41:495–505.
19. Yousif SNAL, Wanik MZC, Mohamed A. Implementation of different passive filter designs for harmonic mitigation. In: *Proceedings of IEEE Conference on Power and Energy*; November 2004. pp. 229–234.
20. Azri M, Rahim NA. Design analysis of low-pass passive filter in single-phase grid-connected transformer less inverter. In: *Proceedings of IEEE Conference on Clean Energy and Technology*, Kuala Lumpur, Malaysia; June 2011. pp. 348–353.
21. Cha H, Vu T-K. Comparative analysis of low pass output filter for single-phase grid connected photovoltaic inverter. In: *Proceedings of IEEE Applied Power Electronics Conference and Exposition*, Palm Springs, CA; February 2010. pp. 1659–1665.
22. Jayaraman M, Sreedevi VT, Balakrishnan R. Analysis and design of passive filters for power quality improvement in stand-alone PV systems. In: *Presented in IEEE International Conference on Engineering*, Ahmedabad, India; November 2013.
23. Saha SS, Suryavanshi R. Power system harmonic mitigation of an offshore oil rig using passive shunt filter. In: *Presented in IEEE India Conference (INDICON)*, Kolkata, India; December 2010.
24. Chellammal N, Subhransu SD, Velmurugan V, Ravitheja G. Power quality improvement using multilevel inverter as series active filter. In: *Proceedings of IEEE International Conference on Emerging Trends in Science, Engineering and Technology*, Tiruchirappalli, Tamilnadu, India; December 2012. pp. 450–455.
25. Singh B, Al-Haddad K, Chandra A. A review of active filters for power quality improvement. *IEEE Trans Ind Electron* 1999;46 (5):960–971.
26. Farahat MA, Zobah A. Active filters for power quality improvement by artificial neural networks technique. In: *Proceedings of IEEE International Universities Power Engineering Conference*, Bristol, UK. Vol. 1; September 2004. pp. 878–883.
27. Wen J, Zhou L, Smedley K. Power quality improvement at medium-voltage grids using hexagram active power filters. In: *Proceedings of Twenty-Fifth Annual IEEE Applied Power Electronics Conference and Expo*, Palm Springs, CA; February 2010. pp. 47–57.
28. Akagi H. New trends in active filters for improving power quality. In: *Proceedings of IEEE International Conference on Power Electronics, Drives and Energy Systems for Industrial Growth*, New Delhi, India. vol. 1; January 1996. pp. 417–425.
29. Jovanovic MM, Jang Y. State-of-the-art single-phase active power factor correction techniques for high power applications—An overview. *IEEE Trans Ind Electron* 2005;52(3):701–708.
30. Chaoui A, Gauberti J-P, Krim F. Power quality improvement using DPC controlled three-phase shunt active filter. *Electric Power Syst Res* 2010;80:657–666.
31. Zahira R, Peer Fathima A. A technical survey on control strategies of active filter for harmonic suppression. *Procedia Eng* 2012:30:686–693.
32. Miller TJE. *Reactive Power Control in Electric Systems*. Toronto, ON: Wiley; 1982.
33. Arrillage J, Bradley DA, Bodger PS. *Power System Harmonics*. San Diego, CA: Wiley; 1985.

34. Paice DA. *Power Electronic Converter Harmonics-Multi Pulse Methods for Clean Power.* New York: IEEE Press; 1996.
35. Mikkili S, Kumar Panda A. Types-1 and -2 fuzzy logic controllers based shunt active filter Id-Iq control strategy with different fuzzy membership functions for power quality improvement using RTDS hardware. *IET Power Electron* 2013;6(3):818–833.
36. Karupanan P, Kanta Mahapatra K. PI and fuzzy logic controllers for shunt active power filter—A report. *ISA Trans* 2012;51:163–169.
37. Sasaki H, Machida T. A new method to eliminate AC-harmonic current by magnetic flux compensation-consideration on basic design. *IEEE Trans Power Appl Syst* 1971;PAS-90:2009–2019.
38. Harashima F, Inaba H, Tsuboi K. A closed-loop control system for the reduction of reactive power required by electronic converters. *IEEE Trans Ind Electron Control Instrum* 1976;IECI-23:162–166.
39. Rajagopalan V, Jacob A, Sevigny A, Nguy TN, Andy L. Harmonic currents compensation-scheme for electrical distribution system. In: *Proceedings of IFAC Control in Power Electronics and Electric Drives Conference*, Lausanne, Switzerland; 1983. pp. 683–690.
40. Alexandrovitz A, Yair A, Epstein E. Analysis of a static VAR compensator with optimal energy storage element. *IEEE Trans Ind Electron* 1984;IE-31:28–33.
41. Kazerani M, Ziogas PD, Joos G. A novel active current wave shaping technique for solid-state input power factor conditioners. *IEEE Trans Ind Electron* 1991;38:72–78.
42. Enjeti P, Shireen W, Pitel I. Analysis design of an active power filter to cancel harmonic currents in low voltage electric power distribution system. In: *Proceedings of IEEE International Conference on Industrial Electronics, Control, Instrumentation, and Automation, Power Electronics and Motion Control*, San Diego, CA; November 1992. pp. 368–373.
43. Duke RM, Round SD. The steady state performance of a controlled current active filter. *IEEE Trans Power Electron* 1993;8:140–146.
44. Nastran J, Cajhen R, Seliger M, Jereb P. Active power filter for nonlinear AC loads. *IEEE Trans Power Electron* 1994;9:92–96.
45. Torrey DA, Al-Zamel AMAM. Single-phase active power filter for multiple nonlinear loads. *IEEE Trans Power Electron* 1995;10:263–272.
46. Wu JC, Jou HL. A new UPS scheme provides harmonic suppression, input power factor correction. *IEEE Trans Ind Electron* 1995;42:629–635.
47. El Shatshat R, Kazerani M, Salama MMA. Power quality improvement in 3-phase 3-wire distribution systems using modular active power filter. *Electr Power Syst Res* 2002;61:185–1894.
48. Chaoui A, Krim F, Gaubert J-P, Rambault L. DPC controlled three-phase active filter for power quality improvement. *Electr Power Energy Syst* 2008;30:476–485.
49. Ghadbane I, Ghamri A, Benchouia MT, Golea A. Three-phase shunt active filter for power quality improvement using sliding mode controller. In: *Presented in IEEE 2nd International Conference on Communications, Computing and Control Applications*, Marseilles, France; 2012.
50. Esfandiari A, Pamiani M, Mokhtari H. Power quality improvement in three phase four-wire system using shunt active filters. In: *Proceedings of IEEE the Ninth International Power Electronics Congress*; 2004. pp. 193–197.
51. Bin L, Minyong T. Control method of the three-phase four leg shunt active power filter. In: *Energy Procedia*; 2012. pp. 1825–1830.
52. Turunen J, Tuusa H. Improvement of the voltage compensation performance of the series active power filter using a simple PI-control method. In: *Presented in IEE European Conference on Power Electronics and Applications*, Aalborg, Denmark; September 2007.
53. Jain SK, Agarwal P, Gupta HO. Fuzzy logic controlled shunt active power filter for power quality improvement. *IEE Proc Electr Power Appl* 2002;149 (5):317–328.
54. Sumner M, Palethorpe B, Thomas David WP. Impedance measurement for improved power quality-part-2: A new technique for stand-alone active shunt filter control. *IEEE Trans Power Deliv* 2004;19 (3):1457–1463.
55. Esfandiari A, Parniani M. Electric arc furnace power quality improvement using shunt active filter and series inductor. In: *IEEE Proceedings of TENCON'2004.* Vol. 4; November 2004. pp. 105–108.
56. Tey LH, So PL, Chu YC. Improvement of power quality using adaptive shunt active filter. *IEEE Trans Power Deliv* 2005;20(2):1558–1568.

57. Saad S, Zellouna L. Fuzzy logic controller for three-level shunt active filter compensating harmonics and reactive power. *Electric Power Syst Res* 2009;79:1337–1341.
58. Leng S, Chung I-Y, Cartes DA. Distributed operation of multiple shunt active power filter considering power quality improvement capacity. In: *IEEE Proceedings of 2nd International Symposium on Power Electron for dg Systems*; June 2010. pp. 543–548.
59. Kabir MA, Mahhub U. Synchronous detection and digital control of shunt active power filter in power quality improvement. In: *Presented in IEEE Power and Energy Conference at Illinois*, Champaign, IL; February 2011.
60. Mikkili S, Panda AK. Real-time implementation of PI and fuzzy logic controllers based shunt active filter control strategies for power quality improvement. *Electr Power Energy Syst* 2012;43:1114–1126.
61. Prenalatha S, Dash SS, Babu PC. Power quality improvement features for a distributed generation system using shunt active power filter. *Procedia Eng* 2013;64:265–267.
62. Mesbahi N, Ouri A, Ould AD, Djamah T, Omeiri A. Direct power control of shunt active filter using high selectivity filter under distorted or unbalanced conditions. *Electr Power Syst Res* 2014;108:113–123.
63. Swarnkar P, Kumar JS, Nema RK. Adaptive control schemes for improving the control system dynamics: A review. *IETE Tech Rev* 2014;31(1):17–33.
64. Sakthivel A, Vijayakumar P, Senthilkumar A, Lakshminarasimman L, Paramasivam S. Experimental investigations on ant colony optimized pi control algorithm for shunt active power filter to improve power quality. *Control Eng Pract* 2015;42:153–169.
65. Hosseini SH, Nouri T, Sabahi M. A novel hybrid active filter for power quality improvement and neutral current cancellation. In: *IEEE Proceedings of International Conference on Electrical and Electronics Engineering*, Bursa, Turkey; November 2009. pp. I-244–8.
66. Singh B, Verma V, Chandra A, Al-Haddad K. Hybrid filters for power quality improvement. *IEE Proc Gener Transm Distrib* 2005;152(May (3)):365–378.
67. Wu JC, Jou H-L, Wu KD, Hsiao H-H. Three-phase four-wire hybrid power filter using a smaller power converter. *Electr Power Syst Res* 2012;87:13–21.
68. Jou H-L, Wu JC, Wu KD, Huang M-S, Lin C-A. A hybrid compensation system comprising hybrid power filter and AC power capacitor. *Electr Power Energy Syst* 2006;28:448–458.
69. Khanna R, Chacko ST, Goel N. Performance and investigation of hybrid filter for power quality improvement. In: *IEEE Proceedings of the PEOCO'2011*, Shah Alam, Selangor, Malaysia; June 2011. pp. 93–97.
70. Rahmani S, Hamadi Ab, Al-Haddad K. A comprehensive analysis of hybrid active power filter for power quality enhancement. In: *IEEE Proceedings of the Annual Conference on Industrial Electron Society*, MontrealQC; October 2012. pp. 6258–6267.
71. Sharanya M, Basavaraja B, Sasikala M. Voltage quality improvement and harmonic mitigation using custom power devices: DVR and hybrid filters. In: *IEEE Proceedings of the Asia Pacific Conference on Postgraduate Research in Microelectronics and Electronics*, Vishakhapatnam, Andhra Pradesh, India; December 2013. pp. 213–218.
72. Forghani M, Afsharnia S. Online wavelet transform based control strategy for UPQC control system. *IEEE Trans Power Deliv* 2007;22 (1):481–491.
73. Peng FZ, Akagi H, Nabae A. A new approach to harmonic compensation in power systems—A combined system of shunt passive and series filters. *IEEE Trans Ind Appl* 1990;26(6):983–990.
74. IEEE Power Engineering Society, FACTs Overview. IEEE Special Publication, 95TP108; 1995.
75. Singh B, Jayaprakash P, Kothari DP, Chandra A, Al Haddad K. Comprehensive study of DSTATCOM configurations. *IEEE Trans Ind Inf* 2014;10(2):854–870.
76. Mahela OP, Shaik AG. A review of distribution static compensator. *Renew Sustain Energy Rev* 2015;50:531–546.
77. Kumar C, Mishra M. A multifunctional DSTATCOM operating under stiff source. *IEEE Trans Ind Electron* 2014;61(7):3131–316. doi:10.1109/TIE.2013.2276778.
78. Mishra M, Ghosh A, Joshi A. Operation of a DSTATCOM in voltage control mode. *IEEE Trans Power Deliv* 2003;18(1):258–264. doi:10.1109/TPWRD.2002.807746.
79. Perera L, Ledwich G, Ghosh A. Multiple distribution static synchronous compensators for distribution feeder voltage support. *IET Gener Transm Distrib* 2012;6(4):285–293. doi:10.1049/iet-gtd.2011.0197.
80. Nijhawan P, Bhatia R, Jain D. Improved performance of multilevel inverter based distribution static synchronous compensator with induction furnace load. *IET Power Electron* 2013;6(9):1939–1947. doi:10.1049/ietpel.2013.0029.

81. Chidurala A, Saha T, Mithulananthan N. Power quality enhancement in unbalanced distribution network using solar-DSTATCOM. In: *Australasian Universities Power Engineering Conference (AUPEC)*; 2013. pp. 1–6. doi:10.1109/AUPEC.2013.6725394.
82. Arya S, Singh B. Implementation of kernel incremental metalearning algorithm in distribution static compensator. *IEEE Trans Power Electron* 2015;30(3):1157–1169. doi:10.1109/TPEL.2014.2315495.
83. Farhoodnea M, Mohamed A, Shareef H, Zayandehroodi H. Optimum DSTATCOM placement using firefly algorithm for power quality enhancement. In: *2013 IEEE 7th International Power Engineering and Optimization Conference (PEOCO)*; 2013. pp. 98–102. doi:10.1109/PEOCO.2013.6564523.
84. Devi S, Geethanjali M. Optimal location and sizing of distribution static synchronous series compensator using particle swarm optimization. *Int J Electr Power Energy Syst* 2014;62(0):646–653. doi: 10.1016/j. ijepes.2014.05.021.
85. Jou H-L, Wu K-D, Wu J-C, Chiang W-J. A three-phase four-wire power filter comprising a three-phase three-wire active filter and a zigzag transformer. *IEEE Trans Power Electron* 2008;23 (1):252–259.
86. Singh B, Jayaprakash P, Kothari DP. Three-leg VSC and a transformer based three-phase four-wire DSTATCOM for distribution systems. In: *Proceedings of the NPSC'08*, Bombay, India; December 2008. pp. 602–607.
87. Singh B, Jayaprakash P, Kothari DP. Three leg VSC with a star/hexagon transformer based DSTATCOM for power quality improvement in three-phase four-wire distribution system. *Int J Emerg Electr Power Syst* 2008;9 (8):1–18, Article 1.
88. Singh B, Jayaprakash P, Somayajulu TR, Kothari DP. Reduced rating VSC with a zig-zag transformer for power quality improvement in three-phase four-wire distribution system. *IEEE Trans Power Del* 2009;24(1):249–259.
89. Singh B, Jayaprakash P, Kothari DP. A T-connected transformer and three-leg VSC based DSTATCOM for power quality improvement. *IEEE Trans Power Electron* 2008;23 (6):2710–2718.
90. Lin B-R, Yang K-T. Active power filter based on NPC inverter for harmonics and reactive power compensation. In: *Proceedings of the IEEE Region 10 Conference TENCON*. vol. 4; November 2004. pp. 93–96.
91. Lin B-R, Chiang H-K, Yang K-T. Shunt active filter with three-phase four-wire NPC inverter. In: *Proceedings of the 47th Midwest Symposium on Circuits and Systems (MWSCAS)*. vol. 2; July 2004. pp. II-281–II-284.
92. Salmeron P, Montano JC, Vazquez JR, Prieto J, Perez A. Compensation in nonsinusoidal, unbalanced three-phase four-wire systems with active power-line conditioner. *IEEE Trans Power Del* 2004;19(4):1968–1974.
93. Dinavahi VR, Iravani MR, Bonert R. Real-time digital simulation and experimental verification of a D-STATCOM interfaced with a digital controller. *Int J Electr Power Energy Syst* 2004;26(November (9)):703–713.
94. Tavakoli Bina M, Eskandari MD, Panahlou M. Design and installation of a D-STATCOM for a distribution substation. *Electr Power Syst Res* 2005;73(3):383–391.
95. Salmeron P, Montano JC, Thomas JP. Analysis of power losses for instantaneous compensation of three-phase four-wire systems. *IEEE Trans Power Electron* 2005;20(4):901–907.
96. Benhabib MC, Saadate S. New control approach for four-wire active power filter based on the use of synchronous reference frame. *Electr Power Syst Res* 2005;73(3):353–362.
97. Singh B, Adya A, Mittal AP, Gupta JRP. Modeling and control of DSTATCOM for three-phase, four-wire distribution systems. *Ind Appl Conf* 2005;4:2428–2434.
98. Soares V, Verdelho P. Digital implementation of a DC bus voltage controller for four-wire active Filters. In: *Proceedings of the 32nd Annual Conference on IEEE Industrial Electronics (IECON)*; November 2006. pp. 2763–2768.
99. De Morais AS, Barbi I. Power redistributor applied to distribution transformers. In: *Proceedings of the 32nd Annual Conference on IEEE Industrial Electronics (IECON)*; November 2006. pp. 1787–1791.
100. De Morais AS, Barbi I. A control strategy for four-wire shunt active filters using instantaneous active and reactive current method. In: *Proceedings of the 32nd Annual Conference IEEE Industrial Electronics (IECON)*; November 2006. pp. 1846–1851.
101. Barrado JA, Grino R, Valderrama H. Standalone self-excited induction generator with a three-phase four-wire active filter and energy storage system. In: *Proceedings of the IEEE International Symposium on Industrial Electronics (ISIE)*; June 2007. pp. 600–605.

102. Zhou J, Wu X-j, Geng Y-w, Dai P. Simulation research on a SVPWM control algorithm for a four-leg active power filter. *J China Univ Mining Technol* 2007;17(4):590–594.
103. Ucar M, Ozdemir E. Control of a 3-phase 4-leg active power filter under non-ideal mains voltage condition. *Electr Power Syst Res* 2008;78(1):58–73.
104. Montero MIM, Cadaval ER, González FB. Comparison of control strategies for shunt active power filters in three-phase four-wire systems. *IEEE Trans Power Electron* 2007:22(1):229–236.
105. Jou HL, Wu KD, Wu JC, Li CH, Huang MS. Novel power converter topology for three-phase four-wire hybrid power filter. *IET Proc Power Electron* 2008;1(1):164–173.
106. Karanki SB, Geddada N, Mishra MK, Kalyan Kumar B. A DSTATCOM topology with reduced DC-link voltage rating for load compensation with nonstiff source. *IEEE Trans Power Electron* 2012;27(3):1201–1211.
107. Ghosh A, Ledwich G. *Power Quality Enhancement Using Custom Power Devices.* Norwell, MA: Kluwer Academic Publishers, 2002.
108. Dugan RC, McGranaghan MF, Beaty HW. *Electric Power Systems Quality.* 2nd ed. New York: McGraw Hill, 2006.
109. Moreno-Munoz A. *Power Quality: Mitigation Technologies in a Distributed Environment.* Berlin, Germany: Springer-Verlag, 2007.
110. Akagi H, Watanabe EH, Aredes M. *Instantaneous Power Theory and Applications to Power Conditioning.* Hoboken, NJ: Wiley, 2007.
111. Padiyar KR. *FACTS Controllers in Transmission and Distribution.* New Delhi, India: New Age International, 2007.
112. Fuchs EF, Mausoum MAS. *Power Quality in Power Systems and Electrical Machines.* Amsterdam, the Netherlands: Elsevier, 2008.
113. Singh B, Jayaprakash P, Kumar S, Kothari DP. Implementation of neural network controlled VSC based DSTATCOM. *IEEE Trans Ind Appl* 2011;47(4):1892–1901.
114. Jayaprakash P, Singh B, Kothari DP. DSP implementation of three-leg VSC based three-phase four-wire DSTATCOM for voltage regulation and power quality improvement. In: *Proceedings of the IECON'09*, Portugal; 2009. pp. 312–318.
115. Singh B, Jayaprakash P, Kothari DP. Star/hexagon transformer and non-isolated three-leg VSC based 3-phase 4-wire DSTATCOM. *Int J Power Energy Convers* 2010;1(2/3):198–214.
116. Singh B, Jayaprakash P, Somayajulu TR, Kothari DP. DSTATCOM with reduced switches using two-leg VSC and zig-zag transformer for power quality improvement in three-phase four-wire distribution system. In: *Proceedings of the IEEE TENCON'08*, Hyderabad, India; November 2008.
117. Singh B, Jayaprakash P, Kothari DP. Three-phase four-wire DSTATCOM with H-bridge VSC and star/delta transformer for power quality improvement. In: *Proceedings of the IEEE INDICON'08*, Kanpur, India; December 2008. pp. 412–7.
118. Singh B, Jayaprakash P, Kothari DP. H-Bridge VSC with a T-connected transformer based three-phase four-wire DSTATCOM for power quality improvement. In: *Proceedings of the NSC'08*, Roorkee, India; 2008, pp. 1–6.
119. Jayaprakash P, Singh B, Kothari DP. Three-phase four-wire DSTATCOM based on H-bridge VSC with a star/hexagon transformer for power quality improvement in distribution systems. In: *Proceedings of the IEEE ICIIS'08*, Kharagpur, India; December 2008. pp. 1–6.
120. Jayaprakash P, Singh B, Kothari DP. Three single-phase voltage source converter based DSTATCOM for three-phase four-wire systems. In: *Proceedings of the ICPS'09*, Kharagpur, India; 2009. pp. 112–118.
121. Jayaprakash P, Singh B, Kothari DP. Digital signal processor implementation of isolated reduced-rating voltage source converter using a zig-zag transformer for three-phase four-wire distribution static compensator. *Electr Power Compon Syst* 2011;39:15–30.
122. Singh B, Jayaprakash P, Somayajulu TR, Kothari DP. Integrated zigzag transformer and 3-leg VSC for power quality improvement in three phase four wire system. In: *Proceedings of the IEEE IECON'08*, Orlando, FL, USA; July 2008. pp. 796–801.
123. Singh B, Jayaprakash P, Kothari DP. A three-phase four-wire DSTATCOM for power quality improvement. *J Power Electron* 2008;8(3):249–255.
124. Singh B, Jayaprakash P, Kothari DP. Three-leg VSC integrated with T-connected transformer as three-phase four-wire DSTATCOM for power quality improvement. *Electr Power Compon Syst* 2009;37(8):817–831.

125. Jayaprakash P, Singh B, Kothari DP. Implementation of isolated three-leg VSC and star/hexagon transformer based three-phase four-wire DSTATCOM. In: *Proceedings of the ICETET 09*, Nagpur, India; 2009. pp. 1–6.
126. Jayaprakash P, Singh B, Kothari DP. Integrated H-bridge VSC with a zig-zag transformer based three-phase four-wire DSTATCOM. Indian patent in process.
127. Singh B, Jayaprakash P, Kothari DP. Three-phase four-wire DSTATCOM with H-bridge VSC and star/delta transformer for power quality improvement. In: *Proc. IEEE Int. Conf. Sustain. Energy Technol (ICSET'08)*, Singapore; November 2008. pp. 366–371.
128. Singh B, Jayaprakash P, Kothari DP. Three-phase four-wire DSTATCOM with reduced switches for power quality improvement. *Asian Power Electron J* 2008;2(2):1–6.
129. Jayaprakash P, Singh B, Kothari DP. Magnetics for neutral current compensation in three-phase four-wire distribution system. In: *Proceedings of the IEEE International Conference on Power Electronics Drives and Energy Systems (PEDES'10)*, New Delhi, India; December 2010. pp. 461–467.
130. Choi S, Jang M. Analysis and control of a single-phase-inverter zigzag transformer hybrid neutral-current suppressor in three-phase four-wire systems. *IEEE Trans Ind Electron* 2007;54(4):2201–2208.

6

Control Topologies for Series Active Filters

6.1 Background

Diode or thyristor rectifiers and cyclo-converters are different nonlinear loads that lead to harmonic pollution. These loads are generally used in industrial applications and transmission/distribution systems, which will pose serious problems to the power system. Owing to a restricted amount of source impedance, distortion in voltage at point of common coupling (PCC) results from harmonic currents formed by these nonlinear loads. Passive filters (PFs) are tuned LC filters and/or high-pass filters are usually used to recover power factor and to attract harmonics in power systems because of their low cost, simplicity, and good efficiency. This tuned LC filter is designed for lower impedance at that tuned harmonic frequency than supply impedance, in order that about every harmonic current at that harmonic frequency enters the LC filter. In principle, filtering uniqueness of a passive filter is resolute through the impedance ratio of the source and the passive filter. So, it is hard for a passive filter fixed in the area of a harmonic-generating load, which is connected to a quite rigid AC source, to meet the above-mentioned criteria. In addition, a passive filter has the following drawbacks:

- The distorted voltage due to non-linear loads at utility supply could overload the passive filter. In the worst scenario, these passive filters might plunge in series resonance with the source impedance.
- For a definite frequency, these passive filters could lead to parallel resonance with the source impedance, so that increase of the harmonic current occurs and the currents in the supply and the passive filter can be extreme.

An active filter (AF) is controlled to produce zero impedance to external circuit at fundamental frequency and an infinite resistance (K) (Ω) at harmonic frequencies. The active filter AC voltage revokes the harmonic voltage across a PF, hence providing a harmonic current branch with zero impedance. This is because all the harmonic currents created here enter the PF, so that no escape for harmonic current. Also, the active filter AC voltage compensates for the other harmonic voltages, consequently blocking the harmonic currents to flow from the supply to the PF.

The harmonic current could be concealed when using a passive or an active power filter (APF). Typically, the passive power filter (PPF) can solve the problems of harmonic current pollution in the industrial power system as it is very cheap. Yet, it has major disadvantages like [2]:

1. Susceptible to the disparity of power system impedance
2. Sensitive to utility frequency variation
3. The danger of series or parallel resonance
4. The filter frequency is set and is difficult to adjust

In the above list, the series or parallel resonance is the main severe drawback. It will result in over-current or overvoltage on the capacitor and inductor leading to damage of the PF. As the system impedance has a major effect on the behavior of the PF, it is tough to get desired filter performance in a practical way. Further, the harmonic current generated by the nonlinear loads can flow into the PPF and consequently overload the PPF.

The active filter can avoid the PF leading to harmonic drain for non-linear loads. Moreover, it will prevent the compensation based on the system impedance. Theoretically, the ideal situation is if the proportionality constant k, linking the output voltage of an active filter and supply current harmonics, has a high value. However, maximum would be an infinite value and the control objective was in practice impossible to achieve. The value of k is generally chosen to be small for avoiding high power active filters and the system instabilities. However, the choice of the correct k value is an uncertain question as it is interrelated to the PF and the supply impedance values. Also, this approach is not feasible for use in systems with variable loads as the reactive power of the PF is constant, and thus, the set compensation apparatus and load has an uneven power factor.

Quick disparity of reactive power produced by non-linear and distorted loads, like arc furnaces and harmonics produced thyristor or by diode rectifiers, is serious. It leads to harmonic interference or flicker in industrial applications and transmission/distribution systems. So far, shunt PFs, which consist of tuned LC filters and/or high-pass filters, have been used to improve power factor and to suppress harmonics in power systems. But, shunt PFs pose the previously mentioned issues so as to discourage industrial applications. A shunt PPF shows lower impedance at the tuned harmonic frequency than the supply impedance to lessen the harmonic currents flowing into the source. The filtering characteristic principle of a shunt PPF is determined by the impedance ratio of the source and shunt PPF. Hence, the shunt PF has subsequent issues like:

- The supply impedance, if unknown and fluctuating with the system design, strongly misleads the filtering behavior of the shunt PPF.
- The shunt PPF performs as a sink to the harmonic current entering from the source. In a worst-case situation, the shunt PPF will be in series resonance with the source impedance.
- For specific frequency, parallel resonance or anti-resonance occurs among the source impedance and the shunt PPF, so-called as harmonic amplification.

The active filter terminates or restrains the low frequency ripple components, which are very difficult to satisfy with a designed single passive low-pass filter. This technique allows a considerable decrease in the size of the PF, with additional benefits like decrease in converter size, cost, and weight. Dynamic voltage restorers (DVRs) or series active power filters (SeAPF) are mostly used to protect sensitive loads from the disturbing electrical network voltage like sags or swells or interruptions and possibly will be used to diminish harmonic distortion of AC source voltages. Voltage sags are sudden fall (between 10% and 99%) in the AC rms voltage value, lasting within 60 s. Distinctive sag depths vary from 50% to 90% of the nominal voltage and last from 10 ms to a few seconds. Voltage sags are of more interest as the wide usage of voltage-sensitive loads like process control equipment, adjustable speed drives and computing devices like computers, etc. Voltage sags could cause wide disruption or interruption to the industrial process sector and are a severe power quality issue.

However, the cost of shunt active power filters (SAPFs) is high, they are difficult to implement on a large scale, and they have lower efficiency than shunt PFs. Therefore, different solutions are being anticipated to enhance the practical use of APFs. One such application is a hybrid system utilizing shunt passive and series active filters. This method allows for designing the active filter for merely a part of the total load power, decreasing costs, and rising overall system effectiveness. Many problems remain to be solved ahead of the specifications to be met. The first problem to be solved is the circuit topology. In recent times, it has been accepted that the three-phase six-valve parallel bridge topology permits either the upper valve or the lower valve to conduct at one time. Using the PWM technique, a strategy for tri-state logic was developed that recent workers have used to produce sharply defined, close to sinusoidal current waveforms as per the specifications during steady state. These tri-state logic methods are designed for static loads and even if the loads are dynamic in nature. Advanced research is still necessary to validate their abilities to suggest dynamic transient states through the tri-logic pulse-width modulation (PWM).

But, the parallel active filters are expensive and hard to realize on a large scale. Series active filters in principle work as isolators, rather than generators of harmonics; hence, they require different control strategies. SeAPFs act as controllable voltage sources and controllable current sources are proposed in the literature. Series active filters enhance the current waveform on the source side of the filter sacrificing cost of voltage-waveform diminishing on the load side of the filter. One solution to this issue is to employ hybrid parallel PFs and series active filters. Still, the power rating of a series active filter is significantly lower than a parallel active filter counterpart, so execution of the large system is still not important for the following reasons: (1) It is at all times understood completely that the input impedance on the source side is entirely inductive, which is practically not true. (2) As complete source voltage appears across the current transformer during a load-side short-circuit, extra protection actions are requisite to protect the system.

The active filter with an rms voltage rated at 5% and with a peak voltage rated at 7.6% enables the diode rectifier to draw three-phase sinusoidal currents from the utility. In addition, it can provide the supplementary value-added function of regulating the common DC-bus voltage to a limited extent of 5%, slightly increasing the rms voltage rating but not increasing the peak voltage rating. The function of achieving not only harmonic compensation, but also DC-bus voltage regulation, justifies the integration of the active filter with the diode rectifier practically and economically.

The large amount of electronic equipment infects electrical distribution system with harmonics, distorting the voltage and the current waveforms. Simultaneously, the outcome for the user is the effect on the electrical installations of extreme heating and accelerated ageing of the electrical equipment. It also involves the functioning of applications, owing to the rate of inappropriate firing and the unfeasibility of feeding sensitive loads. Finally, it has an impact on the electrical power available in the installation as there is an over consumption due to the harmonics. The notable raise of harmonics on system has created the concern and request of numerous norms for enhancing the quality of energy supplied. In the last decade, novel topological promise of voltage filtering has been anticipated, with control strategies, all due to the progress of electronic technology. The mainly frequent use of sinking voltage and current turbulences is the fixing of PFs. This result can create resonances in the filter and impedance in the system, growing the distortions. Application of active filters in parallel is the generally used way of decreasing current disturbances, though this technique needs a PWM current reverser, with huge capacity, reduced losses, and high speed of response. The use of series active filters is exploited to lessen the voltage mismatches or harmonic disturbances. AFs linear models are found by applying averaging techniques or during vital simplifications to the device, letting at same time, the use of standard controllers. The PWM allows use of active filters to control the discrete systems owing to the uniqueness of the method and the control device. Few authors previously used control discrete methods for series-voltage restorers by applying generalized methods to facilitate them to set up a linear model.

Usually, shunt PPFs are used to maintain harmonic pollution within suitable limits. Yet, SPFs have various disadvantages in practice; for example, they might cause impedance of power system, robustly influencing the filtering effects and resonance. To overcome these disadvantages, the APF has been proposed. As the APF can efficiently filter away all harmonics and the system parameters do not involve the filtering effect, the APF suits the key research way of harmonics mitigation. Different APF designs and control methods have been researched during the last two decades. To date, a huge number of SAPF have been applied in practice; however there are still some limitations: (i) superior capacity, the principal investment is too high and the user is hard to allow; (ii) the harmonic flow is created by operation with the shunt PPF collectively; (iii) the APF cannot recover the filter characteristics when the system resonance is formed by PPF.

Series active filters operate as isolators instead of harmonics generators and, thus, they use different control methodologies. So far, series active filters operating as controllable voltage sources are proposed [5]. With this method, the estimation of the series filter reference voltage is necessary. This is generally fairly difficult, because the reference voltage is mostly composed of harmonics and it needs to be assessed through accurate measurements of voltages and/or current waveforms. One more way to obtain the reference voltage for the series active filter is using the "p–q theory." But, this solution has the demerit of requiring a very complex control circuit containing several analog multipliers, operational amplifiers, and dividers.

The increasing quantity of equipment illustrates as nonlinear loads has created a negative impact on the excellence of electric power supply [1,2]. As a result, harmonic and reactive current are responsible for: (i) low energy efficiency, (ii) low power factor, (iii) flicker, (iv) audible noise, and (v) shorter life of galvanic insulation [1,3–5]. The SAPF is popular for dynamic improvement of the harmonic currents between either directly connected non-linear loads or to the PCC [5–7]. Such connection makes the SAPF an appropriate regulator due to its simple repair (i.e., in case of failure) or installation. SAPF development in design comprises several factors such as: (i) control strategy, (ii) converter source type, (iii) converter topology, (iv) source filter type, etc. The converter source types mostly cover current source converter (CSC) [8] otherwise voltage source converter (VSC) [9–14]. VSC is more used than CSC in most applications due to its advantages. From the view of control point, the instantaneous reactive power compensation initiated in [15] has revealed the benefits of a practical system without the necessity of any energy storage apparatus. But, some control strategies by digital signal processors (DSPs) have been explored merging obtainable theories for SAPF [9,11,16,17]. A further significant concern to the SAPF topology is the line filter. Such a module can be an arrangement of L, LC, or LCL passive circuit alternative as the best choice for a few applications [18–20]. A complete state-of-the-art abbreviation of major characteristics associated to SAPF as well as an analysis of main contributions in the literature is explained in [5].

An extremely general SAPF configuration in a three-phase three-wire (3P3W) system is the three-phase full-bridge VSC [1,5,17,21]. But this topology is not appropriate for high or medium power purposes. With the advance of silicon carbide (SiC)-based converters [22–24], the harmonic mitigation by using SAPF with SiC-dependent converter for high-power applications is not established enough and is considered to be restricted by the accessibility in the market [25–27]. Consequently, different SAPF topologies are being anticipated in the methodological literature for high-power application standards, such as six-leg conventional configurations [28] (as in Figure 6.1), multilevel topology [29], hexagram topology [30], interconnected converters [31], and multilevel cascaded H-bridge [32,33]. Each of these topologies has a few advantages and disadvantages. The multilevel cascaded H-Bridge topology has the merits to decrease harmonic mitigation level as well as the rating of power switches. But, the downside is,

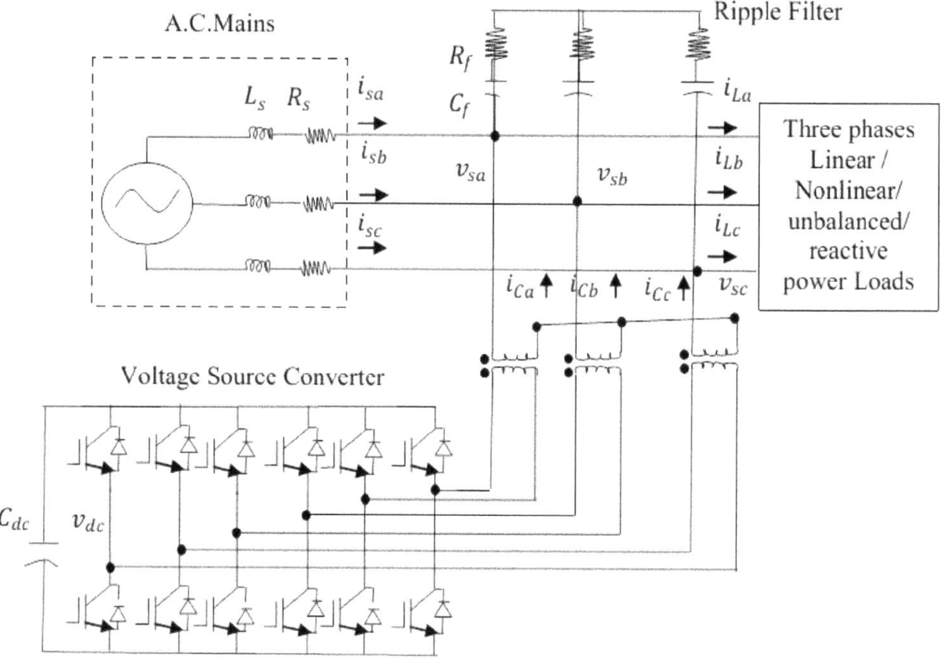

FIGURE 6.1 Conventional shunt active power filter configuration (6L) for MV application. (From Gupta, R. et al., Control of 3-level shunt active power filter using harmonic selective controller, in *Power India Conference, 2006 IEEE*, p. 7, 2006.)

it requires a high number of DC-link capacitors and so, there is an imbalance control issue, which raises the control difficulty. Hexagram configuration is another case of topologies that can decrease the stress at the power switches. Alternatively, it requires 36 power switches and 6 DC-link capacitors with an extra need of circulating current control. The general six-leg configuration is useful as lower devices count and an easy control technique. Such a topology, shown in Figure 6.1, is referenced as 6L here. Then, because of the strong and weak points linked with each topology, the choice depends on particular design criteria.

A series active filter (SeAF), shown in Figure 6.1, can be used to mitigate for both harmonic distortion and imbalance in the source voltage. The voltages, also on the supply side or the load side, can be mitigated as desired. But, this deals with applications where the loads are susceptible to voltage waveform superiority. Thus, the compensation is taken care only on the load side. The compensation voltages, necessary to reduce voltage harmonics and create the balanced system, are injected across the recompensing transformers' secondary windings that are in series with the line. Obviously, the efficacy of such a filter is based on the technique and exactness with which the preferred compensation voltages can be determined and generated.

Harmonic interference issues in power systems are becoming more and more serious due to the wide application of power electronic equipment and non-linear loads in modern years. Usually, shunt passive power filters (SPFs) are used to maintain harmonic pollution within satisfactory limits [34,35]. Still, it has many shortcomings. SPF may lead to resonance and its filtering characteristic is susceptible to the power system impedance. During the last two decades, various APF topologies with their particular control strategies have been proposed [36–43]. A novel theory of fundamental magnetic flux compensation (FMFC) is proposed in [44], which is applied as series hybrid APF (SHAPF) [45,46] and controllable reactor [47]. In the SHAPF, with the appropriate control of PWM inverters, the series transformer (ST) shows very small leakage impedance to fundamental and physically high magnetizing impedance to the harmonics.

Consequently, the harmonic currents are required to flow into the PF branches. The ST performs as a "harmonic isolator." Apart from the traditional SAPF, merely the fundamental current element needs to be identified and tracked. The bandwidth of the filtering system is really lower and the control scheme is easier to be realized. This novel type series hybrid active power filter (NTSHAPF) based on FMFC is very appropriate for filtering the harmonic current. Xie et al. [45,64] explained the analytical model of the PWM inverter and described the filtering characteristic ultimately. A 400 V, 20 kVA practical prototype has been built and the experimental results have verified the legitimacy of the effective theory of the NTSHAPF.

Likewise, to protect the PCC from voltage distortion, using DVR function is advised. The technique is to reduce the effluence of power electronics-based loads connected at the source. While several efforts are made for specific case study a general solution is investigated [46,47]. There are two types of active power devices to prevail over described power quality problems. The foremost category are series active filters (SeAPF) with hybrid ones. They were developed to remove current harmonics formed by non-linear load from the system. SeAPFs are less widespread than SAPFs [48,49]. The merits of SeAPF compared to shunt type are the lower rating of the compensator against load nominal rating [50]. Yet, the difficulty of the configuration and need of an isolation ST have slowed their industrial use in distribution system. The next category was developed in apprehension of answering voltage problems on sensitive loads. Generally known as DVR, they have a similar configuration to SeAPF. These two groups are dissimilar from each other in their control mechanism. This difference depends on the purpose of their application in the system. A hybrid series active filter (HSeAF) was proposed to address the previously mentioned issues with only one combination. Hypothetically, they are capable of compensating current harmonics, ensuring a power factor correction, and eliminating voltage distortions at the PCC [51,52]. These properties make it an appropriate candidate for power quality investments. The three-phase series active filters are well documented [53,54], whereas limited research work has reported on the single-phase applications of series active filters in the literature. In this paper [54], a single-phase transformer-less-HSeAF is proposed and able to clean up the grid-side connection bus bar from current harmonics created by a non-linear load [55]. With a minor rating up to 10%, it could simply restore the shunt active filter [56]. Moreover, it might restore a sinusoidal voltage at the load PCC.

In modern years, there has been a substantial attention in power quality and voltage stability of utility because of the augmented number of non-linear loads. The harmonic pollution in the power system by non-linear loads like rectifiers, inverters, and variable frequency drive (VFD) is bound to be present. Typically, the power quality issues were lessening using PPFs. Recently these issues have been solved with the aid of SAPFs, which are regarded as a type of a controlled current source mitigating the harmonic current due to the non-linear loads. However, the cost of shunt active filters is comparatively higher and they are not chosen for a large-scale system as the power capacity of filter rises with rise in the load current to be compensated [57–59]. In addition, their performance is better in the current-harmonics-created load than in a voltage-harmonics-generating load [60,61].

In the late 80s, SeAF appeared as an alternate result for SAPF [61,62]. The SeAF only is more valuable for mitigating voltage-type harmonic for non-linear loads. In recent years the hybrid active power filter has become very famous as they extend multi-functionality and help to reduce the inverter capability. The SHAPF is capable of mitigating voltage harmonic creating loads as well as current harmonic producing loads [63–65]. The SHAPF works as harmonic isolator rather than a harmonic voltage generator and strengthens the harmonic current to flow through PFs. This topology can also control the load voltage to a preferred value by controlling the inverter output so as to mitigate abnormal utility voltages [66,67].

To get better SHAPF performance, it is significant to choose appropriate reference voltage generation algorithm. The purpose of control strategy for SHAPF is to extract source current harmonics or load voltage harmonics or a combination of both [68]. Two methods exist to extract the required harmonic component of voltage or current; one extends the accepted definition of the power components and in another approach PFs are used. As PF undergo an inherent issue of phase shift due to frequency disparities, the first approach of using instantaneous power theories is highly preferred. A major advance happened when Akagi et al. [69,76] initiated this instantaneous power theory. If the supply voltage has harmonics then the product of current and voltage of same frequencies will result in an average component, which is indivisible from that created by the fundamental frequency component. To circumvent this issue, harmonic removal using synchronous reference frame theory (SRF) was introduced. This method processes only the current to remove its harmonic components but needs synchronization with the grid voltage, which is under non-sinusoidal source voltage conditions, so this might pose serious problems [71–74].

Afterward the instantaneous power theory for polyphase system was developed by Montano and Salmeron [72]. This involves the decomposition of current into two orthogonal components and then multiplying it with voltage to obtain the instantaneous powers. Peng et al. [73] proposed a definition of instantaneous reactive vector based on the outer-product operation of the current and voltage vectors. Dai et al. [74] introduced the concept of generalized instantaneous power multi-vector for a direct expression of instantaneous power quantities with no coordinate transformations.

The topology of the SHAPF is shown in Figure 6.2. The configuration is for a three-phase PWM voltage-source inverter (VSI) connected in series through three single-phase transformers. To permit current harmonic mitigation, a parallel LC filter must be connected between the STs and nonlinear loads (Figure 6.2). Current harmonic and the voltage unbalance mitigation are obtained by creating suitable voltage waveforms with the three-phase PWM VSI. Even though there are a number of papers dealing with the study and design of APFs connected in series, the three-phase SeAPF presented in this paper [74] varies from before discussed methods in the following ways:

(a) it is realized with a three-phase PWM VSI; (b) the APF is appropriate to mitigate three-phase power systems with three or four phases (three phases and a neutral), in order that it can compensate current harmonics generated by three-phase and single-phase loads; (c) the SeAPF is capable to mitigate simultaneous line-to-voltage unbalances at the load terminals and current harmonics; (d) by mitigating the zero-sequence harmonics generated by single-phase loads, the current passing through the neutral wire is considerably reduced, and the total harmonic distortion (THD) of the line currents is enhanced. As the voltage unbalance is based mainly by the fundamental current or voltage components, with the control scheme, the SeAPF can mitigate the negative and zero-sequence components of the load current and voltage harmonics at the same time. Further, zero-sequence current harmonics passing through the neutral are mitigated without sensing the resultant neutral current, thus making the current control scheme easier. The treatment presented in this paper [74] includes a comprehensive steady-state and transient state study of the proposed SeAPF. Also, the design technique and control circuit performance of the proposed SeAPF are reported.

SeAPF compensate current system distortion produced by non-linear loads by impressing a high impedance path to current harmonics. This makes the high frequency currents pass through LC PF connected

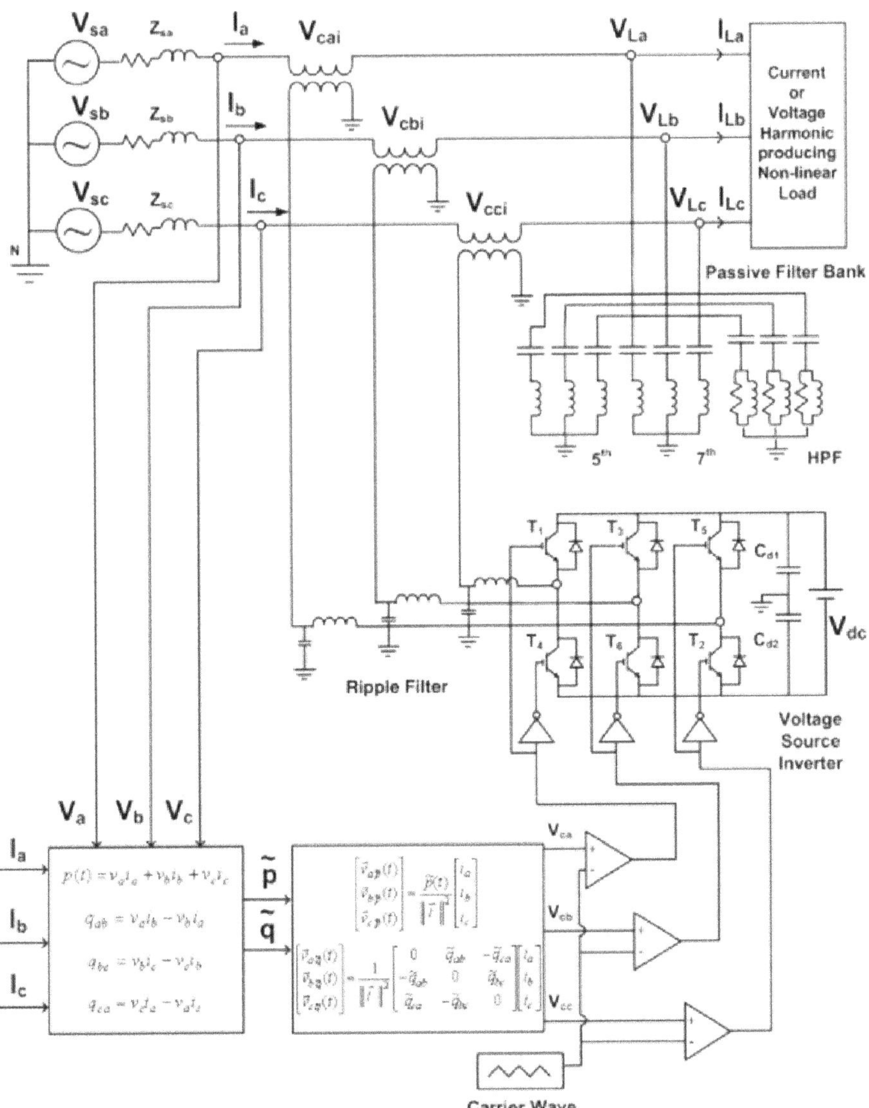

FIGURE 6.2 SHAPF circuit configuration with the control scheme.

in parallel to the load [l]. The high impedance forced by the SeAPF is produced by generating a voltage of the same frequency as the current harmonics that must be eradicated. Voltage unbalance is improved by mitigating the fundamental frequency zero and negative sequence voltage components of the system.

APFs are effective in mitigating non-linear loads [64,75,76]. Shunt topology is the most analyzed topology in which the APF is connected shunt to the load. Its conventional use for the removal of current harmonics created by loads producing disturbances, it is known as harmonic current source (HCS) type loads. But, parallel APF is ineffective in conditions where the load produces voltage harmonics, known as harmonic voltage source loads [77–80]. In this case, series APF configuration was proposed and different control techniques were tried out [81]. In any occasion, mitigating systems create only an APF, either connected in parallel or in series; they do not completely explain the issue of harmonic reduction for all types of loads. Other configurations proposed in [82–89] are combinations of series and parallel topologies with both active and PFs and are called hybrid topologies [113].

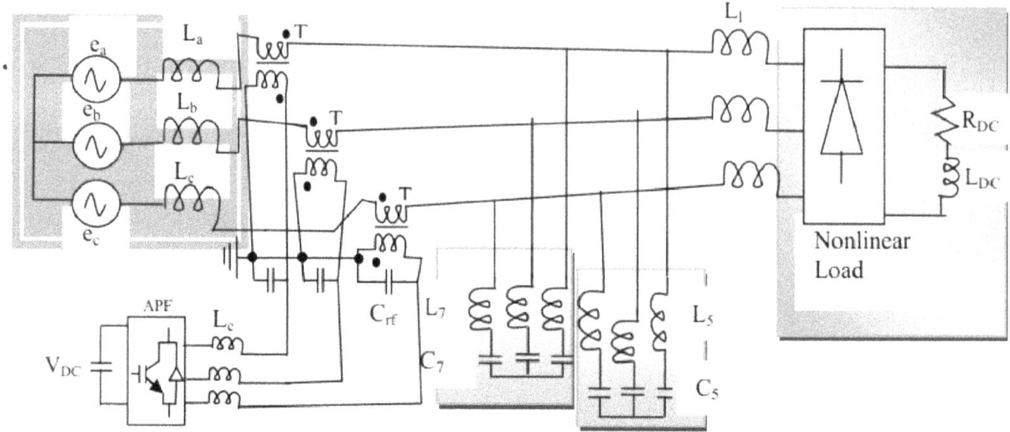

FIGURE 6.3 System with SAPPF filter and load type HCS.

Among the hybrid topologies, the SeAPF with parallel PF (series active and parallel passive filter [SAPPF]) is the main analyzed topology [87–92] and is shown in Figure 6.3. For this configuration, various control strategies are proposed. One of them is taken from early 1990s; the control purpose is based on creating a voltage proportional to the source current harmonics using a proportionality constant *k* [68]. In this case, APF permits the filtering of parallel connection PF to be enhanced. The development of the new strategy was studied from a steady-state model. But, this theoretical advance is not useful for getting the proportionality constant value. Also, this is not a proper way to analyze the system stability. In a similar way, other control techniques for the SeAPF appeared. Thus, in [81] three control methods are studied to observe the performance, while this analysis does not state design criterion for each compensation technique. In this paper [81], the theoretical analysis of a hybrid filter SeAPPF is carried out. This theoretical advance allows the establishment of design strategy for each case. For this, the state model equations are obtained and the performance of mitigation equipment is studied for three different control techniques:

- Source current detection (SCD). In this strategy, the SeAPF creates a voltage proportional to the source current harmonics.
- Load voltage detection (LVD). In this technique, the SeAPF creates a voltage opposite to the load voltage harmonics.
- Current and voltage detection (CVD). This method combines the SCD and the LVD.
- The resulting model lets the control design be recognized in order to get the value of k that guarantees system stability and reduces the source current harmonics to the level of reduction particular for the design.

For this reason, a simulation platform based on MATLAB/Simulink was developed. Each method was studied in the cases of sinusoidal supply and non-sinusoidal supply voltages.

Most modern power converters are equipped with unregulated DC voltage from AC lines using diode rectifiers. Two popularly used strategies of diode rectifier are: the rectifiers with adequately large DC inductor to supply steady DC current to the load and the rectifiers having suitably large capacitance on DC side to supply constant DC voltage to the load. The source current characteristics of diode rectifiers are not gracious to supply lines due to their non-linear nature and hence it has drawn much attention [93–97]. The situation is worst for diode rectifiers accompanied by a large DC capacitive filter. The capacitor filter keeps the DC voltage very close to the peak line-to-line voltage, resulting in highly non-linear source current flow for a brief period.

In addition to source current harmonics, the unbalance of source current is a new primary concern of this design. In field situations, small unbalance is mostly present in source voltage, which may result in major unbalance in source currents. Unbalanced rectifier currents could cause effects such as uneven current distribution in the legs of the rectifier. The raised rms ripple current in the smoothing capacitor and increased total rms line current, harmonics and in particular, non-characteristic triplen (3*n*) harmonics that do not emerge under balanced condition [95–97]. As minute unbalance is forever present in the major industrial and commercial power lines, care must be taken in installing and designing diode rectifiers to maintain the current unbalance at an acceptable point and to circumvent the above objectionable effects.

Compensation of load in power engineering is the method used to get the source currents sinusoidal with balanced waveforms. One method is to integrate power-factor correction (PFC) in the converter topology. Diode rectifier with the continuous-conduction-mode type boost converter [98–100], PWM rectifier [101], and PWM AC choppers [102] are the various topologies for employing PFC. The closed-loop action of static power converter with PFC promises acceptable performance to get high input power factor and control converter output voltage over a broad operating range. Increased complication, electromagnetic interference (EMI), and decreased robustness are distinctive characteristics of these methods [99]. These methods deal with the mitigation of supply current harmonics, but they cannot compensate for supply voltage unbalances. In another approach, APFs are recommended as power electronic solutions for load mitigation or compensation. A variety of APF configurations and control techniques have been analyzed during the last two decades.

Generally used APF configurations are: SAPF, which introduces compensation currents [103]; series APF, which injects compensating voltages through a ST [104]; and SHAPF, which is a hybrid system of shunt PPF and series APF [105–108]. So as to reduce inverter rating and because of multi-functionality, SHAPF is gaining popularity in recent advances. To get efficient SHAPF behavior, it is significant to choose an appropriate reference-based algorithm. The aim of control strategy for SHAPF is to get load voltage harmonics or source current harmonics or a combination of both [104]. Using vector algebra, decomposing voltage vector into quantities is possible to represent various components of power [108,109]. The separated voltage components are useful for getting SHAPF references that compensate for source current harmonics. Further, the reference for source voltages unbalance is added to this reference to mitigate source voltage unbalance.

Particularly on three-phase four-wire (3P4W) distribution systems, the third-harmonic currents are increased. The excessive third harmonic current leads to overheating of the neutral conductors. To reduce the neutral current component, a PF connected in series with the neutral conductor, and the shunt active filters and their control methods, are proposed. The shunt active filters can compensate not only third harmonic currents but also higher order harmonic currents. On the other hand, the voltage drops on both the distribution transformers and the power cables caused by the third-harmonic current result in voltage distortion on utility outlets. In this case, lighting equipment cuts out suddenly due to the third harmonic voltage in the three-phase four-wire distribution system. However, few papers have reported the mitigation methods of third-harmonic voltages on utility outlets.

The APF is restricted to operate only for the third harmonic frequency, not generating any fundamental voltage. Hence, for the fundamental frequency, the APF can be observed as a short circuit. The APF consists of a single-phase VSI in series with neutral conductor. The required power of the APF is less than 10% of the harmonic-producing loads. The APF can compensate either third-harmonic voltage or the neutral current on the 3P4W distribution system.

To solve the above issues, a novel SHAPF is based on controllable harmonic impedance [55]. Through injecting the harmonic current in phase with the primary current to the secondary winding, the harmonic equivalent impedance of primary winding of a ST can be regulated arbitrarily, which forces the harmonic current to flow into the PPF. The self-inductance of the ST in the APF is very small, about 0.01–0.10 in per-unit, and the PWM inverter works as a controlled HSC instead of a harmonic voltage source or a fundamental current source. Owing to the self-inductance of the ST, the proposed APF does not have the weaknesses of resonance. The inverter and secondary winding of the ST need not endure the fundamental current but only endure the injected harmonic currents, the inverter capacity can be sharply reduced, and the LC fundamental tuned bypass can be omitted, so the cost of the proposed APF will be very low.

To segregate the DC bus from the power systems, the SeAPF is connected to the grid across three single-phase inductors as shown in Figure 6.3. This makes the behavior of the series compensator

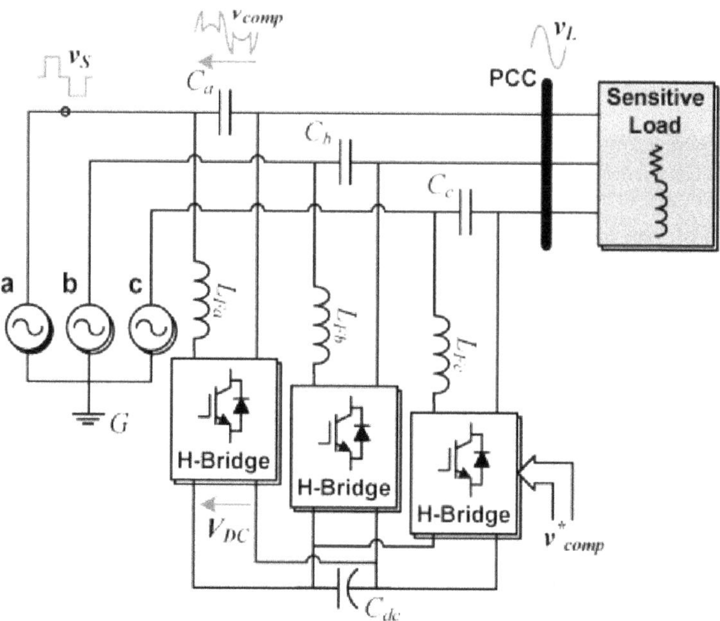

FIGURE 6.4 Transformer-less dynamic voltage restorer (T-DVR).

difficult. Various methods are available to eliminate this expensive transformer-based component of the DVR. The promoting transformer-less configuration can be a reasonable way to substitute shunt AF and UPQCs in the future smart grid. Figure 6.4 shows a simple topology for this new configuration. The exclusion of transformers decreases clearly the production cost.

6.2 Advantages and Comparison with Shunt Active Filters

The parallel active filter functions as a current source, while series active filter (SeAPF) works as a VSC. Inductive or current-source loads or the harmonic current supplies are such loads for shunt active filters. Examples for such loads are, e.g., phase-controlled thyristor-based rectifiers of DC drives. The SeAPF have voltage source loads or capacitive or harmonic voltage sources as loads, e.g., diode rectifiers with direct smoothing capacitor loads for AC drives are examples of SeAPF. The load impedance for the shunt active filter should be high and the active filter must have $|1-G|_h << 1$, whereas for SeAPF, the load impedance should be low and the filter should have $|1-G|_h << 1$. The shunt active filter has characteristics like excellent and independent relation of the source impedance, Z_s for the current-source loads, but based on Z, when the load impedance, Z_L, is low. For SeAPF, excellent and independent of the source impedance, Z_s and the load impedance, Z_L for voltage-source loads, but depends on Z_L, when the loads are a current-source type. The performance for capacitive or inductive load is that the shunt active filter may not behave properly when injected current flows into the load side and may cause over-current when applied to a capacitive or voltage-source load. However, the SeAPF gives low impedance so a parallel branch (parallel PF or power-factor improvement capacitor bank) is necessary when applied to an inductive or current-source load.

6.3 Design of Series Active Filter Components

Usually, shunt passive filters (SPFs) are widely used to keep harmonic pollution between acceptable limits. But, SPFs have numerous disadvantages in practice; for example, they can cause resonance and the impedance of power system strongly affects the filtering effects. In order to overcome these disadvantages, the APF has been addressed, as the APF can efficiently filter out all harmonics and the system elements do not affect the filtering effect. Various APF topologies and control methods have been analyzed during the last two decades. Up to now a large number of SAPF have been applied in practice, but there are still some drawbacks: (i) larger capacity, the basic investment is too high, and it is difficult for the user to accept; (ii) the circulation of harmonics is produced by operating with the SPF together; (iii) the APF may not improve the filter characteristics when the SPF creates the system resonance.

It is shown for a weak AC source the load voltage control using DSTATCOM has great bandwidth and good reduction in source voltage and nonlinear load perturbations. But, the DVR here passes high-frequency load components approximately un-attenuated and produces the presence of notches in the waveforms of the load voltage. For a strong AC source, the DVR has better bandwidth and attenuation properties. The DSTATCOM here in this case cannot control the load bus voltage. The comprehensive converter topology using cascaded multilevel inverter using multi-carrier phase-shifted PWM is used for the control of load voltage of an MV distribution system.

But, the cost of shunt active filters is high, and they are complex to realize on a large scale. Moreover, they have lower efficiency than shunt PFs. Hence, various solutions are proposed to enhance the practicality of active filters. One of them is the use of a combined system of shunt PFs and SeAPF. This technique helps one to design the APF for only a fraction of the total load power, decreasing costs and raising overall system effectiveness. There are many issues that need to be resolved before the provision is met. The issues list starts with the circuit topology. Recently, it has become known that the current source, three-phase six-valve (3P6V) parallel bridge topology allows, at most, only one upper valve and one lower valve to conduct at any time.

However, the shunt active filters are expensive and are difficult to implement on a large scale. In addition, they present lower efficiency than parallel PFs. Series active filters work as isolators instead of generators of harmonics and, hence, they use different control strategies. Series active filters working as controllable voltage sources and controllable current sources have been proposed. Series active filters enhance the current waveform on the line side of the filter at the cost of voltage-waveform degradation on the load side of the filter. One of the solutions to this problem is to use a combined system of parallel PFs and series active filters. Though the power rating of a series active filter is considerably lower than that of a parallel active filter, implementation of the overall system is still not that trivial for the following reasons: (1) It is always assumed implicitly that the input impedance on the line side is purely inductive, which is not always true. (2) Since full line voltage can appear across the current transformer during a load-side short-circuit, additional protection measures are required to protect the system.

The major function of shunt active filter is harmonic compensation, whereas series active filter is harmonic isolation between source and load. The advantages of shunt active filter are reduction in capacity of shunt active conditioner, so conventional shunt active conditioners are applicable. The advantages with series active filter are great reduction in capacity of series active conditioner, so already existing shunt PFs are applicable. The issues with shunt active filters are harmonic current may flow from source to shunt active conditioner and compensating current injected by shunt active conditioner may flow into shunt PF. The series active filter provides isolation and protection of series active conditioner.

The major functions of shunt active power line conditioners are flicker compensation or voltage regulation into the improvement of stability in power systems as the capacity of shunt active power line conditioners becomes larger. The combined system of a small-rated series active power line conditioner and a shunt PF aims at reducing initial costs and improving efficiency, thus giving both practical and economical points of view. This system will be put into practical use in the near future, considered a prospective alternative to shunt active power line conditioners for harmonic compensation of large capacity thyristor or diode rectifiers.

It is evident that the injected harmonic current from a parallel active filter flows into the load side rather than into the source side for a harmonic voltage-source load, and is thus unable to cancel the harmonic current of the source and enlarges the harmonic current of the load instead. To solve the above problems, a large series reactor should be placed on the load side. However, a large series reactor is bulky, increases costs, and causes a fundamental voltage drop, so it is undesirable. However, series active filters are better suited for harmonic compensation of a harmonic voltage source; a series active filter is applied to harmonic compensation of the diode rectifier. In addition, the series active filter and the diode rectifier can share the same DC capacitor (source) by selecting an appropriate turn ratio for the isolation transformer. In this way, the DC voltage control will become very easy. Further, the switching ripple filtering inductor, L, can be incorporated into the transformer to reduce component count.

The parallel active filter will increase harmonic current and may cause over-current of the load when the load is a harmonic voltage source. Instead, it has been verified that the series active filter is better suited for compensation of a harmonic voltage source such as a diode rectifier with smoothing DC capacitor. In some cases, a combined system of parallel active filter and series active filter may be necessary by utilizing the harmonic isolation function of the series active filters. No doubt active filters are superior to PFs if used in their niche applications.

6.4 Topology

The most common harmonically polluting load is a diode or thyristor bridge-rectifier in which the semiconductors operate at line frequency. These are commonplace in industrial variable-speed drives as the first stage of an AC/DC/AC power conversion. They were also widespread in consumer electronics until recently and have had a particularly deleterious effect in offline switching regulators in which the first power conversion stage was diode rectification at mains voltage with no utility-side transformer. Recently, however, at least some of these power supplies have used "power factor corrected" switch-mode rectifiers with near-sinusoidal current waveforms. However, switch-mode rectifiers in consumer electronics or industrial drives are expensive compared with bridge rectifiers and are resisted unless other advantages are apparent (e.g., universal input voltage range or regeneration). APFs have long been touted as the retro-fit solution to harmonic problems of bridge rectifiers [1–4] and are even considered as an option in new-build equipment in preference to an active rectifier.

The case for using an APF is that a relatively small inverter can be used to inject cancellation current (or voltage) to compensate for the distortion produced by a nonlinear load. It is important not to take this argument at face value but to examine in detail the factors that affect the rating of an APF. Ratings are not the only determinant of costs but they are a very important factor. An APF and an active rectifier are technologically similar and so ratings may well be the deciding factor when choosing between them. Both are considerably more complex (in terms of both control and semiconductor characteristics) than a diode rectifier. The APF will not be adopted where it does not offer a significant ratings advantage. Several styles of APF exist and have been usefully categorized and discussed in [5] using ideas of duality to explore the similarity between different approaches. Figure 6.5a shows a SAPF injecting current to cancel the distortion of a HSC. Figure 6.5b shows a series APF injecting a voltage to cancel the distortion of a harmonic voltage source. Figure 6.5c shows a SAPF connected to inject current to cancel current distortion cause indirectly by a harmonic voltage source. Figure 6.5d–f show various hybrid schemes, in which PFs correct some of the distortion and the APF deals with the remainder.

Figure 6.6a shows the system topology of the standard inverter type parallel APF (conventional parallel APF). The conventional parallel APF can do the reactive power compensation harmonic current

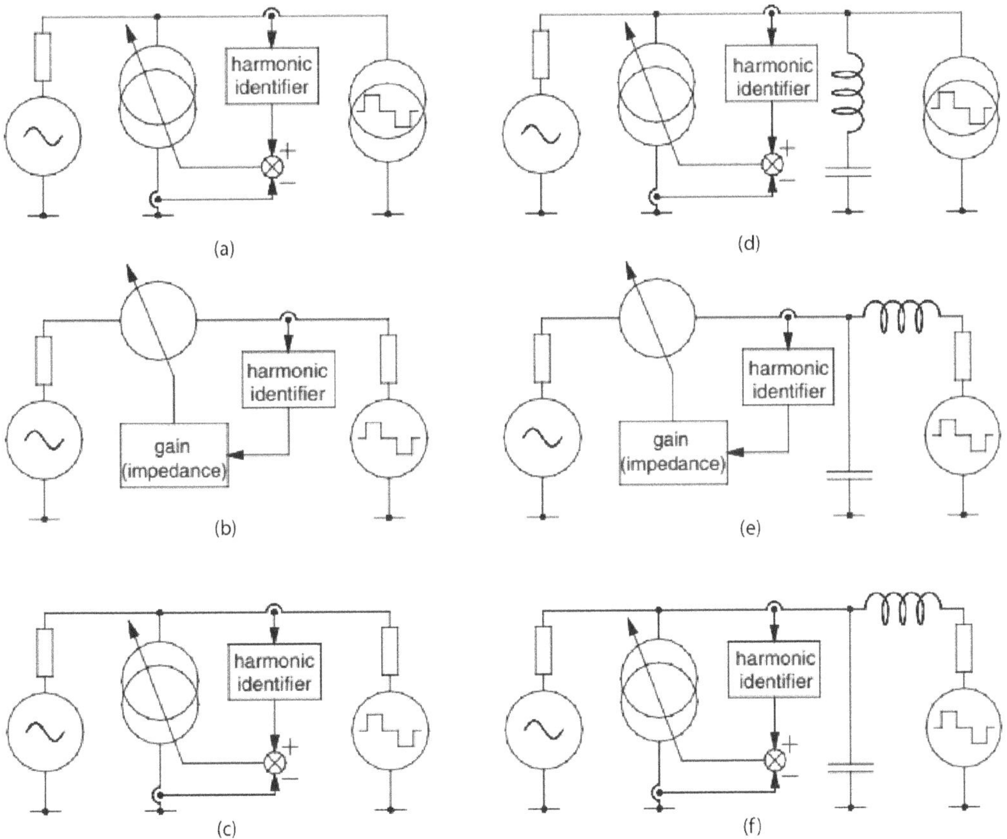

FIGURE 6.5 Example power filters systems (a–c) active filters, (d–f) hybrid active/passive filters.

suppression, and balancing three-phase currents. This filter consists of a voltage-source power converter and a filter inductor connected in series. The role of the filter inductor is used to suppress the high frequency ripple current generated while switching the power electronic devices of the power converter. The inductance of the filter inductor depends on switching frequency, DC voltage, and ripple current limitation. The DC bus voltage must be higher than the peak value of the utility voltage to force the output current of the APF under the command of compensating current in the conventional parallel APF. The use of high DC bus voltage has many disadvantages such as large filter inductance and high voltage rating of DC capacitor and power electronic devices. A larger filter inductor will result in significant power loss, more heat dissipation, bulk dimension, and weight, and it degrades the performance of frequency response. The requirement of high voltage rating of DC capacitor and power electronic devices limits high power application of APFs due to the high-power rating of the power converter and cost. Figure 6.6b shows the system configuration of the series active filter. The major advantages of the series active filter over the parallel APF are that it can maintain the output voltage waveform as sinusoidal and balance the three-phase voltages. However, the series filter is less popular in industrial applications due to the inherent drawbacks of series circuits, namely, it must handle high load currents, which increases their current rating compared with the parallel APFs.

In some applications, the combinations of several types of filters can achieve greater benefits. The major combinations include parallel active filter and series active filter, series active filter and parallel PF, parallel active filter and parallel PF, and active filter in series with parallel PF. Among these configurations, the active filter in series with parallel PF, also known as the hybrid power filter, is more widely discussed in the literature. This configuration is shown in Figure 6.6c where the PF filters the dominant harmonic, and the power converter is used to enhance the filter performance and to protect

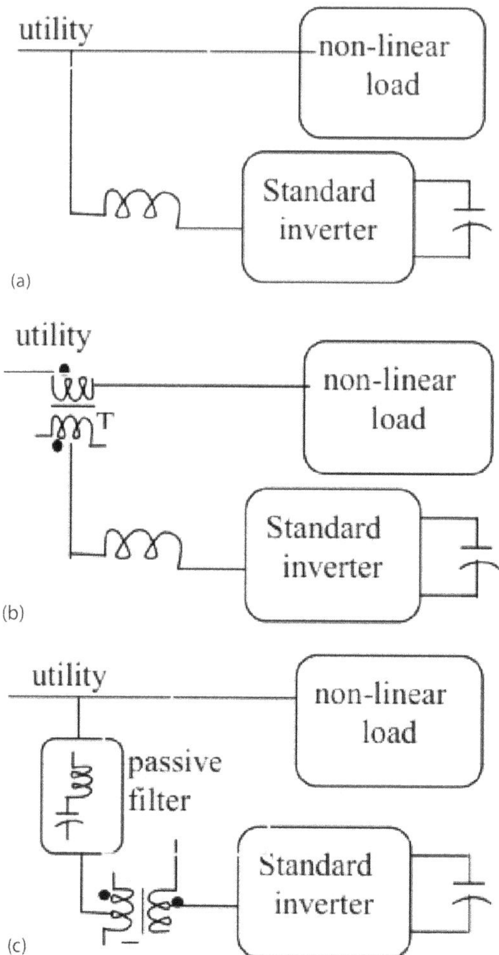

FIGURE 6.6 System configuration of active power filter, (a) standard inverter type parallel active power filter, (b) series active power filter, (c) hybrid power filter.

the PF from power resonance. Hence, the capacity of the power converter is smaller than that of the parallel APF for the same nonlinear load. Besides, the voltage stress applied to the power electronic switches in the power converter is low. As a result, the hybrid filter is suitable to high-power applications. However, the hybrid power filter requires a bulk PPF set and a voltage-matching transformer. Based on the type of compensation, the APF can be divided into reactive power compensation, harmonic compensation, balancing of three-phase systems, and multiple compensations. The conventional parallel APF belongs to multiple compensations, and it can compensate for the harmonic current and reactive power simultaneously. The hybrid power filter belongs to the harmonic compensation, and it only compensates for the harmonic current.

The circuit of Figure 6.7a shows the fundamental topology of the system, which is composed of three 9-level inverters connected in series between the source and the load and a shunt PF tuned at the fifth and seventh harmonics. The PF presents a low-impedance path to load-current harmonics and also helps to partially correct the power factor. Each phase of the nine-level series APF comprises Two H Bridges connected at the same DC-link capacitor. The two bridges are connected to the AC line using independent transformers scaled in powers of three, as shown in Figure 6.7b. More than nine levels increases hardware complexity and does not improve noticeably the filtering characteristics.

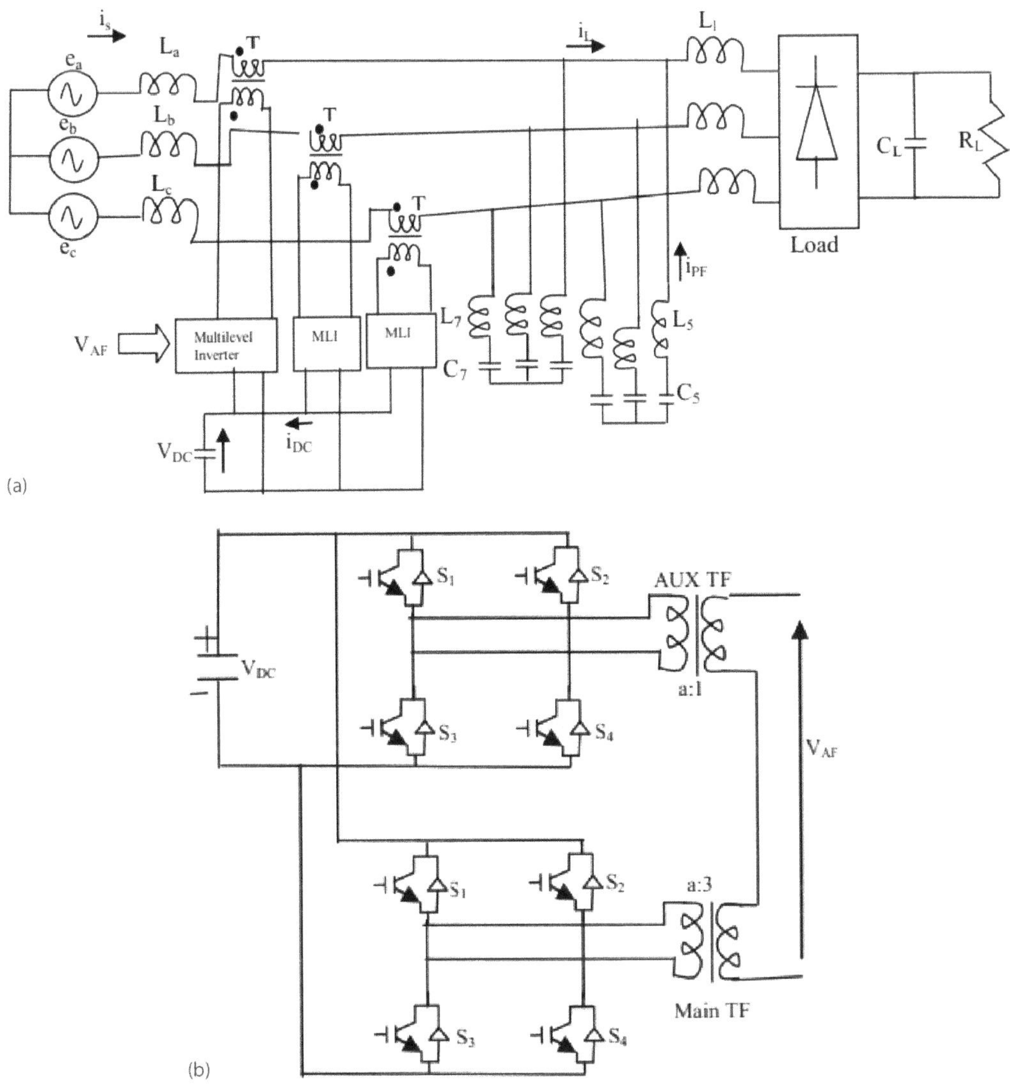

FIGURE 6.7 (a) Main components of the hybrid-series APF, (b) multilevel inverter topology (one phase).

The H-bridge converter is able to produce three levels of voltage at the AC side: +vdc, −vdc, and zero. The outputs of the modules are connected through transformers whose voltage ratios are scaled in powers of three, allowing $3n$ levels of voltage. Then, with only two converters per phase ($n = 2$), nine different levels of voltage are obtained: four levels of positive values, four levels of negative values, and zero. The transformer located at the bottom of the figure has the highest voltage ratio and, with its corresponding H Bridge, is called the main converter. The other transformer defines the auxiliary converter (Aux). The main converter manages most of the power but works at the lowest switching frequency, which is an additional advantage of this topology. Amplitude modulation is used to determine the output level of the inverter, rounding the reference signal to the nearest integer between the nine possible levels.

A hybrid-series AFP based on a low-rated multilevel inverter, acting as high-harmonic impedance, and a shunt PF acting as a harmonic-current path, were developed and tested. All the control tasks were programmed in an industrial controller, which was adapted for this particular application. With the control algorithm and taking advantage of the multilevel topology of the active filter, almost purely sinusoidal currents and voltages were achieved, without the usual high frequency content present in PWM inverters.

Two topologies into a unified AC and DC active filter component are investigated in [110]. This component will mitigate ripple and assist in dynamic response of the DC bus while ensuring a smooth load current of the test bed. This combination will also highlight issues for consideration in adding active loads to the MVDC amplifier. Initially, [111,112] explored design issues in the MVDC amplifier system. However, in the case of [111], a Thevenin-equivalent circuit was used in place of a thyristor model. This provides a similar voltage waveform but ignored the complexities caused by the CSC, thyristors. Further, [112] looked at purely resistive loads and voltage reference changes. For clarity, the DCAF and hybrid ACAF topologies are represented in Figure 6.8a and b, respectively. These two active filters have been

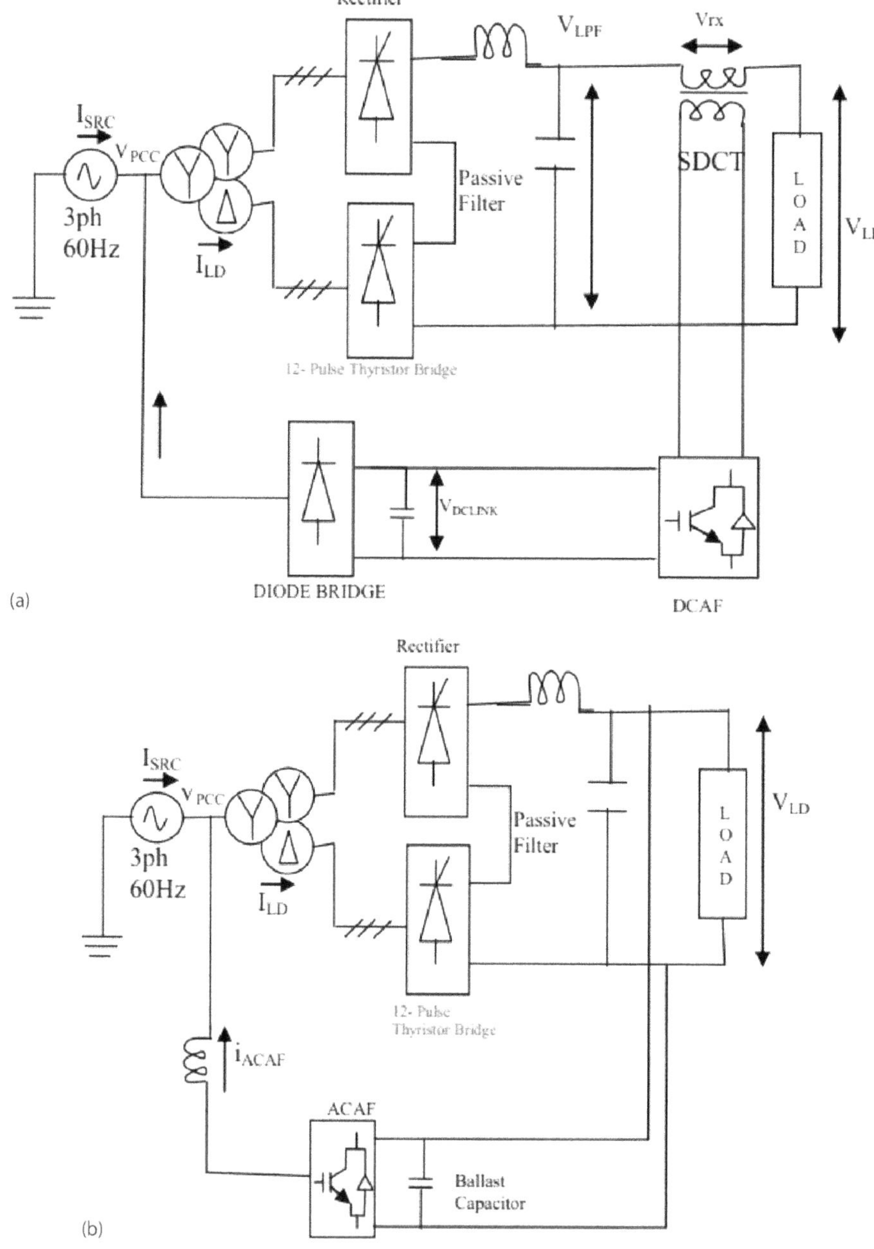

FIGURE 6.8 (a) DC active filter with series DC transformer in MVDC test bed, (b) hybrid AC active filter in MVDC test bed.

tested in isolation with good results. Therefore, there is a motivation to combine these two systems into a unified platform. This paper will first investigate adding active loads and necessary filters to the MVDC amplifier. Next, two active filter topologies will be explored, first the superposition of the two circuits and second, a back-to-back topology with the two filters DC bus linked. Finally, load profile will be shown with the back-to-back filters.

This section has presented design issues involved in the thyristor-based medium voltage DC amplifier with active loads. In particular, the DC bus suffers from static ripple and slow dynamic response. Similarly, the load current of the MVDC amplifier is a nonlinear load causing significant harmonic distortion in the source current. Recent work has shown that active filters can compensate for the drawbacks of the thyristor rectifier.

Two active filter topologies were presented that work to mitigate the AC current distortion and DC voltage ripple. The dual active filter with hybrid AC active filter shows the two filters decoupled. The hybrid AC active filter introduces a low impedance path that can destabilize the DC bus. Thus, a back-to-back topology is proposed that couples the two active filters. The back-to-back topology shows further promise by removing the low impedance path and providing dynamic DC link voltage to the DC active filter.

By combining passive and active filters into a single device, their individual disadvantages can be mitigated. These devices are referred to as hybrid APFs (HAPF). The purpose of the active part of a hybrid filter is not a standalone operation, but improving operational characteristics of the passive part. This allows us to considerably reduce the necessary power ratings of the active part, which is typically up to 10% of the rating of the passive part [3].

The hybrid filter presented in this paper is composed of a three-phase voltage-source converter connected in parallel with the PF inductor (Figure 6.9a). The main advantage of this structure is that the voltage drop on the capacitor reduces the VSC voltage ratings, while the inductor conducts the fundamental reactive current. In this paper, the rating requirements are analyzed and compared with the most commonly used series HAPF topology (Figure 6.9b).

The MVDC supply system main circuit is shown in Figure 6.10 in [114]. It consists of four parts: a twelve-pulse thyristor-bridge front-end, a PPF, transformer-coupled series DC APF, and load. The PF immediately after the thyristor-bridge filters out some of the high frequency ripple voltages and currents. The DC-APF can nearly remove all of the undesired ripples left after passive filtering. As can be seen in the power circuit diagram, series DC active filter deploys a special fourth-order output filter to eliminate

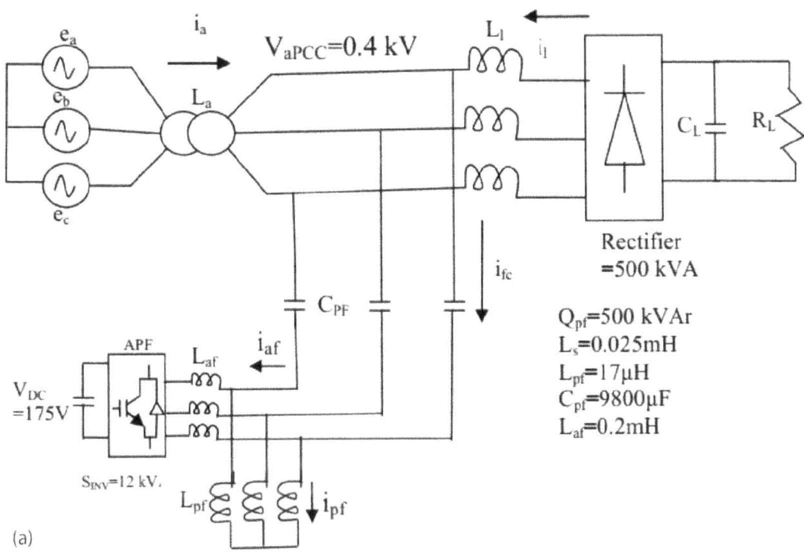

FIGURE 6.9 (a) Basic circuit of the parallel HAPF. *(Continued)*

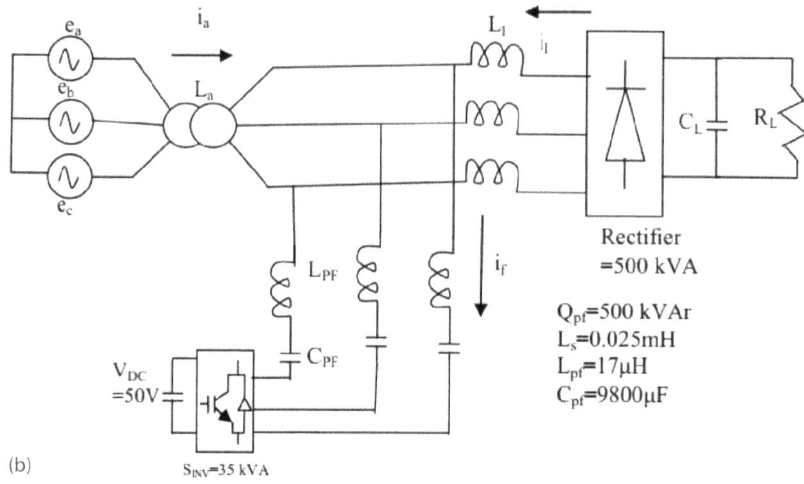

FIGURE 6.9 (Continued) (b) basic circuit of the series HAPF.

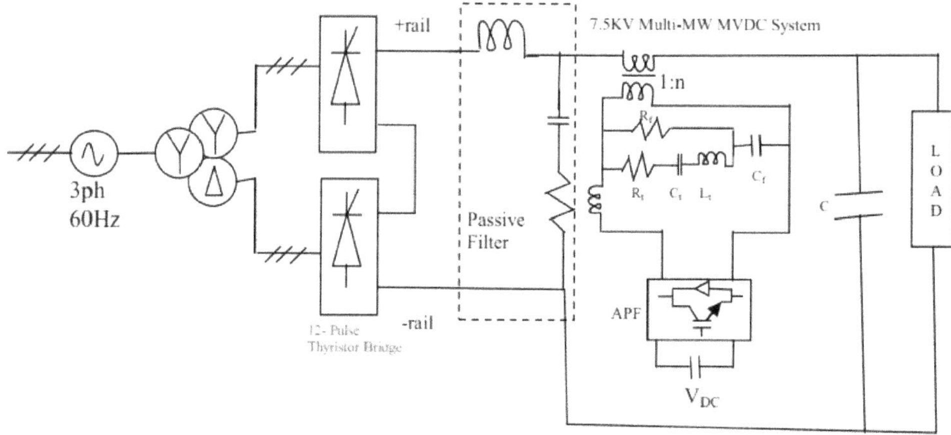

FIGURE 6.10 The 7.5 kV multi-MW medium-voltage DC (MVDC) system based on twelve-pulse thyristor-bridge front-end technology.

switching frequency ripple and damp out unwanted harmonic frequencies between converter controller bandwidth and switching frequency. The primary side of the series coupling transformer is connected to the SADF's output filter and the secondary is connected in series with medium-voltage DC bus.

Figure 6.11 shows system power stage and control block diagram. The dynamic performance of the system can be affected by the following system components: thyristor-bridge low-pass filter, angle control unit, voltage buck or boost rate limit and quantization, dynamic saturation, and system controllers. Each of these components are briefly explained but series DC active filter control will be the main scope of this section.

6.4.1 Thyristor Bridge Low-Pass Filter

The LPF filter parameters are restricted to a certain range of values required by the filtering performance and thyristor bridge predictable behavior. For the thyristor bridge to behave predictably, the load that it sees on the DC-side should be dominantly inductive. This requires a minimum amount of inductance on the DC-side and therefore, the capacitor and damping resistor are chosen accordingly. Due to filtering requirements, the filter resonant frequency can only be varied over a narrow frequency band. In this case, it can vary from 500–1000 Hz.

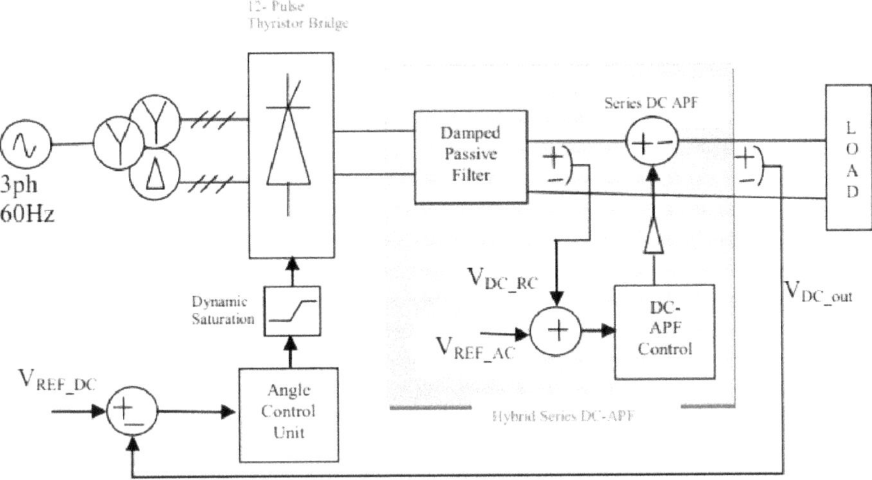

FIGURE 6.11 MVDC system component and control block diagram.

6.4.2 Angle Control Unit

An angle control PI controller makes sure that the total DC-bus voltage is following the reference value. It affects the transient and steady-state error but is limited by the voltage-boost or -buck rate limit and dynamic saturation limits, which vary according to the commanded reference DC-bus and load current.

6.4.3 Voltage-Boost or Buck Rate Limit and Quantization

This is the kind of limitation that is forced by the hardware that is firing the thyristors. The maximum rising and falling edge rate considered for this work is 4/ms and it is quantized with a sampling frequency of 2.5 kHz.

6.4.4 Dynamic Saturation

The dynamic saturation helps to control and hold the voltage excursions within a certain predefined band around the commanded reference. This is actually done by dynamically limiting the firing angle within a band around the reference value and it can be done in a symmetric or non-symmetric fashion.

6.4.5 Series DC Active Filter Controller

The controller structure and block diagrams for series DC active filter (SDAF) [111]. The system consists of three major control loops in order to perform the desired functions. The final goal is to produce a voltage ripple at the output terminals of SDAF's transformer, which is as close as possible in all its signal attributes to the ripple component at the output of the front-end rectifier but which possess an opposite phase in order to cancel each other out. With this, the final output DC-bus voltage after SDAF is smooth and free of ripples and disturbances. To accomplish this goal, the following control actions are recognized and implemented:

a. *DC Error Elimination*: This is to make sure that DC voltage error caused during transient load condition will not lead to coupling transformer saturation.
b. *Harmonic Extraction*: This block is considered the most important block in the controller structure since it is responsible for extracting the ripple component out of the front-end rectifier output voltage and for conditioning it for SDAF inverter.

c. *Band-Pass Damping*: The reference voltage to the PWM modulator of SDAF inverter is required to produce a specific ripple component at its output. Due to hardware limitations such as limited switching frequency, limited bandwidth, improper switching, dead-time effect, etc., not all of harmonic components are injected as the reference given to SDAF. This controller, in a closed-loop feedback system, generates a reference voltage signal to the PWM modulator such that series DC active filter injects harmonics only up to a specific frequency.

The first topology is presented in Figure 6.12a [106]. In this topology, the SHAPF consists of a SAPF connected between the supply and the load using coupling transformer and two passive LC-shunt circuits tuned for 5th and 7th harmonic frequencies. An LC-filter is connected between the SAPF and the coupling transformer to filter high frequency ripple caused by the active filter switching. The LC shunt circuits are connected in parallel with the load. The function principle of the SHAPF can he examined using single-phase equivalent circuit shown in Figure 6.12b.

The second topology is presented in Figure 6.12c [115]. This filter also consists of an SAPF and *two* passive LC shunt circuits. In this topology, the SAPF is connected in series with two LC shunt circuits. The LC shunt circuits are tuned for fifth and seventh harmonic frequencies. Although the main circuit is different compared to topology I, the operating principle of this topology control system is similar because its purpose is to produce a compensation voltage reference that is inversely proportional to the harmonic components of the supply current. The single-phase equivalent circuit of topology II is presented in Figure 6.12d.

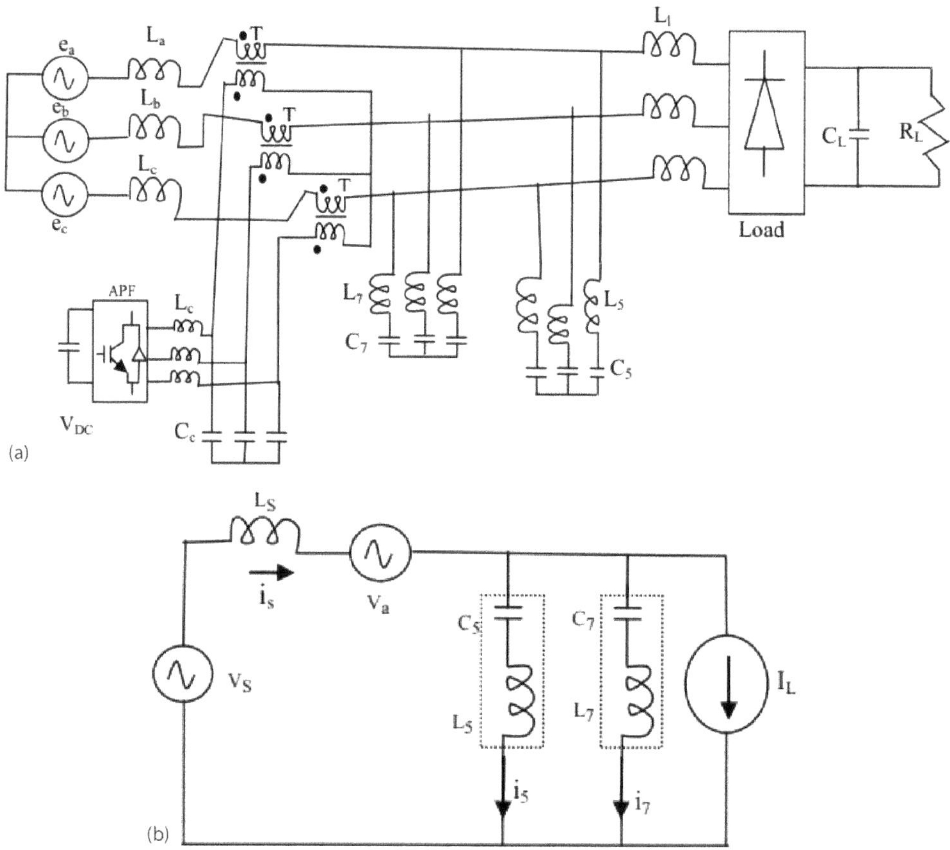

FIGURE 6.12 (a) SHAPF topology I, (b) single-phase equivalent circuit of SHAF topology I. *(Continued)*

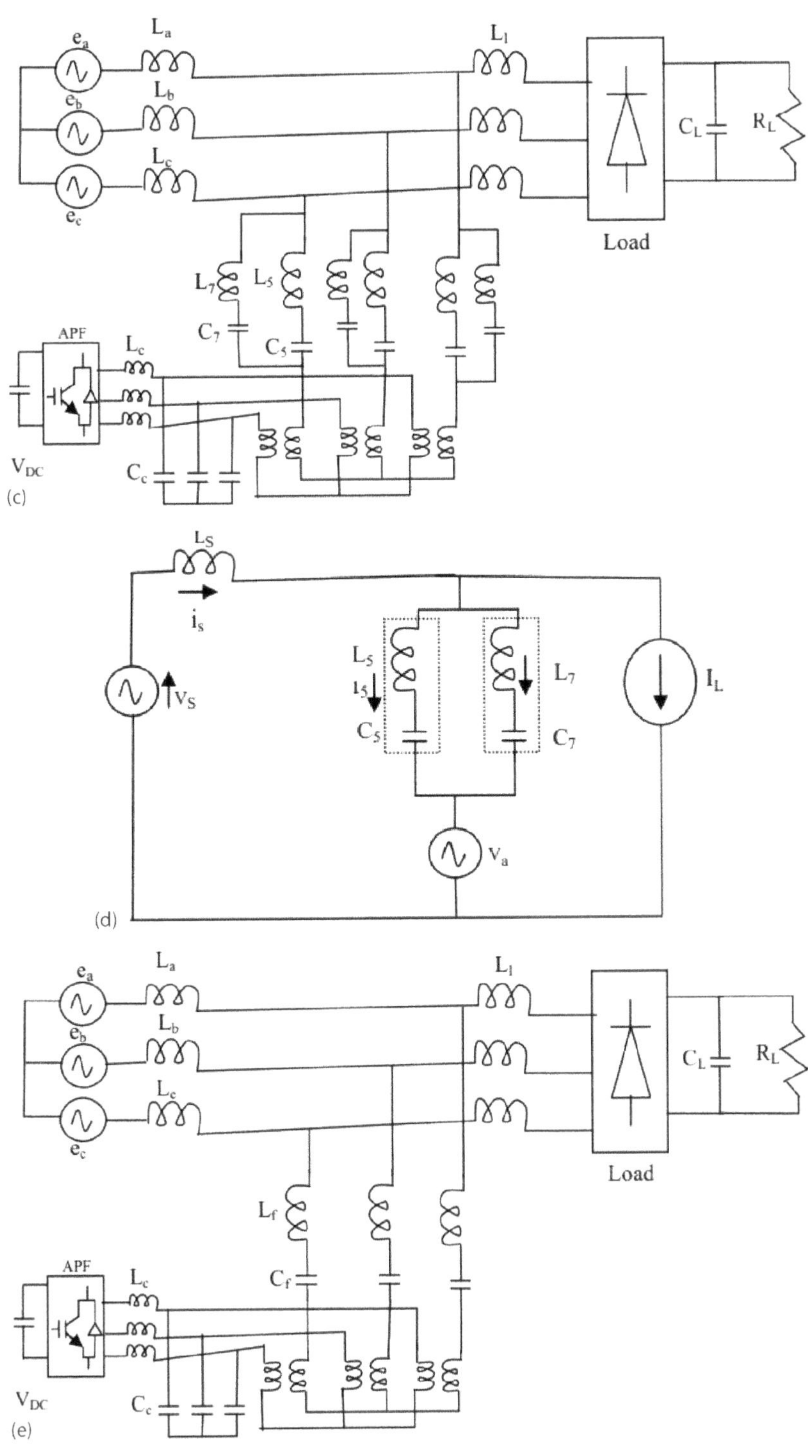

FIGURE 6.12 (Continued) (c) SHAPF topology II, (d) single-phase equivalent circuit of SHAF topology II, (e) hybrid filter topology III.

(*Continued*)

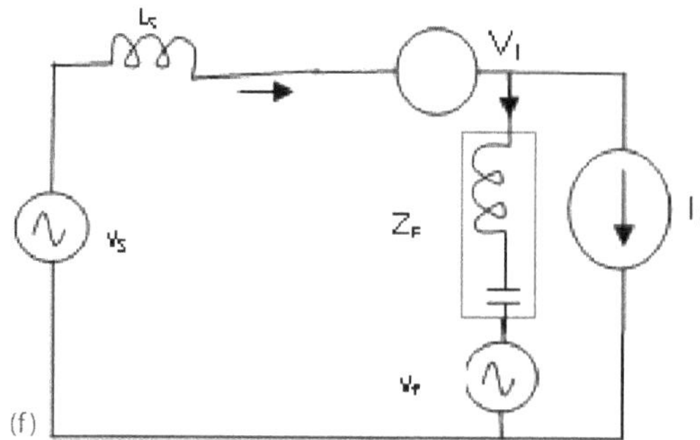

FIGURE 6.12 (Continued) (f) single-phase equivalent circuit of SHAPF topology III.

The third topology is presented in [116,107]. The main circuit is presented in Figure 6.12e. The main circuit of this filter is very similar to the previous topology. The difference is that in this topology there is only one LC shunt circuit. The resonance frequency of the LC shunt circuit is not specifically defined, as it is the task of the SAPF to electrically tune the resonance frequency of the LC shunt circuit. Besides this, SAPF forms also simultaneously an active resistance in the supply branch just like was presented in the case of topology II. The single-phase equivalent circuit of the SHAPF is shown in Figure 6.12f. The electrical tuning of the LC shunt circuit is done by generating an active inductance L_a to the system by the SAPF.

Based on the simulations presented in [107], it can be stated that SHAPF topology I resulted in good current filtering result in both simulation cases. However, to reach this goal, quite a small transformation ratio of the coupling transformer had to be used. Because of this, the power rating of the SAPF had to be large compared to the power of the load.

Topology II required the smallest transformation ratio of the coupling transformer. The reason for this was the DC-link voltage control that produced large voltage component at fundamental frequency. In the case of smaller load, this meant that the power rating of the SAPF had to be quite large, and in the case of higher-powered load this problem with DC-link voltage control resulted in poor current filtering.

Based on these simulations, SAPF topology III seemed to require the smallest power rating of the SAPF. Unfortunately, this topology was sensitive to the variations in supply impedance, which resulted in poor filtering result when only active inductance control was used.

6.4.5.1 The Traditional SHAPF

The traditional SHAPF topology and its equivalent one phase circuit are shown in Figure 6.13a, where Z_s is the source impedance and Z_F is the equivalent impedance of the shunt PFs. As is shown in Figure 6.13b, the active filter is controlled as the controlled $V_c = KI_{sh}$. Let's assume that a source V_s is sinusoidal, the filter characteristic of the traditional SHAPF can be expressed as:

$$I_{sh} = \frac{Z_F}{Z_s + Z_F + K} I_{lh}$$

where:

I_{lh} is the load harmonic current
I_{sh} is the source harmonic current

If the K impedance is much larger than the source impedance $s\ Z$ and the equivalent impedance Z_F, it will not have any harmonic source current and not be influenced by the variations in the source impedance and excellent harmonic isolation effect, i.e., no harmonic current flowing from the load side into the

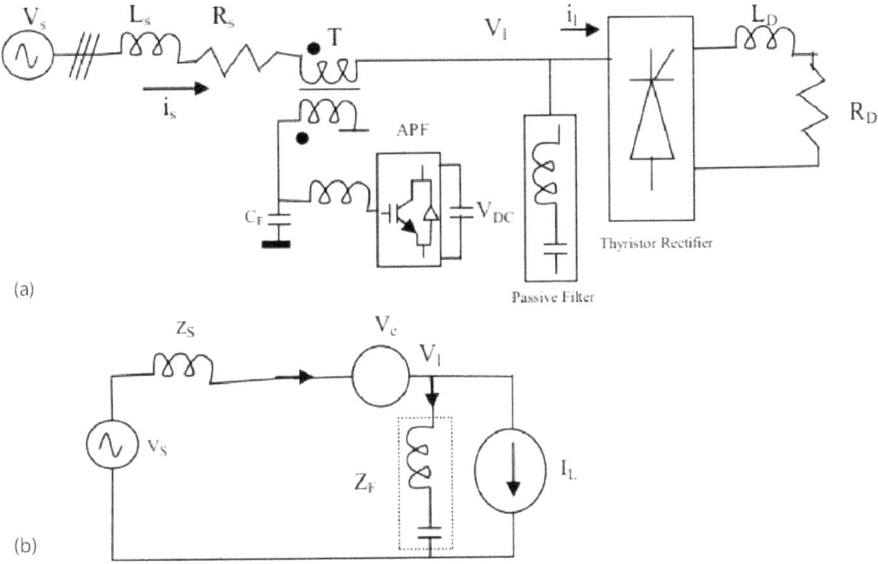

FIGURE 6.13 The traditional SHAPF: (a) topology and (b) the equivalent one-phase circuit.

source side or from the source side into the load or into the shunt PFs. The series active filter also acts as damping impedance, which can eliminate the parallel resonance and harmonic sink problems inherent to the shunt PFs.

However, with the increment value of the *K* the traditional SHAPF system will oscillate and even lead to system instability, so the filter performance is confined to its close-loop control capability. Additionally, the series active filter endures the whole source current and harmonic voltage, which will cause the inverter rating larger. So, many APF topologies are developed to improve or solve the problems; the series-in SHAPF is one of the topologies as shown in Figure 6.14.

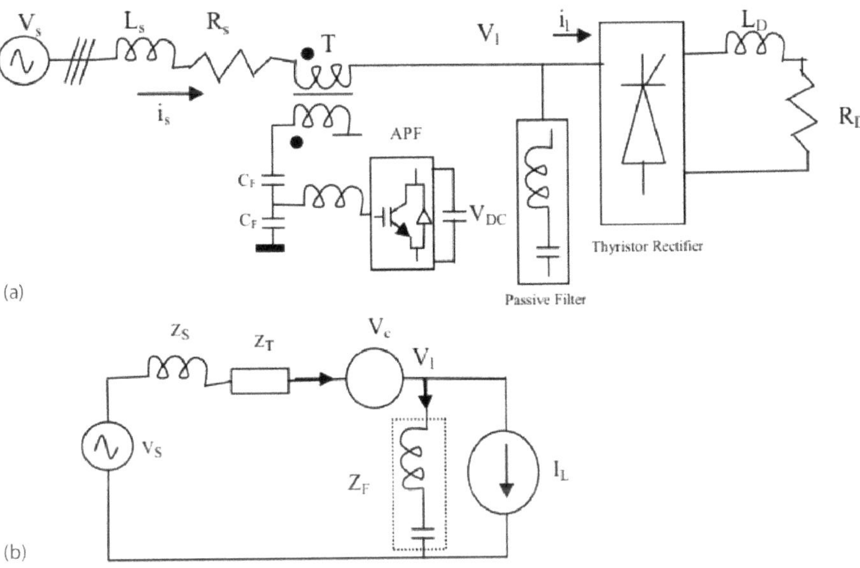

FIGURE 6.14 The series-in SHAPF (a) topology and (b) the equivalent one-phase circuit.

6.4.5.2 The Series-in SHAPF

The series-in SHAPF topology is in Figure 6.14a and its equivalent one-phase circuit is shown in Figure 6.14b; its filter characteristic is given by

$$I_{sh} = \frac{Z_F}{Z_s + Z_F + K + Z_1} I_{lh}$$

$$Z_1 = \frac{\left(j\omega L_1 - \frac{j}{\omega C_1} + R_1 \right)}{n^2}$$

Here n is the turn ratio of the couple transformer. The L_1C_1 fundamental frequency series resonance circuit presents the zero impedance at the fundamental frequency and the linear increment impedance characteristic at the harmonic frequency, especially very high impedance at high frequency. So, it can increase the loop harmonic impedance gain for the harmonic load current and the source harmonic voltage suppression. For it will endure the most harmonic voltage of the series part, the harmonic voltage of the inverter can be decreased. Additionally, it also can increase harmonic impedance gain of the main circuit and avoid the parameter variation of the line and PFs. But the larger rating of the passive part and the higher peak value at the lower resonant frequency when only the passive part is on work are inevitably disadvantages for the series-in SHAPF.

6.4.5.3 The SHAPF

To decrease the rating of the passive part, suppress the resonant peak value, and retain all the advantages of the series in SHAPF, a new series-in SHAPF is shown in Figure 6.15. Its main idea is

- Unite the two main parts of the series-in SHAPF, the L_1 C_1 fundamental frequency series-resonance circuit and the output filter, which is applied to suppress the switching current and voltage ripple.
- Make L_f and C_1 be the fundamental frequency resonant circuit, remove L_1 and combine C_1 and L_f to be the output filter, as shown in Figure 6.3.

Figure 6.16 presents a schematic diagram of the 18-pulse diode rectifier characteristic for parallel operation of three three-phase bridge converters, which operate at the leading load [11,19]. The three-phase supply network is represented by the source voltage e and the line impedance Z_s. Moreover, reactors of inductance L_d are connected in series to the supply source, to reduce, in general, the amplitude of higher harmonics of the currents, which are taken from the network by the 18-pulse rectifier. The basic assumption in the adopted concept of the presented rectifier system is the construction of three three-dimensional

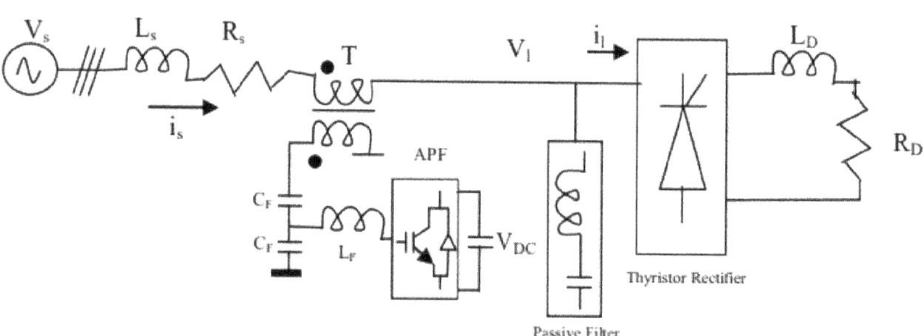

FIGURE 6.15 The series-in SHAPF.

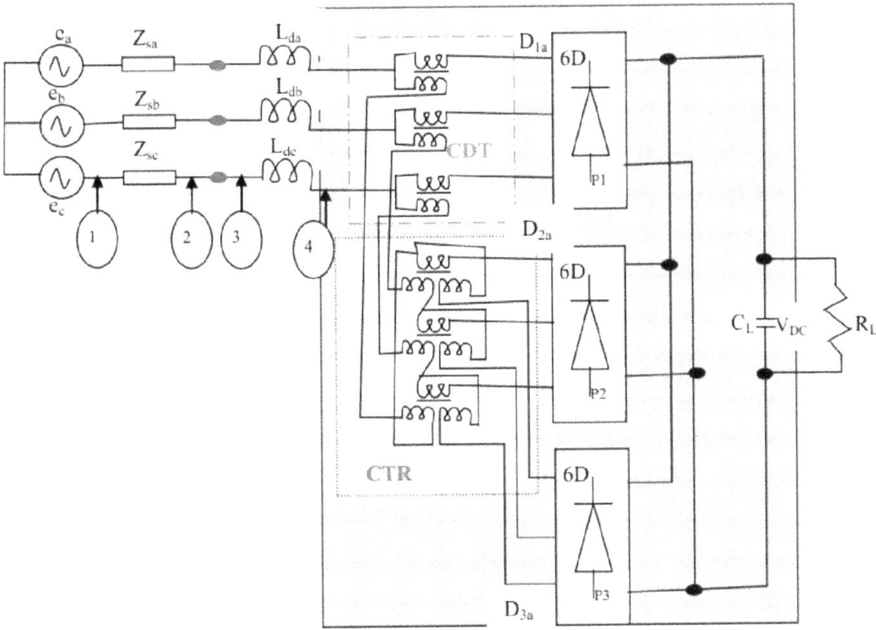

FIGURE 6.16 Supply system based on the 18-pulse diode rectifier with coupled three-phase reactors (CDT&CTR).

vectors of three-phase voltages supplying the bridge rectifiers, which are shifted by 20° with respect to each other. When this condition is met, the three-phase supply voltage v_K, which has the sinusoidal shape at no-load running, takes the 18-step shape at nominal load. The voltages v_K measured between the terminals K and the neutral point 0 can be interpreted as the quantities created as a result of cyclic switching of the current voltage V_{DC} on (Figure 6.1) via the diodes of three bridge rectifiers. The 18-pulse waveform of the voltages v_K (Figure 6.16) is only obtained when the diode conduction angles in particular rectifiers are equal to π, and when the conduction intervals (18 intervals) are symmetrically distributed along the supply voltage period. In this case, each of the three-phase bridge rectifiers generates a three-level voltage at the alternating-current terminals. The phase voltages measured at terminals *D*2 (with respect to the neutral point 0) are by 20° ahead of the relevant phase voltages measured at terminals *D*1. In turn, the phase voltages measured at terminals *D*3 are phase-delayed by 20° with respect to the relevant phase voltages measured at terminals *D*1. As a consequence, CTR executes 40° phase shift between two output voltage systems. The applied electromagnetic systems CDT and CTR make it possible to obtain the required division of the electric current, taken from the supply source, on particular bridge rectifiers. The three three-phase current systems, which supply these rectifiers, reveal the same rms value of the phase currents and an approximately sinusoidal shape. As for voltages, the system currents *P*2 are by 20° ahead of the system currents *P*1, while the system currents *P*3 are phase-delayed by 20° with respect to the rectifier *P*1.

The effect of the action of the coupled three-phase reactors is converting the three-phase voltage of the supply source into the nine-phase voltage. The reactors comprise six separate magnetic cores, on which relevant windings are wound. Selection of the number of winding turns and their connections results from the required voltage phase shift (CTR system) and required preliminary electric current division (CDT system). The application of the 18-pulse converter with the system of coupled reactors cooperating with a low power SeAPF seems to be an interesting solution for the problem of clean AC-DC energy conversion, due to:

- Simplification of the converter system resulting from possible elimination of the network reactor
- Small overall power of the both systems, decreasing their costs
- Remarkable reduction of the content of higher harmonics in the supply current waveform

- Small susceptibility of the system to supply asymmetry and load changes
- Potential for construction of low-cost and high-reliability supply systems

The advantages resulting from the application of the presented filter system refer to efficient minimization of the negative action of the multi-pulse converter onto the supply network, in various supply conditions. We can assume that the 18-pulse rectifier in the configuration with a SeAPF compose a converter system working in conditions close to CPC.

6.5 Comparison of Topologies

A few important topologies—like six-leg conventional topology, multilevel configurations and hexagram topology, interconnected converters, and multilevel cascaded H-bridge—are discussed in detail in the literature. The multilevel cascaded H-Bridge configuration has the advantage to reduce harmonic distortion level as well as the rating of power switches. However, the downside is that it needs a high number of DC-link capacitors and there is an imbalance control problem, which increases the control complexity. Hexagram topology is another example of a topology that can reduce the stress at the power switches. On the other hand, it needs 36 power switches and 6 DC-link capacitors with an additional need of circulating current control. The conventional six-leg configuration has the advantages of lower devices count and a simple control strategy.

The series active filter alone is more effective for compensating voltage-type harmonic producing non-linear loads. In recent development the HAPF has become very popular as they extend multi-functionality and they help reducing the inverter capacity. The SHAPF is capable of compensating voltage harmonic producing loads as well as current harmonic producing loads [63–65]. The SHAPF works as a kind of harmonic isolator rather than a harmonic voltage generator and forces the harmonic current to flow through PFs. This arrangement can also regulate the load voltage to a desired value by controlling the inverter output so as to compensate for abnormal utility voltages.

The advantage of series active filter compared to shunt type is the inferior rating of the compensator versus load nominal rating [50]. However, the complexity of the configuration and necessity of an isolation ST has decelerated their industrial application in distribution systems. The second category was developed in concern of addressing voltage issues on sensitive loads. Commonly known as DVRs, they have a similar configuration as series active filters. These two categories are different from each other in their control principle. This difference relies on purpose of their application in the system. The HSeAF in Figure 6.2 was given to address the previously mentioned issues with only one combination. Hypothetically, they are capable of compensating current harmonics, ensuring a power factor correction, and eliminating voltage distortions at the PCC [51,52]. These properties make it an appropriate candidate for power quality investments.

The configuration is based on a three-phase PWM VSI connected in series with the power lines through three single-phase transformers. In order to allow current harmonic compensation, a parallel LC filter must be connected between the nonlinear loads and the STs.

To isolate the DC bus from the power systems in Figure 6.4, the series active filter is connected to the grid across three single-phase transformers. This complicates the behavior of the series compensator. Different approaches are available to eradicate this pricey component of the DVR. Promoting the transformer-less configuration could be an affordable solution to substitute shunt AF and UPQCs in future smart grids.

The design issues involved in the thyristor-based medium voltage DC amplifier with active loads is shown in Figure 6.8a. In particular, the DC bus suffers from static ripple and slow dynamic response. Similarly, the load current of the MVDC amplifier is a nonlinear load causing significant harmonic distortion in the source current. Recent work has shown that active filters can compensate for the drawbacks of the thyristor rectifier.

TABLE 6.1
Disturbance Reduced or Eliminated and Specifications

Line Voltage Disturbances	Load Current Disturbances
Permanent rms deviations	Low and high frequency harmonics (permanent current distortions)
Erratic slow variations (flicker)	Erratic fast variations
Transient under and over-voltages	Power factor
Erratic fast variations	Phase current unbalance
Harmonics (permanent voltage distortions)	
Phase-voltage unbalance	
Short sustained interruptions	

The hybrid filter is composed of a three-phase voltage-source converter connected in parallel with the PF inductor (Figure 6.8b). The main advantage of this structure is that the voltage drop on the capacitor reduces the VSC voltage ratings, while the inductor conducts the fundamental reactive current. The rating requirements are analyzed and compared with the most commonly used series HAPF topology.

MVDC consists of four parts: twelve-pulse thyristor-bridge front-end, PPF, transformer-coupled series DC APF, and load. The PF immediately after the thyristor-bridge filters out some of the high frequency ripple voltages and currents. The DC-APF can nearly remove all of the undesired ripples left after passive filtering. As can be seen in the power circuit diagram, series DC active filter deploys a special fourth-order output filter to eliminate switching frequency ripple and damp out unwanted harmonic frequencies between converter controller bandwidth and switching frequency. The primary side of the series coupling transformer is connected to the SADF's output filter and the secondary is connected in series with medium-voltage DC bus.

In this topology in Figure 6.9, the SHAPF consists of a SAPF connected between the supply and the load using coupling transformer and two passive LC-shunt circuits tuned for fifth and seventh harmonic frequencies. An LC-filter is connected between the SAPF and the coupling transformer to filter high frequency ripple caused by the active filter switching. The LC shunt circuits are connected in parallel with the load. The disturbances considered for the elimination or reduction is listed in Table 6.1

6.6 Summary

Shunt configuration has been the most studied topology where the APF is connected parallel to the load. Its traditional use is the elimination of current harmonics produced by loads generating disturbances, known as HCS loads. However, parallel APF is not effective in situations where the load generates voltage harmonics, which are called harmonic voltage source loads. In any event, compensation systems composed only of an APF, whether in parallel or in series, connections do not completely solve the problem of harmonic elimination for all load types. To this end, other configurations have been proposed that are combinations of series and parallel topologies with active and passive filters. These are called hybrid topologies. Among the hybrid topologies, the series APF with parallel PF (series active-parallel passive filter [SAPPF]) is the most studied topology. For this configuration, different control strategies have been proposed. One of them has its origin in the early 1990s; the control objective was based on generating a voltage proportional to the source current harmonics through a proportionality constant k. In this instance, APF allowed the filtering features of parallel connection PF to be improved. The functionality of the new strategy was analyzed from a steady-state model. However, this theoretical development is not helpful for determining the proportionality constant.

The typical series active filter (SeAF) topology essentially contains a VSI, an injection transformer connected between the AC voltage line and the sensitive load, and a DC energy storage device. An alternative SeAF topology with a PWM auto-transformer can minimize the number of switches, reducing the costs and allowing, nevertheless, the rising of the THD output voltage at the sensitive load.

The series-connected transformer secondary winding injects the compensating voltages generated by the VSI to mitigate voltage sags of the AC line. Multilevel VSI converters such as diode-clamped converters, multilevel flying capacitor converters, or multilevel cascaded H-bridge converters are able to synthesize voltage waveforms with lower harmonic content than two-level converters and able to operate at higher DC voltages.

The controlled VSI converts the stored energy (DC source) into AC compensating voltages UDVR, which are filtered by a second-order LC low-pass filter. The filter capacitor can be placed either on the converter side or on the line side of the ST winding. By inserting the filtering capacitor on the converter side, the high-frequency switching harmonics are filtered locally, but the bandwidth of the injected compensating voltage is reduced, and the capacitor cannot aid in the compensation of the load displacement power factor, since, dynamically, an LC tank circuit is in series with the AC power supply. On the other hand, connecting the filter capacitor on the ST line side does not limit the bandwidth of the injected compensating voltage, and the capacitor can aid in the displacement power factor compensation in spite of sustaining the full line voltage, since the AC power supply sees a parallel LC circuit. A crucial performance issue of the DVR is the need for fast detection of the main AC voltage amplitude and the generation of its fundamental frequency and phase, even under unbalanced and distorted AC voltages.

Among all the compensation alternatives, the hybrid topologies appear very attractive in the distribution networks where some passive compensation is already installed. In particular, the hybrid shunt active filter formed with the connection of a low-rate active filter in series with one or several PFs is gaining attention [4,16–19]. Such a combination between active and passive filters allows for significantly reducing the rating of the active filter. Its main tasks are to improve the filtering performance and to avoid the resonance problems introduced by PFs. Moreover, no extra components are required to filter the ripple caused by the power inverter. It constitutes a simple and cheap solution for harmonics in a power distribution network.

The performance of the DVR is strongly influenced by the accuracy of the AC voltage synchronization. To inject a series voltage to dynamically restore the nominal AC voltages, it is mandatory to detect the sag depth. Several methods were reported to obtain the sag depth information: The peak detection of the network AC voltage provides the sag depth, but it can take up to half a cycle to obtain this information; the detection of direct voltage component v_d gives the information about sag depth in a balanced system, but components with a frequency of 100 Hz occur in unbalanced voltage systems, which can take up to a half a cycle before reaching the minimum value of v_d; locking a narrow band-pass filter can track changes in the network AC voltage phase, but it cannot directly return information regarding sag depth; and applying the Fourier transform to each phase returns the magnitude and phase of each frequency component, but filtering out harmonics other than the fundamental (50 Hz) introduces transient delays. The filtering capacitor is placed on the AC side to improve the mitigation capability of the DVR in a wide frequency range of network AC voltage perturbations. This line-side capacitor topology is compared to the converter-side capacitor placement.

An optimal predictive phase quadrature synchronizer that computes the phase shift correction factor in real time to quickly detect the phase of the network AC voltage, even with balanced and unbalanced sags and swells, high-order THD distortion, and short interruptions, is used here on a multilevel DVR to generate the fundamental frequency. The DVR performance in steady state and in transient mode also depends on the control laws, which shape the series compensating voltage to add to network AC voltage to compensate the sag and to compensate the AC load voltage THD. Known DVR system fundamental control strategies include the presag, in-phase, and minimal energy. Recent examples of successful control laws use a P+ resonant, having high gains around ±50 Hz, to achieve good positive and negative sequence fundamental voltage controls. The H∞ control method is used to achieve robust control of the DVR voltage. A virtual inductance DVR controller can ensure zero real power absorption during DVR compensation, thus minimizing the stress on the DVR DC power supply.

Control strategies can reduce the distortion of load voltage and save the DC-link capacitor energy. A DVR with a series converter on the source side connected through a DC-link capacitor to a back-to-back shunt converter on the load side is used to reduce the capacitor energy storage and compared with a DVR topology with a series converter on the load side and the shunt converter on the source side.

Despite the control strategies and capable results in reducing the impact of voltage sags to sensitive loads, most control processes do not deal with the reduction of the voltage THD at sensitive loads, mostly due to the injected compensation voltage. They are not concerned about voltage short interruptions, neither with the sharp notches or overshoots that occur mostly at the beginning and at the end of the sag.

REFERENCES

1. Akagi, H.: "New trends in active filters for power conditioning," *IEEE Trans. Ind. Appl.*, 1996, 32, 1312–1322.
2. F II, I.: "IEEE recommended practices and requirements for harmonic control in electrical power systems," *IEEE Std 519-1992*, 1993.
3. Harmonics, P.: "Power system harmonics: An overview," *IEEE Trans. Power Appl.*, 1983, PAS-102, 2445–2460.
4. Cavallini, A. and Montanari, G.: "Compensation strategies for shunt active-filter control," *IEEE Trans. Power Electron.*, 1994, 9, 587–593.
5. Singh, B., Al-Haddad, K. and Chandra, A.: "A review of active filters for power quality improvement," *IEEE Trans. Ind. Electron.*, 1999, 46, 960–971.
6. Limongi, L., Roiu, D., Bojoi, R. and Tenconi, A.: "Analysis of active power filters operating with unbalanced loads," in *Energy Conversion Congress and Exposition, 2009. ECCE 2009. IEEE*, pp. 584–591, 2009.
7. Asiminoaei, L., Lascu, C., Blaabjerg, F. and Boldea, I.: "Performance improvement of shunt active power filter with dual parallel topology," *IEEE Trans. Power Electron.*, 2007, 22, 247–259.
8. Terciyanli, A., Avci, T., Yilmaz, I., Ermis, C., Kose, K., Acik, A., Kalaycioglu, A., Akkaya, Y., Cadirci, I. and Ermis, M.: "A current source converter based active power filter for mitigation of harmonics at the interface of distribution and transmission systems," *IEEE Trans. Ind. Appl.*, 2012, 48, 1374–1386.
9. Bhattacharya, S., Frank, T., Divan, D. and Banerjee, B.: "Active filter system implementation," *IEEE Ind. Appl. Mag.*, 1998, 4, 47–63.
10. Akagi, H., Nabae, A. and Atoh, S.: "Control strategy of active power filters using multiple voltage-source PWM converters," *IEEE Trans. Ind. Appl.*, 1986, IA-22, 460–465.
11. Saetieo, S., Devaraj, R. and Torrey, D.: "The design and implementation of a three-phase active power filter based on sliding mode control," *IEEE Trans. Ind. Appl.*, 1995, 31, 993–1000.
12. Khadem, S., Basu, M. and Conlon, M.: "Harmonic power compensation capacity of shunt active power filter and its relationship with design parameters," *IET Power Electron.*, 2014, 7, 418–430.
13. Pereira, R., da Silva, C., da Silva, L., Lambert-Torres, G. and Pinto, J.: "New strategies for application of adaptive filters in active power filters," *IEEE Trans. Ind. Appl.*, 2011, 47, 1136–1141.
14. Asiminoaei, L., Blaabjerg, F., Hansen, S. and Thogersen, P.: "Adaptive compensation of reactive power with shunt active power filters," *IEEE Trans. Ind. Appl.*, 2008, 44, 867–877.
15. Akagi, H., Kanazawa, Y. and Nabae, A.: "Instantaneous reactive power compensators comprising switching devices without energy storage components," *IEEE Trans. Ind. Appl.*, 1984, IA-20, 625–630.
16. Kanjiya, P., Khadkikar, V. and Zeineldin, H.: "Optimal control of shunt active power filter to meet IEEE std. 519 current harmonics constraints under non-ideal supply condition," *IEEE Trans. Ind. Electron.*, 2014, 99, 1.
17. Ribeiro, R., Rocha, T., de Sousa, R., Junior, E. and Lima, A. "A robust DC-link voltage control strategy to enhance the performance of shunt active power filters without harmonic detection schemes," *IEEE Trans. Ind. Electron.*, 2014, 99, 1.
18. Liu, Q., Peng, L., Kang, Y., Tang, S., Wu, D. and Qi, Y.: "A novel design and optimization method of an LCL filter for a shunt active power filter," *IEEE Trans. Ind. Electron.*, 2014, 61, 4000–4010.
19. Asiminoaei, L., Aeloiza, E., Kim, J., Enjeti, P., Blaabjerg, F., Moran, L. and Sul, S.: "Parallel interleaved inverters for reactive power and harmonic compensation," in *Power Electronics Specialists Conference, 2006. PESC'06. 37th IEEE*, pp. 1–7, 2006.
20. Chaoui, A., Gaubert, J.-P., Krim, F. and Rambault, L.: "On the design of shunt active filter for improving power quality," in *Industrial Electronics, 2008. ISIE 2008. IEEE International Symposium on*, pp. 31–37, 2008.

21. Yi, H., Zhuo, F., Zhang, Y., Li, Y., Zhan, W., Chen, W. and Liu, J.: "A source-current-detected shunt active power filter control scheme based on vector resonant controller," *IEEE Trans. Ind. Appl.*, 2014, 50, 1953–1965.
22. Ericsen, T., Hingorani, N. and Khersonsky, Y.: "Power electronics and future marine electrical systems," *IEEE Trans. Ind. Appl.*, 2006, 42, 155–163.
23. Ozpineci, B. and Tolbert, L.: "Smaller, faster, tougher," *IEEE Spectr.*, 2011, 48, 45–66.
24. Wang, Z., Shi, X., Tolbert, L., Wang, F., Liang, Z., Costinett, D. and Blalock, B.: "A high temperature silicon carbide MOSFET power module with integrated silicon-on-insulator-based gate drive," *IEEE Trans. Power Electron.*, 2015, 30, 1432–1445.
25. Asiminoaei, L., Aeloiza, E., Enjeti, P. and Blaabjerg, F.: "Shunt active power-filter topology based on parallel interleaved inverters," *IEEE Trans. Ind. Electron.*, 2008, 55, 1175–1189.
26. Franquelo, L., Rodriguez, J., Leon, J., Kouro, S., Portillo, R. and Prats, M.: "The age of multilevel converters arrives," *IEEE Ind. Electron. Mag.*, 2008, 2, 28–39.
27. Hochgraf, C., Lasseter, R., Divan, D. and Lipo, T.A.: "Comparison of multilevel inverters for static VAr compensation," in *Industry Applications Society Annual Meeting, 1994. Conference Record of the 1994 IEEE*, pp. 921–928, vol. 2, 1994.
28. Gupta, R., Ghosh, A. and Joshi, A.: "Control of 3-level shunt active power filter using harmonic selective controller," in *Power India Conference, 2006 IEEE*, p. 7, 2006.
29. Kouro, S., Malinowski, M., Gopakumar, K., Pou, J., Franquelo, L., Wu, B., Rodriguez, J., Perez, M. and Leon, J.: "Recent advances and industrial applications of multilevel converters," *IEEE Trans. Ind. Electron.*, 2010, 57, 2553–2580.
30. Wen, J., Zhou, L. and Smedley, K.: "Power quality improvement at medium-voltage grids using hexagram active power filter," in *Applied Power Electronics Conference and Exposition (APEC), 2010 Twenty-Fifth Annual IEEE*, pp. 47–57, 2010.
31. Jacobina, C., Fabricio, E., Menezes, A., Correa, M. and Carlos, G.: "Shunt compensator based on three-phase interconnected converters," in *Energy Conversion Congress and Exposition (ECCE), 2013 IEEE*, pp. 5222–5228, 2013.
32. Rani, K. and Porkumaran, K.: "Multilevel shunt active filter based on sinusoidal subtraction methods under different load conditions," in *Computational Technologies in Electrical and Electronics Engineering (SIBIRCON), 2010 IEEE Region 8 International Conference on*, pp. 692–697, 2010.
33. Ortuzar, M., Carmi, R., Dixon, J. and Moran, L.: "Voltage-source active power filter based on multilevel converter and ultracapacitor DC link," *IEEE Trans. Ind. Electron.*, 2006, 53, 477–485.
34. El-Saadany, E.F., Salama, M.M.A. and Chikhani, A.Y.: "Passive filter design for harmonic reactive power compensation in single-phase circuits supplying nonlinear loads," *IEE Proc. Gener. Transm. Distrib.*, 2000, 147(6), 373–380.
35. Chang, G.W., Wang, H.-L., Chuang, G.-S. and Chu, S.-Y.: "Passive harmonic filter planning in a power system with considering probabilistic constraint," *IEEE Trans. Power Deliver*, 2009, 24(1), 208–218.
36. Kim, S. and Enjeti, P.N.: "A new hybrid active power filter (APF) topology," *IEEE Trans. Power Electron.*, 2002, 17(1), 48–54.
37. Jou, H.-L., Wu, J.-C., Chang, Y.-J. and Feng, Y.-T.: "A novel active power filter for harmonic suppression," *IEEE Trans. Power Deliver*, 2005, 20(2), 1507–1513.
38. Wu, L.H., Zhuo, F., Zhang, P.B., Li, H.Y. and Wang, Z.A.: "Study on the influence of supply-voltage fluctuation on Shunt active power filter," *IEEE Trans. Power Deliver*, 2007, 22(3), 1743–1749.
39. Yang, H.Y. and Ren, S.Y.: "A practical series-shunt hybrid active power filter based on fundamental magnetic potential self-balance," *IEEE Trans. Power Deliver*, 2008, 23(4), 2089–2096.
40. Shuai, Z.K., Luo, A., Shen, J. and Wang, X.: "Double closed-loop control method for injection-type hybrid active power filter," *IEEE Trans. Power Electron.*, 2011, 26(9), 2393–2403.
41. Li, D.Y., Chen, Q.F., Jia, Z.C. and Ke, J.X.: "A novel active power filter with fundamental magnetic flux compensation," *IEEE Trans. Power Deliver*, 2004, 19(2), 799–805.
42. Li, D.Y., Chen, Q.F., Jia, Z.C. and Zhang, C.Z.: "A high-power active filtering system with fundamental magnetic flux compensation," *IEEE Trans. Power Deliver*, 2006, 21(2), 823–830.
43. Sheng, J.K., Chen, Q.F., Ke, J.X. and Jia, Z.C.: "A novel principle of magnetic flux compensation and its application in power systems," *Int. J. Electron.*, 2003, 90(11–12), 707–720.
44. Zhang, Y., Chen, Q.F., Tian, J., Li, X. and Li, J.H.: "Controllable reactor based on transformer winding current regulating," *Proc. CSEE*, 2009, 29(18), 113–118.

45. Xie, B.R., Chen, Q.F., Tian, J., Chen, Y.D. and Yu, H.: "Analysis on filtering characteristic of series hybrid active power filter based on fundamental magnetic flux compensation," *Autom. Electr. Power Syst.*, 2007, 31(20), 75–79.
46. Akagi, H. and Isozaki, K.: "A hybrid active filter for a three-phase 12-pulse diode rectifier used as the front end of a medium-voltage motor drive," *IEEE Trans. Power Deliver*, 2012, 27, 69–77.
47. Zobaa, A.F.: "Optimal multi objective design of hybrid active power filters considering a distorted environment," *IEEE Trans. Ind. Electron.*, 2014, 61, 107–114.
48. Sixing, D., Jinjun, L. and Jiliang, L.: "Hybrid cascaded H-bridge converter for harmonic current compensation," *IEEE Trans. Power Electron.*, 2013, 28, 2170–2179.
49. Hamad, M.S., Masoud, M.I. and Williams, B.W.: "Medium-voltage 12-pulse converter: Output voltage harmonic compensation using a series APF," *IEEE Trans. Ind. Electron.*, 2014, 61, 43–52.
50. Liu, J., Dai, S., Chen, Q. and Tao, K.: "Modelling and industrial application of series hybrid active power filter," *IET Power Electron.*, 2013, 6, 1707–1714.
51. Javadi, A., Fortin Blanchette, H. and Al-Haddad, K.: "An advanced control algorithm for series hybrid active filter adopting UPQC behavior," in *IECON 2012–38th Annual Conference on IEEE Ind. Electron. Society*, Montreal, Canada, pp. 5318–5323, 2012.
52. Senturk, O.S. and Hava, A.M.: "Performance enhancement of the single-phase series active filter by employing the load voltage waveform reconstruction and line current sampling delay reduction methods," *IEEE Trans. Power Electron.*, 2011, 26, 2210–2220.
53. Goharrizi, A.Y., Hosseini, S.H., Sabahi, M. and Gharehpetian, G.B.: "Three-phase HFL-DVR with independently controlled phases," *IEEE Trans. Power Electron.*, 2012, 27, 1706–1718.
54. Abu-Rub, H., Malinowski, M. and Al-Haddad, K.: *Power Electronics for Renewable Energy Systems, Transportation, and Industrial Applications*. Chichester, UK: Wiley Inter Science, 2014.
55. Rahmani, S., Al-Haddad, K. and Kanaan, H.: "A comparative study of shunt hybrid and shunt active power filters for single-phase applications: Simulation and experimental validation," *J. Math. Comput. Simul. (IMACS)*, 2006, 71(4–6), 345–359.
56. Nogueira Santos, W.R., Cabral da Silva, E.R., Brandao Jacobina, C., de Moura Fernandes, E., Cunha Oliveira, A., Rocha Matias, R. et al., "The transformerless single-phase universal active power filter for harmonic and reactive power compensation," *IEEE Trans. Power Electron.*, 2014, 29, 3563–3572.
57. Liu, J., Yang, J. and Wang, Z.: "A new approach for single-phase harmonic current detecting and its application in a hybrid active power filter," in *Proceedings of the 25th Annual Conference IEEE Industrial Electronics Society (IECON'99)*, pp. 849–854, vol. 2, 1999.
58. Fujita, H. and Akagi, H.: "A practical approach to harmonic compensation in power systems-series connection of passive and active filters," *IEEE Trans. Ind. Appl.*, 1991, 27(6), 1020–1025.
59. Grady, W.M., Samoty, M.J. and Noyola, A.H.: "Survey of active line conditioning methodologies," *IEEE Trans. Power Deliver*, 1990, 5, 1536–1542.
60. Akagi, H., Watanabe, E.H. and Aredes, M.: *Instantaneous Power Theory and Applications to Power Conditioning*. IEEE Press, 2007.
61. Peng, F.Z., Akagi, H. and Nabae, A.: "A new approach to harmonic compensation in power systems—A combined system of shunt passive and series active filters," *IEEE Trans. Ind. Appl.*, 1990, 26, 983–990.
62. Moran, L., Pastorini, I., Dixon, J. and Walace, R.: "Series active power filter compensates current harmonics and voltage unbalance simultaneously," *Proc. IEE Gener. Transm. Distrib.*, 2000, 147(1), 31–36.
63. Singh, B., Verma, V., Chandra, A. and Al-Haddad, K.: "Hybrid filters for power quality improvement," *IEE Proc. Gener. Transm. Distrib.*, 2005, 152(3), 365–378.
64. Tian, J., Chen, Q., Xie, B.: "Series hybrid active power filter based on controllable harmonic impedance," *IET J. Power Electron.*, 2012, 5(1), 142–148.
65. Salmerón, P. and Litrán, S.P.: "A control strategy for hybrid power filter to compensate four-wires three-phase systems," *IEEE Trans. Power Electron.*, 2010, 25(7), 1923–1931.
66. Campos, A., Joos, G., Ziogas, P.D. and Lindsay, J.F.: "Analysis and design of a series voltage unbalance compensator based on a three-phase VSI operating with unbalanced switching functions," *IEEE Trans. Power Electron.*, 1994, 9, 269–274.
67. Lee, G.-M., Lee, D.-C. and Seok, J.-K.: "Control of series active power filters compensating for source voltage unbalance and current harmonics," *IEEE Trans. Ind. Electron.*, 2004, 51(1), 132–139.

68. Wang, Z., Wang, Q., Yao, W. and Liu, J.: "A series active power filter adopting hybrid control approach," *IEEE Trans. Power Electron.*, 2001, 16(3), 301–310.
69. Akagi, H., Ogasawara, S. and Kim, H.: "The theory of instantaneous power in three-phase four-wire systems: A comprehensive approach," in *Conf. Record IEEE-IAS Annual Meeting*, pp. 431–439, 1999.
70. Bhattacharya, S., Divan, M. and Benerjee, B.: "Synchronous reference frame harmonic isolator using series active filter," in *Proc. Fourth European Power Electronics Conf.*, Florence, Italy, pp. 30–35, vol. 3, 1991.
71. Li, G.Y.: "Definition of generalized instantaneous reactive power in dqO coordinates and its compensation." *Proc. Chin. Soc. Electr. Eng.*, 1996, 16, 176–179.
72. Montano, J.C. and Salmeron, P.: "Instantaneous and full compensation in three-phase systems," *IEEE Trans. Power Deliver*, 1998, 13(4), 1342–1347.
73. Peng, F.Z., Oh, G.W. and Adams, J.D.J.: "Harmonic and reactive power compensation based on the generalized reactive power theory for three-phase four-wire systems," *IEEE Trans. Power Electron.*, 1998, 13(6), 1174–1181.
74. Dai, X., Liu, G. and Gretsch, R.: "Generalized theory of instantaneous reactive quantity for multiphase power system," *IEEE Trans. Power Deliver*, 2004, 19(3), 965–972.
75. Peng, F.Z. and Adams, D.J.: "Harmonics sources and filtering approaches," *Proc. Ind. Appl. Conf.*, 1999, 1, 448–455.
76. Akagi, H.: "Active harmonic filters," *Proc. IEEE*, 2005, 93(12), 2128–2141.
77. Tey, L.H., So, P.L. and Chu, Y.C.: "Adaptive neural network control of active filters," *IEEE Trans. Power Electron.*, 2005, 74(1), 37–56.
78. Montero, M.I.M., Cadaval, E.R. and Gonzalez, F.B.: "Comparison of control strategies for shunt active power filters in three-phase four-wire systems," *IEEE Trans. Power Electron.*, 2007, 22(1), 229–236.
79. Uyyuru, K.R., Mishra, M.K. and Ghosh, A.: "An optimization-based algorithm for shunt active filter under distorted supply voltages," *IEEE Trans. Power Electron.*, 2009, 24(5), 1223–1232.
80. Herrera, R.S. and Salmerón, P.: "Instantaneous reactive power theory: A reference in the nonlinear loads compensation," *IEEE Trans. Ind. Electron.*, 2009, 56(6), 2015–2022.
81. Jou, H.-L., Wu, K.-D., Wu, J.-C., Li, C.-H. and Huang, M.S.: "Novel power converter topology for three phase four-wire hybrid power filter," *IET Power Electron.*, 2008, 1(1), 164–173.
82. Shuai, Z., Luo, A., Tu, C. and Liu, D.: "New control method of injection-type hybrid active power filter," *IET Power Electron.*, 2011, 4(9), 1051–1057.
83. Milanés-Montero, M.I., Romero-Cadaval, E. and Barrero-González, F.: "Hybrid multiconverter conditioner topology for high-power applications," *IEEE Trans. Ind. Electron.*, 2011, 58(6), 2283–2292.
84. Akagi, H. and Kondo, R.: "A transformerless hybrid active filter using a three-level pulse width modulation (PWM) converter for a medium-voltage motor drive," *IEEE Trans. Power Electron.*, 2010, 25(6), 1365–1374.
85. Luo, A., Zhao, W., Deng, X., Shen, Z.J. and Peng, J.C.: "Dividing frequency control of hybrid active power filter with multi-injection branches using improved i_p–i_q algorithm," *IEEE Trans. Power Electron.*, 2009, 24(10), 2396–2405.
86. Rodriguez, P., Candela, J.I., Luna, A. and Asiminoaei, L.: "Current harmonics cancellation in three-phase four-wire systems by using a four-branch star filtering topology," *IEEE Trans. Power Electron.*, 2009, 24(8), 1939–1950.
87. Zhao, W., Luo, A., Shen, Z.J. and Wu, C.: "Injection-type hybrid active power filter in high-power grid with background harmonic voltage," *IET Power Electron.*, 2011, 4(1), 63–71.
88. Rivas, D., Moran, L., Dixon, J. and Espinoza, J.: "A simple control scheme for hybrid active power filter," *IEE Proc. Gener. Transm. Distrib.*, 2002, 149(4), 485–490.
89. Hamadi, A., Rahmani, S. and Al-Haddad, K.: "A hybrid passive filter configuration for VAR control and harmonic compensation," *IEEE Trans. Ind. Electron.*, 2010, 57(7), 2419–2434.
90. Luo, A., Shuai, Z., Zhu, W., Fan, R. and Tu, C.: "Development of hybrid active power filter based on the adaptive fuzzy dividing frequency-control method," *IEEE Trans. Power Deliver.*, 2009, 24(1), 424–432.
91. Ribeiro, E.R. and Barbi, I.: "Harmonic voltage reduction using a series active filter under different load conditions," *IEEE Trans. Power Electron.*, 2006, 21(5), 1394–1402.
92. Fujita, H., Yamasaki, T. and Akagi, H.: "A hybrid active filter for damping of harmonic resonance in industrial power systems," *IEEE Trans. Power Electron.*, 2000, 15(2), 215–222.

93. Rice, D.E.: "A detailed analysis of six-pulse converter harmonic currents," *IEEE Trans. Ind. Appl.*, 1994, 30, 294–304.
94. Sakui, M., Fujita, H. and Shioya, M.: "A method for calculating harmonic currents of a three-phase bridge uncontrolled rectifier with DC filter," *IEEE Trans. Ind. Electron.*, 1989, 36, 434–440.
95. Bauta, M. and Grotzbach, M.: "Noncharacteristic line harmonics of AC/DC converters with high DC current ripple," *IEEE Trans. Power Deliver*, 2000, 15, 1060–1066.
96. Sakui, M. and Fujita, H.: "An analytical method for calculating harmonic currents of a three-phase diode-bridge rectifier with DC filter," *IEEE Trans. Power Electron.*, 1994, 9(6), 631–637.
97. Jeong, S.-G. and Choi, J.-Y.: "Line current characteristics of three-phase uncontrolled rectifiers under line voltage unbalance condition," *IEEE Trans. Power Electron.*, 2002, 17(6), 935–945.
98. Chen, Y. and Smedley, K.M.: "Parallel operation of one-cycle controlled three-phase PFC rectifiers," *IEEE Trans. Ind. Electron.*, 2007, 54(6), 3217–3224.
99. Singh, B., Singh, B.N., Chandra, A., Al-Haddad, K., Pandey, A. and Kothari, D.P.: "A review of single-phase improved power quality AC-DC converters," *IEEE Trans. Ind. Electron.*, 2003, 50(5), 962–981.
100. Jovanovic, M.M. and Jang, Y.: "State-of-the-art, single-phase, active power-factor-correction techniques for high-power applications–An overview," *IEEE Trans. Ind. Electron.*, 2005, 52(3), 701–708.
101. Rodríguez, J.R., Dixon, J.W., Espinoza, J.R., Pontt, J. and Lezana, P.: "PWM regenerative rectifiers: State of the art," *IEEE Trans. Ind. Electron.*, 2005, 52(1), 5–22.
102. Kolar, J.W., Friedli, T., Rodriguez, J. and Wheeler, P.W.: "Review of three-phase PWM AC–AC converter topologies," *IEEE Trans. Ind. Electron.*, 2011, 58(11), 4988–5006.
103. Peng, F.Z. and Lai, J.S.: "Generalized instantaneous reactive power theory for three phase power systems," *IEEE Trans. Instrum. Meas.*, 1996, 45(1), 293–297.
104. Peng, F.Z., Akagi, H. and Nabae, A.: "Compensation characteristics of the combined system of shunt passive and series active filters," *IEEE Trans. Ind. Appl.*, 1993, 29(1), 144–152.
105. Bhattacharya, S., Cheng, P.-T. and Divan, D.M.: "Hybrid solutions for improving passive filter performance in high power applications," *IEEE Trans. Ind. Appl.*, 1997, 33(3), 732–747.
106. Bhattacharya, S. and Divan, D.M.: "Hybrid Series Active/Parallel Passive Power Line Conditioner with Controlled Harmonic Injection," US Patent No. 5, 465, 203, 1995.
107. Mulla, M.A., Chudamani, R. and Chowdhury, A.: "A novel control scheme for series hybrid active power filter for mitigating source voltage unbalance and current harmonics," in *Presented at the Seventh International Conference on Industrial and Information Systems (ICIIS-2012) Held at Indian Institute of Technology Madras*, Chennai, India, August 6–9, 2012.
108. Menti, A., Zacharias, T. and Milias-Argitis, J.: "Geometric algebra: A powerful tool for representing power under nonsinusoidal conditions," *IEEE Trans. Circuits Syst.*, 2007, 54(3), 601–609.
109. European standard EN 50160: "Voltage characteristics of electricity supplied by public distribution systems," CENELEC, November 1994.
110. Beddingfield, R., Davis, A., Mirzaee, H. and Bhattacharya, S.: "Investigation of series DC active filter and hybrid AC active filter performance in medium voltage DC amplifier," in *Electric Ship Technologies Symposium (ESTS), 2015 IEEE. IEEE*, pp. 161–166, 2015.
111. Mirzaee, H., Dutta, S. and Bhattacharya, S.: "A medium-voltage DC (MVDC) with series active injection for shipboard power system applications," in *Energy Conversion Congress and Exposition (ECCE), 2010 IEEE*, pp. 2865–2870, 12–16 September 2010.
112. Mirzaee, H., Bhattachary, S. and Bala, S.: "Design issues in a medium voltage DC amplifier with multi-pulse thyristor bridge front-end," in *Energy Conversion Congress and Exposition (ECCE), 2012 IEEE*, pp. 603–609, 15–20 September 2012.
113. Herman, L., Blažič, B. and Papič, I.: "Comparison of circuit configuration and filtering performance between parallel and series hybrid active filter," in *Proceedings of 22nd International Conference and Exhibition on Electricity Distribution*, Stockholm, Sweden, pp. 1–4, 2013.
114. Mirzaee, H., Parkhideh, B. and Bhattacharya, S.: "Design and control of series DC active filter (SDAF) for shipboard medium-voltage DC power system," in *Electric Ship Technologies Symposium (ESTS), 2011 IEEE. IEEE*, pp. 452–458, 2011.
115. Fuji, H. and Akagi, H.: "A practical approach to harmonic compensation in power systems–A series connection of passive and active filled," *IEEE Trans. Ind. Appl.*, 1991, 27, 1020–1025.
116. Turunen, J., Salo, M. and Tuusa, H.: "A new approach for harmonic filtering in high power applications," in *Proc. Cons Power Electronics and Drive System.*, Singapore, pp. 1500–1505, 2003.

7

Control Strategies for Active Filters

7.1 Background

A power quality problem is an occurrence manifested as a nonstandard voltage, current, or frequency that results in a failure or faulty operation of end-use equipment. The presence of harmonics in power lines poses a power quality problem and results in greater power losses in the distribution system, interference problems in communication systems and, sometimes, in operation failures of electronic equipment. Active filters are special equipment that use power electronic converters to compensate for current and/or voltage harmonics originated by non-linear loads, or to avoid those harmonic voltages that might be applied to sensitive loads.

Passive filters are widely used in electrical system for power quality improvements. Their first installations date from the 1940s and their advantages make them an attractive and standard solution up to the current day. However, passive filters have their filtering characteristics deteriorated due to parameter variations caused by aging or temperature. Additionally, a capacitor bank for power factor correction is designed for specific loads and may not supply the right amount of reactive power when loads keep being added or changed. When these issues make the passive filter and the capacitor bank incapable of keeping the system operating within acceptable levels of power quality, an inconvenience arises and a solution must be provided. A common one is to replace both of them either by new elements or by active power compensators. However, replacing the passive filter and the capacitor bank may not be economically feasible because they belong to a past investment. Hence, combined passive and active filters are being used for techno-economic feasibility.

In all of them, it has been demonstrated underbalanced and sinusoidal AC voltage conditions; the strategies such as the so-called $p–q$ theory and synchronous reference frame (SRF) theory provide similar performances. Differences arise when these are used under distorted and unbalanced AC voltages. In addition, it has been proved in the literature [1] that the supply voltage fluctuations have influence on the DC link voltage and the expression signifying the influence of the change in supply voltage in the change in DC link voltage has also been derived for the shunt active filter (ShAF). A conventional PI controller used for DC link voltage control suffers from high overshoot and undershoot during the supply voltage fluctuations. However, no attempt is made on adaptive fuzzy logic controllers (FLCs) to regulate the DC link voltage along with other parts of the algorithm in case of sudden load changes. An adaptive FLC is used and is compared with a conventional PI controller to reduce the DC link voltage variations in case of supply voltage fluctuations along with the sudden load changes. An active power filter (APF) requires accurate harmonic current identification to compensate harmonics in power system distribution. In recent years, the concept of decentralizing power generation through the deployment of distributed generators (DGs) has been widely accepted and applied, driven by the growing market of renewable energy sources. These DGs are normally equipped with a switching power interface, acting as front-end with the grid. These are also sources of non-sinusoidal or polluted sine waveform of the power supply connected at the grid terminal.

Rapid growth of non-linear loads in distribution systems has attracted power system engineers' attention from a power quality point of view. Connection of the non-linear loads deteriorates power quality in the distribution system by introducing current harmonics. These current harmonics, when circulated in the electric network, interact with the system impedance and generate voltage harmonics. These current and voltage harmonics together can affect other consumers connected in the distribution network. Some typical non-linear loads, such as electric arc furnace (EAF), are an inherent source of voltage harmonics that give rise to voltage flicker, which can cause large voltage fluctuation in the connected distribution system. This chapter presents performance evaluation of a hybrid active filter (HAF) for power quality improvement of an (EAF)-connected distribution system. The performance of HAF is evaluated in terms of harmonic and voltage flicker mitigation capability. EAF is one of the causes for deteriorating power quality in the distribution network by introducing harmonics, propagating voltage flicker, and causing unbalance in voltages and currents. The HAF consists of a shunt passive filter (PF) connected with a lower rated voltage source PWM converter-based series active filter (SeAF) or ShAF. Various control strategies for SeAF and ShAF control are surveyed in the literature.

In recent times, due to the increased use of reactive power, it is important to bill the electrical power consumption based on both the active and reactive power components. Reactive power is a very significant factor in electric power systems since it affects the efficiency of these systems. For this reason, the measurement of reactive energy is also important. Several methods are used to measure reactive power. Reactive energy meters provide accurate measurement only if the input is fundamental, because meters are constructed for operating under sinusoidal conditions. Due to the widespread use of nonlinear loads and natural disturbances such as lightning, the power quality of the waveforms is considerably deteriorated. As the corrupted waveforms are composed of non-stationary components, in order to estimate the true reactive power of the power system fundamental component, a time-frequency analysis is required.

Current harmonic compensation is the main concern in improving power quality. In order to maintain good power factor, the current must be harmonic free. For current harmonic compensation, a voltage source inverter (VSI)-based ShAF is used, whose operation is entirely governed by a pulse-generation algorithm. The main aim of the control algorithm is to extract the fundamental component of the current. This component is then subtracted from the load current to compute the reference current for the pulse generation circuit.

In general, harmonic elimination is done using the combined topology of SeAF and shunt PF. The passive filter commonly eliminates the current harmonics produced by the load, but it has some limitations that result in sub-optimal operation. On the other hand, the SeAF improves the filtering characteristics of the passive filter as well as compensating voltage harmonics produced by the load and appearing on the source side (source currents), which improves the performance of the power system. Also, it reduces the rating of the active filter, which results in a practical and economical system. The SeAF is realized using conventional inverter topology with an appropriate control scheme to minimize the effects of the harmonics and make the desired waveform sinusoidal.

7.1.1 Control Strategy

Control strategy plays a vital role in overall performance of the compensating device. The control of a compensating device is realized in three stages. In the first stage, the essential voltage and current signals are sensed using power transformers (PTs), CTs, Hall-effect sensors, and isolation amplifiers to gather accurate system information. In the second stage, compensating commands in terms of current or voltage levels are derived based on different control methods and device configurations. In the third stage of control, the gating signals for the solid-state devices of the compensating devices are generated either in open loop or closed loop.

Among open loop the most popular schemes are PWM [2–7,8] and SPWM [9–12], while for closed loop, hysteresis controls [13–16] are the most common form of tracking control for lower order systems. For second and higher order systems sliding-mode control [5,17,18], linear quadratic regulator (LQR) [19,20,16], pole shift control [21], dead beat control [22–25], and Kalman filter [26–28] are used. Nowadays, with the help of microprocessors, microcontrollers, and digital signal processors (DSPs), it is possible to implement complex algorithms like fuzzy logic [6,29], neural networks [15,16,23,29,30], and genetic algorithms [31] for improving the dynamic and steady state performance of these devices.

7.1.1.1 Signal Conditioning

Voltage signals are sensed using either PTs or Hall-effect voltage sensors or isolation amplifiers. Current signals are sensed using CTs and/or Hall-effect current sensors. The voltage and current signals are sometimes filtered to avoid noise problems. The filters are either hardware-based (analog) or software-based (digital) with low-pass, high-pass, or band-pass characteristics. These signals are also useful to monitor, measure, and record various performance indexes, such as total harmonic distortion (THD), power factor, active and reactive power, crests factor, etc. The typical voltage signals are AC terminal voltages, DC-bus voltage of the custom power devices, and voltages across series elements. The current signals to be sensed are load currents, supply currents, compensating currents, and DC-link current of the custom power device.

7.1.1.2 Derivation of Compensating Signals

Development of compensating signals either in terms of voltages or currents is the important part of compensating devices for their control and affects their rating and transient, as well as steady-state, performance. The control strategies to generate compensation commands are based on frequency domain or time-domain correction techniques. Apart from this, control strategies for custom power devices can also be categorized on the basis of linear and non-linear, classical and modern, hard computing and soft computing, on-line and offline, but for the sake of brevity, they are not discussed here.

a. ***Compensation in Frequency Domain***: Control strategy in the frequency domain is based on Fourier analysis [14,32–34], wavelet analysis [32–35], and infinite impulse response [36,37]. Using Fourier transformation or wavelet transformation (WT), compensating harmonic components are separated from the harmonic-polluted signals and combined to generate compensating commands. The on-line application of Fourier transforms or WT (solution of a set of nonlinear equations) is a cumbersome computation and results in a large response time. This makes it difficult for real time application with dynamically varying loads.
b. ***Compensation in Time Domain***: Control methods of the compensating devices in the time domain are based on instantaneous derivation of compensating commands in the form of either voltage or current signals from distorted and harmonic-polluted voltage or current signals. There are large numbers of control methods in the time domain, which are known as instantaneous "*p–q*" theory ($\alpha\beta$ transformation) [37–40], proposed by Akagi et al. and then revised by Marshal et al. and Nabae, to make it applicable to eliminate neutral current of three-phase four-wire distribution systems. These are known as "modified *p–q* theory" [40], "extended *p–q* theory" [41], and *p–q–r* theory [20,42–44]. In addition to this SRF theory [45–49], symmetrical component transformation [12,50,51,18] and unit vector control [52–55] are other popular control schemes to extract reference signals in the time domain. Apart from this, control strategies based on no reference [56,57] are also reported in the literature.

7.1.1.3 Generation of Gating Signals to Compensating Devices

The third stage of control of the compensating devices is to generate gating signals for the solid-state devices of the compensating device based on the derived compensating commands, in terms of voltages or currents.

7.2 Control Strategy for Shunt Active Three-Phase Three-Wire System

The comparison of a three-phase three-wire shunt active power filter (SAPF) and its control strategies are studied, analyzed, and simulated [58]. Passive power filters are used in AC systems for simplicity of operation and implemented with low cost, but the major drawback of a passive filter is its performance dependency on the system parameter and its resonance problem. To mitigate the problem of a passive filter, the implementation of an SAPF is mandatory. The main issue consideration in this analysis is harmonics and reactive power compensation, and this is because of nonlinear load. Nonlinear load is the key point to disturbing a power system. To mitigate the effects from nonlinear loads SAPF is placed in shunt with the line. The control techniques of SAPF are performed based on *p–q* theory. A hysteresis current control (HCC) scheme is used for IGBT switching control and PWM generation.

Adaptive linear neuron (ADALINE) is a powerful estimation technique that has been used in power system applications, such as fault detection and classification, and SAPF [59]. In this paper, an ADALINE-based approach for control of three-phase three-wire unified power quality conditioner (UPQC) is presented. Two ADALINEs are used in the UPQC control strategy, one for load current fundamental extraction and the other for supply voltage fundamental extraction.

Three-Wire Compensating Devices: Three-phase three-wire non-linear loads, such as ASDs, are major applications of solid-state power converters. A large number of publications have appeared on three-phase DSTATCOM [8,15,16,29,40,43,44,48,49,57,60–71], DVR [62,72–108] and UPQC [109–123], with three wires on the AC side and two wires on the DC side. At medium voltage level these devices are developed with voltage-fed type having single-stage VSI [63,64] and for enhanced voltage and power handling capacity multilevel VSI [65,76,124,109], cascaded multilevel VSI [73,74,125], cascaded H-bridge [48,68,72,77,109], and multipulse VSI [48,66,78,110] are used. A schematic diagram of a VSC-based three-phase DSTATCOM is shown in Figure 7.1 and reference supply currents estimation using STF-based IRPT control algorithm are given in Figure 7.2. A controller block diagram is given in Figure 7.3. A DSTATCOM control algorithm diagram is given in Figure 7.4. Figure 7.5a–c shows the controller based on SAPF. Figure 7.6a shows the block diagram of IRPT-based control of DSTATCOM and Figure 7.6b depicts the block diagram of SRFT-based control of DSTATCOM.

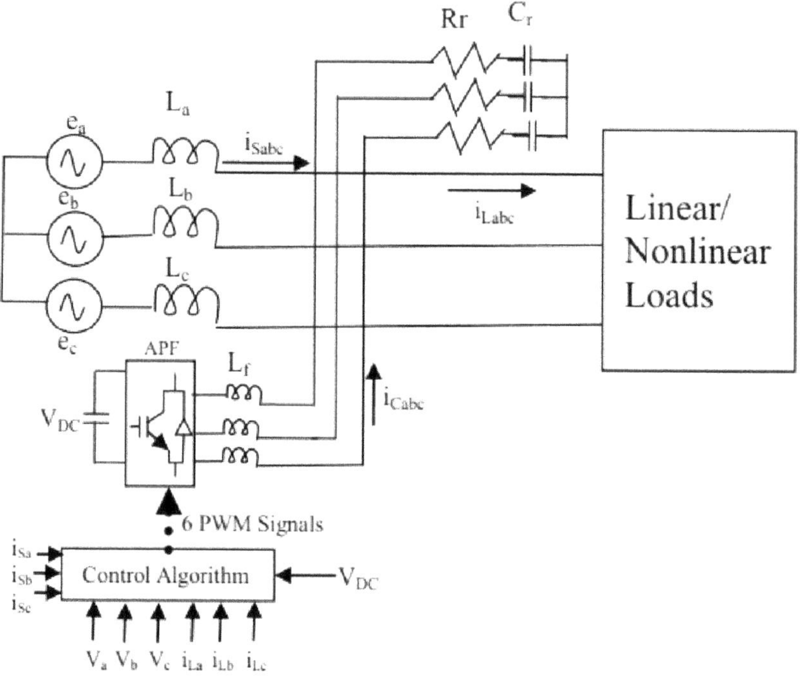

FIGURE 7.1 Schematic diagram of VSC-based three-phase DSTATCOM.

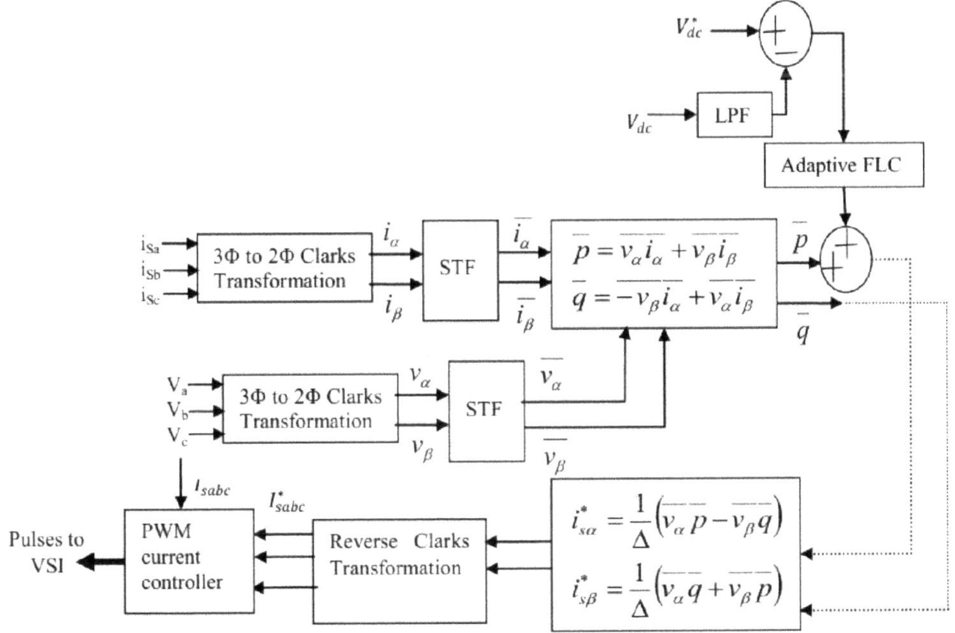

FIGURE 7.2 Reference supply currents estimation using STF-based IRPT control algorithm.

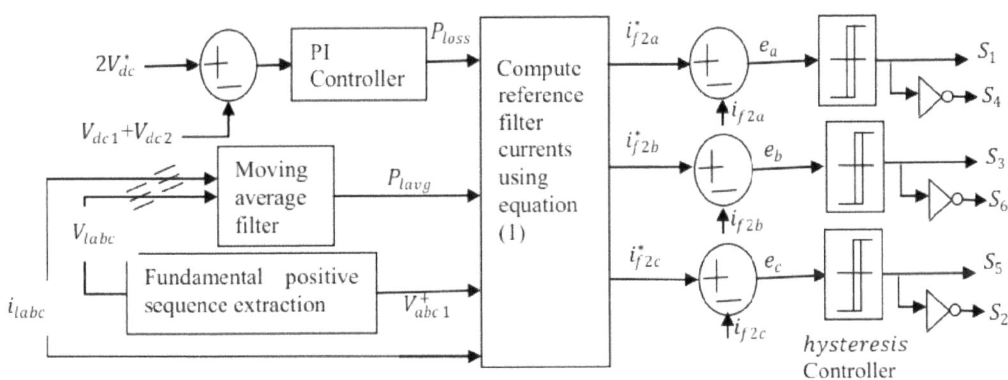

FIGURE 7.3 Controller block diagram.

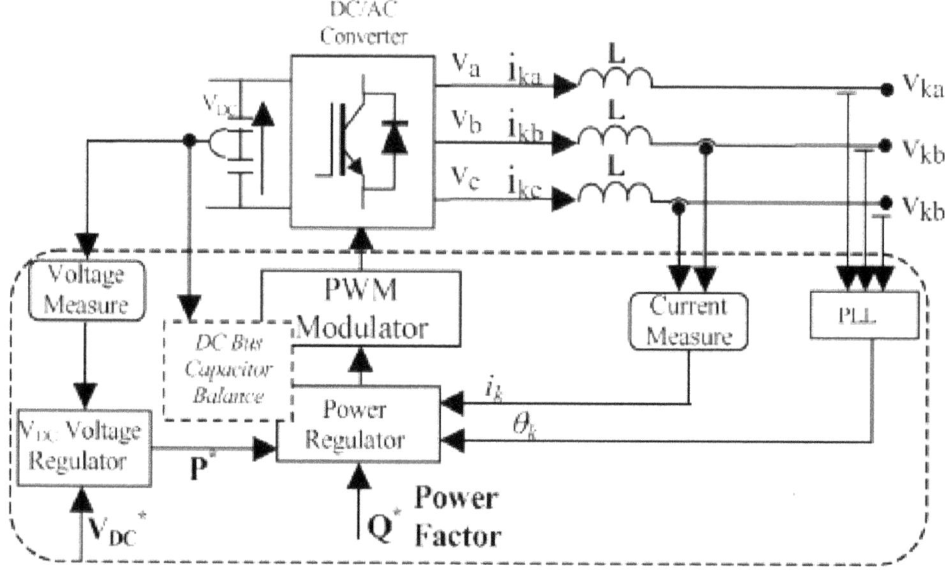

FIGURE 7.4 DSTATCOM control algorithm.

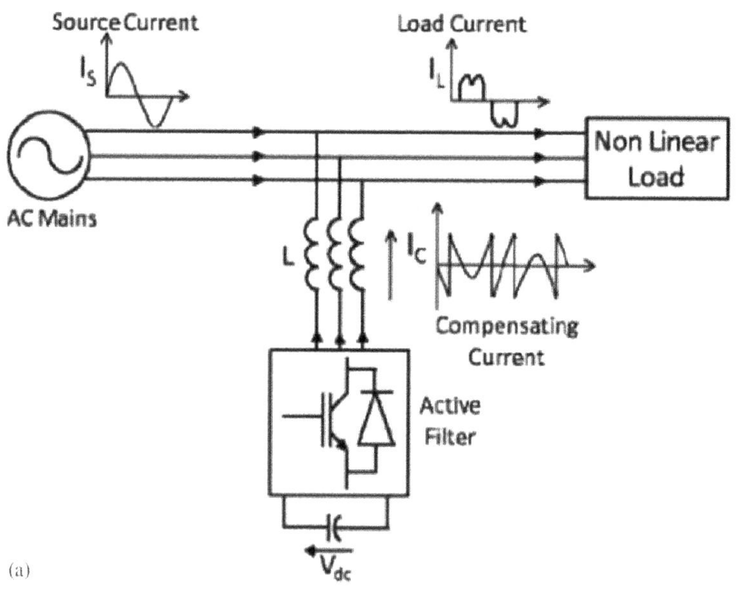

FIGURE 7.5 (a) SAPF.

(Continued)

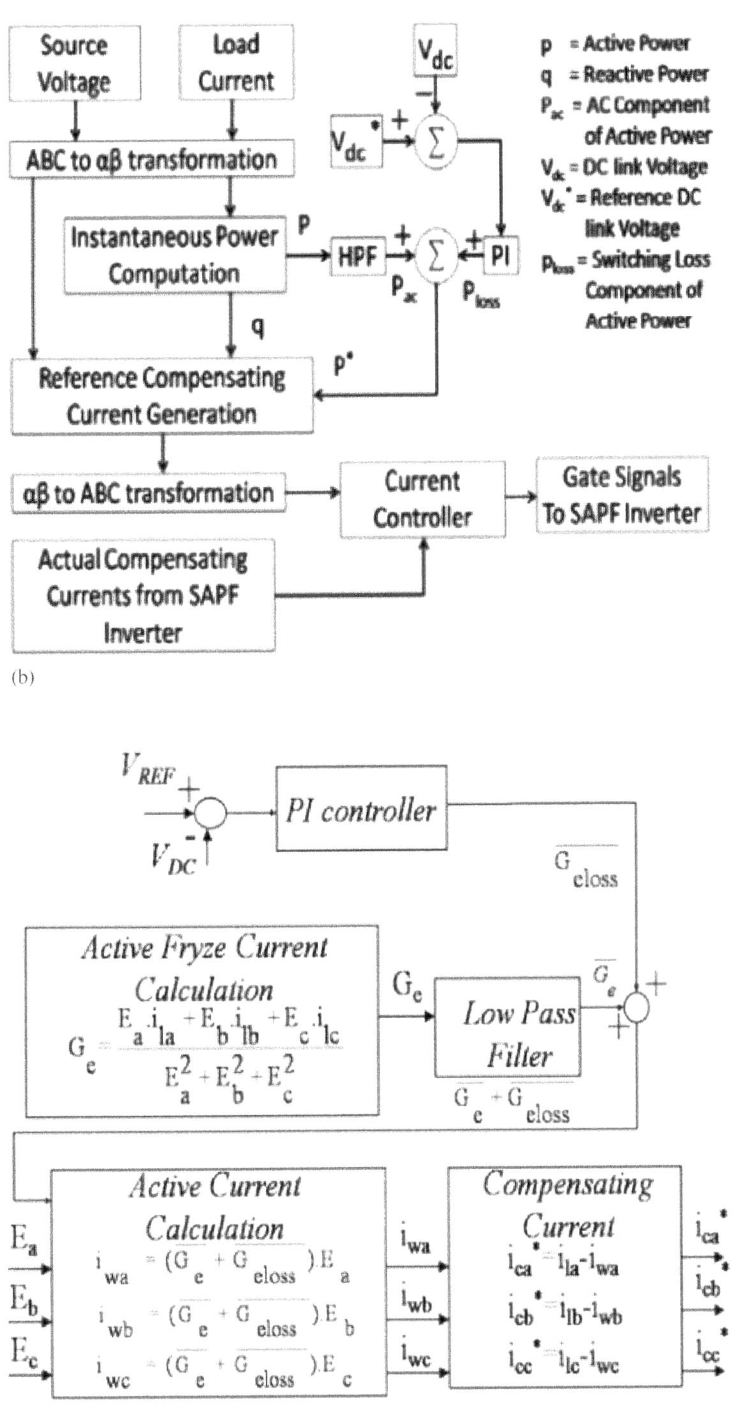

FIGURE 7.5 (Continued) (b) Control scheme for IRP theory, and (c) Control scheme for Fryze current computation technique.

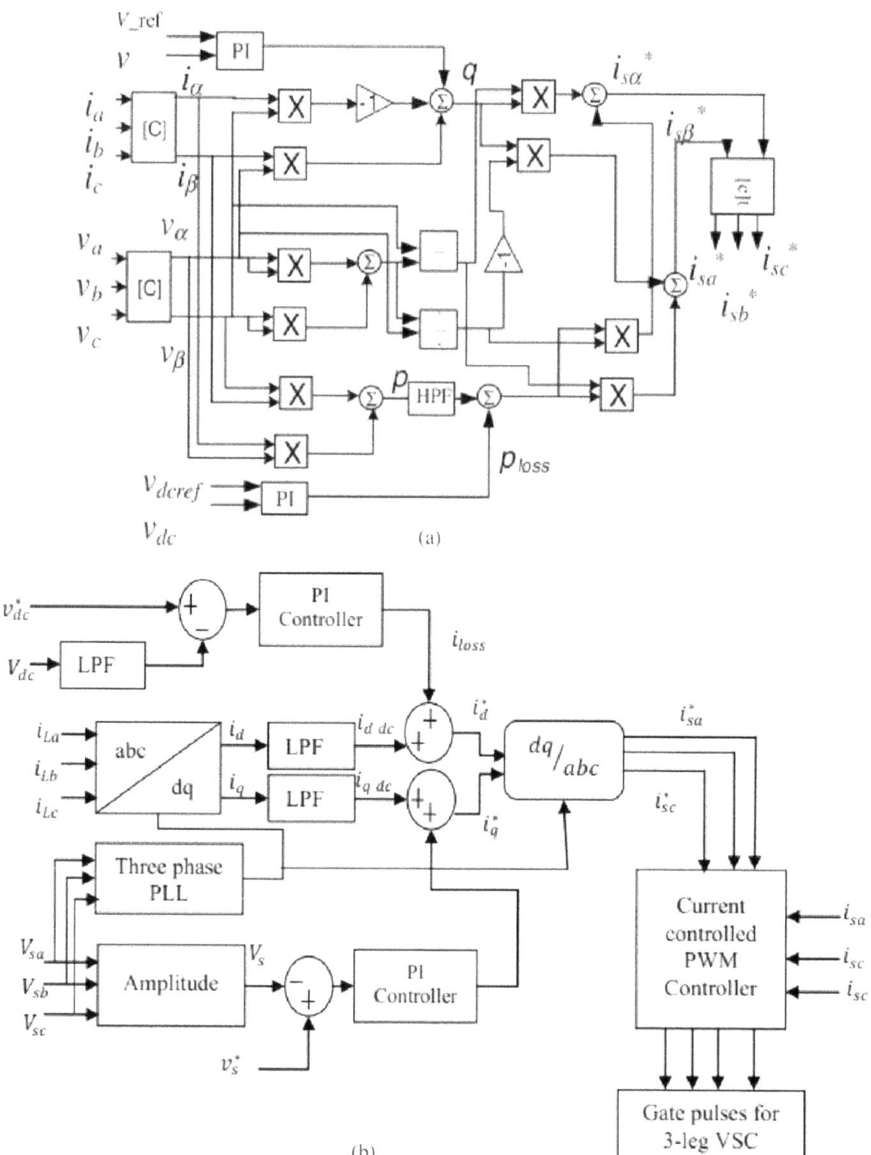

FIGURE 7.6 (a) Block diagram of IRPT-based control of DSTATCOM and (b) block diagram of SRFT-based control of DSTATCOM.

7.3 Three-Phase Four-Wire Shunt Active Filter

A large number of single-phase loads may be supplied from three-phase mains with neutral conductor [11]. They cause excessive neutral current, harmonics and reactive power burden, and unbalance. To reduce these problems, four-wire compensating devices have been attempted. These devices have been developed as DSTATCOM [126,127], DVR [128–131], and UPQC [109,132–137]. To develop these devices for three-phase four-wire distribution systems, different configurations of VSI having capacitor mid-point [31,126,128,129,135,137], three single-phase H-bridge VSI configurations [15,47,79,80,132,135], and four-pole VSI [60,61,81,115,128,130,133] are used. The capacitor mid-point VSI configuration is used for small ratings as the entire neutral current flows through DC bus capacitor, while in the four-pole VSI configuration; the fourth pole is used to stabilize the neutral of a particular compensating device. The three single-phase H bridge scheme is popular as it allows the proper voltage matching for solid state devices and enhances the reliability of devices. Figure 7.7 shows the system configuration of reactive current compensator. Figure 7.8 shows the reference current calculation. Figure 7.9 depicts the two- and three-level VSIs for the APF. Table 7.1 illustrates the components and parameters of various DSTACOM topologies. Table 7.2 shows the comparison of DSTATCOM control techniques and topologies with respect to its performance. Table 7.3 details the DSTATCOM compensating variables with respective strategies and Table 7.4 elucidates the overall comparative analysis.

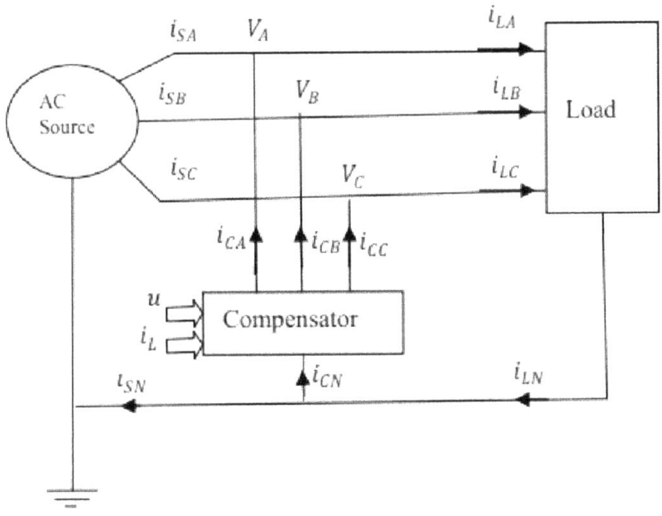

FIGURE 7.7 System configuration of reactive current compensator.

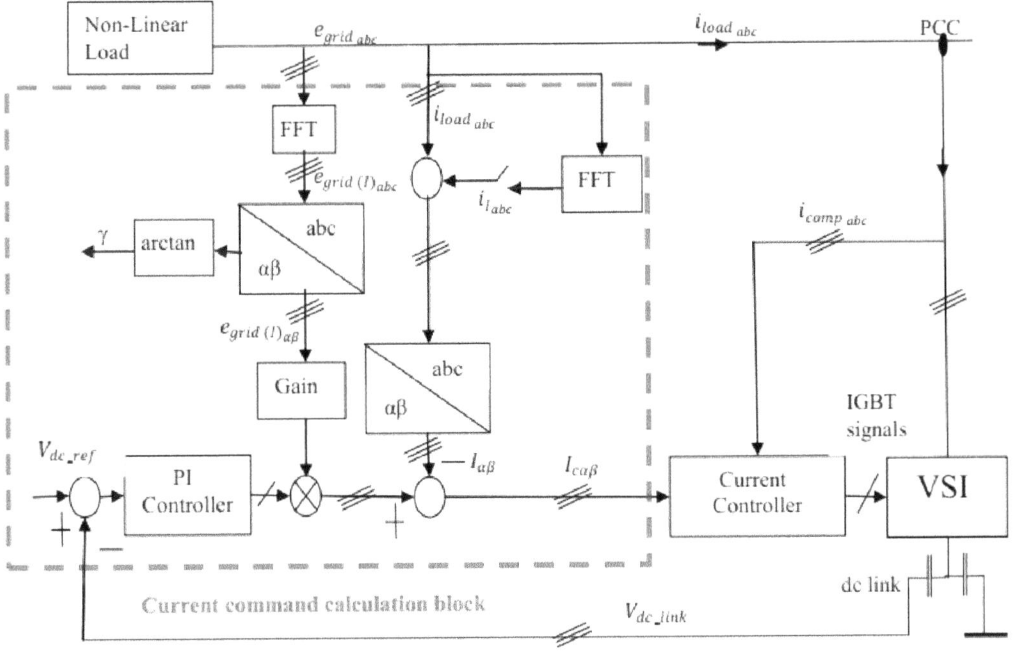

FIGURE 7.8 Reference current calculation.

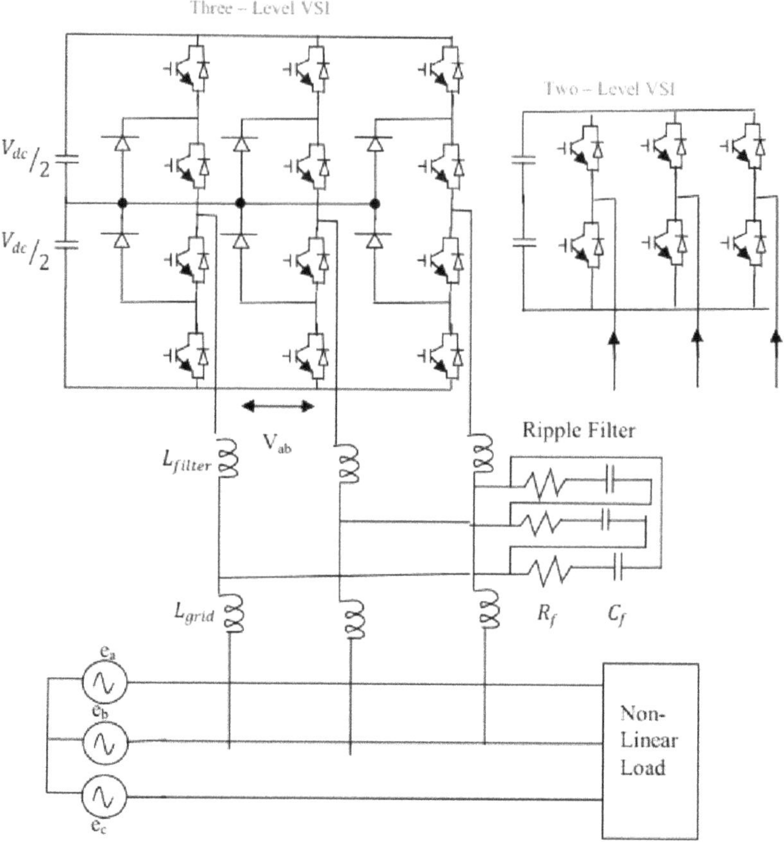

FIGURE 7.9 Two- and three-level VSIs for the APF.

TABLE 7.1

Components and Parameters of DSTATCOM (shunt active power filter)

S. No	Topology	Transformer	Interfacing Inductance L (mH)	Capacitance C (uH)	DC Bus Voltage V_{dc} (V)	Isolation	kVA Rating of Transformer	Semiconductor Devices
1	Three-leg VSC	Not required	2.5	3000	700	No	Nil	6
2	Three-leg VSC with split capacitor	Not required	7	5000	1400	No	Nil	4
3	Four-leg VSC	Not required	3.5	2200	700	No	Nil	8
4	Non-isolated zig-zag transformer with three-leg VSC	Zig-zag transformer	3.5	2200	700	No	5	6
5	Non-isolated star-delta transformer with two-leg VSC	Star-delta transformer	7	5000	1400	No	8	4
6	Three single-phase VSC	Not required	7	2200	400	Yes		12
7	Isolated T-connected transformer with three-leg VSC	T-connected transformer	2.3	6600	400	Yes	12	6
8	Isolated star-hexagon transformer with two-leg VSC	Star-hexagon transformer	3.5	6600	400	Yes	12	4

TABLE 7.2

Comparison of DSTATCOM Control Techniques

		Performance of Control Techniques						
S. No	**Attributes**	**IRP**	**SRF**	**SC**	**AUPF**	**PI Controller**	**NN**	**SMC**
1	Reactive power compensation	Partial	Good	Excellent	Excellent	Partial	Good	Excellent
2	Harmonic mitigation	Good	Good	Better	Excellent	Good	Good	Excellent
3	Load balancing	Excellent	Average	Excellent	Good	Good	Excellent	Excellent
4	Source neutral current elimination	Excellent	Good	Good	Good	Average	Good	Good
5	Computational complexity	High	Average	Simpler	High	Average	Simpler	High

TABLE 7.3

Details of DSTATCOM Compensating Variables with Respective Strategies

Compensated variables	Inverter topology	Current control method	Modulation strategy	References
Voltage regulation and unbalance	Two-level PWM VSI	Wavelet transform	SPWM	[138]
	Two-level PWM VSI	Synchronous reference frame	SPWM	[139]
Voltage regulation, current unbalance, and neutral current	3HB p–q theory	Hysteresis-based	PWM	[140]
Voltage sag and unbalance	Seven-level CHB-MLI Two-level PWM VSI	Synchronous reference frame Synchronous reference frame	— Hysteresis-based PWM	[141] [142]
Voltage sag, swell, and unbalance	Single-phase VSI	Phase space	Carrier PWM	[143]
Current harmonics, unbalance, and neutral current	p–q, cross vector, vectorial, global, p–q–r, unity power factor	SRF	Instantaneous symmetrical components	[144]
Current harmonics, unbalance and neutral current hysteresis-based	3HB VSI two-level PWM VSI two-level PWM VSI two-level PWM VSI	Instantaneous symmetrical component Synchronous reference frame NN Single-phase p–q theory	PWM SPWM — —	[145] [146] [147] [148]
Current harmonics and unbalances	five-level FC-MLI	Instantaneous symmetrical components	Hysteresis-based PWM	[149]
	3HB VSI ins.	Symmetrical component	Hysteresis-based PWM	[150]
	two-level PWM VSI	Unity power factor	Hysteresis-based PWM	[151]
	two-level PWM VSI	Composite observer-based	SPWM	[152]
	two-level PWM VSI	LMS-based ADALINE	Hysteresis-based PWM	[153]
	3HB	p–q theory	Hysteresis-based PWM	[154]

(*Continued*)

TABLE 7.3 (*Continued*)

Details of DSTATCOM Compensating Variables with Respective Strategies

Compensated variables	Inverter topology	Current control method	Modulation strategy	References
Reactive power, current harmonic, current unbalances, and voltage control	Two-level PWM VSI	p–q theory	Hysteresis-based PWM	[155]
Reactive power, current harmonics, and neutral current	Two-level PWM VSI Five-level CHB-MLI	Synchronous reference frame p–q theory	— PS-PWM	[156] [157]
Dual-mode voltage sag or current harmonics	Three-level FC-MLI	Synchronous reference frame	PS-PWM	[158]
Current harmonics, unbalance and neutral current or voltage regulation, current harmonics, unbalance	4 L VSI	Instantaneous symmetrical component	—	[159]
Current harmonics and unbalance or voltage regulation, current	Two-level PWM VSI	Harmonics adaptive theory	—	[152]
Voltage unbalance or current harmonics, unbalances, and neutral current	4 L VSI Two-level PWM VSI	p–q theory admittance-based	Hysteresis-based PWM —	[159] [160]
Reactive power and current harmonics or voltage regulation	Two-level PWM VSI	p–q theory	SPWM	[161]
Load balancing and current harmonics or voltage regulation	Two-level PWM VSI Two-level PWM VSI Two-level PWM VSI Two-level PWM VSI Two-level PWM VSI	NN-based adjustable step LMS NN-based anti-Hebbian leaky LMS adaptive filter Back-propagation p–q theory	— — — SPWM —	[162] [162] [163] [164] [165]
Reactive power, current harmonics, neutral current, and load balance or voltage regulation, current harmonics, neutral current, and load balance	4 L VSI 2 leg VSI	Synchronous reference frame Synchronous reference frame	SPWM SPWM	[166] [167]
Reactive power, current harmonics, and load balance or voltage regulation, current harmonics, and load balance	Two-level PWM VSI Two-level PWM VSI	Synchronous detection algorithm Adaptive filter	— —	[168] [169]
Reactive power, current harmonics, and load balance or reactive power, voltage regulation	Two-level PWM VSI	Peak detection	—	[170]

(*Continued*)

TABLE 7.3 (*Continued*)

Details of DSTATCOM Compensating Variables with Respective Strategies

Compensated variables	Inverter topology	Current control method	Modulation strategy	References
Single-mode reactive power	Three-level NPC-MLI	Time-varying Fourier	SV-PWM	[171]
	Two-level PWM VSI	SRF	Hysteresis-based PWM	[172]
	Two-level PWM VSI	Synchronous reference frame	SPWM	[173]
	Five-level DC-MLI	Instantaneous symmetrical component	LQR	[174]
	Three-level NPC-MLI	Instantaneous symmetrical component	Hysteresis-based PWM	[175]
	Two-level PWM VSI	Adaptive based on AIS	SPWM	[176]
	Five-level CHB-MLI	Sliding mode hysteresis-based	PWM	[177]
	Two-level PWM VSI	Sliding mode carrier	PWM	[178]
	Two-level PWM VSI	p–q theory	Dead-beat	[179]
Voltage sag	Two-level PWM VSI	Insttantaneous symmetrical component	SPWM	[180]
	84-pulses VSI	Synchronous reference frame	—	[181]
	Two-level PWM VSI	Synchronous reference frame	—	[182]
Voltage flicker	Two-level PWM VSI	SRF	SPWM	[183,184]
Voltage unbalance	Three-HB VSI	p–q theory	Pole-shift technique	[185]
Current harmonics	Two-level PWM VSI	Fortescue with RDFT	Dead-beat	[186]
Voltage sag and flicker	Three-HB VSI	Kalman filter	Hysteresis-based PWM	[187]
Voltage and frequency	Two-level PWM VSI	SRF	Carrier PWM	[188]
Reactive power and voltage control	11-level CHB-MLI	Synchronous reference frame	PS-PWM	[189]
	—	Synchronous reference frame	—	[190]
	Two-level PWM VSI	Synchronous reference frame	SPWM	[191]
	Two-level PWM VSI	Synchronous reference frame	SPWM	[192]
	Five-level CHB-MLI	Synchronous reference frame	SPWM	[193]
	Two-level PWM VSI	Synchronous reference frame	SV-PWM	[194]
Reactive power and unbalance load	Two-level PWM VSI	Synchronous reference frame	Hysteresis-based PWM	[195]
	Two-level PWM VSI	ADALINE-based	Hysteresis-based PWM	[196]
Reactive power and current harmonics	Five-level CHB-MLI Two-level PWM VSI Two-level PWM CSI	Synchronous reference frame Synchronous reference frame Fourier transform	LS-PWM, PS-PWM predictive SHE	[197] [142] [198]
Voltage and voltage harmonic	Two-level PWM VSI CHB-MLI Single-phase VSI	Synchronous reference frame Synchronous reference frame Sliding mode	SHE — Hysteresis-based PWM	[199] [170] [200]
Voltage sag and swell with phase angle control	Two-level PWM VSI	Two-loop feedback control	Dead-beat	[201]

(*Continued*)

TABLE 7.3 (*Continued*)

Details of DSTATCOM Compensating Variables with Respective Strategies

Compensated variables	Inverter topology	Current control method	Modulation strategy	References
Voltage support and balance with angles control, balance the active and reactive power	Two-level PWM VSI	p–q theory	SPWM	[148]
Current harmonic and current unbalances	2C-VSI 3HB VSI Three-level NPC 4 L VSI 3HB VSI Two-level PWM VSI	Instantaneous symmetrical component Instantaneous symmetrical component Instantaneous symmetrical component Instantaneous symmetrical component Instantaneous symmetrical components Synchronous reference frame	Hysteresis-based PWM Hysteresis-based PWM Hysteresis-based PWM Hysteresis-based PWM LQR —	[202] [203] [204] [205] [206] [207]
Current harmonics and neutral current	2C-VSI	Instantaneous symmetrical component	Hysteresis-based PWM	[208]

TABLE 7.4

Comparative Analysis

Compensating Variable	Phase and Wire System	Inverter Topology	Aim/ Technique	Control Method	Modulation Strategy	References/ author
Unbalanced- neutral current	Three-phase four-wire	Two-level PWM VSI	Mitigate the flow of zero-sequence harmonic components	Modified synchronous reference frame controller (MSRFC)	PWM	da Silva et al. [209]
Mitigate voltage sags and swells, reduce voltage harmonics, and eliminate voltage unbalance	Three single-phase IGBT-based voltage source inverters (VSI)	Two-level PWM VSI	Shunt rectifier uses a PI controller with two closed loops	Improved direct voltage comparison method	PWM	Ran Cao et al. [210]
Reduction of 5th and 7th harmonics with reduced number of devices	Single-phase quasi-square wave inverter	Neutral point clamped three-level leg	Selective harmonic elimination	Algebraic harmonic elimination approach	PWM	Vivek et al. [211]
Unbalances, harmonics and flicker and damp possible resonances involving the passive filters and the system impedance	Three-phase three-wire systems	Two-level PWM VSI	Designed to drain harmonic current-components generated by the load	Instantaneous power theory	PWM	Pinto et al. [212]

(*Continued*)

TABLE 7.4 (*Continued*)
Comparative Analysis

Compensating Variable	Phase and Wire System	Inverter Topology	Aim/ Technique	Control Method	Modulation Strategy	References/ author
Uninterruptible power supply (UPS)	Single-phase phase-locked loop (PLL) systems	Single-phase	Fictitious quadrature voltage	Three-phase instantaneous active power	PWM	Silva et al. [213]
Interactive uninterruptible power supply and reactive power and to eliminate harmonic currents	Three-phase three-wire systems	Two-level PWM VSI	Compensating fundamental zero-sequence component of the neutral current,	SRF method	PWM	Da Silva and Modesto [214]
Compensate medium voltage power distribution systems	Neutral point clamped (NPC) multilevel converters					[215]
Voltage sags and total harmonic distortion	Three-phase PWM	Two-level PWM VSI	Controlled voltage source injects the compensation voltage	—	PWM	[216]
Reduction of the inverter ratings harmonic distortion and/or imbalance and attaining the unity displacement power factor	Multi-converter conditioner three-phase four-wire systems	Auto adjustable SRF PWM	Strategies and the tracking techniques are based on estimating the load current	Collaborative control strategies	PWM	[217]
Restore sags and short interruptions while reducing the THD	Multilevel converter and three-phase neutral point clamped	Synchronous frame and stationary frame with PI and PIR controllers	Dynamically inject a compensation voltage vector in series with the line voltage, through series-connected transformer secondary windings	Optimal predictive controller and suitable quadratic weighed cost functional	PWM	[218]
Compensation voltages	Three-phase PWM	Two-level PWM VSI	Transformed to p–q–r coordinates without time delay	PQR power theory	PWM	[219]
Line-interactive uninterruptible power supply	Three-phase four-wire system	Two-level PWM VSI	Obtaining the reference currents, both conventional Butterworth filters, and moving-average filters (MAF)	SRF method and SPS-MAF and ISPS-MAF	PWM	[220]

(*Continued*)

TABLE 7.4 (*Continued*)
Comparative Analysis

Compensating Variable	Phase and Wire System	Inverter Topology	Aim/ Technique	Control Method	Modulation Strategy	References/ author
Voltage sags	Transformer-less D VR	Two-level PWM VSI	Respond very fast on the request from abruptly changing reference signals	Fuzzy logic control	PWM	[221]
Voltage sags/swells, harmonic compensation, or reactive power compensation	Multilevel series compensator (MSC);	Three-phase bridge with cascaded single-phase transformers	Generalization for K-stages in which K-transformers are coupled with K-TPB	—	PWM strategy	[222]
Voltage sag problem with phase angle jump	Three-phase PWM	Two-level PWM VSI	Optimum voltage injection angle based on minimum energy compensation	Instantaneous reactive power theory and the low-pass filter theory	PWM	[223]
Voltage sags, fault protection	Three-phase PWM	Two-level PWM VSI	*p*–*q*–*r* coordinates are then inversely transformed to the original a-b-c	PQR instantaneous power theory	PWM	[224]
Block the propagation of harmonic current	Three-phase PWM	Two-level PWM VSI	Ability to extract only harmonic component from supply current	Adapted control strategy based on Instantaneous Power Theory	PWM	[225]
Selective harmonic compensation	Three-phase PWM	Two-level PWM VSI	Control of steady-state power through the DVR	Selective harmonic feedback control strategy	PWM	[226]

7.4 Capacitor Charging in Active Filters

7.4.1 Design of VSC

The designs of some topologies of DSTATCOM are reported in the literature. However, the component selection for a three-leg VSC used as three-phase three-wire DSTATCOM shown in Figure 7.1b is given here for a sample calculation. The DSTATCOM is designed for a rating of 12 kVA for the reactive power compensation of the load given in the Appendix. The DC bus voltage, DC bus capacitor, and interfacing inductor are selected as follows.

1. **DC Capacitor Voltage**: The minimum DC bus voltage should be greater than twice the peak of the phase voltage of the system [101]. The DC bus voltage is calculated as

$$V_{dc} = \frac{\sqrt{2}V_{LL}}{\sqrt{3}m}$$

where m is the modulation index and is considered as 1. Thus, V_{dc} is selected as 700 V for a V_{LL} of 415 V.

2. **DC Bus Capacitor**: The value of the DC capacitor (C_{dc}) depends on the instantaneous energy available to the DSTATCOM during transients [101]. The principle of energy conservation is applied as

$$\frac{1}{2}C_{dc}\left[\left(V_{dc}^2\right)-\left(V_{dc1}^2\right)\right]=3V(aI)t$$

where V_{dc} is the reference DC voltage and V_{dc1} is the minimum voltage level of DC bus, "a" is the over loading factor, V is the phase voltage, I is the phase current, and t is the time by which the DC bus voltage is to be recovered. Considering the minimum voltage level of DC bus, V_{dc1} = 690 V, and V_{dc} = 700 V, V = 239.6 V, I = 27.82 A, t = 350 µs, a = 1.2, the calculated C_{dc} value is approximately 3000 µF.

3. **AC Inductor**: The selection of the AC inductance (L_f) depends on the current ripple [101], switching frequency, and DC bus voltage (V_{dc}), and is given as

$$L_f=\frac{\left(\sqrt{3}mV_{dc}\right)}{\left(12af_s i_{cr(pp)}\right)}$$

where m is the modulation index and a is the over load factor. Considering $i_{cr,p-p}$ = 5%, f_s is 10 kHz, m = 1, V_{dc} = 700 V, and a = 1.2, the L_f value is approximately 2.5 mH.

7.5 Applications of Compensating Devices

A brief criterion for selection of an appropriate compensating device for a specific application is discussed in this section.

A. **Current-Based Compensation**

In addition to the current harmonics mitigation, DSTATCOM is most widely used for power factor correction [53], reactive power compensation [227], and load balancing/load compensation [9,53,128,228] when connected at the load terminals. It can also perform voltage regulation [53], mitigate voltage sag/swell [26,129,130], voltage dip [68,69,131], voltage flicker [5,26], and voltage fluctuations [72] when connected to a distribution bus.

B. **Voltage-Based Compensation**

Voltage-based compensation is categorized as voltage harmonics compensation, improving voltage regulation, voltage balancing, voltage flicker reduction, and removing voltage sags, swells and dips. Apart from voltage harmonics compensation, DVR finds its application for reactive power compensation [126], voltage regulation [20,73–75], voltage sags/swells [31,32,47,229,230], voltage dips [127], and flicker [18].

C. **Voltage and Current-Based Compensation**

Many applications require a compensation of a combination of voltage- and current-based problems, a few of them being interrelated. The UPQC (combination of active series with active shunt filters) is an ideal choice for such mixed compensation. It is considered as a most versatile device that can inject current in shunt and voltage in series simultaneously in a dual control mode. In addition to the current- and voltage-based harmonics mitigation it can perform load compensation/load balancing [54,76], reactive power compensation [43,44,57,77], power factor correction [28,78], voltage regulation [79–81], voltage sag/swells [8,35,54,55,78], voltage dip [68], and voltage flicker [12,43] at the same time.

7.5.1 Reference Signal Extraction Techniques

The development of real-time methods for the detection and analysis of disturbances is a major concern to evaluate the quality of supply voltage and to prevent the harmful effects on equipment. The performance of a DSTATCOM strictly depends on its reference signal generation technique. In general, frequency-domain and time-domain methods are used to generate the reference signal. The time-domain methods are faster and easier to implement than the frequency-domain methods but they present worse detection performance than the frequency-domain methods.

These harmonic compensation approaches include instantaneous reactive power theory, instantaneous power theory based on symmetrical components, generalized instantaneous reactive power theory, SRF method, synchronous detection method, and a novel approach in the *a–b–c* reference frame based on symmetrical components. It is shown in the literature that the six approaches are all effective on compensating reactive and harmonic components of load currents under ideal source voltages. However, when the source voltages are either unbalanced or non-sinusoidal, each approach presents inconsistent results on maintaining sinusoidal source current.

7.5.1.1 Frequency-Domain Methods

Frequency-domain methods are suitable for both single- and three-phase systems. They are mainly derived from the Fourier analysis and include the following three subdivisions.

7.5.1.1.1 Fast Fourier Transform

A fast Fourier transform (FFT) is used to compute the discrete Fourier transform (DFT) and its inverse. A Fourier transform converts functions from time to frequency domains and vice versa. FFT computes such transformations by factorizing the DFT matrix into a product of sparse factors. In DSTATCOMs, FFT is used to extract the harmonic components from the harmonic polluted signals. Owing to excessive computation in on-line application of FFT, it has high response time [90,91]. Generally, in order to detect current harmonics and voltage sags, the conventional detection algorithms such as RMS-based algorithm and FFT-based algorithm have been introduced to achieve the detection task. In this paper [92], an alternative algorithm, the adaptive predictive algorithm, is adopted to do the same task as the conventional algorithms. The salient feature of the adaptive algorithm is that it can detect the event as fast as the conventional algorithm can. Thus, it would be appropriate to detect events like voltage sags, because voltage sags of even very short duration may result in malfunction to sensitive loads.

This paper [93] proposes a new algorithm to generate a reference signal for an APF using a sliding-window FFT operation to improve the steady-state performance of the APF. In the proposed algorithm the sliding-window FFT operation is applied to the load current to generate the reference value for the compensating current. The magnitude and phase angle for each order of harmonics are respectively averaged for 14 periods. Furthermore, the phase-angle delay for each order of harmonics passing through the controller is corrected in advance to improve the compensation performance. The steady-state and transient performance of the proposed algorithm was verified through computer simulations and experimental work with a hardware prototype. A single-phase APF with the proposed algorithm can offer a reduction in THD from 75% to 4% when it is applied to a non-linear load composed of a diode bridge and a RC circuit. The APF with the proposed reference generation method shows accurate harmonic compensation performance compared with previously developed methods, in which the THD of the source current is higher than 5%.

7.5.1.1.2 Kalman Filter

The Kalman filter is a recursive optimal estimator and requires a state variable model for the parameters to be estimated and a measurement equation that relates the discrete measurement to the state variables. The Kalman filter uses a mathematical model of the states to be estimated and is suitable for real time applications. If the harmonic contents have a time-varying amplitude, a Kalman filter-based algorithm tracks the time variation after the initialization period [231–233]. The parameter tuning is based on a combined pole placement and an optimal estimation based on the Kalman–Bucy filter [94]. This paper

[95] deals with the harmonic detection that is decoupled from the operation of an APF. A Kalman filter for harmonic detection based on a stochastic state-space model is proposed. However, it is a challenging task in large time-varying systems to know the process and noise covariance matrices Q and R. In this APF application, the current sensors TLC277CD and ADC LTC1403A, which introduce load current measurement inaccuracies, are analyzed to decide a rough R.

Because R is exactly known, two adaptive Kalman filter algorithms to scale Q are proposed. One of the adaptive Kalman methods switches two basic Q matrices depending on the system in transient or steady state. The other Kalman algorithm tunes an optimal Q at each step by using the information of innovations sequence. This study [96] presents a new reference current estimation method using proposed robust extended complex Kalman filter (RECKF) together with model predictive current (MPC) control strategy in the development of a three-phase SAPF. A new exponential function embedded into the RECKF algorithm helps in the estimation of the in-phase fundamental component of voltage (v_h) at the point of common coupling (PCC) considering grid perturbations such as distorted voltage, measurement noise, and phase angle jump and also for the estimation of fundamental amplitude of the load current (i_h). The estimation of these two variables (v_h, i_h) is used to generate reference signals for MPC. The proposed RECKF-MPC needs fewer voltage sensors and resolves the difficulty of gain tuning of proportional-integral (PI) controller. The proposed RECKF-MPC approach is implemented using MATLAB/SIMULINK and also Opal-RT was used to obtain the real-time results. The results obtained using the proposed RECKF together with different variants of Kalman filters (Kalman filter (KF), extended KF (EKF), and extended complex KF (ECKF)) and PI controller are analyzed in both the steady state and transient state conditions. From the above experimentation, it was observed that the proposed RECKF-MPC control strategy outperforms PI controller and other variants of Kalman filtering approaches in terms of reference tracking error, power factor distortion, and percentage THD in the SAPF system. This paper [97] presents a robust model-based control in natural frame for a three-phase SAPF. For the proposed control method, a linear converter model is deduced. Then, this model is used in a Kalman filter in order to estimate the system state-space variables.

Even though the state estimations do not match the variables of the real system, it has allowed for design of three sliding mode (SM) controllers providing the following features to the closed-loop system: a) robustness due to the fact that control specifications are met independently of any variation in the system parameters; b) noise immunity, since a Kalman filter is applied; c) a lower THD of the current delivered by the grid compared with the standard solution using measured variables; d) the fundamental component of the voltage at PCC is estimated even in the case of a distorted grid; and e) a reduction in the number of sensors. Thanks to this solution the sliding surfaces for each controller are independent. This decoupling property of the three controllers allows using a fixed switching frequency algorithm that ensures a perfect current control.

7.5.1.1.3 Wavelet Transform Base Algorithm

Harmonic detection technology is one of the key technologies of APF. Its development decides the development of APF technology. Due to the grid harmonic inherent characters, like non-linear randomness, distribution, non-stationary and factors influencing the complexity of other features, it is difficult to accurately detect the harmonic. Therefore, the research method for detecting the harmonics in the power system is very important. Wavelet technology helps in detecting the harmonics effectively. This method is based on the definition of the active and reactive power in the time-frequency domain using the complex WT.

The voltage and current signals are transformed to the time-frequency domain using the complex wavelet with scaling and translation parameters to set the frequency range and localize the frequency, respectively [138,232,234]. A novel method [98] to determine the reference filter current for an active power conditioner filtering harmonic current components and/or current contributing to reactive power is presented. It is based on definitions of active and reactive power quantities in the time-frequency domain associated to complex wavelets, allowing a better time localization than Fourier-based quantities. Its behavior is illustrated in a well-chosen single-phase example. This paper [99] advances a novel method for power quality monitoring by using integer lifting wavelet transform (ILWT). ILWT-based wavelet uses time domain for its integer calculation and could lessen calculation complexity due to its structure, compared with modified wavelet transform (MWT) and WT, which depends on floating point coefficients. The whole method is tested over sample disturbances like voltage sag, voltage swell, and harmonics.

7.5.1.2 Time-Domain Methods

The following time-domain approaches are mainly used for three-phase systems.

7.5.1.2.1 *p–q* Theory

The problem of harmonics due to non-linear loads can be reduced by series APF. *p–q* Theory is used as the control algorithm in the proposed series APF. The performance of the filter is evaluated by monitoring the reduction of THD and the improvement of power factor. In *p–q* theory, voltages/currents of 3P3 W system are converted into two-phase voltage/current components by Clarke transformation on orthogonal α–β coordinates, thus the instantaneous active and reactive powers can be determined without any time delay. *p–q* Theory provides a theoretical validation that the instantaneous active and reactive powers are uniquely related with the instantaneous active and reactive currents, respectively, in 3P3 W systems. *p–q* Theory does not follow power conservation and conflicts with the general understanding of power in that the zero-sequence instantaneous reactive power cannot be defined by this theory in 3P4 W systems [235–237].

(ii) Synchronous detection theory: This technique is similar to *p–q* theory and comprises three approaches, equal power, equal current, and equal resistance criterion. The average power is calculated and divided equally between the three phases. In synchronization process, the compensation signals are synchronized with relative utility grid phase voltage. It is easy theory to implement but it is affected by voltage harmonics [238,239]. The controller based on a novel combination of SRF and *p–q* theory [100] to extract voltage and current harmonics and unbalances is developed in the paper. To produce gate switching signals for the three-phase THSeAF, a conventional PI controller, is used to produce reference waveforms produced earlier to rectify harmonic currents initiated from typical nonlinear loads. The device ensures a reliable and dynamically restored power supply on the load's PCC by means of three auxiliary DC supplies. The purpose of the paper [111] is to present a comparative study of two techniques for harmonics currents identification based on *p–q* theory and ADALINE neural networks [101]. The harmonic current can be identified from powers or currents. The first method is based on the instantaneous powers taking advantage of the relationship between load currents and power transferred from the supply source to the loads. The second method concerns the artificial neural networks based on the least mean square (LMS) algorithm. This approach adjusts the weights by iteration and provides more flexibility to perform the compensation.

Several publications are reported and compared the performances of different reference current generation strategies under balanced, sinusoidal, unbalanced, or distorted AC voltages conditions [86–88]. In all of them, authors have demonstrated that under balanced and sinusoidal AC voltages conditions, the strategies such as the so-called *p–q* theory and SRF theory provide similar performances. Differences arise when these are used under distorted and unbalanced AC voltages. In addition, it has been proved in the literature [89] that the supply voltage fluctuations have influence on the DC link voltage and the expression signifying the influence of the change in supply voltage in the change in DC link voltage has also been derived for the ShAF. A conventional PI controller used for DC link voltage control suffers from high overshoot and undershoot during the supply voltage fluctuations. However, there no attempt is made on adaptive FLCs to regulate the DC link voltage along with other parts of the algorithm in case of sudden load changes. In this paper, adaptive FLC is used and is compared with conventional PI controller to reduce the DC link voltage variations in case of supply voltage fluctuations along with the sudden load changes.

7.5.1.2.2 Cross Vector Theory

In cross vector theory, the instantaneous active and reactive power is defined by scalar/vector product of the voltage and the current space vectors in a 3P4W system. Cross vector theory identifies one instantaneous active power and three instantaneous reactive powers. However, the three components of instantaneous reactive powers are linearly dependent on each other. In the presence of a zero-sequence voltage, the neutral line current cannot be eliminated completely by compensating the instantaneous reactive power [144,236,237,240,241]. The theory, which applies to four-wire systems, is shown to contain mostly the same limitations as the original instantaneous reactive power theory, namely that the instantaneous active current is not the lowest rms current in all circuit conditions and could contain higher levels of harmonic distortion than the supply voltage. Also, most recommended compensation strategies based on the theory require the supply voltage to be approximately sinusoidal and symmetrical to avoid erroneous compensation. Under these conditions, the compensation strategies produce the desired results. However, they require calculation of average power which is not instantaneous, and energy storage is necessary if the oscillating component of instantaneous power is compensated.

7.5.1.2.3 Global Theory

Since the reference compensation currents are determined in the *A–B–C* reference frame, there is no reference-frame transformation requirement. Therefore, the global theory gives less complexity in realizing the control circuit of the DSTATCOM. By using this theory, DSTATCOM is able to compensate reactive power and suppress harmonic/neutral currents of the imbalanced/distorted load without supplying or consuming real power [144,242,243].

7.5.1.2.4 Vectorial Theory

In this method, vectorial formulation does not need to undergo any kind of coordinates translation. Vectorial theory utilizes the same power variables as *p–q* theory and identifies the instantaneous reactive power in phase coordinates. The current vector is split into three components. The first one is collinear with respect to the modified voltage vector, and it transports the instantaneous real power. The second one is collinear with respect to the zero-sequence voltage vector, and it transports the instantaneous zero-sequence power. The last one is normal with respect to the modified and zero-sequence voltage vectors, and it transports the instantaneous imaginary power [34,144,244,245]. One of the control strategies based on the dual formulation of the electric power vectorial theory is implemented in [102] for balance and resistive load. In this paper an attempt is made to apply the same control theory for unbalanced and non-sinusoidal voltage conditions of the distribution system. A state-space averaging model of HAF constructed to analyze its system stability by traditional control strategy, considering the effect of the time delay. Performance of the HAF is compared and analyzed with that of the (PF) to improve power quality at point of common coupling PCC as per the IEEE standards. This paper presents mathematical modeling of composite filter (CF) for power quality improvement of EAF distribution network. The CF consists of a shunt LC passive filter connected with a lower rated voltage source PWM converter-based series APF. The control strategy adopted for CF operation is based on simultaneous detection of source current and load voltage harmonic based on the vectorial theory dual formulation [246] of instantaneous reactive power. A state-space averaging model of a CF constructed to analyze its system stability by traditional control strategy, considering the effect of the time delay.

7.5.1.2.5 p–q–r Theory

The *p–q–r* theory takes advantage of both *p–q* theory and cross vector theory. The defined instantaneous powers follow power conservation. Both instantaneous active and reactive powers can be defined in the zero-sequence circuit in three-phase four-wire systems. The three power components are linearly independent of each other. In the presence of a zero-sequence voltage, the neutral line current can be eliminated completely by applying the *p–q–r* theory [144,237]. This paper [102] discusses how to calculate the compensation voltages in dynamic voltage restorers (DVR) by use of PQR power theory. Directly sensed three-phase voltages are transformed to *p–q–r* coordinates without time delay, then the reference voltages in *p–q–r* coordinates have very simple form: DC values. The controller in *p–q–r* coordinates is very simple and clear and has better steady state and dynamic performance than conventional controllers. The controlled variables in *p–q–r* coordinates are then inversely transformed to the original *a–b–c* coordinates instantaneously, generating reference compensation voltages to a DVR. The control algorithm can be used for various kinds of series compensators such as DVRs, SeAFs, synchronous static series compensators (SSSC), and bootstrap variable inductances (BVI).

The paper directly [247] calculates compensation voltages without time delay and compensates for the faulted voltages dynamically in the time domain. Using the PQR transformation, the reference voltage in p–q–r coordinates become simple in form: a single DC value. Thus, the controllers will be very simple and are expected to have a better steady state and dynamic performance. Voltages can be sensed on the source side, the load side, or both sides according to applications. When the system is in sinusoidal balance, only the p-axis voltage is needed to control. Thus, the number of control variable becomes one in sinusoidal balance systems. The control algorithm can ride through any kind of voltage faults on the source side and can compensate also for the back-ground noise when the source voltages are distorted by the harmonics.

If sinusoidal balanced voltages are assumed in a–b–c coordinates, the reference waves in 0–α–β coordinates will consist only of sinusoidal and orthogonal. Using the reference waves, in the mapping matrix, the voltages in 0–α–β coordinates can be transformed to p–q–r coordinates, given by the following equation [84,85].

$$\begin{bmatrix} v_p \\ v_q \\ v_r \end{bmatrix} = \begin{bmatrix} 0 & \dfrac{v_{\alpha\text{REF}}}{v_{\alpha\beta\text{REF}}} & \dfrac{v_{\beta\text{REF}}}{v_{\alpha\beta\text{REF}}} \\ 0 & -\dfrac{v_{\beta\text{REF}}}{v_{\alpha\beta\text{REF}}} & \dfrac{v_{\alpha\text{REF}}}{v_{\alpha\beta\text{REF}}} \\ 1 & 0 & 0 \end{bmatrix} \begin{bmatrix} v_0 \\ v_\alpha \\ v_\beta \end{bmatrix}$$

where $V_{\alpha\beta\text{REF}} = \sqrt{V_{\alpha\text{REF}}^2 + V_{\beta\text{REF}}^2}$

The physical meaning of the proposed PQR transformation is shown in Figure 7.1. The reference voltage space vector V^{REF} rotates on the α–β plane with the angle of

$$\theta(t) = \tan^{-1}\left(\frac{V_{\beta\text{REF}}(t)}{V_{\alpha\text{REF}}(t)}\right)$$

from the α-axis. The p-axis is aligned with the reference voltage space vector V^{REF}. The q-axis is perpendicular to the p-axis on the α–β plane. The r-axis is equal to the 0-axis in 0–α–β coordinates. Thus, the p-axis and q-axis rotate on the α–β plane with the angle of $\theta(t)$ from the α-axis and β-axis, respectively. The r-axis becomes the shaft of rotating p–q–r coordinates. Figure 7.10 shows the simple case of the

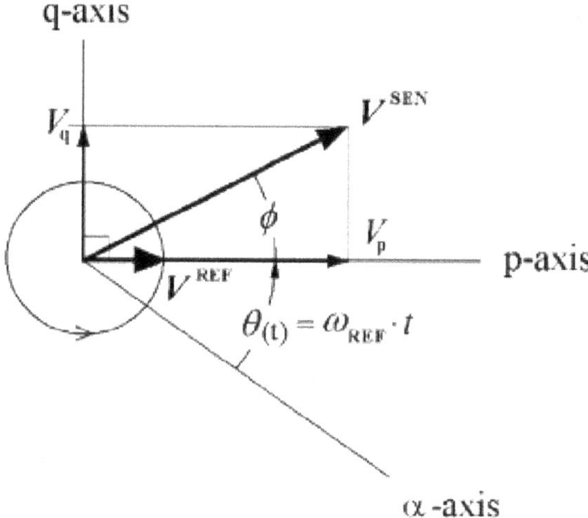

FIGURE 7.10 Simple case of the sensed three-phase voltages in p–q–r coordinates when the normal grid voltages are assumed in constant angular frequency without the zero-sequence component.

sensed three-phase voltages in p–q–r coordinates when the normal grid voltages are assumed in constant angular frequency without the zero-sequence component.

7.5.1.2.6 Synchronous Frame-Based Theory

This algorithm is based on the transformation of the three-phase system into synchronously rotating frame to extract the direct, quadrature, and zero-sequence components of signals. The active and reactive components of the system are defined by the direct and quadrature components, respectively. The high-order harmonics still remain in the signal; however, they are modulated at different frequencies. These are the undesired components to be eliminated from the system and they represent the reference harmonic current. The system is very stable since the controller deals mainly with DC quantities. The computation is instantaneous but causes a time delay in filtering the DC quantities [138,144,172,194–197].

7.5.1.2.7 Instantaneous Symmetrical Component Theory

This instantaneous theory is valid for steady and transient states and for generic voltage and current waveforms. Some examples explaining the physical meaning of the new concepts are presented. By using the concepts of symmetrical components together with the new theory, the powers in an unbalanced system are analyzed, including the zero-sequence instantaneous power. In instantaneous symmetrical components method, a symmetrical voltage and current are transformed by symmetrical components to obtain positive sequence, negative sequence, and zero-sequence components of three phase variables. The instantaneous symmetrical component theory has advantages such as: it is simple in formulation, computationally less intensive for reference currents generation thus ensuring fast dynamic response, and avoids interpretation of various definitions of instantaneous reactive powers and complex transformations. Numerous studies focused on symmetrical components theory were used for extraction the reference signal in [144,145,150,202,206,248]. In this algorithm the theory of instantaneous symmetrical component is used to extract a positive sequence of supply voltage to generated reference compensator currents under the assumptions of unbalanced and non-stiff source voltages. The APF is implemented with pulse-width modulation-based current controlled VSI. The switching signals for VSI are generated through proposed three-level hysteresis current controllers.

7.5.1.2.8 Unity Power Factor Theory

This is another technique, except the fact that it forces the instantaneous current signal to track the voltage-reference waveform. This implies that the power factor would be fixed to unity and the system would only be suitable for the combined system of VAr and current harmonic compensation [144,238,249,151].

Neural network (NN)-based theory: NN-based algorithms are used to extract required information after processing of signals by learning or training and activation function. Arya and Singh [147] proposed an algorithm-based on load conductance estimation using NN and implemented on a three-phase DSTATCOM. Its structure is reflected as Kohonen learning or Kohonen feature maps. It is used for extraction of the fundamental component of load currents in terms of conductance and susceptance. Arya et al. [162] proposed a NN-based Anti-Hebbian control algorithm for PQ improvement under linear/non-linear type consumer loads, which is used for extraction of fundamental active and reactive power components of load currents in terms of weighted signals. Reference [163] used a NN-based adjustable-step LMS for signal extraction. It uses autocorrelation time mean estimate error signal for updating the step size in place of simple error signal. This theory is motivated by the necessity to correct the deteriorated power factor caused by the increase of consumer electronic equipment. The APF is designed to compensate for the ldquonon-activerdquo power of the loads. The current compensation is-based on the concept of non-active current. This definition permits the power factor to become unity, even under non-sinusoidal source voltage conditions, and balances the active power per phase.

7.5.1.2.9 Back Propagation-Based Theory

The back propagation (BP) algorithm includes three steps: the feed-forward of the input signal training, calculation and BP of the error signals, and upgrading of training weights. Continuity, differentiability, and non-decreasing monotony are the main characteristics of this algorithm. It is based on a mathematical equation and does not need special features of function in the learning process. It also has smooth variation on weight correction because of batch updating features on weights. In the training process, it is slow because of more learning steps, but after the training of weights, this algorithm generates very fast trained output response [164].

7.5.1.2.10 Learning Vector Quantization-Based Theory

Learning vector quantization (LVQ) is a standard statistical clustering technique that is also known as special case of competitive network. The desired values are extracted through training of weighed values of load currents using the gradient descent method. In the training process, the desired signals are at the position of the learning stage. After training, an LVQ network classifies the supply current vector by assigning it to the same class as the output stage. It has its weighted vector closest to the input vector. In the LVQ network, each unit has a known value or elements and used supervised learning, which differed from the Kohonen self-organizing map [250,251]. A four-leg VSI-based active filter is implemented in [106] for elimination of neutral current and harmonics compensation with load balancing. It is controlled through an algorithm based on LVQ under current-fed type nonlinear load. This double stage supervisory control algorithm is used for extraction of reference supply currents.

7.5.1.2.11 Adaptive-Based Theory

Adaptive-based theory is a closed-loop controller that can adjust system behavior in terms of response to disturbances. An area of adaptive control provides an automatic adjustment of the controller gains and parameters in real time, to achieve a desired level of performance. Characteristics of these control algorithms are the ability to extract required information from real online data to tune the controller, and also used for grid synchronization. Based on this control theory, many control algorithms are also reported in available literature such as adaptive nature for synchronous extraction [169], adaptive theory-based improved linear sinusoidal tracer [152], adaptive control strategy based on artificial immune system [176], and leaky LMS adaptive filter [163]. This paper [107] presents hybrid SeAF based on an adaptive predictive algorithm to compensate voltage sags and current harmonics for residential equipment. Generally, in order to detect current harmonics and voltage sags, the conventional detection algorithms such as RMS-based algorithm and FFT-based algorithm have been introduced to achieve the detection task. In this paper, an alternative algorithm, the adaptive predictive algorithm, is adopted to do the same task as the conventional algorithms. The salient feature of the adaptive algorithm is that it can detect the event as fast as the conventional algorithm can. Thus, it would be appropriate to detect events like voltage sags. This is because even very short-duration voltage sags may result in malfunction to sensitive loads.

7.5.1.2.12 Composite Observer-Based Theory

Composite observer (CO) is used to extract individual harmonics from repetitive signals. The settling period and the bandwidth of the observer depend on how far the observer poles have been placed from the origin of the S-plane or the Z-plane. The errors in the magnitude and phase of the extracted components, because of the deviation of the signal from the central frequency of the observer, are made very small by providing an integrated phase-locking arrangement. Further improvement in the accuracy, particularly in the extracted higher harmonics, is because of the introduction of multi-rate sampling. Advantages of this algorithm are that it is less sensitive with supply frequency variation, low distortion in the extracted signal without leakage of harmonics, and so on [152,163]. A standalone micro grid system consisting of a solar photovoltaic (SPV) array, diesel engine driven permanent magnet synchronous generator (PMSG), and a battery energy storage system (BESS) is analyzed with a CO-based control methodology [108,252]. The PV array-fed boost converter is controlled for providing the maximum power of the SPV array at any given temperature and insolation level. The BESS and diesel generator (DG) are used to control the transfer of power in the system. The CO-based control technique is used for estimation of active and reactive powers of the standalone system. A three-leg voltage source converter (VSC) with a BESS is used to provide load balancing, reactive power compensation, and harmonics elimination capabilities under different loading conditions.

7.5.1.2.13 Conservative Power Theory

Conservative power theory (CPT) is studied in [246,253–260] for active power filtering and power factor correction, specifically under distorted and asymmetric voltage regimes. It is demonstrated how the CPT can selectively neutralize current non-idealities, under harsh voltage conditions. The CPT framework provides means of achieving Fryze currents, with unity power factor, transporting real power at the minimum line current RMS value. The CPT current decomposition includes, beside the active, reactive, and void components, an unbalanced component. This component makes it possible for the active filter to compensate the current unbalance, if desired. This is particularly interesting when the voltage is also unbalanced because of the load unbalance or not. In this case, the compensation will eliminate the voltage asymmetry due to the load current unbalance, but if the grid impedances are not equal on all the three phases, the compensated sinusoidal and balanced current will unavoidably produce asymmetric voltages in the PCC. Moreover, due to the CPT definition of the active current, the grid voltage asymmetry will produce a load current unbalance. The current decomposition based on the CPT was analyzed on various case studies, for a combination of typical nonlinear balanced and unbalanced industrial loads. A shunt APF using a multilevel inverter operating in low switching frequency is used in [257] to improve the power quality in electric metro power supply systems. The proposed topology is a three level/phase voltage source neutral point clamped (NPC) and the CPT is used to extract the harmonic currents. A current control schema sensing only the grid-side voltages and currents is used and a capacitor voltage controller that balances the natural point voltage is implemented. The CPT can selectively neutralize current non-idealities under harsh voltage conditions. In this technique in [259], each current component is weighted by compensation coefficients (k_i), which are adjusted instantaneously and independently, in any percentage, by means of load conformity factors (λ_i), thus providing online flexibility with respect to the objectives of compensation and injection of active power. The CPT and the linear programming (simplex algorithm) are used in [132] to optimize the compensation of reactive power, harmonic distortion and unbalanced load applied to distributed electronic power processors, for example, APFs and/or photovoltaic grid-connected inverters, especially when their power capacity is limited.

7.5.1.3 Other Algorithms

There are numerous optimization and estimation techniques such as ADALINE [159], LMS-based ADALINE [132,261], differential evolution [186], time-varying Fourier coefficient series [171], Fortescue decomposition with recursive DFT [186], and peak detection [170,262]-based algorithms used to extract the reference signal. This paper develops a waveform reconstruction method (WRM) [103–106] for high accuracy and bandwidth signal decomposition of voltage-harmonic-type three-phase diode rectifier load voltage into its harmonic and fundamental components, which are utilized in the SeAF control algorithms. The SAF-compensated system utilizing WRM provides high-performance load harmonic voltage isolation and load voltage regulation at steady-state and during transients compared to the system utilizing the synchronous reference-frame-based signal decomposition. In addition, reducing the line current sampling delay in the discrete-time implementation enhances the stability of the SAF.

7.5.2 Current Control Techniques

The current control of shunt power filters is critical since poor control can reinforce existing harmonic problems. Various control strategies have been proposed by many researchers. Generation of suitable switching signal is the most significant part of DSTATCOM's control algorithm and has a high influence on the compensation performance [263]. PWM is the most reliable way of reconstructing a desired output voltage waveform. The frequency of the switching should be significantly higher than that of the desired signal for a reliable signal representation [264]. PWM methods are often categorized as open loop (feed-forward) and closed loop (feed-back) methods. The open loop method is subdivided into SPWM and space vector PWM (SVM). The closed loop method is classified into HCC and linear current control involving ramp comparison, state feedback, synchronous vector, predictive, deadbeat, SM, linear quadratic regulator, and pole shift controllers. Apart from these methods the selective harmonic elimination technique is also used for generation of a proper switching signal. This paper [134] presents a new and efficient control scheme that adopted the VSI to decrease current harmonics generated by the nonlinear load.

The SM control is used in the current control loop to achieve fast dynamics and a simple PI controller is adopted in the outer voltage control loop to achieve slow dynamics. The proposed scheme implements simplified control algorithms based on direct current control (DCC) and indirect current control (ICC) for designing trajectories in SM control-based SAPF. Compared to the conventional DCC scheme, the ICC scheme using SM control can achieve high power factor and current harmonics elimination in SAPF. This paper [135] therefore proposes a novel fuzzy logic variable hysteresis band current control technique of SAPF, to compensate the harmonics and/or the reactive power generated by nonlinear loads. In this technique, the interference between phases due to the construction of the circuits is first eliminated, and the hysteresis band is adjusted to ensure nearly constant switching frequency, based on the fuzzy logic control of the reference currents and system parameters. This method allows the advantages of quick response, good current tracking accuracy of APF, and minimal current ripple in three-phase systems. The conventional adaptive hysteresis concept is hybridized with FLCs, which facilitates discarding of uncertainty in the system. In fact, conventional PI controllers for ShAF are based on a linearized model that fails to react under transient events [136]. On the other side, FLC has widened its applicability to many engineering fields and offers satisfactory results for a wide variety of operating conditions. It helps in fulfilling the need for perfection, such as stability and robustness for every system. All this motivated to adopt FLC for SAPF applications. By incorporating an adaptive fuzzy hysteresis band, APF gains outstanding compensation ability under steady-state and transient conditions.

7.5.2.1 Open Loop PWM Methods

SPWM: In SPWM, a sinusoidal reference signal is compared with a triangular carrier waveform to generate switching signals. The multi-carrier SPWM control methods were used to increase the performance of inverter, especially in multilevel inverter-based DSTATCOMs. The multi-carrier SPWM can be categorized according to vertical or horizontal arrangements of carrier signal. The vertical multi-carrier SPWM techniques are identified as level-shifted (LS-PWM), which includes phase disposition (PD-PWM), phase opposition disposition (POD-PWM), and alternative phase opposition disposition, while horizontal multi-carrier SPWM is defined as phase-shifted (PS) control technique [138,141,146,157,161,164,165,173,180,184,192,193,197,265–269]. Compared with the traditional multilevel converter, multilevel voltage source cascade converter (ML-VSCC) using a carrier phase-shifted SPWM (CPS-SPWM) technique converter utilizes fewer power switches and has a balanced load, a good linearity between input and output, and a perfect control feature. A novel modulation strategy for ML-VSCC, sample time staggered space vector modulation (STS-SVM), is adopted in [137]. The equivalent switching frequency is improved and the harmonic component is greatly decreased under this technique. Multi-modular SAPF is widely used in high voltage and high-power fields to promote the power quality. However, the traditional multi-modular SAPF needs to handle harmonic and reactive current, so the algorithm is complex, and the harmonic compensation performance is deteriorated. A multi-modular SAPF based on DCC and frequency doubling carrier phase-shifted (FDCPS-SPWM) control scheme is proposed in the paper. There is no need to sense load current, and it can realize higher equivalent switch frequency with lower real switch frequency devices. Thus, it has better harmonic compensation performance. Double closed loop control scheme for the DC voltage and in-grid current is constructed, and then the small-signal model and stability conditions are studied [109]. Sinusoidal pulse-width modulation (SPWM) technique has fixed switching frequency and provides advantage over others.

Space vector modulation (SVM): The goal of the SVM is to find the proper switching combinations and their duty ratios according to certain modulation scheme. SVM uses the control variable given by the control system and identifies each switching vector as a point in complex space of (α, β). SVM operates in a complex plane divided in the six sectors separated by a combination of conducting or non-conducting switches in the power circuit. Although with the good reliability and strong anti-jamming of digital control technique, SVM is of low speed of response caused by the inherent calculation delay [270]. The main feature of the SVM switching strategy is that it enables control of the DC voltage and can balance voltages of the DC capacitors in multilevel inverter, with no requirement of additional controls or auxiliary devices [271,272]. With SVM method, it is also possible to reduce THD and to solve the voltage unbalance problem [171,273]. SVM was recently much more developed for the power quality improvement [110–119]. With the conventional SVM scheme, AC drives generate high-peak common-mode currents (CMCs), particularly at a very low modulation index. This occurs because of the low dwell times on the active voltage vectors, resulting in nearly simultaneous switching on all three phases. High-peak CMCs present several issues-high-voltage spikes from the DC bus to ground, a pump-up of the DC bus voltage, and increased conducted emissions that can affect the reliability and performance of the AC drive. The solutions for addressing this phenomenon include placing capacitors to ground on the DC bus, which cannot be applied when the power system is not solidly grounded (e.g., high-resistance grounding), and filters at the output of the drive, which are bulky, expensive, and may not always be effective. Three-level three-dimensional space vector modulation technique in the $\alpha\beta o$ axes (three-level 3D SVM) is proposed to generate the switching signals of this SAPF for power quality improvements under balanced and unbalanced loads. In order to improve the output voltage forms, lower switching losses, fix switching frequency, minimize the harmonics of source currents, reduce the magnitude of neutral current, eliminate the zero-sequence current, and compensate the reactive power in the four-wire distribution network, and for a good dynamic, the SRF theory in the $dq0$-axes for generating and extracting the reference currents, a three-level NPC four leg inverter is used. The HCC is a popular modulation technique used in APF applications owing to its simplicity of implementation, fast current response, and peak current limiting capability. Despite these advantages, it has a major disadvantage when applied to three-phase isolated neutral point (INP) systems: inter-phase dependency causes very high-switching frequencies. To overcome this problem, space-vector hysteresis current control (SVHCC) modulation for APFs can be applied to three-phase INP systems. The space-vector-based method allows for the efficient use of zero-voltage vectors and prevents high switching frequencies caused by phase interference. The approach comprises initially performing two hysteresis bands around the error vector in the $\alpha\beta$ stationary reference frame (SRF) and then selecting the next voltage vector based on the present location of the error vector to reduce the next error vector. The control algorithm in [113] is composed of indirect current loop for active and reactive power control, predictive current loop for harmonic filtering control and independent control for auxiliary balancing circuits. Among these modules, the indirect current along with the predictive current control will generate a combined voltage reference to the module of space vector pulse-width modulation, which could output optimal modulation signals to the voltage control of a diode-clamped multilevel converter. Compared with traditional DC control for two-level APF, this system takes advantages of ICC and multilevel circuits, including improvement in waveform quality, increase in system power rate, and reduction in voltage stress and switching frequency on power semiconductors. a new constant frequency space-vector hysteresis-band current control (CF-SVHCC) in the SRF for three-level APF applications when applied to INP systems. CF-SVHCC is designed based on two recognized modulation methods: (1) space-vector modulation and (2) adaptive HCC. The proposed technique consists of a simple circular hysteresis strategy around the current-error vector in SRF with the purpose of employing the zero- and nonzero-voltage vectors of the three-level VSI. CF-SVHCC continuously estimates an adaptive outer hysteresis-band in the SRF using the inverter switching signals by a simple and fast artificial neural network method called the adaptive linear neuron algorithm. The main part of CF-SVHCC is a supervisory control unit that operates in the SRF to avoid interphase dependency and systematically uses the voltage vectors associated with the estimated outer hysteresis-band to prevent a high switching frequency and, in turn, maintain the switching frequency constant. CF-SVHCC retains most benefits of the conventional HCC and also introduces additional advantages, including a constant switching frequency and the interphase independency in three-phase INP systems; it is proposed in [114]. The implemented switching techniques are: periodic-sampling (PS), triangular carrier pulse-width modulation (TC-PWM) and SVM [115]. The comparison between them is made in terms of the compensated currents THD%, implementation complexity, necessary CPU time, and SAPF efficiency. The SVM method provides compensation for harmonics and reactive power and has an excellent dynamic performance. The APF's current reference is obtained by subtracting the measured load current from the supply current reference. In PWM rectifier control strategy, direct power control (DPC) has been used more and more widely in order to solve the problem of complex rotating coordinate transformation of vector control. The traditional DPC method with variable switching frequency is applied based on instantaneous power theory and switching table. This approach requires more voltage and current sensors to calculate active and reactive power, leading to low anti-jamming capability. The use of bang-bang control results in large power fluctuation. For these problems, an improved DPC strategy is proposed in [119] with fixed switching frequency based on virtual flux and SVM. This method simplifies the circuit structure by omitting the voltage sensors and obtains continuous control by SVM instead of switching table.

7.5.2.2 Closed Loop PWM Methods: Hysteresis Controller

Hysteresis control is a widely used current control technique in the DSTATCOM applications because of its ease of implementation, fast dynamic response, and inherent peak current-limiting capability [175]. The basic principle of current hysteresis control technique is that the switching signals are derived from the comparison of the current error signal with a fixed width hysteresis band. To bring the switching frequency to acceptable level for practical devices, hysteresis logic with suitable width (h) is required. The various switching schemes of hysteresis controller to achieve the desired switching frequency are two-level switching [273,274], basic three-level switching [273,274], and forced switching [274].

7.5.2.2.1 Second and Higher Order Systems

Ramp comparison controller: The ramp comparison current controller also called as the stationary controller uses three proportional-integrator controllers to provide a high DC gain, which eliminates steady-state errors and provides a controlled high-frequency response. In this scheme, comparison with the triangular carrier signal generates the control signals (switching functions) [195,275,276].

Predictive and deadbeat controllers: This technique predicts the current error vector on the basis of the present error and the AC side parameters at the beginning of each sampling period. The voltage vector to be generated by PWM during the next sampling period is determined to minimize the forecast error [275]. Usually, predictive current controllers [277] have been known as the deadbeat controller with a high bandwidth. Most predictive current control methods have shown high performances in RL-based inverter model. The grid-connected inverter needs to interface with LCL filter, but it is very sensitive in terms of resonance problem. In the proposed controller, the current predictive algorithm is used to reduce the overshoot phenomenon of the input current caused by the delay controller. The compensating current reference that is necessary for the deadbeat control is estimated by using the current predictive algorithm. This method in [120] can improve the accuracy of compensation and ensure the fast, dynamic response of SAPF. In this paper [121], a robust predictive current control scheme is proposed for the grid-connected inverter with LCL filter. Three-stage deadbeat control is designed to decouple the discrete inverter model. In order to minimize the resonant behavior during the grid connection, the predictive active damping method is added with the phase and gain compensations. In addition, a DQ-reference frame filler-based harmonics prediction provides enhanced current regulation under the grid distortion. Finally, the proposed control method is able to obtain fast controller responses as well as robust harmonic compensation through predictive approach.

1. **Constant Switching Frequency Predictive Algorithm**: In this case, the predictive algorithm calculates the voltage vector commands once every sample period. This will force the current vector according to its command. The inverter voltage and EMF voltage of the load are assumed to be constant over the sample period. The calculated voltage vector is then implemented in the PWM modulator algorithm, for example, space vector. Note that, while the current ripple is variable, the inverter switching frequency is fixed. The disadvantage of this algorithm is that it does not guarantee the inverter peak current limit [195,276,278–280]. The predictive control method of APF [122] is based on the DPC algorithm and the current error vector minimization criteria. This non-linear method with vector modulation ensures constant switching frequency and grid current error vector minimization without deterioration of very good dynamic properties. Constant switching frequency provides a simple way to eliminate (using LC passive filter) the current higher harmonics associated with switching transistors.

2. **Deadbeat Controllers**: When the choice of the voltage vector is made to null the error at the end of the sample period, the predictive controller is often called a deadbeat controller. Among the additional information given to the controller, non-available state variables can be included. Their determination can require the use of observers or other control blocks, which often may be shared with the control of the entire scheme [186,201,276,278,281].
3. **Linear Quadratic Regulator (LQR)**: The basic principle of LQR involves choosing the positive definite state and control input matrices, Q and R that provide satisfactory closed-loop performance. The closed-loop eigenvalues are related to these weighting matrices. The LQR control not only ensures the stability of a closed loop system, but it is also a controller with robustness features, because of the wide gain and phase margins. Theoretically, it can provide a gain margin between −6 and +∞ dB and a phase margin between −60° and +60°. Besides, it provides a low sensibility of the complete system. Another important feature of the LQR is that it can tolerate the input non-linearities [206,174,279,282,283].
4. **Sliding Mode (SM) Controller**: SM can be used to achieve robust system performance against parameter variations and load disturbances. SM control scheme performs a high-speed switching control law to drive the state trajectory of the plant onto a specified surface in the state space, and to keep the state trajectory on this surface for all subsequent time. When sliding on the sliding surface, the structure of the system is changed discontinuously according to the instantaneous values of the system states evaluated along the trajectory. Owing to the discontinuous change of the structure of the system, SM control system is insensitive to the parameter changes of the plant and the external disturbances. With the inherent robustness and switching characteristics of SM control, it is especially suitable in the applications of closed-loop regulation of DSTATCOM [267,280,284].
5. **Pole Shift Controller**: This is a discrete-time control technique in which the open-loop system poles are radially shifted towards the more stable locations to form the closed-loop poles. The pole-shift control algorithm utilizes the on-line updated auto-regressive moving average model parameters to determine the new closed-loop poles of the system that are always inside the unit circle in the z-plane. It achieves regulation of the system to a constant set-point in the shortest interval of time [185,264,285].
6. **Resonant Current Controllers**: A new current control scheme for selective harmonic compensation is proposed for SAPFs. The method employs an array of resonant current controllers, one for the fundamental and one for each harmonic, implemented in fundamental reference frame in order to reduce the overall computational effort. The proposed controller design is based on the pole-zero cancellation technique, considering the load transfer function at each harmonic frequency. Two design methods are provided, which give controller transfer functions with superior frequency response. The complete current controller is realized as the superposition of all individual harmonic controllers. The frequency response of the entire closed loop control is optimal with respect to filtering objectives, i.e., the system provides good overall stability and excellent selectivity for interesting harmonics. The resonant controllers [123] are applied for active filters to enhance the harmonic elimination and making waveform sinusoidal.

7.5.2.3 Selective Harmonic Elimination PWM

SHE-PWM is based on fundamental frequency switching method and realized to eliminate the defined harmonic orders. This method defines the switching angles of harmonic orders to eliminate and obtain the Fourier series expansion of output voltage. Basically, in SHE-PWM, the harmonic components of the predefined switched waveform with the unknown switching angles are brought to zero for those undesired harmonics, while the fundamental component is kept to the desired reference amplitude. SHE-PWM is a very attractive option for multilevel inverter applications, because the equipment requires operating at a very low switching frequency to decrease the power switches losses [198,199,276,286–290]. Much deeper analysis about selected harmonic reduction is available in the next chapter.

7.5.3 Main Circuits

A. Voltage-Source Active Power Filter

The main circuit of a voltage-source ShAF was presented in Figure 7.11. The PWM bridge consists of six controllable switches (IGBTs) with antiparallel diodes. The current stresses of the semiconductor devices are limited to the peak value of the compensating current and the devices have to withstand the unipolar DC link voltage as the switching voltage [11]. A typical DC link voltage in active filter applications used in 400-V mains is approximately 700–750 V, while the theoretical minimum for the voltage is the peak of the supply line-to-line voltage, i.e., 565 V.

The voltage-source topology requires the filter to be placed between the supply and the PWM bridge. The filter has usually either first order (L) or third-order (LCL) structure. Moreover, the filter makes it possible to control the currents. In the DC link there is an electrolytic capacitor with a DC voltage as energy storage. The limited lifetime of the electrolytic capacitor can be considered a disadvantage. The DC link voltage should be so high that the filter currents can be controlled to draw the load current harmonics through the supply filter.

B. Current-Source Active Power Filter

The current-source active filter (Figure 7.12) PWM bridge is built with six controllable unidirectional switches. They have to withstand the active filter DC link current. The semiconductor devices are under bipolar voltage stresses and the maximum values of these are the peak value of the supply filter capacitor line-to-line voltage [11], which in steady state is nearly equal to supply line-to-line voltage. This is 565 V in the case of 400-V mains. Figure 7.1b shows the antiparallel diodes of the commercial IGBT power modules. Because of these and the very low reverse voltage blocking capability of the IGBTs, additional diodes have to be connected in series with the transistors. Instead of the series connection the use of the RB–IGBTs discussed previously would also be possible. The switching devices have to be protected against over voltages with a separate clamp circuit. This is also shown in Figure 7.11.

The overvoltage spikes are caused by supply network and supply filter when the supply voltages are switched on or when the control signals are removed from the active filter, by the DC link inductance when the control signals are removed from the active bridge, and by the

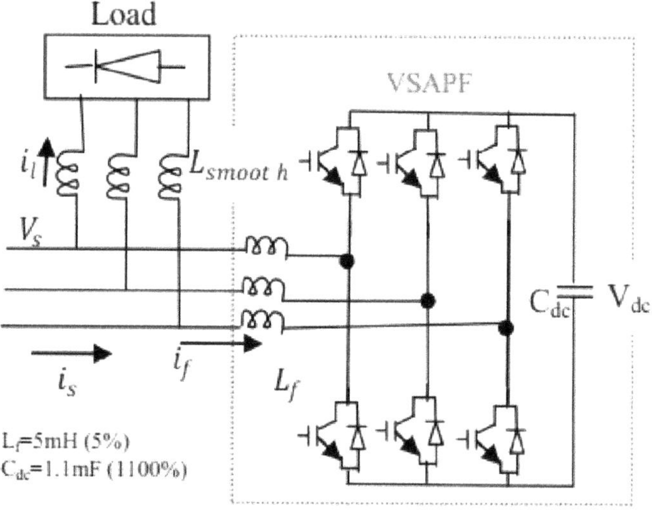

FIGURE 7.11 Voltage-source shunt active filter.

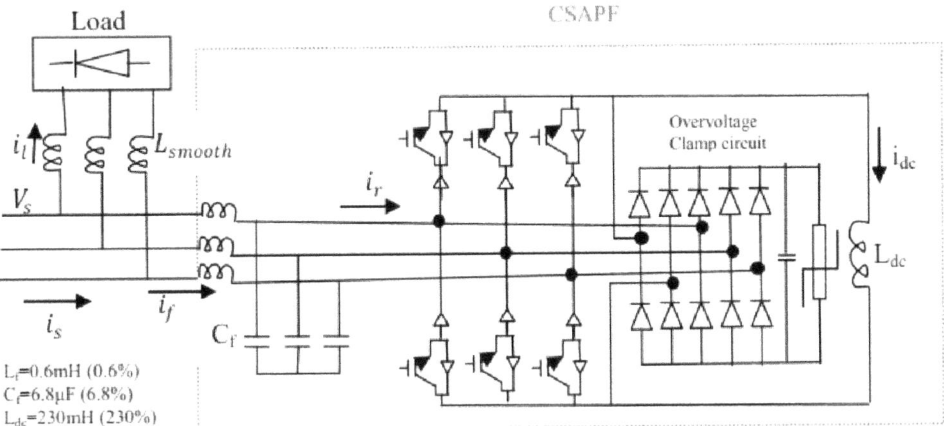

FIGURE 7.12 Current-source active filter.

inductances of the bridge circuit themselves when the current is commutating from one phase leg to another. The PWM bridge of the current-source APF is connected to mains through the second order filter, which filters the carrier frequency components from the PWM currents. As the energy storage of the current-source APF there is an inductor L_{dc} with a DC current flowing through it. The current should be at least as high as the peak value of the compensating current. The theoretical minimum of the current is zero. Although the lifetime of the DC coil is not limited compared to that of the electrolytic capacitor in the VSAPF, the coil is a bulky and heavy component.

7.5.3.1 Space-Vector Modulation

Various vector modulation methods for voltage and current source PWM systems have been presented, e.g., in [12,13]. This section discusses the methods used in this study. In the figures presented, the following notations are used: "+++" refers to the on-state of the upper switch and "---" to the on-state of the lower switch while "0" indicates that both of the switches in a phase are in off-state and "++-" that both of the switches in a phase are on. Sectors are labeled with Roman numerals.

A. ***Modulation of a Voltage-Source PWM Bridge***

The voltage vectors of the voltage-source PWM bridge are shown in Figure 7.13. Since the bridge is controlled so that three switches are simultaneously in on-state, it is possible to realize six active vectors and two zero vectors. The switching combinations to produce each vector are also presented in Figure 7.13a. The modulation method applied is called asymmetric regular sampled space-vector modulation [12]. In the method, the voltage reference vector is produced on average during a modulation period by using two active vectors and both zero vectors. To ensure precise operation, the switching durations are calculated and updated twice in the PWM carrier period. In the modulation strategy applied, the two active vectors next to the reference vector and both zero vectors are used during each period. Every carrier period begins with the zero vectors and this is followed by the active vector that is adjacent to the reference vector in the sector. The active vector is chosen so that the state of switches in only one phase needs to be changed at a time. After that the other active vector in the sector is applied and finally the zero vector. Then, the new switching times are calculated and in the latter part of the carrier period the vectors are applied in inverse sequence. Figure 7.13a illustrates the switching sequence if the reference vector lies in sector I. To prevent the short-circuiting of the DC link capacitor both upper and lower switches of the phase are not allowed to be simultaneously in the on-state. In practical implementations there

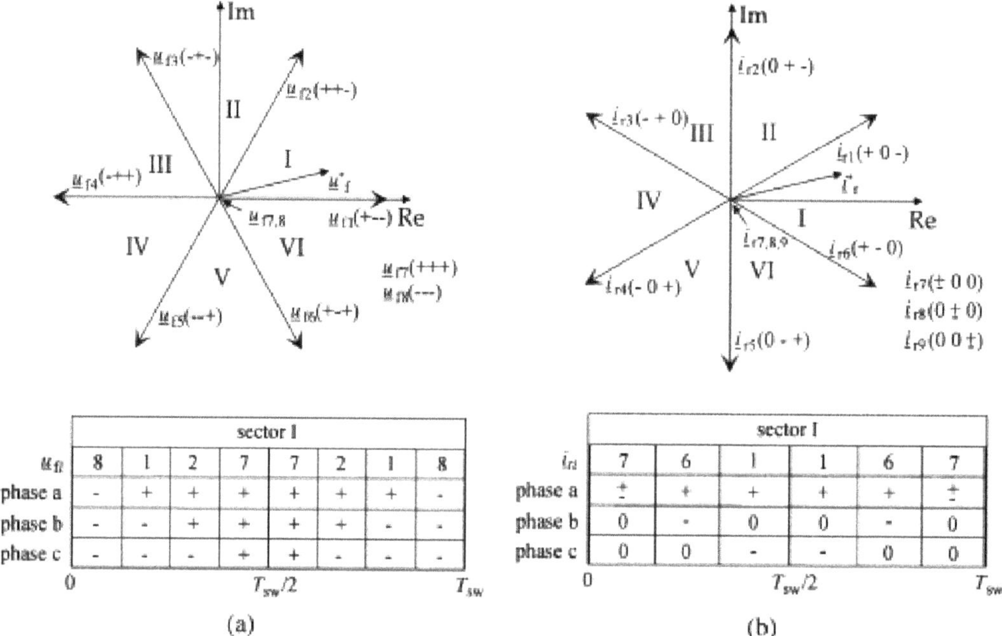

FIGURE 7.13 Switching vectors and applied sequence of vectors when the reference vector lies in sector I in the case of (a) voltage-source PWM bridge and (b) current-source PWM bridge.

is a short time period called dead time between the turn-off of one of the two switching devices in a phase and the turn-on of the other device in the phase. Figure 7.14 shows the control system for (a) the voltage-source APF and (b) the current-source APF.

B. ***Modulation of a Current-Source PWM Bridge***

The current-source PWM bridge modulates unidirectional DC current. The bridge is controlled so that one of the upper switches and one of the lower switches are in on-state at a time. This results in six possible active vectors and three zero vectors. The zero vectors mean that although the DC current continues to flow, this is not circulating through the mains. This is done by short-circuiting the DC link coil by turning on both the upper and the lower switch in a phase. In the so-called halfwave symmetrical space-vector modulation technique [13] applied, the current reference vector is realized on average during a modulation period. This is done using two active switching vectors and one of the zero vectors. The active vectors applied are adjacent to the reference vector and the zero vector is chosen so that one of the switches in the bridge is always in on-state during the half of the carrier period. This way the switching losses can be reduced. As in the modulation of the voltage-source bridge, the new switching times are calculated every. The PWM bridge is not allowed to break the DC current path. To ensure this, instead of the dead time used in the voltage source bridge, overlapping of the modulation signals is needed in practice.

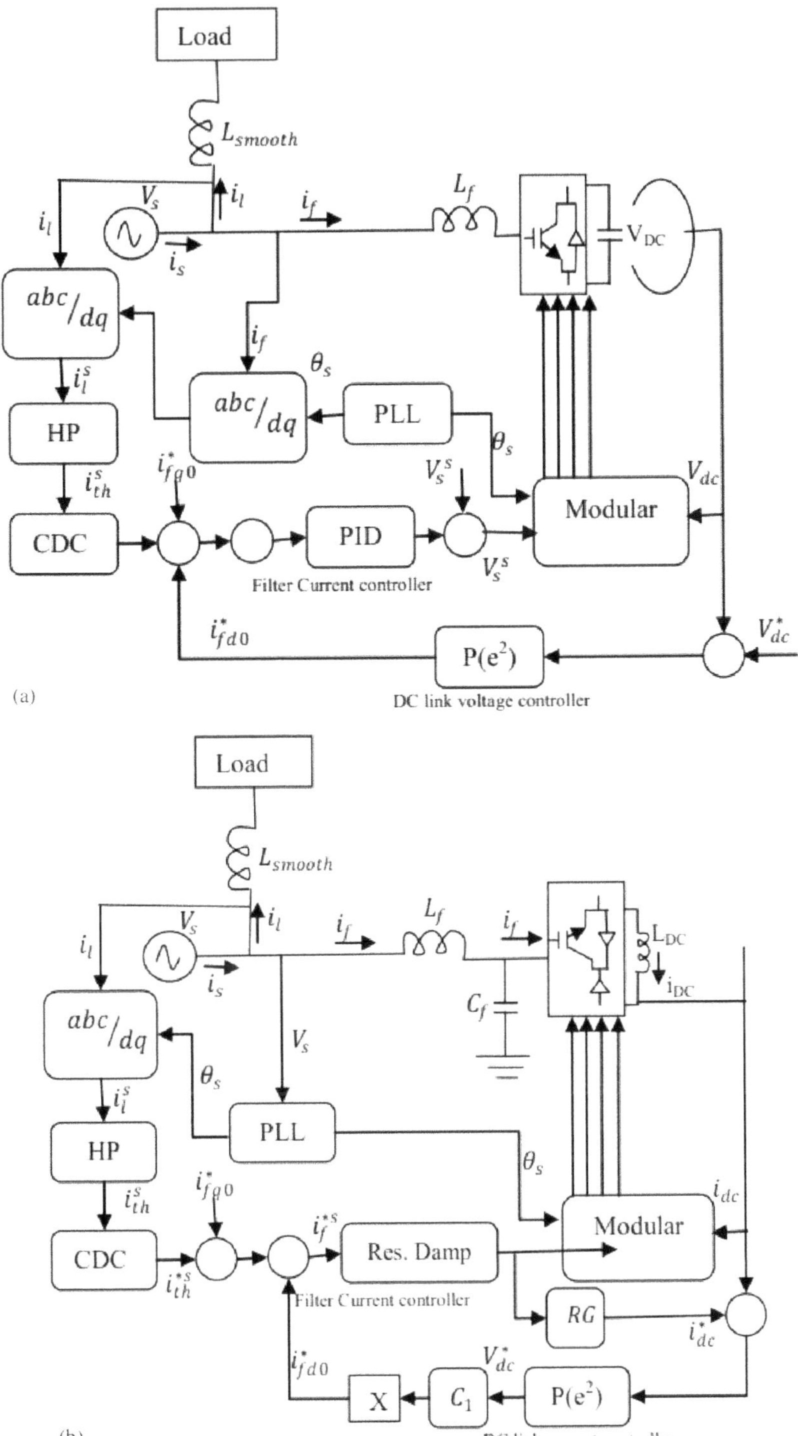

FIGURE 7.14 Control system for (a) the voltage-source Active Power Filter and (b) the current-source Active Power Filter.

7.5.4 Control Systems

The main characteristics of the control systems used are presented in this section. A detailed explanation of the systems used can be found in [14,17]. Block diagrams of the control systems for voltage and current-source active filters are presented. They are implemented in the supply voltage vector-oriented rotating SRF. The reference frame angle is determined with a phase-locked loop (PLL) from the supply voltages. The underlined variables refer to space-vectors, the superscripts refers to the SRF and reference values, respectively. Further, subscripts s, l, f, and r refer to supply, load, active filter, and CSAPF rectifier bridge variables, respectively, and d and q to direct and quadrature axis components of a space-vector in the SRF. Both of the control systems are based on the feed-forward of the load currents. The three-phase load currents (i_l) are measured and transformed into the rotating reference frame with the block "abc/dq" in Figure 7.3(a) and (b). Since the fundamental current component can be seen as a DC quantity in the reference frame, the load current harmonics ($i_{lh}{}^{s}$) are extracted with high-pass filters (blocks "HP"). The control delay compensation block ("CDC") compensates the delay caused by the digital control system. This is done by modifying the signal produced with the load current feed-forward. The method was originally presented in [17]. The output of the block is the reference for the harmonics compensating filter current ($i_{fh}{}^{*s}$).

A. ***Control of a Voltage-Source Active Power Filter***

The active filter current reference vector is the sum of the harmonic compensating current reference vector, the fundamental reactive power controlling axis reference, and the fundamental active current reference. The axis current reference is the output of a non-linear *P* voltage controller used to maintain the DC link voltage. The controller output is proportional to the square of the error [14] and allows a slight deviation between the reference and the actual voltage without affecting the supply currents. The active filter currents are realized with a closed-loop control. The output of the current controller is the reference value for the voltage vector over the filter inductor. The voltage vector reference for the active filter is then calculated by subtracting the voltage vector from the supply voltage vector. Finally, the voltage vector reference is given to the space-vector modulator.

B. **Control of a Current-Source Active Power Filter**

In the current-source active filter the compensating currents are produced by pulse-width modulating of the DC current. As in the voltage-source active filter a constant DC link voltage was maintained; the current-source active filter has to control the DC link current. This is done with a non-linear PI controller [291], which gives the reference value for the pulse-width modulated DC link voltage. The controller used behaves like a conventional PI controller, but the proportional gain varies as a function of the error [291]. The controller allows a slight deviation between the reference and the actual DC current without influence to the supply currents. Since the control system is implemented in the SRF, the fundamental active filter current component is a DC quantity. Thus, the reference for the current is then calculated from this assuming that AC power equals the DC power [17]:

$$i_{fd0}^{*} = \frac{2}{3u_{sd}} u_{dc}^{*} i_{dc} = c_1 u_{dc}^{*} i_{dc}.$$

The active filter current control is performed in open loop manner; thus, no measurement of filter currents is needed. As with the voltage-source APF, the current-source APF current reference vector consists of a harmonic compensating part, plus fundamental reactive and active current references. Since the open loop control of the active filter currents may cause resonance in the filter, this is also damped in an open loop manner by considering the filter effect from the bridge current to the active filter current. This is done in the block "Res. Damp.," which gives the reference for the current vector produced with the PWM bridge [17]. Furthermore, the DC link current reference is generated so that the DC link current is greater than the peak value of the current reference. This is done in the block "RG."

7.6 Summary

A comprehensive literature review on different control strategies of the series and ShAFs is carried out. A flexible control technique for power electronic converters is necessary, which can function as an active power filter, local power supply interface, or both functions simultaneously. Thus, it can compensate for current disturbances while simultaneously injecting active power into the electrical grid, transforming the power converter into a multifunctional device. The main objective is to use all the capacity available in the electronic power converter to maximize the benefits when it is installed in the electricity grid. This study proposes a flexible APF controller operating selectively to satisfy a set of desired load performance indices defined at the source side. The definition of such indices, and of the corresponding current references, is based on the orthogonal instantaneous current decomposition and conformity factors provided by the CPT. This flexible approach can be applied to single- or three-phase APFs or other grid-tied converters, as those interfacing DGs in smart grids. Optimizing the performance of power system networks using conventional methods is quite difficult because of the complex nature of systems that are highly non-linear and non-stationary. Due to the increased usage of loads with power electronic control, utility service providers may enforce more strict power factor and harmonic standards in the future. One of the solutions towards this is to employ APFs.

A comparative evaluation has been made with the aim of exploring and exposing the design, model, and simulation. The compensation of reactive power for power factor correction or voltage regulation, harmonics elimination, load balancing, and neutral current compensation has been demonstrated for three-phase three-wire and three-phase four-wire series and SAPF. The performances of control strategies of series and SAPFs selected from each category have been demonstrated to validate the designed series and parallel active filter system.

According to the developed review, it can be concluded that DSTATCOM or DVR is an effective tool for PQ improvement in distribution systems. The commonly used DSTATCOM/DVR topologies are isolated and non-isolated 3P3W, isolated two-leg and three-leg 3P4W, and non-isolated three-leg/two-leg with and without transformer. The commonly used control techniques are SRF, IRP, SC, PI controller, SMC, NN, and AUPF. A comparative study presented will help the users in selecting the particular topology and control technique of DSTATCOM/DVR that suits a specific application. It is hoped that this review on active filters will be beneficial to the users, designers, manufacturers, researchers, and power engineers for enhancing the quality of power.

Control circuits constitute a minor portion of the total cost of active filters. This is because the new generation of microcontrollers and DSPs can operate at extremely high frequencies and at very low cost. The number of instructions and operations performed per second is phenomenal. Thus, the most complex control requirements can be incorporated without a great deal of concern about this part of the cost for any system. To facilitate understanding and selection of particular configuration and control techniques for a given application, the classification is based on five main criteria. The power-circuit configurations of active filters and the ratings of the compensated systems define the two broad categories. The other three classification criteria are based on the control strategies, control techniques, and reference-estimation methods generally employed. The review also considers the criteria for selecting passive components, and the switching frequencies and losses for the various configurations are also discussed.

It is impossible to guarantee sinusoidal compensated currents without the use of a PLL circuit or a voltage detector that extract, at least, the frequency and the phase angle of the fundamental positive-sequence component of the system voltage. A simple example where the system voltage and load current were already sinusoidal, but unbalanced due to a negative-sequence component, together with a zero-sequence component, was sufficient to prove this necessity.

Topologies and control strategies used in both 3P3W and 3P4W distribution systems are analyzed critically and a comparative study of different types of control strategies is presented. The compensation of neutral currents without the need of energy storage elements introduces harmonics in the source currents that may not be present in the load currents. Hence, the usefulness of this control method in practical cases may become dubious. The control techniques such as IRP, SRF, SC, PI controller, SMC, NN, and AUPF are analyzed and their performance is presented. A comparative study of transformers used in the DSTATCOM/DVR controllers is also presented. Selection considerations of DSTATCOM and DVR control techniques for specific applications have also been outlined.

A strategy to eliminate the neutral current without the need of energy storage elements and based on the p–q theory was proposed. This was useful to demonstrate that the reference power control method based on the p–q–r theory reduces to the same results, with the advantage of avoiding the time-consuming $\alpha\beta 0 \leftrightarrow pqr$ transformations.

Development of new control strategies and execution of multifunctional compensation capability are the main research trends to perform mitigation of various PQ disturbances using active filters. Series or shunt active filters with passive filters arrangement are essentially candidates for the future utility grid and industrial facilities to perform high quality, reliable, and efficient electricity supply. To achieve this goal, various topologies, controllers, and structures should be employed to increase the capacity of installed active filters.

In summary, this chapter evaluates various techniques for current and voltage harmonic reduction, reactive power control, etc. As expected, no single topology or controller is ideal for all applications. However, knowing the application requirements and the costs of the various components, this chapter enables the selection of the best topology/strategy. The review and classification of published work in this field shows that there has been a significant increase in interest in active filters and associated control methods. This is due to increasing concern about power quality and the availability of suitable power-switching devices at affordable prices.

REFERENCES

1. Singh B, Solanki J. A comparative study of control algorithms for DSTATCOM for load compensation. In: *Proceedings of the IEEE International Conference on Industrial Technology*, 2006, pp. 1492–1497.
2. Yang X-p, Zhang Y-x, Zhong Y-r. Three-phase four-wire DSTATCOM based on a three-dimensional PWM algorithm. In: *Proceedings of the International Conference on Electric Utility Deregulation and Restructuring and Power Technologies*, IEEE Publishers, April 2008, pp. 2061–2066.
3. Sao CK, Lehn PW, Iravani MR, Martinez JA. A benchmark system for digital time-domain simulation of a pulse-width-modulated D-STATCOM. *IEEE Trans Power Deliver* 2002;17:1113–1120.
4. Kimura N, Morizane T, Taniguchi K, Nishida Y. Modified double-modulation signal PWM control for D-STATCOM using five level double converter. In: *Proceedings of the EPE'2007*, IEEE publishers, September 2007, pp. 1–10.
5. Sun J, Czarkowski D, Zabar Z. Voltage flicker mitigation using PWM-based distribution STATCOM. In: *Proceedings of the IEEE PES Summer Meeting,* July 2002, Vol. 1, pp. 616–621.
6. Jurado F and Valverde M. Fuzzy logic control of a dynamic voltage restorer. In: *Proceedings of the IEEE International Symposium*, May 2004, Vol. 2, pp. 1047–1052.
7. Fernandes DA and Naidu SR. A novel PWM scheme for the 4-leg voltage source converter and its use in dynamic voltage restoration. In: *Proceedings of the IEEE PES General Meeting*, 2007, pp. 1–5.
8. Hong-Je R, Geun-Hie R, Tae-Jin K, Kisck DO. Digital-controlled single-phase unified power quality conditioner nonlinear and voltage sensitive load. In: *Proceedings of the IEEE-IECON'04*, November 2004, Vol. 1, pp. 24–29.
9. Xu S, Song Q, Zhu Y, Liu W. Development of a D-STATCOM prototype based on cascade inverter with isolation transformer for unbalanced load compensation. In: *Proceedings of the IEEE International Conference on Industrial Technology*, December 2005, pp. 6.

10. Changjiang Z, Arulampalam A, Jenkins N. Four-wire dynamic voltage restorer based on a three-dimensional voltage space vector PWM algorithm, *IEEE Trans Power Electron* 2003;18:1093–1120.
11. Changjiang Z, Ramachandaramurthy VK, Arulampalam A, Fitzer C, Kromlidis S, Bames M, Jenkins N. Dynamic voltage restorer based on voltage-space-vector PWM control. *IEEE Trans Ind Appl* 2001;37:1855–1863.
12. Elnady A, El-Khattam W, Salama MMA. Mitigation of AC arc furnace voltage flicker using the unified power quality conditioner. In: *Proceedings of the IEEE PES Winter Meeting*, January 2002, Vol. 2, pp.735–739.
13. Shukla A, Ghosh A, Joshi A. A hysteresis current controlled flying capacitor multilevel inverter based DSTATCOM. In: *Proceedings of the IEEE Power Engineering Society General Meeting*, 2005, Vol. 1, pp. 857–864.
14. Jowder FAL. Modeling and simulation of dynamic voltage restorer (DVR) based on hysteresis voltage control. In: *Proceedings of the IEEE Conference on Industrial Electronics Society*, 2007, pp. 1726–1731.
15. Tey LH, So PL, Chu YC. Unified power quality conditioner for improving power quality using ANN with hysteresis control. In: *Proceedings of the PowerCon 2004*, November 2004, Vol. 2, pp. 1441–1446.
16. Li S-H, Liaw C-M. On the DSP-based switch-mode rectifier with robust varying-band hysteresis PWM scheme. *IEEE Trans Power Electron* 2004;19(6):1417–1425.
17. Gosh A, Jindal AK, Joshi A. Inverter control using output feedback for power compensating devices. In: *Proceedings of the TENCON 2003*, Bangalore, India; IEEE publishers, October 2003, vol. 1, pp. 48–52.
18. Errabelli RR, Kolhatkar YY, Das SP. Experimental investigation of DVR with sliding mode control. In: *Proceedings of the IEEE Conference on Power India*, 2006, pp. 5.
19. Ledwich G. Linear switching controller convergence. *IEE Proc Control Theory Appl* 1995;142(4):329–334.
20. Hyosung K, Jun-Keun J, Jang-Hwan K, Seung-Ki S, KyungHwan K. Novel topology of a line interactive UPS using PQR instantaneous power theory. In: *Proceedings of the IEEE Industry Applications Conference*, 2004, October 2004, Vol. 4, pp. 2232–2238.
21. Rai D, Gokaraju R, Faried SO. Adaptive control using constrained RLS and dynamic pole-shift technique for TCSCs. *IEEE Trans Power Del* 2014;29(1):224–234.
22. Mishra MK, Ghosh A, Joshi A. Operation of a DSTATCOM in voltage control mode, *IEEE Trans Power Deliver* 2003;1 8:258–264.
23. Deng H, Oruganti R, Srinivasan D. A multi-layer neural network controller for single-phase inverters. In: *Proceedings of the IEEE PEDS'2003*, pp. 370–375.
24. Ghosh A, Jindal AK, Joshi A. Design of a capacitor-supported dynamic voltage restorer (DVR) for unbalanced and distorted loads. *IEEE Trans Power Deliver* 2004;19:405–413.
25. Kamran F, Habetler TG. Combined deadbeat control of a series-parallel converter combination used as a universal power filter. *IEEE Trans Power Electron* 1998;13(1):160–168.
26. Beides HM, Heydt GT. Dynamic state estimation of power system harmonics using kalman filter methodology. *IEEE Trans Power Del* 1991;6(4):1663–1670.
27. Gonzalez M, Cardenas V, Alvarez R. Detection of sags, swells, and interruptions using the digital RMS method and Kalman filter with fast response. In: *Proceedings of IEEE IECON'06*, November 2006, pp. 2249–2254.
28. Kwan KH, So PL, Chu YC. A harmonic selective unified power quality conditioner using MVR with Kalman filters. In: *Proceedings of the IPEC 2007*, December 2007, pp. 332–337.
29. Chakraborty S, Weiss MD, Simoes MG. Distributed intelligent energy management system for a single-phase high-frequency AC microgrid. *IEEE Trans Ind Electron* 2007;54(1):97–109.
30. Singh B, Solanki J, Verma V. Neural network based control of DSTATCOM with rating reduction for three-phase four-wire system. In: *Proceedings of the Conference on IEEE-PEDS'05*, November 2005, Vol. 2, pp. 920–925.
31. Zhang Y, Milanovic JV. Application of niching genetic algorithms in system-wide voltage sag mitigation studies. In: *Proceedings of the IEEE Lausanne on Power Technology*, July 2007, pp. 1515–1521.
32. Fitzer C, Barnes M, Green P. Voltage sag detection technique for a dynamic voltage restorer. *IEEE Trans Ind Appl* 2004;40:203–212.
33. Yanlei Z. Design and implementation of inverter in dynamic voltage restorer based on selective harmonic elimination PWM. In: *Proceedings of the International Conference on Electric Utility Deregulation and Restructuring and Power Technologies*, 2008, pp. 2239–2244.

34. Chang GW, Shee TC. A novel reference compensation current strategy for shunt active power filter control. *IEEE Trans Power Deliver* 2004;19(4):1751–1758.
35. Forghani M, Afsharnia S. Online wavelet transform-based control strategy for UPQC control system. *IEEE Trans Power Deliver* 2007;22:481–491.
36. Changjiang Z, Ramachandaramurthy VK, Arulampalam A, Fitzer C, Barnes M, Jenkins N. Universal custom power conditioner (UCPC) with integrated control. In: *Proceedings of the IEEE Power Engineering Society Winter Meeting*, 2001, February 2001, Vol. 3, pp. 1039–1044.
37. Reviriego P, Ruano O, Antonio Maestro J. Implementing concurrent error detection in infinite-impulse-response filters. *IEEE Trans Circuits Syst II: Exp Briefs* 2012;59(9):583–586.
38. Zhuo F, Choi San S, Sng Eng Kian K, Wang T, Wang Z. Study on the dynamic voltage restorer use in an isolated power supply system. In: *Proceedings of the International Conference Power Electronics and Motion Control Conference*, 2004, Vol. 3, pp. 1207–1212.
39. Zhi LX, Yin ZD, Ding H, Han JB, Hu FX. Study on energy saving strategies for dynamic voltage restorer. In: *Proceedings of the International Conference on Power System Technology*, 2006, pp. 1–5.
40. Correa JM, Farret FA, Simoes MG. Application of a modified single-phase *P–Q* theory in the control of shunt and series active filters in a 400 Hz microgrid. In: *Proceedings of the IEEE Conference on Power Electronics Specialists Conference*, 2005, pp. 2585–2591.
41. Haque MT, Ise T, Hosseini SH. A novel control strategy for unified power quality conditioner (UPQC). In: *Proceedings of the IEEE on PESC'02*, June 2002, Vol. 1, pp. 94–98.
42. Lee S-J, Kim H, Sul S-K, Blaabjerg F. A novel control algorithm for static series compensators by use of PQR instantaneous power theory. *IEEE Trans Power Electron* 2004;19(3):814–827.
43. Fan N, Man-Chung W, Ying-Duo H. Analysis and control of UPQC and its DC-link power by use of *p–q–r* instantaneous power theory. In: *Proceedings of the Power Electronics Systems & Application,* 2004, pp. 43–53.
44. Zhili T, Xun L, Jian C, Yong K, Shanxu D. A direct control strategy for UPQC in three-phase four-wire system. In: *Proceedings of the IEEE Conference on Power Electronics and Motion Control*, 2006, Vol. 2, pp. 1–5.
45. Xiangyun F, Jianze W, Yanchao J. A novel control method for DSTATCOM under unbalanced conditions. In: *Proceedings of the International Conference on Power System Technology,* 2006, October 2006, pp. 1–6.
46. Songcen W, Guangfu T, Kunshan Y, Jianchao Z. Modeling and control of a novel transformer-less dynamic voltage restorer based on H-bridge cascaded multilevel inverter. In: *Proceedings of the International Conference on Power System Technology*, 2006, pp. 1–9.
47. Boonchiam P, Mithulananthan N. Dynamic control strategy in medium voltage DVR for mitigating voltage sags/swells. In: *Proceedings of the International Conference on Power System Technology*, 2006, pp. 1–5.
48. Reyes JR, Espinoza JR, Sepulveda CA. Operating region of single-phase UPQCs. In: *Proceedings of the IEEE PESC*, 2005, pp. 1726–1731.
49. Hamadi A, Al-Haddad K, Rahmani R. Series active filter to mitigate power quality for medium size industrial loads. *IEEE Int Conf Ind Electron* 2006;2:1510–1515.
50. Mishra MK, Karthikeyan K, Linash PK. A development and implementation of DSP based DSTATCOM to compensate unbalanced nonlinear loads. In: *Proceedings of the IEEE Power India Conference 06*, April 2006, pp. 8.
51. Marei MI, El-Saadany EF, Salama MMA. A new approach to control DVR based on symmetrical components estimation. *IEEE Trans Power Deliver* 2007;22:2017–2024.
52. Vadirajacharya K, Agarwal P, Gupta HO. A simple control strategy for unified power quality conditioner using current source inverter. In: *Proceedings of the International Conference IPEC 2007*, December 2007, pp. 1219–1223.
53. Singh B, Adya A, Mittal AP, Gupta JRP. Analysis, simulation and control of DSTATCOM in three-phase, four-wire isolated distribution systems. In: *Proceedings of the IEEE Power India Conference*, April 2006.
54. Khadkikar V, Chandra A, Barry AO, Nguyen TD. Application of UPQC to protect a sensitive load on a polluted distribution network. In: *IEEE Power & Energy Society General Meeting*, June 2006.

55. Khadkikar V, Chandra A, Barry AO, Nguyen TD. Analysis of power flow in UPQC during voltage sag and swell conditions for selection of device ratings. In: *Canadian Conference on Electrical and Computer Engineering*, May 2006, pp. 867–872.
56. Chen G, Chen Y, Smedley KM. Three-phase four-leg active power quality conditioner without references calculation. In: *Proceedings of the IEEE APEC '04*, Anaheim, CA, USA, February 2004, IEEE Publisher, Vol. 1, pp. 587–593.
57. Liu L, Li H, Xue Y, Liu W. Reactive power compensation and optimization strategy for grid-interactive cascaded photovoltaic systems. *IEEE Trans Power Electron* 2015;30(1):188–202.
58. Chaudhari KR, Trivedi TA. Analysis on control strategy of shunt active power filter for three-phase three-wire system. In: *Transmission & Distribution Conference and Exposition: Latin America (PES T&D-LA)*, 2014 IEEE PES, Medellin, 2014, pp. 1–6.
59. Qasim M, Khadkikar V. ADALINE based control strategy for three-phase three-wire UPQC system. In: *2014 16th International Conference on Harmonics and Quality of Power* (*ICHQP*), Bucharest, 2014, pp. 586–590.
60. Prasai A, Divan D. Zero energy sag correctors–Optimizing dynamic voltage restorers for industrial applications. In: *Proceedings of the IEEE IAS Annual Meeting*, September 2007, pp. 1585–1592.
61. Woodley NH, Morgan L, Sundaram A. Experience with an inverter-based dynamic voltage restorer. *IEEE Trans Power Deliver* 1999;14:1181–1186.
62. Peel MK, Sundaram A, Woodley N. Power quality protection using a platform-mounted SCD-demonstration project experience. In: *Proceedings of the IEEE Industrial and Commercial Power Systems Technical Conference*, May 2000, pp. 133–139.
63. Vilathgamuwa DM, Wijekoon HM, Choi SS. Interline dynamic voltage restorer: A novel and economical approach for multiline power quality compensation. *IEEE Trans Ind Appl* 2004;40:1678–1685.
64. Chan K. Technical and performance aspects of a dynamic voltage restorer. In: *IEE Proceedings of the Dynamic voltage Restorers–Replacing Those Missing Cycles*, February 1998, pp. 5/1–5/25.
65. Karshenas HR, Moradlou M. Design strategy for optimum rating selection in interline DVR. In: *CCECE 2008*, May 2008, pp. 1919–1924.
66. Mahesh SS, Mishra MK, Kumar BK, Jayashankar V. Rating and design issues of DVR injection transformer. In: *Proceedings of the IEEE APEC*, February 2008, pp. 449–455.
67. Kisck DO, Navrapescu V, Kisck M. Single-phase unified power quality conditioner with optimum voltage angle injection for minimum VA requirement. In: *Proceedings of the IEEE PESC*, 2007, pp. 574–579.
68. Khoor MS, Machmoum M. A novel single-phase reduced parts on-line UPS with power quality conditioning capability. In: *Proceedings of the IEEE PESC*, 2007, June 2007, pp. 1170–1175.
69. Renders B, De Gusseme K, Ryckaert WR, Stockman K, Vandevelde L, Bollen MHJ. Distributed generation for mitigating voltage dips in low-voltage distribution grids. *IEEE Trans Power Del* 2008;23(3):1581–1588.
70. Faranda R, Valade I. UPQC compensation strategy and design aimed at reducing losses. In: *Proceedings of the IEEE ISIE'02*, Vol. 4, pp. 1264–1270.
71. Jiang X, Xiao X, Liu H, Ma Y. The output spectrum analysis of high-power multilevel voltage source converters using double Fourier series. In: *IEEE IPES Conference and Exhibition on Transmission and Distribution, Asia and Pacific*, 2005, pp. 1–5.
72. Hill JE. A practical example of the use of distribution static compensator (D-STATCOM) to reduce voltage fluctuations. In: *Proceedings of the IEE Colloquium on Power Electronics for Renewable Energy* (Digest No: 1997/1 70), June 1997, pp. 7/1–7/5.
73. Krasselt P, Weck S, Leibfried T. Voltage-based harmonic compensation using MCCF state estimation. In: *Proceedings of the IEEE Innovative Smart Grid Technologies: Asia, Kuala Lumpur*, Malaysia, May 20–23, 2014.
74. Ghosh A, Ledwich G. Compensation of distribution system voltage using DVR. *IEEE Trans Power Deliver* 2002;17:1030–1036.
75. Jindal AK, Ghosh A, Joshi A. Voltage regulation using dynamic voltage restorer for large frequency variations. In: *Proceedings of the IEEE Power Engineering Society General Meeting*, 2005, Vol. 1, pp. 850–856.

76. Basu M, Das SP, Dubey GK. Performance study of UPQC-Q for load compensation and voltage sag mitigation. In: *Proceedings of the IEEEIECON02*, November 2002, Vol. 1, pp. 698–703.
77. Jayanti NG, Basu M, Michael Conlon F, Gaughan K. Optimising the rating of the UPQC for applying to the fault ride through enhancement of wind generation. In: *Proceedings of the UPEC '06*, September 2006, Vol. 1, pp. 123–127.
78. Basu M, Das SP, Dubey GK. Performance study of UPQC-Q for load compensation and voltage sag mitigation. In: *Proceedings of the IEEEIECON02*, November 2002, Vol. 1, pp. 698–703.
79. Bojoi RI, Limongi LR, Roiu D, Tenconi A. Enhanced power quality control strategy for single-phase inverters in distributed generation systems. *IEEE Trans Power Electron* 2011;26(3):798–806.
80. Ghosh A, Jindal AK, Joshi A. A unified power quality conditioner for voltage regulation of critical load bus. In: P*roceedings of the IEEE Power & Energy Society General Meeting,* June 2004, Vol. 1, pp. 471–476.
81. Jindal AK, Ghosh A, Joshi A. Interline unified power quality conditioner. *IEEE Trans Power Deliver* 2007;22:364–372.
82. Singh B, Verma V, Chandra A, Al-Haddad K. Hybrid filters for power quality improvement. *Proc IEE Gener Transmission Distrib* 2005;152:365–378.
83. Senini ST, Wolfs PJ. Systematic identification and review of hybrid active filter topologies. In: *Proceedings of the IEEE PESC*, 2002, pp. 394–399.
84. Guide for Application of Power Electronic Guide for Application of Power Electronics for Power Quality Improvement on Distribution Systems Rated kV through 38 kV, IEEE P1409 Distribution Custom Power Task Force, 2003.
85. Custom Power-State of the Art Cigre WG14.31, 2000.
86. Montero M, Cadaval ER, Gonzalez F. Comparison of control strategies for shunt active power filters in three-phase four-wire systems. *IEEE Trans Power Electron* 2007;22:229–236.
87. George S, Agarwal V. A DSP based optimal algorithm for shunt active filter under non sinusoidal supply and unbalanced load conditions. *IEEE Trans Power Electron* 2007;22:593–601.
88. Chang GW, Yeh C.M. Optimisation-based strategy for shunt active power filter control under non-ideal supply voltages. *IEE Proc Electr Power Appl* 2005;152:182–90.
89. Longhui W, Fang Z, Pengbo Z, Hongyu L, Zhaoan W. Study on the influence of supply-voltage fluctuation on shunt active power filters. *IEEE Trans Power Deliver* 2007;22(3):1743–1749.
90. Heydt GT, Fjeld PS, Liu CC, Pierce D, Tu L, Hensley G. Applications of the windowed FFT to electric power quality assessment. *IEEE Trans Power Del* 1999;14(4):1411–1416.
91. Massoud AM, Finney SJ, Williams BW. Review of harmonic current extraction techniques for an active power filter. In: *11th International Conference on Harmonics Quality Power*, September 12–15, 2004, pp. 154–159.
92. Puengsungwan S, Kumhom P, Chamnongthai K, Chaisawadi A, Lasseter RH. Adaptive-based hybrid series active filter to compensate voltage sags and current harmonics for residential equipment. In: *TENCON 2006–2006 IEEE Region 10 Conference*, Hong Kong, 2006, pp. 1–3.
93. Lee J-H, Jeong J-K, Han B-M, Bae B.Y. New reference generation for a single-phase active power filter to improve steady state performance. *J Power Electron* 2010;10(4):412–418.
94. Ramos Fuentes GA, Cortés-Romero JA, Zou Z, Costa-Castelló R, Zhou K. Power active filter control based on a resonant disturbance observer. *IET Power Electron* 2015;8(4):554–564.
95. Wang H, Liu S. Adaptive Kalman filter for harmonic detection in active power filter application. In: *Electrical Power and Energy Conference (EPEC)*, 2015 IEEE, London, UK, 2015, pp. 227–232.
96. Panigrahi R, Subudhi B, Panda PC. Model predictive-based shunt active power filter with a new reference current estimation strategy. *IET Power Electron* 2015;8(2):221–233.
97. Guzman R, de Vicuna L, Morales J, Castilla M, Miret J. Model-based control for a three-phase shunt active power filter. *IEEE Trans Ind Electron* 2016;63(7):3998–4007.
98. Driesen J, Belmans R. Active power filter control algorithms using wavelet-based power definitions. In: *Harmonics and Quality of Power, 10th International Conference*, 2002, Vol. 2, pp. 466–471.
99. Chandrasekar P, Kamaraj V. Integer lifting wavelet transform based hybrid active filter for power quality improvement. In: *Electrical Energy Systems (ICEES), 2011 1st International Conference*, Newport Beach, CA, 2011, pp. 103–107.
100. Javadi A, Hamadi A, Al-Haddad K. Three-phase power quality device tor weak systems based on SRF and p-q theory controller. In: *Industrial Electronics Society, IECON 2015 – 41st Annual Conference of the IEEE*, Yokohama, Tokyo, Japan, 2015, pp. 000345–000350.

101. Merabet L, Saad S, Omeiri A, Ould Abdeslam D. A comparative study of harmonic current identification for active power filter. In: *2012 First International Conference on Renewable Energies and Vehicular Technology*, Hammamet, TN, 2012, pp. 366–371.
102. Bhonsle AD, Bhonsle DC. Performance evaluation of hybrid active filter for power quality improvement of electrical distribution system. In: *Smart Energy Grid Engineering (SEGE), 2015 IEEE International Conference*, Oshawa, ON, 2015, pp. 1–9.
103. Bhonsle DC, Kelkar RB. Mathematical modeling of composite filter for power quality improvement of electric arc furnace distribution network. In: *2014 IEEE 6th India International Conference on Power Electronics (IICPE)*, Kurushetra, India, 2014, pp. 1–5.
104. Senturk OS, Hava AH. High-performance harmonic isolation and load voltage regulation of the three-phase series active filter utilizing the waveform reconstruction method. *IEEE Trans Ind Appl* 2009;45(6):2030–2038.
105. Senturk OS, Hava AM. High-performance harmonic isolation and load voltage regulation of the three-phase series active filter utilizing the waveform reconstruction method. *IEEE Trans Ind. Appl* 2009;45(6):2030–2038.
106. Senturk OS, Hava AM. Performance enhancement of the single-phase series active filter by employing the load voltage waveform reconstruction and line current sampling delay reduction methods. *IEEE Trans Power Electron* 2011; 26(8):2210–2220.
107. Han B-M, Bae B-Y, Ovaska SJ. Reference signal generator for active power filters using improved adaptive predictive filter. *IEEE Trans Ind Electron* 2005;52(2):576–584.
108. Singh B, Arya SR, Jain C, Goel S, Chandra A, Al-Haddad K. Four leg VSI based active filter in distribution system. In: *Power Systems Conference (NPSC), 2014 Eighteenth National*, Guwahati, India, 2014, pp. 1–6.
109. Suresh Y, Panda AK, Suresh M. Real-time implementation of adaptive fuzzy hysteresis-band current control technique for shunt active power filter. *IET Power Electron.* 2012;5(7):1188–1195.
110. Jianlin L, Changsheng H, Liqiao W, Zhongchao Z. APF based on multilevel voltage source cascade converter with carrier phase shifted SPWM [active power filter]. In: *TENCON 2003. Conference on Convergent Technologies for the Asia-Pacific Region*, 2003, Vol. 1, pp. 264–267.
111. Yafang W, Juping G, Ruixiang C, Ling Q, Juan C. The multi-modular shunt APF based on direct current control and frequency doubling carrier phase-shifted SPWM. In: *ECCE Asia Downunder (ECCE Asia)*, 2013 IEEE, Melbourne, VIC, 2013, pp. 867–871.
112. Fereidouni A, Masoum MAS. A new space-vector hysteresis current-control modulation for active power filter applications. In: *PowerTech, 2015 IEEE Eindhoven*, Eindhoven, the Netherlands, 2015, pp. 1–5.
113. Chebabhi A, Fellah MK, Kessal A, Benkhoris MF. Power quality improvement using a three dimensional space vector modulation with SRF theory for three level neutral point clamped four leg shunt active power filter controlling in *dq*0 axes. In: *2015 4th International Conference on Electrical Engineering (ICEE)*, Boumerdes, Algeria, 2015, pp. 1–6.
114. Nguyen TD, Patin N, Friedrich G. Extended double carrier PWM strategy dedicated to RMS current reduction in DC link capacitors of three-phase inverters. *IEEE Trans Power Electron* 2014;29(1):396–406.
115. Zhu H, Shu Z, Gao F, Qin B, Gao S. Five-level diode-clamped active power filter using voltage space vector-based indirect current and predictive harmonic control. *IET Power Electron* 2014;7(3):713–723.
116. Fereidouni A, Masoum MAS, Smedley KM. Supervisory nearly constant frequency hysteresis current control for active power filter applications in stationary reference frame. *IEEE Power Energy Technol Syst J* 2016;3(1):1–12.
117. Araújo Â, Pinto JG, Exposto B, Couto C, Afonso JL. Implementation and comparison of different switching techniques for shunt active power filters. In: *IECON 2014-40th Annual Conference of the IEEE Industrial Electronics Society*, Dallas, TX, 2014, pp. 1519–1525.
118. Yin LJ, Zhao XL, Xin ZQ, Luo HW. A novel control method for single-phase shunt active power filter. In: *Intelligent Control and Automation (WCICA), 2014 11th World Congress.* Shenyang, 2014, pp. 4111–4116.
119. Saidi S, Abbassi R, Chebbi S. Power quality improvement using VF-DPC-SVM controlled three-phase shunt active filter. In: *Systems, Signals & Devices (SSD), 2015 12th International Multi-Conference*, Mahdia, TN, 2015, pp. 1–5.
120. Tallam RM, Leggate D, Kirschnik DW, Lukaszewski RA. Reducing common-mode current: A modified space vector pulse width modulation scheme. *IEEE Ind Appl Mag* 2014;20(6):24–32.

121. Yu G, Wang Y, Ma S, Zhang L. Research of direct power control of PWM rectifier based on virtual flux. In: *Electrical Machines and Systems (ICEMS), 2014 17th International Conference*, Hangzhou, China, 2014, pp. 775–779.
122. Kim B-J, Choi S-C, Bae S-H, Won C-Y. An improved deadbeat control method based on current predictive algorithm for shunt active power filter. In: *Future Energy Electronics Conference (IFEEC), 2015 IEEE 2nd International*, Taipei, Taiwan, 2015, pp. 1–6.
123. Heo HS, Choe GH, Mok HS. Robust predictive current control of a grid-connected inverter with harmonics compensation. In: *Applied Power Electronics Conference and Exposition (APEC), 2013 Twenty-Eighth Annual IEEE*, Long Beach, CA, 2013, pp. 2212–2217.
124. Vilathgamuwa DM, Wijekoon HM, Choi SS. Investigation of resonance phenomena in a DVR protecting a load with PF correction capacitor. In: *Proceedings of the International Conference on Power Electronics and Drive Systems*, November 2003, Vol. 1, pp. 811–815.
125. Choi SS, Li BH, Vilathgamuwa DM. A comparative study of inverter- and line-side filtering schemes in the dynamic voltage restorer. In: *Proceedings of the IEEE PES Winter Meeting*, January 2000, pp. 2967–2972.
126. Papic I, Zunko P, Krajnc A, Povh D, Weinhold M, Zurowski R. 300 kW battery energy storage system using an IGBT converter. In: *Proceedings of the IEEE PES Summer Meeting*, July 1999, Vol. 2, pp. 1214–1218.
127. Zhan C, Barnes M, Ramachandaramurthy VK, Jenkins N. Dynamic voltage restorer with battery energy storage for voltage dip mitigation. In: *EE Proceedings of the Power Electronics and Variable Speed Drives*, 2000, September 2000, pp. 360–365.
128. Singh B, Solanki J. A comparative study of control algorithms for DSTATCOM for load compensation. In: *Proceedings of the IEEE International Conference on Industrial Technology*, 2006, pp. 1492–1497.
129. Woo SM, Kang DW, Lee WC, Hyun DS. The distribution STATCOM for reducing the effect of voltage Sag and Swell. In: *Proceedings of the IEEE Industrial Electronics*, pp. 1132–1137, 2001.
130. Iyer S, Ghosh A, Joshi A. Operation of a controlled rectifier supported dynamic voltage regulator. In: *Proceedings of the IEEE Power Engineering Society General Meeting*, 2005, Vol. 1, pp. 843–849.
131. Cai R, Bongiomo M, Sannino A. Control of D-STATCOM for voltage dip mitigation. In: *Proceedings of the International Conference Future Power Systems*, 2005, November 2005, pp. 1–6.
132. Bonaldo JP, Morales Paredes HK, Pomilio JA. Control of single-phase power converters connected to low-voltage distorted power systems with variable compensation objectives. *IEEE Trans Power Electron* 2016;31(3):2039–2052.
133. Marafão FP, Brandão DI, Costabeber A, Paredes HKM. Multi-task control strategy for grid-tied inverters based on conservative power theory. *IET Renew Power Gener* 2015;9(2):154–165.
134. Brandao DI, Guillardi H, Pomilio JA, Paredes HKM. Optimized compensation based on linear programming applied to distributed electronic power processors. In: *2015 IEEE 24th International Symposium on Industrial Electronics (ISIE)*, Buzios, Brazil, 2015, pp. 373–378.
135. Brandao DI, Paredes HKM, Costabeber A, Marafão FP. Flexible active compensation based on load conformity factors applied to non-sinusoidal and asymmetrical voltage conditions. *IET Power Electron* 2016;9(2):356–364.
136. Patjoshi RK, Mahapatra KK. Performance comparison of direct and indirect current control techniques applied to a sliding mode based shunt active power filter. In: *2013 Annual IEEE India Conference (INDICON)*, Mumbai, India, 2013, pp. 1–5.
137. Li X, Du T, Tang S. A fuzzy logic variable hysteresis band current control technique for three phase shunt active power filter. In: *Control, Automation and Systems Engineering (CASE), 2011 International Conference*, Singapore, 2011, pp. 1–4.
138. Lu X, Iyer KLV, Mukherjee K, Kar NC. A wavelet/PSO based voltage regulation scheme and suitability analysis of copper and aluminum-rotor induction machines for distributed wind power generation. *IEEE Trans Smart Grid* 2012;3(4):1923–1934.
139. Luo A, Fang L, Xu X, Peng S, Wua C, Fang H. New control strategy for DSTATCOM without current sensors and its engineering application. *Int J Electr Power Energy Syst* 2011;33(2):322–331.
140. Pramila E, Babu CHS, Rao SS, Dinesh L. Enhancement of a power quality by DSTATCOM with *pq* theory. *Int J Eng Trends Eng Dev* 2012;4(2):735–752.

141. Sano K, Takasaki M. A transformerless DSTATCOM based on a multivoltage cascade converter requiring no DC sources. *IEEE Trans Power Electron* 2012;27(6):2783–2795.
142. Moon GW. Predictive current control of distribution static compensator for reactive power compensation. *IEE Proc Gener Transm Distrib* 1999;146(5):515–520.
143. Shokri A, Shareef H, Mohamed A, Farhoodnea M, Zayandehroodi H. A novel single-phase phase space-based voltage mode controller for distributed static compensator to improve voltage profile of distribution systems. *Energy Convers Manage* 2014;79:449–455.
144. Kummari NK, Singh AK, Kumar P. Comparative evaluation of DSTATCOM control algorithms for load compensation. In: *15th IEEE Conference Harmonics and Quality of Power*, Hong Kong, June 17–20, 2012, pp. 299–306.
145. Zaveri T, Bhalja B, Zaveri N. Comparison of control strategies for DSTATCOM in three-phase, four-wire distribution system for power quality improvement under various source voltage and load conditions. *Electr Power Energy Syst* 2012;43:582–594.
146. Singh B, Jayaprakash P, Kothari DP. A T-connected transformer and three-leg VSC based DSTATCOM for power quality improvement. *IEEE Trans Power Electron* 2008;23(6):2710–2718.
147. Arya SR, Singh B. Neural network based conductance estimation control algorithm for shunt compensation. *IEEE Trans Ind Inf* 2014;10(1):569–577.
148. Arya SR, Singh B, Chandra A, Al-Haddad K. Power factor correction and zero voltage regulation in distribution system using DSTATCOM. In: *IEEE International Conference Power Electronics, Drives, Energy System*, Bengaluru, India, 16–19 December 2012, pp. 1–6.
149. Shukla A, Ghosh A, Joshi A. Hysteresis current control operation of flying capacitor multilevel inverter and its application in shunt compensation of distribution systems. *IEEE Trans Power Deliver* 2007;22(1):396–405.
150. Ramesh J, Sudhakaran M. Enhancement of power quality using three phase DSTATCOM for variable load. In: *International Conference on Emerging Trends Electrical Engineering Energy Management*, Chennai, India, 13–15 December 2012, pp. 88–92.
151. Kannan VK, Rengarajan N. Investigating the performance of photovoltaic based DSTATCOM using ICOSF algorithm. *Electr Power Energy Syst* 2014;54(1):376–386.
152. Singh B, Arya SR. Adaptive theory-based improved linear sinusoidal tracer control algorithm for DSTATCOM. *IEEE Trans Power Electron* 2013;28(8):3768–3778.
153. Singh B, Solanki J. Load compensation for diesel generator-based isolated generation system employing DSTATCOM. *IEEE Trans Ind Appl* 2011;47(1):238–244.
154. Zaveri T, Bhalja BR, Zaveri N. Load compensation using DSTATCOM in three-phase, three-wire distribution system under various source voltage and delta connected load conditions. *Electr Power Energy Syst* 2012;41:34–43.
155. Sukanth T, Srinivas D, Lazarus MZ, Satsangi KP. Comparative study of different control strategies for DSTATCOM. *Int J Adv Res Electr Electron Inst Eng* 2012;1(5):362–368.
156. Kannan VK, Rengarajan N. Photovoltaic based distribution static compensator for power quality improvement. *Electr Power Energy Syst* 2012;42:685–692.
157. Sreenivasarao D, Agarwal P, Das B. Performance enhancement of a reduced rating hybrid DSTATCOM for three-phase, four-wire system. *Electr Power Syst Res* 2013;97:158–171.
158. Karmiris G, Tsengenes G, Adamidis G. A multifunction control scheme for current harmonic elimination and voltage sag mitigation using a three phase three level flying capacitor inverter. *Simul Modell Pract Theory* 2012;24:15–34.
159. Singh B, Jayaprakash P, Kothari DP. New control approach for capacitor supported DSTATCOM in three-phase four wire distribution system under non-ideal supply voltage conditions based on synchronous reference frame theory. *Electr Power Energy Syst* 2011;33:1109–1117.
160. Singh B, Arya SR. Admittance based control algorithm for DSTATCOM in three phase four wire system. In: *Second International Conference Power, Control Embedded Systems*. Allahabad, India, 17–19 December 2012, pp. 1–8.
161. Singh B, Arya SR. Software PLL based control algorithm for power quality improvement in distribution system. In: *Fifth IEEE India International Conference on Power Electronics*, Delhi, India, December 6–8, 2012, pp. 1–6.

162. Arya SR, Singh B, Chandra A, Al-Haddad K. Control of shunt custom power device based onanti-Hebbian learning algorithm. In: 38th IEEE Annual Conference Industrial Electronics Society, Montreal, Canada, 25–28 October 2012, pp. 1246–1251.
163. Arya SR, Singh B. Performance of DSTATCOM using leaky LMS control algorithm. *IEEE J Emerg Sel Top Power Electron* 2013;1(2):104–113.
164. Singh B, Arya SR. Back-Propagation control algorithm for power quality improvement using DSTATCOM. *IEEE Trans Ind Electron* 2014;61(3):1204–1212.
165. Singh B, Arya SR. Implementation of single-phase enhanced phase locked loop-based control algorithm for three-phase DSTATCOM. *IEEE Trans Power Deliver* 2013;28(3):1516–1524.
166. Wolfs P, Oo AMT. Improvements to LV distribution system PV penetration limits using a DSTATCOM with reduced DC bus capacitance. In: *Power Energy Society General Meeting*, Vancouver, Canada, July 21–25, 2013, pp. 1–5.
167. Singh B, Jayaprakash P, Kothari DP. Three-phase four-wire DSTATCOM with H-bridge VSC and star/delta transformer for power quality improvement. In: *IEEE Annual India Conference*, December 11–13, 2008, Vol. 2, pp. 412–417.
168. Fazeli SM, Ping HW, Bin Abd Rahim N, Ooi BT. Individual-phase control of 3-phase 4-wire voltage–source converter. *IET Power Electron* 2014;7(9):2354–2364.
169. Singh B, Arya SR, Chandra A, Al-Haddad K. Implementation of adaptive filter based control algorithm for distribution static compensator. In: *Industry Applications Society Annual Meeting*, Las Vegas, NV, October 7–11, 2012, pp. 1–8.
170. Singh B, Arya SR, Jain C. Simple peak detection control algorithm of distribution static compensator for power quality improvement. *IET Power Electron* 2014;7(7):1736–1746.
171. Qi Q, Yu C, Wai CK, Ni Y. Modeling and simulation of a STATCOM system based on 3-level NPC inverter using dynamic phasors. In: *IEEE Power Engineering Society General Meeting*, Denver, CO, 10 June 2004, vol. 2, pp. 1559–1564.
172. Reddy VK, Veni KK, Das GT, Pulla S. Performance analysis of DSTATCOM compensator using control techniques for load compensation. *Int J Electr Electron Eng Res* 2011;1(2):149–171.
173. Suvire GO, Mercado PE. Combined control of a distribution static synchronous compensator/flywheel energy storage system for wind energy applications. *IET Gener Transm Distrib* 2012;6(6):483–492.
174. Shukla A, Ghosh A, Joshi A. State feedback control of multilevel inverters for DSTATCOM applications. *IEEE Trans Power Deliver* 2007;22(4):2409–2418.
175. Srikanthan S, Mishra MK. DC capacitor voltage equalization in neutral clamped inverters for DSTATCOM application. *IEEE Trans Ind Electron* 2010;57(8):2768–2775.
176. Sensarma PS, Padiyar KR, Ramanarayanan V. Analysis and performance evaluation of a distribution STATCOM for compensating voltage fluctuations. *IEEE Trans Power Deliver* 2001;16(2):259–264.
177. Gupta R, Ghosh A, Joshi A. Multiband hysteresis modulation and switching characterization for sliding-mode-controlled cascaded multilevel inverter. *IEEE Trans Ind Electron* 2010;57(7):2344–2353.
178. Gupta R, Ghosh A, Joshi A. Characteristic analysis for multisampled digital implementation of fixed-switching-frequency closed-loop modulation of voltage-source inverter. *IEEE Trans Ind Electron* 2009;56(7):2382–2392.
179. Luo A, Ma F, Shuai Z, Shuai Z, Wang Y. Distribution static compensator based on an improved direct power control strategy. *IET Power Electron* 2014;7(4):957–964.
180. Masdi H, Mariun N, Bashi SM, Mohamed A, Yusuf S. Construction of a prototype DSTATCOM for voltage sag mitigation. *Eur J Sci Res* 2009;30(1):112–127.
181. Valderrábano A, Ramirez JM. DStatCom regulation by a fuzzy segmented PI controller. *Electr Power Syst Res* 2010;80(6):707–715.
182. Geddada N, Karanki SB, Mishra MK. Synchronous reference frame based current controller with SPWM switching strategy for DSTATCOM applications. In: *IEEE International Conference on Power Electronics Drives and Energy Systems*, Bengaluru, India, 16–19 December 2012, pp. 1–6.
183. Kumar P, Kumar N, Akella AK. Modeling and simulation of different system topologies for DSTATCOM. *Conf Parallel Distrib Comput Syst* 2013;5:249–261.
184. Muñoz JA, Espinoza JR, Baier CR, Morán LA, Guzmán JI, Cárdenas V.M. Decoupled and modular harmonic compensation for multilevel STATCOMs. *IEEE Trans Ind Electron* 2014;61(6):2743–2753.

185. Shahnia F, Ghosh A, Ledwich G, Zare F. Voltage unbalance improvement in low voltage residential feeders with rooftop PVs using custom power devices. *Electr Power Energy Syst* 2014;55:362–377.
186. Jazebi S, Hosseinian SH, Vahidi B. DSTATCOM allocation in distribution networks considering reconfiguration using differential evolution algorithm. *Energy Convers Manage* 2011;52(7):2777–2783.
187. Elnady A, Salama MMA. Unified approach for mitigating voltage sag and voltage flicker using the DSTATCOM. *IEEE Trans Power Deliver* 2005;20(2):992–1000.
188. Singh B, Murthy SS, Chilipi RR, Madishetti S, Bhuvaneswari G. Static synchronous compensator-variable frequency drive for voltage and frequency control of small-hydro driven self-excited induction generators system. *IET Gener Transm Distrib* 2014;8(9):1528–1538.
189. Yang K, Cheng X, Wang Y, Chen L, Chen G. PCC voltage stabilization by D-STATCOM with direct grid voltage control strategy. In: *IEEE International Symposium on Industrial Electronics*, Hangzhou, China, 28–31 May 2012, pp. 442–446.
190. Chen CS, Lin CH, Hsieh WL, Hsu CT, Ku TT. Enhancement of PV penetration with DSTATCOM in Tai-power distribution system. *IEEE Trans Power Syst* 2013;28(2):1560–1567.
191. Mahmuda MA, Pota HR,Hossain MJ. Nonlinear DSTATCOM controller design for distribution network with distributed generation to enhance voltage stability. *Int J Electr Power Energy Syst* 2013;53:974–979.
192. Suvire GO, Mercado PE. DSTATCOM with flywheel energy storage system for wind energy applications: Control design and simulation. *Electr Power Syst Res* 2010;80(3):345–353.
193. Gawande SP, Khan S, Ramteke MR. Voltage sag mitigation using multilevel inverter based distribution static compensator (DSTATCOM) in low voltage distribution system. In: *Fifth IEEE India International Conference on Power Electronics*, Delhi, 6–8 December 2012, pp. 1–6.
194. Mahfouz MMA, El-Sayed MAH. Static synchronous compensator sizing for enhancement of fault ride-through capability and voltage stabilisation of fixed speed wind farms. *IET Renew Power Gener* 2014;8(1):1–9.
195. Aggarwal M, Gupta SK, Singh M. Integration of wind generation system in low voltage distribution system. In: *IEEE Fifth India International Conference. on Power Electronics*, Delhi, 6–8 December 2012, pp. 1–6.
196. Singh B, Solanki J. A comparison of control algorithms for DSTATCOM. *IEEE Trans Ind Electron* 2009;56(7):2738–2745.
197. Reddy JGP, Reddy KR. Design and simulation of cascaded H-bridge multilevel inverter based DSTATCOM for compensation of reactive power and harmonics. In: *First International Conference on Recent Advanced Information Technology*, Dhanbad, India, March 15–17, 2012, pp. 737–743.
198. Bilgin HF, Ermis M, Kose KN, et al. Reactive power compensation of coal mining excavators by using a new-generation STATCOM. *IEEE Trans Ind Appl* 2007;43(1):97–110.
199. Maswood AI, Neo TG, Rahman MA. An online optimal approach to PWM-SHE gating signal generation. *IEEE Power Eng Rev* 2001;21(3):61–62.
200. Gupta R, Ghosh A. Frequency-domain characterization of sliding mode control of an inverter used in DSTATCOM application. *IEEE Trans Circuits Syst* 2006;53(3):662–676.
201. Shahgholian G, Shafaghi P, Moalem S, Mahdavian M. Analysis and design of a linear quadratic regulator control for static synchronous compensator. In: *2009 Second International Conference on Computer and Electrical Engineering*, Volume 1, 2009, pp. 65–69.
202. Karanki SB, Geddada N, Mishra MK, Kumar BK. A DSTATCOM topology with reduced DC-Link voltage rating for load compensation with non-stiff source. *IEEE Trans Power Electron* 2012;27(3):1201–1211.
203. Kishore PV, Reddy SR, Kishore PV. Modeling and simulation of 14 bus system with DSTATCOM for power quality improvement. *Indian J Sci Res* 2012;3(1):73–79.
204. Kumar C, Mishra MK. A modified DSTATCOM topology with reduced VSI rating, DC link voltage, and filter size. In: *International Conference Clean Electrical Power*, Alghero, 11–13 June 2013, pp. 325–331.
205. Mishra MK, Karthikeyan K. A fast-acting DC-link voltage controller for three-phase DSTATCOM to compensate AC and DC loads. *IEEE Trans Power Deliver* 2009;24(4):2291–2299.
206. Ghosh A, Ledwich G. Load compensating DSTATCOM in weak AC systems. *IEEE Trans Power Deliver* 2003;18(4):1302–1309.
207. Mitra P, Venayagamoorthy GK. An adaptive control strategy for DSTATCOM applications in an electric ship power system. *IEEE Trans Power Electron* 2010;25(1):95–104.

208. Shukla A, Ghosh A, Joshi A. Control schemes for DC capacitor voltages equalization in diode-clamped multilevel inverter-based DSTATCOM. *IEEE Trans Power Deliver* 2008;23(2):1139–1149.
209. da Silva CH, da Silva VF, da Silva LEB, Lambert-Torres G, Takauti EH. Optimizing the series active filters under unbalanced conditions acting in the neutral current. In: *2007 IEEE International Symposium on Industrial Electronics*, Vigo, 2007, pp. 943–948.
210. Cao R, Zhao J, Shi W, Jiang P, Tang G. Series power quality compensator for voltage sags, swells, harmonics and unbalance. In: *Transmission and Distribution Conference and Exposition, 2001 IEEE/PES*, Atlanta, GA, 2001, vol. 1, pp. 543–547.
211. Vivek G, Nair MD, Barai M. Online reduction of fifth and seventh harmonics in single phase quasi square wave inverters. In: *2015 Annual IEEE India Conference (INDICON)*, New Delhi, India, 2015, pp. 1–6.
212. Pinto JG, Pregitzer R, Monteiro LFC, Couto C, Afonso JL. A combined series active filter and passive filters for harmonics, unbalances and flicker compensation. In: *2007 International Conference on Power Engineering, Energy and Electrical Drives*, Setubal, Portugal, 2007, pp. 54–59.
213. da Silva SAO, Campanhol LBG, Goedtel A, Nascimento CF, Paiao D. A comparative analysis of p-PLL algorithms for single-phase utility connected systems. In: *Power Electronics and Applications, 2009. EPE '09. 13th European Conference*, Barcelona, 2009, pp. 1–10.
214. Da Silva S, Modesto RA. Active power line compensation applied to a three-phase line interactive UPS system using SRF method. In: *2005 IEEE 36th Power Electronics Specialists Conference*, Recife, 2005, pp. 2358–2362.
215. Acuña PF, Morán LA, Weishaupt CA, Dixon JW. An active power filter implemented with multilevel single-phase NPC converters. In: *IECON 2011-37th Annual Conference on IEEE Industrial Electronics Society*, Melbourne, Vic, 2011, pp. 4367–4372.
216. Libano FB, Muller SL, Marques Braga RA, Rossoni Nunes JV, Mano OS, Paranhos IA. Simplified control of the series active power filter for voltage conditioning. In: *2006 IEEE International Symposium on Industrial Electronics*, Montreal, Canada, 2006, pp. 1706–1711.
217. Milanés-Montero MI, Romero-Cadaval E, Barrero-González F. Hybrid multiconverter conditioner topology for high-power applications. *IEEE Trans Ind Electron* 2011;58(6):2283–2292.
218. Barros JD, Silva JF. Multilevel optimal predictive dynamic voltage restorer. *IEEE Trans Ind Electron* 2010;57(8):2747–2760.
219. Kim H, Lee S-J, Sul SK. A calculation for the compensation voltages in dynamic voltage restorers by use of PQR power theory. In: *Applied Power Electronics Conference and Exposition, 2004. APEC '04. Nineteenth Annual IEEE*, 2004, pp. 573–579.
220. da Silva SAO, Modesto RA. A comparative analysis of SRF-based controllers applied to active power line conditioners. In: *Industrial Electronics, 2008. IECON 2008. 34th Annual Conference of IEEE*, Orlando, FL, 2008, pp. 405–410.
221. Rajasekaran D, Dash SS, Teja DRA. Dynamic voltage restorer based on fuzzy logic control for voltage sag restoration. In: *Sustainable Energy and Intelligent Systems (SEISCON 2011), International Conference*, Chennai, India, 2011, pp. 26–30.
222. de Almeida Carlos GA, Jacobina CB. Series compensator based on cascaded transformers coupled with three-phase bridge converters. In: *2015 IEEE Energy Conversion Congress and Exposition (ECCE)*, Montreal, Canada, 2015, pp. 3414–3421.
223. Wu FZ, Pei SP. The research and implementation of dynamic voltage restorer with power factor correction. In: *Intelligent Systems and Applications (ISA), 2011 3rd International Workshop*, Wuhan, China, 2011, pp. 1–4.
224. Newman MJ, Holmes DG, Nielsen JG, Blaabjerg F. A dynamic voltage restorer (DVR) with selective harmonic compensation at medium voltage level. *IEEE Trans Ind Appl* 2005;41(6):1744–1753.
225. Tung NX, Fujita G, Horikoshi K. An adapted control strategy for dynamic voltage restorer to work as series active power filter. In: *Power Electronics Conference (IPEC), 2010 International*, Sapporo, 2010, pp. 2283–2287.
226. Newman MJ, Holmes DG, Nielsen JG, Blaabjerg F. A dynamic voltage restorer (DVR) with selective harmonic compensation at medium voltage level. *IEEE Trans Ind Appl* November–December 2005;41(6):1744–1753.

227. Escobar G, Stankovic AM, Mattavelli P. Reactive power, imbalance and harmonics compensation using D-STATCOM with a dissipativity-based controller. In: *Proceedings of the IEEE Decision and Control*, December 2000, Vol. 4, pp. 3051–3055.
228. Sharmeela C, Uma G, Mohan MR. Multi-level distribution STATCOM for voltage sag and swell reduction. In: *Proceedings of the IEEE Power Engineering Society General Meeting*, June 2005, Vol. 2, pp. 1303–1307.
229. Vilathgamuwa DM, Perera AA, Choi SS. Voltage sag compensation with energy optimized dynamic voltage restorer. *IEEE Trans Power Deliver* 2003;18:928–936.
230. Lam C.-S, Wong M-C, Han Y-D. Voltage swell and overvoltage compensation with unidirectional power flow controlled dynamic voltage restorer. *IEEE Trans Power Deliver* 2008;23(4):2513–2521.
231. Gruzs TM. A survey of neutral currents in three-phase computer power systems. *IEEE Trans Ind Appl* 1990;26(4):719–725.
232. Lin C-H, Wang C-H. Adaptive wavelet networks for power-quality detection and discrimination in a power system. *IEEE Trans Power Del* 2006;21(3):1106–1113.
233. Girgis AA, Chang WB, Makram EB. A digital recursive measurement scheme for online tracking of power system harmonics. *IEEE Trans Power Deliv* 1991;6(3):1153–1160.
234. Kumar S, Singh B. Control of 4-Leg VSC based DSTATCOM using modified instantaneous symmetrical component theory. In: *3th International Conference Power Systems*, Kharagpur, India, December 27–29, 2009, pp. 1–6.
235. George V, Mishra MK. Design and analysis of user-defined constant switching frequency current-control-based four-leg DSTATCOM. *IEEE Trans Power Electron* 2009;24(9):2148–2158.
236. Yepes AG, Vidal A, López O, Doval-Gandoy J. Evaluation of techniques for cross-coupling decoupling between orthogonal axes in double synchronous reference frame current control. *IEEE Trans Ind Electron* 2014;61(7):3527–3531.
237. Hyosung K, Blaabjerg F, Bak-Jensen B, Jaeho C. Instantaneous power compensation in three-phase systems by using *p–q–r* theory. *IEEE Trans Power Electron* 2002;17(5):701–710.
238. El-Habrouk M, Darwish MK, Mehta P. Active power filters: A review. *IEE Proc Electr Power Appl* 2000;147(5):403–413.
239. Nair D, Nambiar A, Raveendran M, Mohan NP, Sampath S. Mitigation of power quality issues using DSTATCOM. In: *International Conference on Emerging Trends Electrical Engineering Energy Management*, Chennai, India, December 13–15, 2012, pp. 65–69.
240. Peng FZ, Lai JS. Generalized instantaneous reactive power theory for three-phase power systems. *IEEE Trans Instrum Meas* 1996;45(1):293–297.
241. Peng FZ, Ott GW, Adams DJ. Harmonic and reactive power compensation based on the generalized instantaneous reactive theory for three-phase four-wire systems. *IEEE Trans Power Electron* 1998;13(6):1174–1181.
242. Shukla A, Ghosh A, Joshi A. State feedback control of multilevel inverters for DSTATCOM applications. *IEEE Trans Power Deliver* 2007;22(4):2409–2418.
243. Selvajyothi K, Janakiraman PA. Extraction of harmonics using composite observers. *IEEE Trans Power Deliver* 2008;23(1):31–40.
244. Herrera RS, Salmeron P, Hyosung K. Instantaneous reactive power theory applied to active power filter compensation: Different approaches, assessment, and experimental results. *IEEE Trans Ind Electron* 2008;55(1):184–196.
245. Herrera RS, Salmeron P. Instantaneous reactive power theory: A reference in the nonlinear loads compensation. *IEEE Trans Ind Electron* 2009;56(6):2015–2022.
246. Suru CV, Patrascu CA, Popescu M, Bitoleanu A. Conservative power theory application in shunt active power filtering under asymmetric voltage. In: *2014 International Conference on Optimization of Electrical and Electronic Equipment (OPTIM)*, IEEE, 2014, pp. 647–654.
247. Lee SJ, Kim H, Sul SK, Blaabjerg F. A novel control algorithm for static series compensators by use of PQR instantaneous power theory. *IEEE Trans Power Electron* 2004;19(3):814–827.
248. Wang H, Chang X. A method based on improved instantaneous symmetrical components and three-point algorithm for synchronized phasor measurement. In: *Fifth International Conference on Critical Infrastructure*, Beijing, China, 20–22 September 2010, pp. 1–5.

249. Jayaprakash P, Singh B, Kothari DP. Icosϕ algorithm based control of zig-zag transformer connected three phase four wire DSTATCOM. In: *IEEE International Conference on Power Electronics, Drives Energy Systems*, Bengaluru, India, 16–19 December 2012, pp. 1–6.
250. Arya SR, Singh B. Implementation of distribution static compensator for power quality enhancement using learning vector quantisation. *IET Gener Transm Distrib* 2013;7(11):1244–1252.
251. Arya SR, Chandra A, Al-Haddad K. Control of DSTATCOM using adjustable step least mean square control algorithm. In: *Fifth IEEE Power India Conference*, Murthal, India, 19–22 December 2012, pp. 1–6.
252. Singh B, Arya SR. Back-Propagation control algorithm for power quality improvement using DSTATCOM. *IEEE Trans Ind Electron* 2014;61(3):1204–1212.
253. Wessels C, Hoffmann N, Molinas M, Fuchs FW. StatCom control at wind farms with fixed-speed induction generators under asymmetrical grid faults. *IEEE Trans Ind Electron* 2013;60(7):2864–2873.
254. Srikanthan S, Mishra MK, Rao RKV. Improved hysteresis current control of three-level inverter for distribution static compensator application. *IET Power Electron* 2009;2(5):517–526.
255. Iyer S, Ghosh A, Joshi A. Inverter topologies for DSTATCOM applications—A simulation study. *Electr Power Syst Res* 2005;75:161–170.
256. Tenti P, Morales Paredes HK, Mattavelli P. Conservative power theory, a framework to approach control and accountability issues in smart microgrids. *IEEE Trans Power Electron* 2011;26(3):664–673
257. Philip J, Kant K, Jain C, Singh B, Mishra S. A simplified configuration and implementation of a standalone microgrid. In: *2015 IEEE Power & Energy Society General Meeting*, Denver, CO, 2015, pp. 1–5.
258. Rosa RB, Vahedi H, Godoy RB, Pinto JOP, Al-Haddad K. Conservative power theory used in NPC-based shunt active power filter to eliminate electric metro system harmonics. In: *Vehicle Power and Propulsion Conference (VPPC), 2015 IEEE*, Montreal, QC, 2015, pp. 1–6.
259. Haugan TS, Tedeschi E. Active power filtering under non-ideal voltage conditions using the conservative power theory. In: *2015 IEEE 13th Brazilian Power Electronics Conference and 1st Southern Power Electronics Conference (COBEP/SPEC)*, Fortaleza, Brazil, 2015, pp. 1–6.
260. Busarello TDC, Pomilio JA, Simoes MG. Passive filter aided by shunt compensators based on the conservative power theory. *IEEE Trans Ind Appl* 2016;52(4):3340–3347
261. Singh B, Arya SR. Composite observer-based control algorithm for distribution static compensator in four-wire supply system. *IET Power Electron* 2013;6(2):251–260.
262. Borisov K, Ginn HL. Multifunctional VSC based on a novel Fortescue reference signal generator. *IEEE Trans Ind Electron* 2010;57(3):1002–1007.
263. Melin PE, Espinoza JR, Zargari NR, Moran LA, Guzman JI. A novel multi-level converter based on current source power cell. In: *IEEE Conference Power Electronics Special*, Rhodes, 15–19 June 2008, pp. 2084–2089.
264. Ghosh A, Ledwich G. *Power Quality Enhancement Using Custom Power Devices*. Boston, MA: Kluwer; 2002.
265. Hatami H, Shahnia F, Pashaei A, Hosseini SH. Investigation on D-STATCOM and DVR Operation for voltage control in distribution networks with a new control strategy. In: *IEEE Power Tech*, Lausanne, 1–5 July 2007, pp. 2207–2212.
266. Anaya-Lara O, Acha E. Modeling and analysis of custom power systems by PSCAD/EMTDC. *IEEE Trans Power Deliver* 2002;17(1):266–272.
267. Sreenivasarao D, Agarwal P, Das B. A T-connected transformer based hybrid DSTATCOM for three-phase, four-wire systems. *Electr Power Energy Syst* 2013;44(1):964–970.
268. SajediHir M, Hoseinpoor Y, Ardabili PM, Pirzadeh T. Analysis and simulation of a DSTATCOM for voltage quality improvement. *Aust J Basic Appl Sci* 2011;5:864–870.
269. Huang J, Shi H. Reducing the common-mode voltage through carrier peak position modulation in an SPWM three-phase inverter. *IEEE Trans Power Electron* 2014;29(9):4490–4495.
270. Kouro S, Malinowski M, Gopakumar K, et al. Recent advances and industrial application of multilevel converters. *IEEE Trans Ind Electron* 2010;57(8):2553–2580.
271. Singh B, Al-Haddad K, Chandra A. A review of active filters for power quality improvement. *IEEE Trans Ind Electron* 1999;46(5):960–971.
272. Chen D, Xie S. Review of the control strategies applied to active power filters. In: *IEEE International Conference Electric Utility Deregulation, Restructure Power Technology*, 5–8 April 2004, vol. 2, pp. 666–670.

273. Saeedifard M, Nikkhajoei H, Iravani R. A space vector modulated STATCOM based on a three-level neutral point clamped converter. *IEEE Trans Power Deliver* 2007;22(2):1029–1039.
274. Morcillo JD, Burbano D, Angulo F. Adaptive Ramp technique for controlling chaos and subharmonic oscillations in DC–DC power converters. *IEEE Trans Power Electron* 2016;31(7):5330–5343.
275. Sood VK. *HVDC and FACTS Controllers–Applications of Static Converters in Power Systems*. Boston, MA: Kluwer, 2004.
276. Colak I, Kabalci E, Bayindir R. Review of multilevel voltage source inverter topologies and control schemes. *Energy Convers Manage* 2011;52(2):1114–1128.
277. Grodzki R, Kuźma A. Control of the active power filter using the predictive current control algorithm. *Przeglad Elektrotechniczny* 2013;89(9):30–33.
278. Malesani L, Tomasin P. PWM current control techniques of voltage source converters-A survey. In: *International Conference on Industrial Electronics and Control Instrumentation*, Maui, HI, 15–19 November 1993, vol. 2, pp. 670–667.
279. Ledwich G, Ghosh A. A flexible DSTATCOM operating in voltage or current control mode. *IEE Proc Gener Transm Distrib* 2002;149(2):215–224.
280. Jung SL, Tzou YY. Sliding mode control of a closed-loop regulated PWM inverter under large load variations. In: *24th IEEE Annual Conference Power Electronics Specialist*, Seattle, WA, 20–24 June 1993, pp. 616–622.
281. Kazmierkowski MP, Dzieniakowski MA. Review of current regulation techniques for three-phase PWM inverters. In: *20th International Conference Industrial Electronics and Control Instrumentation*, Bologna, 5–9 September 1994, vol. 1, pp. 567–575.
282. Kanai MM, Nderu JN, Hinga PK. Optimizing LQR to control buck converter by mesh adaptive search algorithm. In: *Proceedings of 2010 JKUAT Scientific Technological and Industrialization Conference*, 2010.
283. Landaeta LM, Sepulveda CA, Espinoza JR, Baier CR. A mixed LQRI/PI based control for three-phase UPQCs. In: *32nd IEEE Annual Conference Industrial Electronics*, Paris, France, 6–10 November 2006, pp. 2494–2499.
284. Mahieddine-Mahmoud S, Chrifi-Alaoui L, Assche VV, Bussy P. Sliding mode control with unmatched disturbance and implementation for a three-tanks system. In: *IEEE International Conference Industry and Technology*, Mumbai, India, 15–17 December 2006, pp. 675–680.
285. Ramakrishna G, Malik OP. Adaptive PSS using a simple on-line identifier and linear pole-shift controller. *Electr Power Syst Res* 2010;80(4):406–416.
286. Lascu C, Asiminoaei L, Boldea I, Blaabjerg F. High performance current controller for selective harmonic compensation in active power filters. *IEEE Trans Power Electron* 2007;22(5):1826–1835.
287. Bilgin HF, Ermis M. Design and implementation of a current-source converter for use in industrial applications of DSTATCOM. *IEEE Trans Power Electron* 2010;25(8):1943–1957.
288. Bilgin HF, Ermis M. Current source converter based STATCOM: Operating principles, design and field performance. *Electr Power Syst Res* 2011;81:478–487.
289. Rodríguez J, Franquelo LG, Kouro S, et al. Multilevel converters: An enabling technology for high-power applications. *IEEE Proc* 2009;97(11):1786–1817.
290. Cetin A. Ermis M. VSC-based D-STATCOM with selective harmonic elimination. *IEEE Trans Ind Appl* 2009;45(3):1000–1015.
291. Ray PK. Power quality improvement using VLLMS based adaptive shunt active filter. *CPSS Trans Power Electron and Appl* 2018;3(2):154–162.

8

An Active Power Filter in Phase Coordinates for Harmonic Mitigation

8.1 Active Power Filter

In power systems, harmonics appear on the current signals, due to nonlinear loads present in the system. Current harmonics, which affect the power quality, are harmful for vulnerable loads. To eliminate current, harmonics filters are used. They are active and passive filters.

Active power filters (Figure 8.1) are simply power electronic converters that are specifically designed to inject harmonic current into the system. Active power filters have capabilities such as the following:

1. Eliminating current and voltage harmonics
2. Reactive power compensation
3. Regulating terminal voltage
4. Compensating the voltage flickering

In this contribution, an efficient active power filter (APF) technique in phase coordinates is suggested that can mitigate the harmonics generated by nonlinear components and loads. It avoids the transformation between different frameworks and allows a control circuit design to be relatively simpler and easier to implement.

8.2 Synchronous Current Detection

Based on the synchronous current detection method (SCD), an APF model in phase coordinates is given in Figure 8.2. The reference currents are calculated as

$$i_{fk}^{*}(t) = i_k(t) - \left(\frac{2p_T}{V_T}\right)\left(\frac{v_k(t)}{V_k}\right) \quad k = a,b,c \tag{8.1}$$

where i_k are the line distorted currents, p_T is the sum of the average real power delivered to the load in each phase, V_K are peak phase voltages and V_T is their sum, v_k are the source voltages, and $k = a, b, c$.

The APF technique can handle distorted currents and unbalanced source voltages. In addition, the APF maintains good performance in the determination of the reference filtering currents. The voltage waveforms are treated according to the techniques.

8.3 Least-Squares Fitting

The curve-fitting method used is based on the assessment of the minimum square error to find the magnitude and phase of the fundamental component. However, the method is usually applicable to any other frequency, thus making a recursive solution to obtain the fundamental frequency component

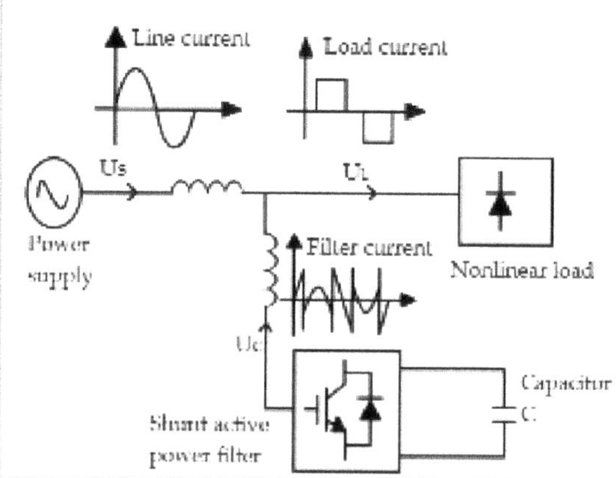

FIGURE 8.1 Active power filter.

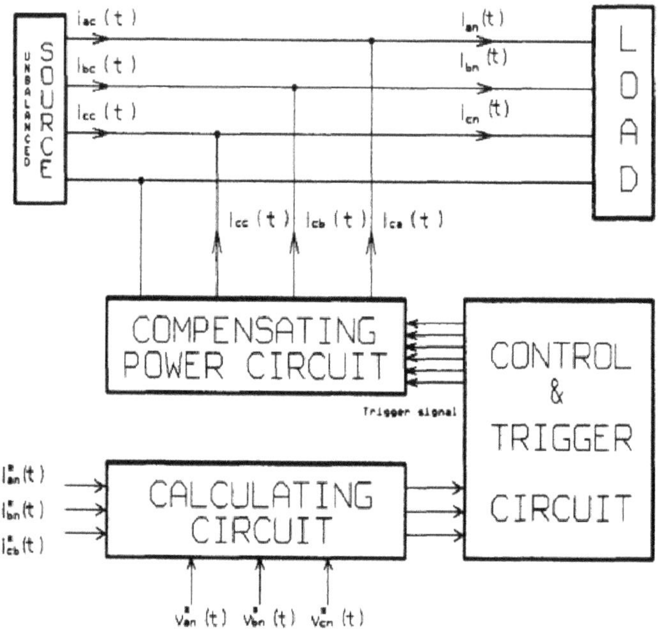

FIGURE 8.2 APF based on SCD.

feasible. A curve-fitting method is adequate for periodic waveforms, as it is usually the case in electric fields, where the fundamental component is, in general, dominant and the data measurement (or truncation) after each period is not necessary, as required by the Fourier transform. Once the source voltage waveforms v_{k1} $k = a, b, c$ at the fundamental frequency are obtained, these are incorporated by substituting v_k by v_{k1}, thus yielding

$$i_{fk}^{*}(t) = i_k(t) - \left(\frac{2p_T}{V_T}\right)\left(\frac{v_{k1}(t)}{V_k}\right), k = a, b, c \tag{8.2}$$

This procedure maintains ideal line currents under the presence of harmonic distortion in the source voltages and in the fundamental frequency amplitude-unbalanced source voltages.

8.4 Phase-Lock Technique

When the fundamental frequency voltages are phase unbalanced, the line currents are, in turn, phase unbalanced and this produces relatively high neutral current in a three-phase four-wire system. To solve this inconvenience, let us assume that a reference phase is selected (e.g., the phase a); thus, the phase angle of a at the fundamental frequency is taken as the reference phase angle; the phase angle for b is obtained as (phase of a–120) and for c as (phase of a+120). This procedure ensures that the line currents to be determined will be symmetrical. This procedure can be seen as a phase-lock technique.

8.5 Determination of Phase Reference Currents

Figure 8.3 shows the APF method which use the reference voltages obtained through the procedures described in the previous sections and to substitute these reference voltages in the above equation; thus, the reference currents are then obtained as

$$\begin{aligned} i_{fa}^{*}(t) &= i_a(t) - \frac{2p_T}{V_T}\sin(\omega t + \phi_{a1}) \\ i_{fb}^{*}(t) &= i_b(t) - \frac{2p_T}{V_T}\sin\left(\omega t + \phi_{a1} - \frac{2\pi}{3}\right) \\ i_{fc}^{*}(t) &= i_c(t) - \frac{2p_T}{V_T}\sin\left(\omega t + \phi_{a1} + \frac{2\pi}{3}\right) \end{aligned} \tag{8.3}$$

8.6 Simulation Model

Figure 8.4 shows the simulation circuits, Figure 8.5 shows their corresponding output voltage waveforms, and Figure 8.6 displays the source-side voltage and current with filter. Input of current harmonics has been eliminated using active filter. The method reduces input current harmonics in a grid-connected system.

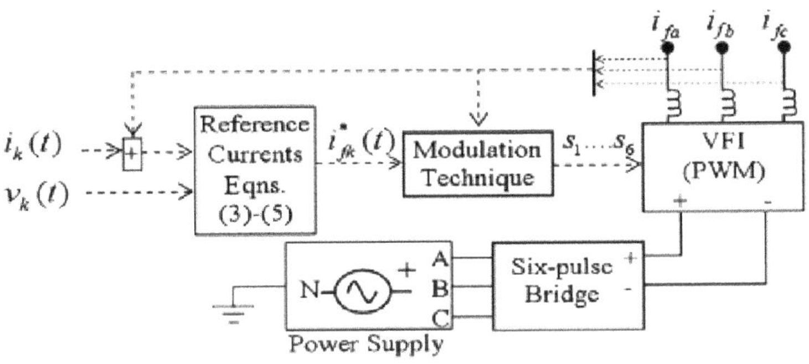

FIGURE 8.3 APF system.

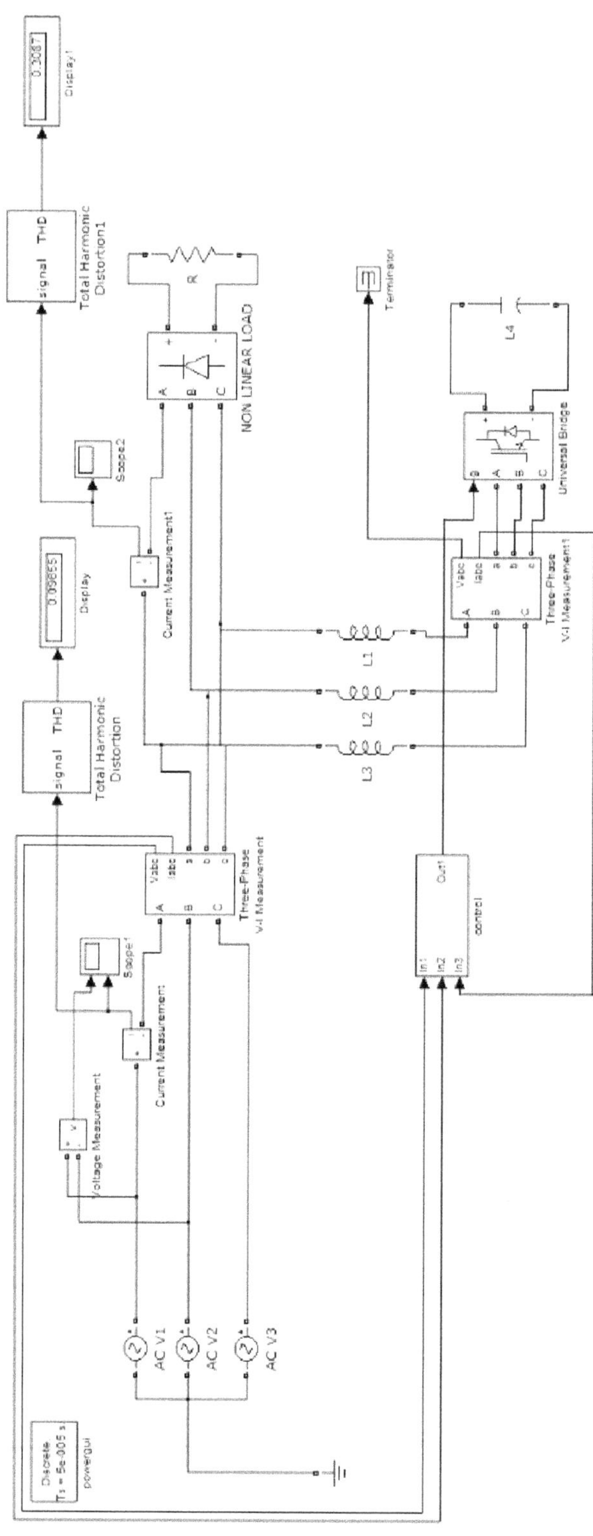

FIGURE 8.4 Simulation circuit.

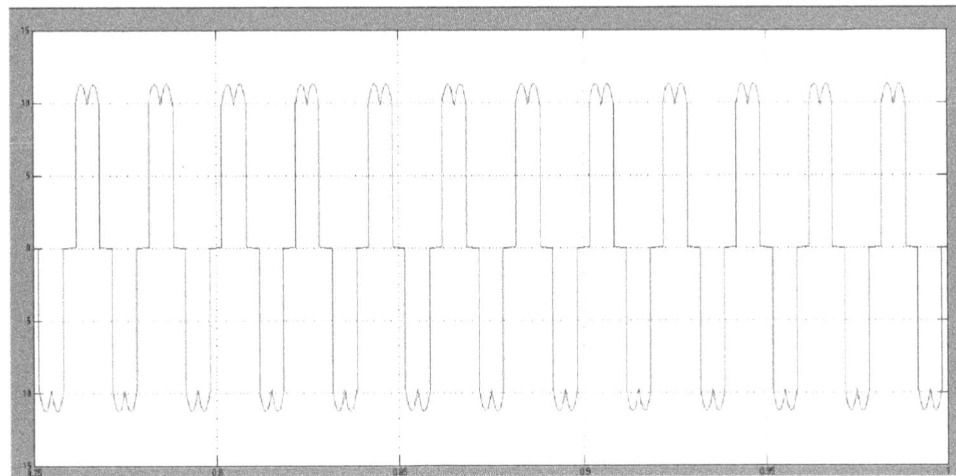

FIGURE 8.5 Output voltage waveform.

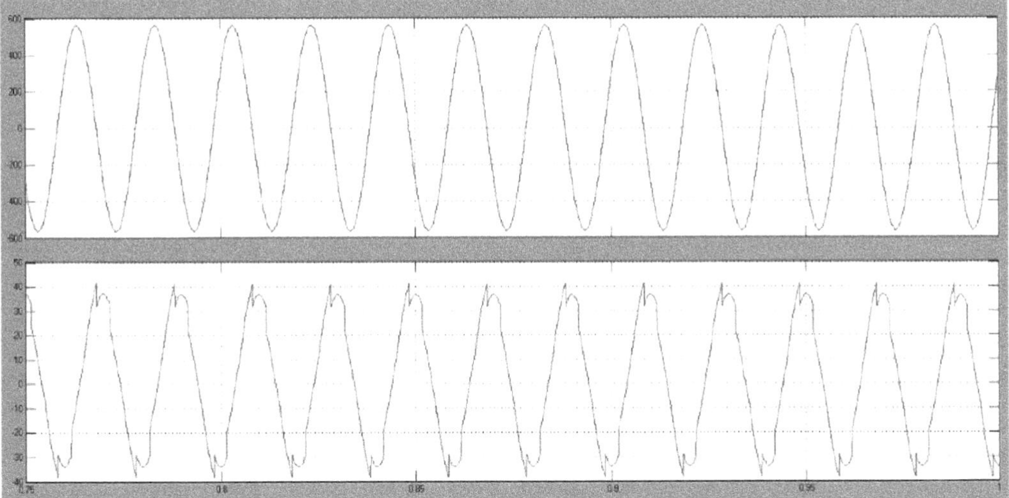

FIGURE 8.6 Source-side voltage and current waveform with filter.

9

Line Harmonics Reduction in High-Power Systems

9.1 Introduction

A rectifier is an electrical device that converts alternating current (AC), which periodically reverses direction, to direct current (DC), which flows in only one direction. The process is known as rectification. Physically, rectifiers take a number of forms, including vacuum tube diodes, mercury-arc valves, copper and selenium oxide rectifiers, semiconductor diodes, silicon-controlled rectifiers, and other silicon-based semiconductor switches.

Historically, even synchronous electromechanical switches and motors have been used. Early radio receivers, called crystal radios, used a "cat's whisker" of fine wire pressing on a crystal of galena (lead sulfide) to serve as a point-contact rectifier or "crystal detector." Rectifiers have many uses but are often found serving as components of DC power supplies and high-voltage DC power transmission systems. Rectification may serve in roles other than to generate DC for use as a source of power. As noted, detectors of radio signals serve as rectifiers. In gas heating systems flame rectification is used to detect presence of a flame. Because of the alternating nature of the input AC sine wave, the process of rectification alone produces a DC current that, though unidirectional, consists of pulses of current.

Many applications of rectifiers, such as power supplies for radio, television, and computer equipment, require a steady constant DC current (as would be produced by a battery). In these applications the output of the rectifier is smoothed by an electronic filter (usually a capacitor) to produce a steady current. More complex circuitry that performs the opposite function, converting DC to AC, is called an inverter. The full-wave rectifier circuit's purpose is to create maximum efficiency from electric current flow in order to deliver maximum power. While it may seem complicated at first glance, the concept is fairly simple. All circuits have a source that generates a voltage, which is needed in order for them to work.

There are two types of voltage sources: those that use DC and those that use AC. One problem with this type of waveform is that only part of it is in the positive direction (above the x-axis) while the rest is in the negative direction (below that x-axis). The negative parts of the waveform work against the positive parts, making the overall power supply much less than what we might want.

For example, if you were to sail a boat against the wind, the wind would slow you down since it is blowing in the opposite, or negative, direction. This makes the boat's overall distance over a period of time less than it ought to be. However, what if there was a way to change the direction of the wind so it blew in the same direction as the boat's direction of travel, so that it would actually increase your speed and overall distance covered in the same timeframe? In terms of power transform, that is the exact purpose of the full-wave rectifier. The circuit takes all the negative parts of the waveform and flips them to positive, so that all of the waveform is working in the same direction. This maximizes the efficiency of the power transferred. The major objective is to design a dominant harmonic active filter (DHAF). Square wave inverters are used to supply the rated line current.

The source will be the three-phase AC source. It will be fed to the three-phase non-linear load. Non-linear load is used to inject the harmonic into that power system. The midpoint passive filter is fed to the star-to-star transformer that will be connected to the voltage source inverter. Figure 9.1 shows the block diagram of the system.

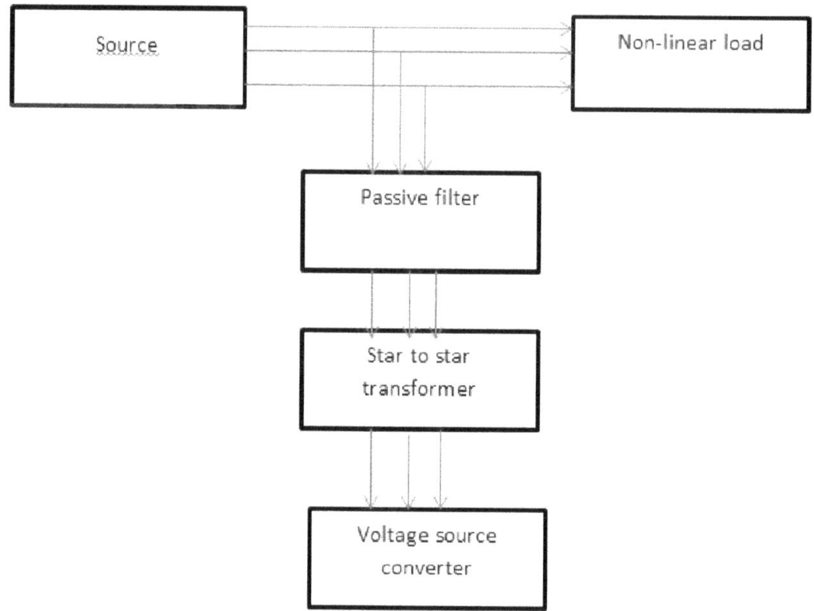

FIGURE 9.1 Block diagram of the system.

9.2 Square Wave Inverter

An inverter or power inverter is a device that converts the DC sources to AC sources. Inverters are used in applications such as adjustable-speed AC motor drivers, uninterruptible power supplies (UPS), and AC appliances run from an automobile battery.

Power inverters produce one of three different types of wave output:

1. Square wave
2. Modified square wave (modified sine wave)
3. Pure sine wave (true sine wave)

The three different wave signals represent three different qualities of power output. Square wave inverters result in uneven power delivery, which is inefficient for running most devices. Square wave inverters were the first type of inverters made.

9.3 Modified Sine Wave

Modified sine wave inverters were the second generation of power inverter. The modified sine wave inverter provides a cheap and easy solution to powering devices that need AC power. Modified sine wave inverters approximate a sine wave and have low enough harmonics that do not cause problems with household equipment. They do have some drawbacks as not all the devices work properly on a modified

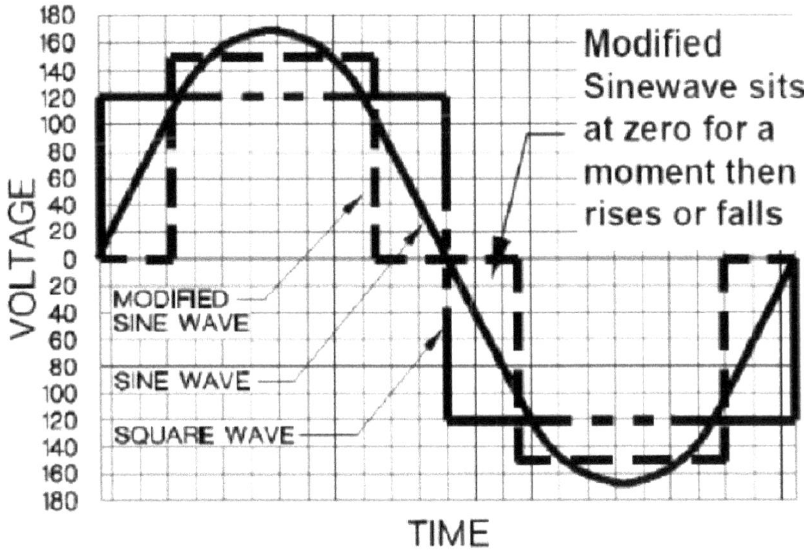

FIGURE 9.2 Square, modified, and pure sine wave.

sine wave; products such as computers and medical equipment need pure sine wave inverters. The main disadvantage of the modified sine wave inverter is that peak voltage varies with the battery voltage. Figure 9.2 shows the square, modified, and pure sine wave.

9.4 Pure Sine Wave

The pure sine wave inverter represents the latest inverter technology. The waveform produced by these inverters is the same as or better than the power delivered by the utility. Usually, sine wave inverters are more expensive than modified sine wave inverters due to their added circuitry.

9.5 Pulse-Width Modulation

Pulse-width modulation (PWM) is a powerful technique for controlling analog processors with digital outputs. There is a wide variety of PWM applications, ranging from measurement and communications to power control and conversion. In PWM inverter harmonics will be much higher frequencies than for a square wave, making filtering easier.

In PWM, the amplitude of the output voltage can be controlled with the modulating waveforms. Reduced filter requirements to decrease harmonics and the control of the output voltage amplitude are two distinct advantages of PWM. Disadvantages include more complex control circuits for the switches and increased losses due to more frequent switching. Control of the switches for sinusoidal PWM output requires (1) a reference signal, sometimes called a modulating or control signal, which is a sinusoidal in this case; and (2) a carrier signal, which is a triangular wave that controls the switching frequency. Figure 9.3 shows the full-bridge inverter circuit.

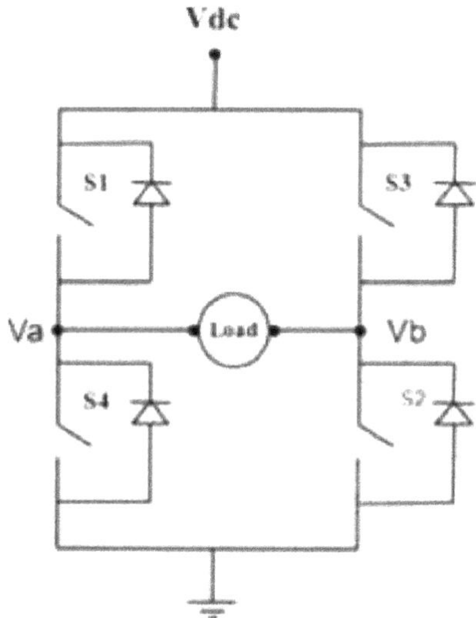

FIGURE 9.3 A full-bridge inverter.

9.6 Bipolar Switching

When the instantaneous value of the sine reference is larger than the triangular carrier, the output is at $+V_{dc}$, and when the reference is less than the carrier, the output is at $-V_{dc}$.

$$V_0 = +V_{dc} \text{ for } V_{sin} > V_{tri}$$
$$V_0 = -V_{dc} \text{ for } V_{sin} < V_{tri} \tag{9.1}$$

This version of PWM is bipolar because the output alternates between plus and minus the DC power supply voltage.

From Figure 9.4, we can see that

S1 and S2 are on when $V_{sin} > V_{tri}$
S3 and S4 are on when $V_{sin} < V_{tri}$

9.7 Unipolar Switching

In a unipolar switching scheme as shown in Figure 9.5 for PWM, the output is switched from either high to zero or low to zero, rather than between high and low, as in bipolar switching. For unipolar switching, control is as follows:

S1 is on when $V_{sin} > V_{tri}$
S2 is on when $-V_{sin} < V_{tri}$
S3 is on when $-V_{sin} > V_{tri}$
S4 is on when $V_{sin} < V_{tri}$
(S1, S4) and (S2, S3) are complementary.

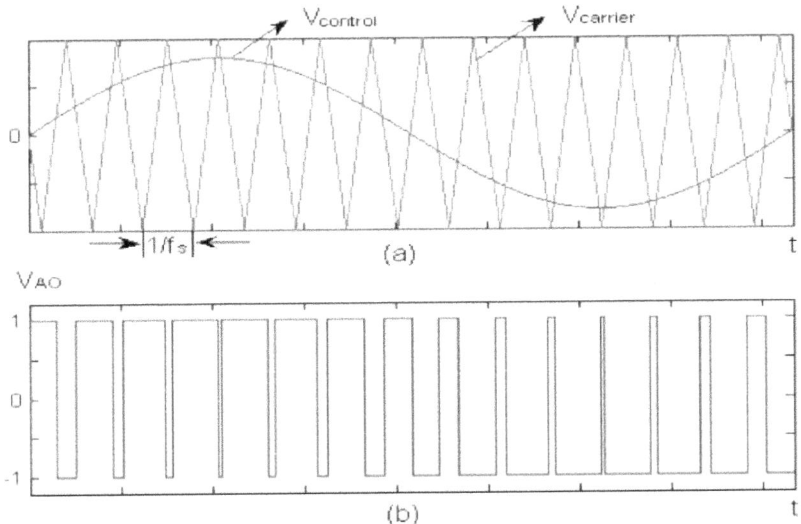

FIGURE 9.4 Bipolar pulse-width modulation. (a) Sinusoidal reference and triangular carrier. (b) Output is $+V_{dc}$ when $V_{sin} > V_{tri}$ and is $-V_{dc}$ when $V_{sin} < V_{tri}$.

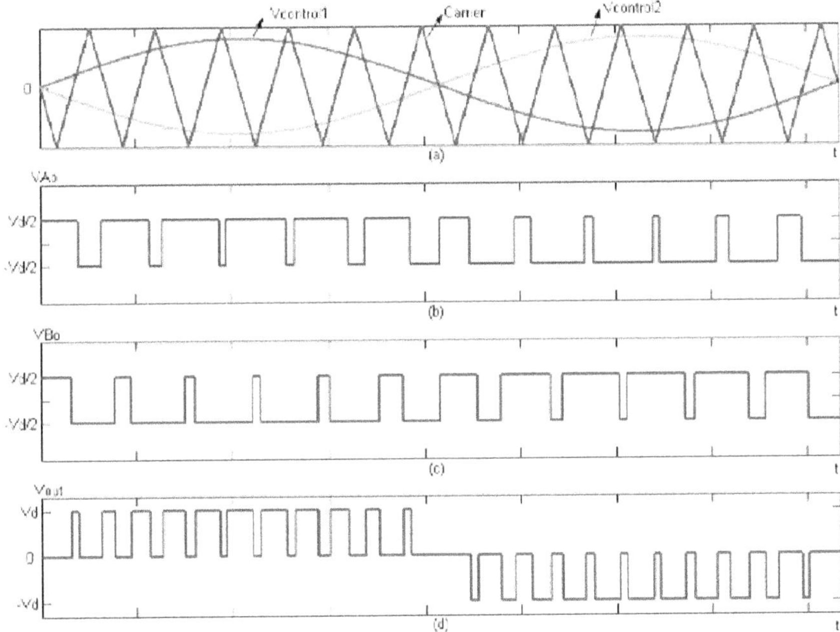

FIGURE 9.5 Unipolar PWM: (a) comparison of carrier and reference, (b) pulses for switches 1,2, (c) pulses for switches 3,4, (d) output voltage waveform.

9.8 Modified Unipolar Switching

In the modified unipolar PWM approach, two arms switch at different frequencies: one is at fundamental frequency while the other one is at the carrier frequency, thus having two high frequency switches and two low frequency switches. It also produces a unipolar output voltage waveform changing between 0 and $+V_{dc}$ or between 0 and $-V_{dc}$. Figure 9.6 depicts the Modified Unipolar PWM scheme and output voltage.

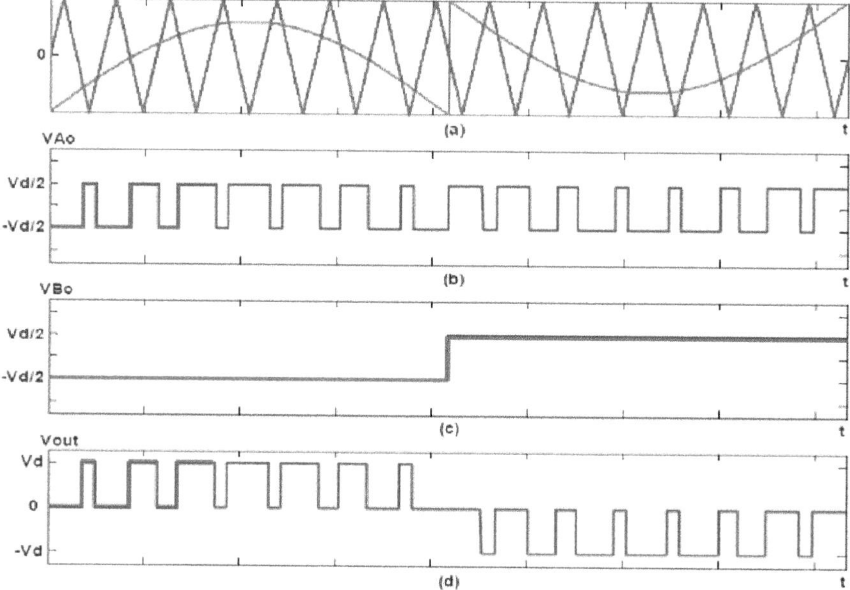

FIGURE 9.6 Modified unipolar PWM: (a) comparison of carrier and reference, (b) pulses for switches 1,2, (c) pulses for switches 3,4, (d) output voltage waveform.

In this switching scheme,

S1 is on when $V_{sin} > V_{tri}$ (high frequency)
S4 is on when $V_{sin} < V_{tri}$ (high frequency)
S2 is on when $V_{sin} > 0$ (low frequency)
S3 is on when $V_{sin} < 0$ (low frequency)

9.9 Voltage Source Inverter

The main objective of static power converters is to produce an AC output waveform from a DC power supply. These are the types of waveforms required in adjustable-speed drives (ASDs), UPSs, static var compensators, active filters, flexible AC transmission systems (FACTS), and voltage compensators, which are only a few applications. For sinusoidal AC outputs, the magnitude, frequency, and phase should be controllable.

According to the type of AC output waveform, these topologies can be considered as voltage source inverters (VSIs), where the independently controlled AC output is a voltage waveform. These structures are the most widely used because they naturally behave as voltage sources as required by many industrial applications, such as ASDs, which are the most popular application of inverters. Similarly, these topologies can be found as current source inverters (CSIs), where the independently controlled AC output is a current waveform. These structures are still widely used in medium-voltage industrial applications, where high-quality voltage waveforms are required. Static power converters, specifically inverters, are constructed from power switches and the AC output waveforms are therefore made up of discrete values. This leads to the generation of waveforms that feature fast transitions rather than smooth ones.

For instance, the AC output voltage produced by the VSI of a standard ASD is a three-level waveform. Although this waveform is not sinusoidal as expected (Figure 9.1b), its fundamental component behaves as such. This behavior should be ensured by a modulating technique that controls the amount of time and the sequence used to switch the power valves on and off. The modulating techniques most used are the carrier-based technique (e.g., sinusoidal pulse-width modulation, SPWM), the space-vector (SV) technique, and the selective-harmonic-elimination (SHE) technique.

9.10 Three-Phase Voltage Source Inverter

Single-phase VSIs cover low-range power applications and three-phase VSIs cover the medium- to high-power applications. The main purpose of these topologies is to provide a three-phase voltage source, where the amplitude, phase, and frequency of the voltages should always be controllable. Although most of the applications require sinusoidal voltage waveforms (e.g., ASDs, UPSs, FACTS, VAR compensators), arbitrary voltages are also required in some emerging applications (e.g., active filters, voltage compensators).

The standard three-phase VSI topology is shown in Figure 9.4 with eight valid switch states. As in single-phase VSIs, the switches of any leg of the inverter (S1 and S4, S3 and S6, or S5 and S2) cannot be switched on simultaneously because this would result in a short circuit across the DC link voltage supply. Similarly, in order to avoid undefined states in the VSI, and thus undefined AC output line voltages, the switches of any leg of the inverter cannot be switched off simultaneously as this will result in voltages that will depend upon the respective line current polarity. Of the eight valid states, two of them (7 and 8) produce zero AC line voltages. In this case, the AC line currents freewheel through either the upper or lower components. The remaining states (1 to 6) produce non-zero AC output voltages. In order to generate a given voltage waveform, the inverter moves from one state to another. Thus, the resulting AC output line voltages consist of discrete values of voltages that are Vi, 0, and –Vi for the topology shown in the selection of the states. The modulating technique was used to generate the given waveform, which should ensure the use of only the valid states.

9.11 Simulation and Results

Figure 9.7 shows the Simulink model of nonlinear load. Figure 9.8 shows the current waveform of the bridge rectifier and Figure 9.9 shows the voltage waveform. Figure 9.10 shows the Simulink model of the SPWM pulse. Figure 9.11 displays the output of the scope of comparing the carrier wave and the sine wave and Figure 9.12 shows the Simulink output SPWM pulse.

Figure 9.13 shows the Simulink model of the DHAF harmonic active filter system and Figure 9.14 depicts the output of the DHAF harmonic active filter system.

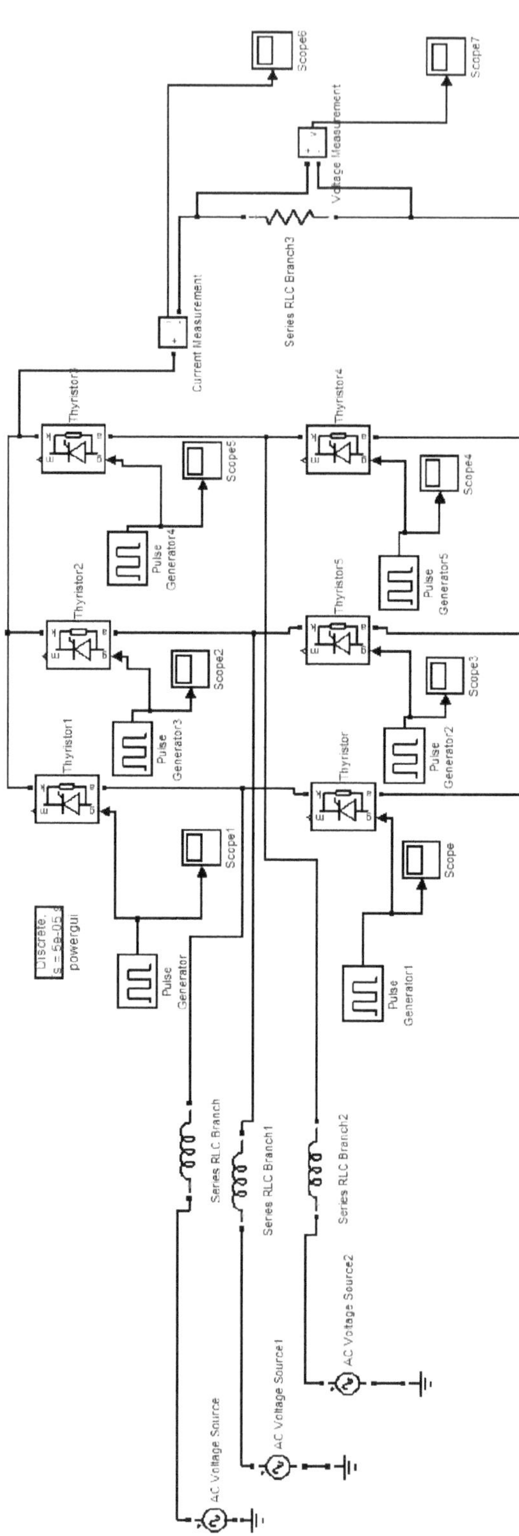

FIGURE 9.7 Simulink model of nonlinear load.

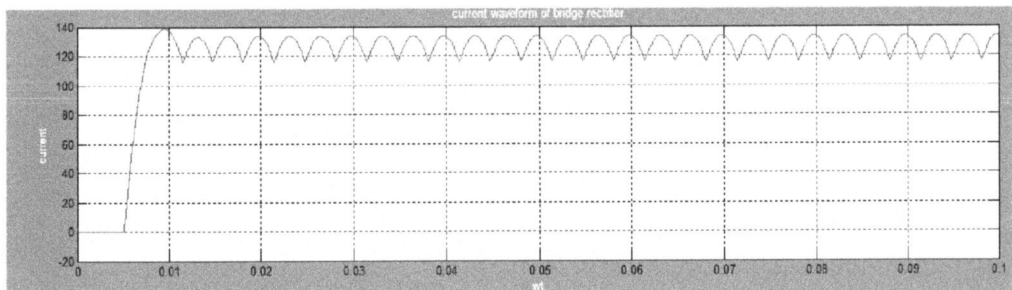

FIGURE 9.8 Current waveform of bridge rectifier.

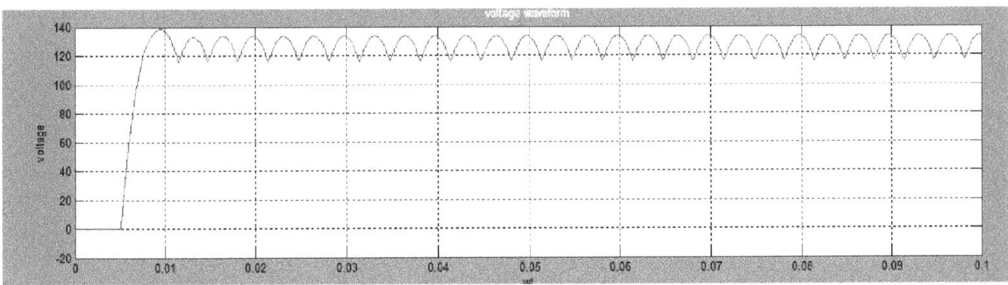

FIGURE 9.9 Voltage waveform.

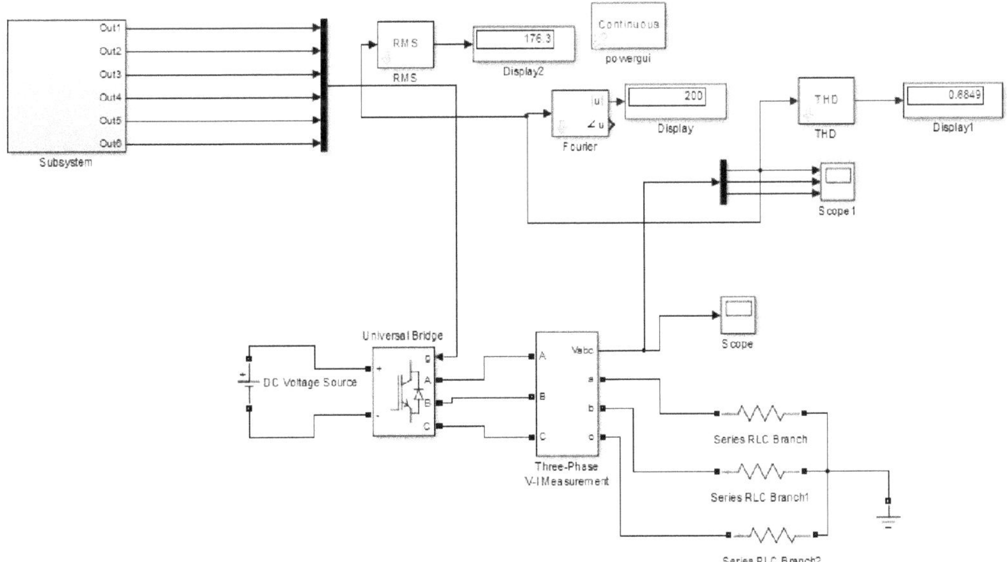

FIGURE 9.10 Simulink model of SPWM pulse.

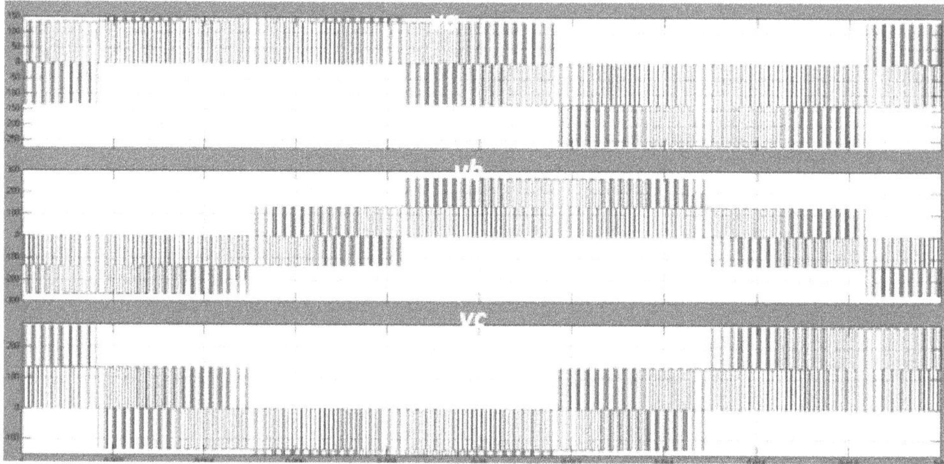

FIGURE 9.11 Output waveform comparing carrier wave and sin wave.

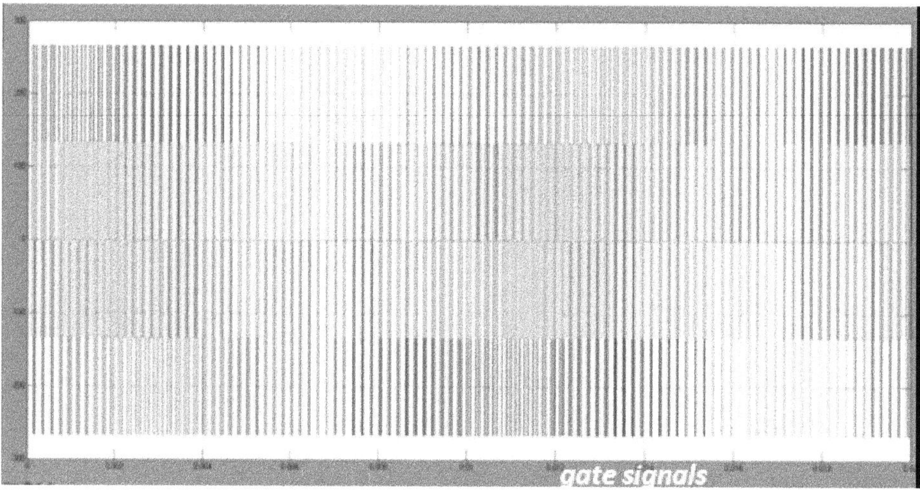

FIGURE 9.12 SPWM pulse waveforms.

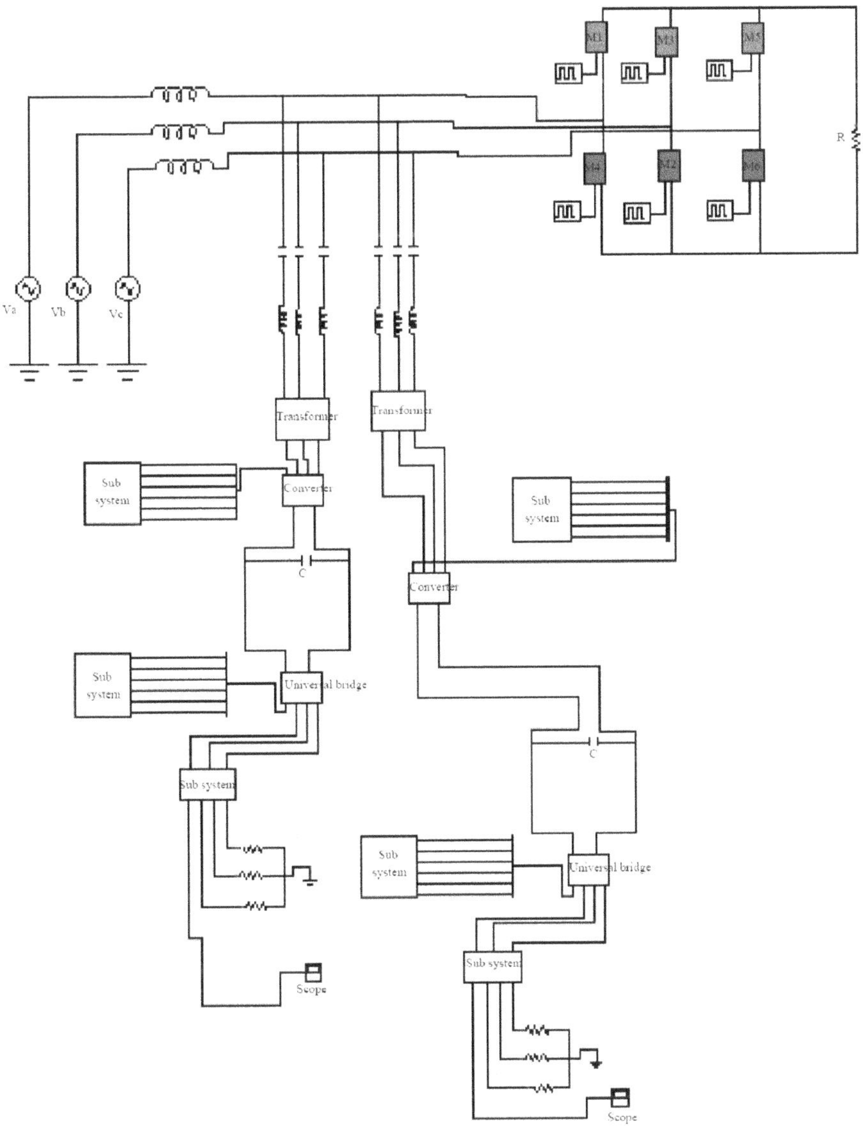

FIGURE 9.13 Simulink model of DHAF harmonic active filter system.

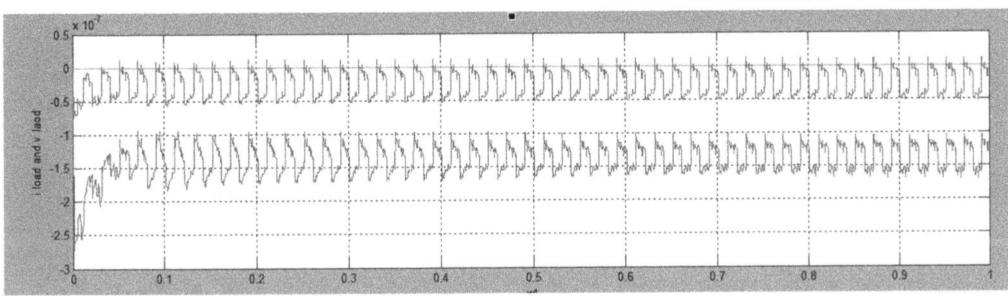

FIGURE 9.14 Output waveform of DHAF harmonic active filter system.

10

AC–DC Boost Converter Control for Power Quality Mitigation

10.1 Introduction

Power factor correction is a mandatory functionality of electronic products in the industrial and commercial market in order to mitigate grid harmonics and operate a power system economically. Since numerous unidirectional AC–DC converters can be connected with AC power systems, existing commercial converters possess the ability to improve substantially the stability of AC power systems by compensating harmonic current and reactive power. The load characteristics of most PFC applications such as home appliances, battery chargers, switched mode power supplies, and other digital products support unidirectional power flow.

The general AC–DC boost converter with step-up chopper is considered a popular topology. This is because it is low in cost, simple, and its performance is well-proven. Its main task inside the system is to maintain DC-link voltage constantly in order to feed loads at different power ratings. In addition, it is necessary to control input current with a pure sinusoidal waveform in phase with input.

Active power filters are another approach capable of improving grid power quality. Conventionally, topologies with bidirectional power flow are used for APF applications. Despite their excellent performance, they may not be the best solution to improve the power quality of an entire power system due to high capital and operating costs related to space and installation, as well as their intrinsic power losses.

Since numerous unidirectional AC–DC converters are connected with AC power systems, existing unidirectional AC–DC boost converters can possess the ability to improve substantially the stability of AC power systems if they can compensate harmonic current and reactive power while fulfilling their basic function of furnishing constant DC-bus voltage.

10.2 Unidirectional AC to DC Boost Converter

The unidirectional AC–DC boost converter is realized as a bridgeless PFC converter. The configuration is given in Figure 10.1. There are a few commercial power modules including IGBTs, gate circuits, and protection circuits already available in the market, which accelerates the application of this topology to home appliances and digital products. The control algorithms are almost the same as the conventional AC–DC converter using a diode rectifier and step-up chopper, except that the bridgeless converter controls AC input current while the conventional one controls rectified AC current.

- The local loads can be considered separately as a non-linear load with harmonic current.
- The linear load with a poor power factor, which is an indicator of poor power quality.

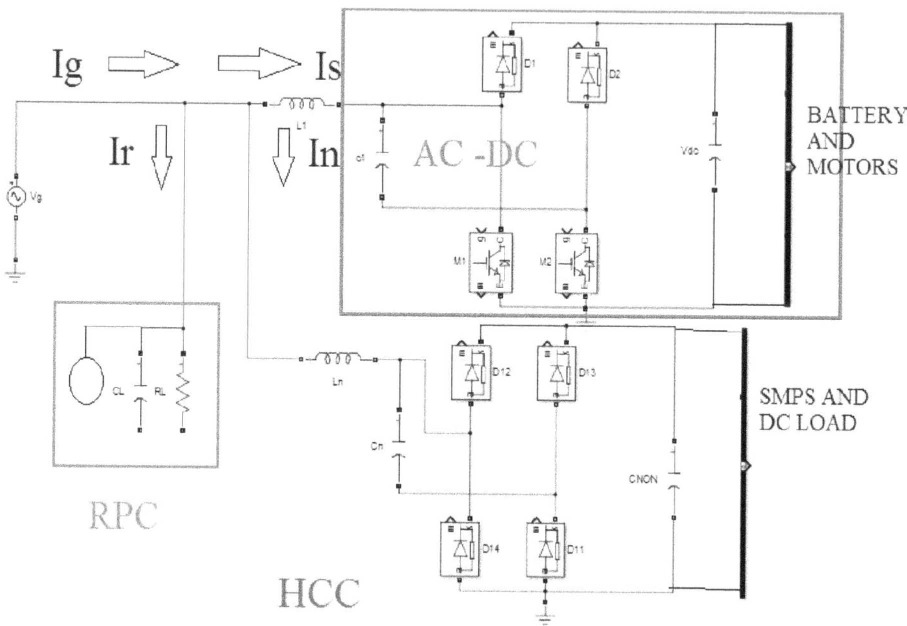

FIGURE 10.1 Example of proposed configuration.

Conventional converters consider the input current to be a purely sinusoidal waveform in phase with the input voltage. The control method can ameliorate harmonic current and reactive power for improved grid power quality as well as regulation of DC-bus voltage. The unidirectional AC–DC converter has three operation modes, i.e., power factor correction (PFC), harmonic current compensation (HCC), and reactive power compensation (RPC). Also, both HCC and RPC can be simultaneously used to improve the distortion and the displacement factors of the grid current. The general block is given in Figure 10.2.

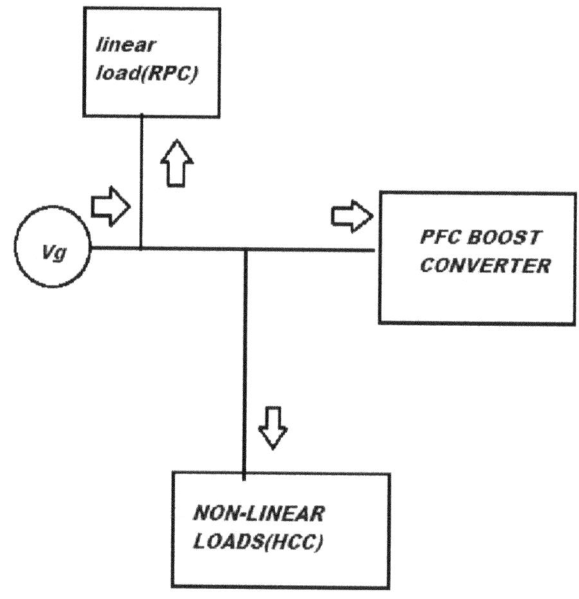

FIGURE 10.2 General block.

10.3 PFC Control

The dual-boost PFC converter is often called the bridgeless PFC converter. Conventional-boost PFC converters are composed of a full-bridge AC to DC diode rectifier followed by a boost converter. The bridge rectifier suffers from high conduction losses, which lowers the system efficiency. A rectifier with the boost PFC topology has been analyzed. The power processors usually consist of some power conversion stages where the operation of these stages is decoupled on an instantaneous basis by means of energy storage elements such as capacitors and inductors. Therefore, the instantaneous power input does not have the equal instantaneous power output. Thus, a converter is a basic module of power electronic systems. It utilizes power semiconductor devices controlled by signal electronics and possibly energy storage elements such as inductors and capacitors.

Most electronic equipment are supplied by 50 or 60 Hz utility power supplies. Almost all power systems of the equipment are processed through some kinds of power converters such as AC to DC, DC to DC, or DC to AC. AC to DC power conversion of electric power is commonly used in adjustable-speed drives, switch-mode power supplies, uninterrupted power supplies for communication system equipment, test equipment, etc. Most of the power supplies have input bridge rectifier with filter capacitor and odd harmonics. However, this harmonic is bad and causes problems in power systems. Moreover, this harmonic content in the pulsating current causes losses. It creates dielectric stresses in capacitors and power distribution cables. As the power factor deteriorates, the distribution current become higher. If there is no power factor correction circuit in the devices then the capacitive filter draws pulsating currents from the AC mains. So, the power quality deteriorates and the higher harmonic content in the power creates bad effects for other users fed from the same line.

Higher harmonics in the line affect the utility grid and other users' appliances as well. A power factor correction circuit used between line and nonlinear load can give both stable output DC voltage and input-side high power factor. This smart feature makes the PFC circuit a more attractive and smart option for power supplies of different power conversions or in different appliance power supplies. Therefore, different kinds of topology and control processes have been developed and evaluated to make the power quality better and improve the power factor. PFC boost topology has been used to improve the power quality and an improved control method is shown Figure 10.3.

10.4 Control Strategy of PFC Control

A simple circuit diagram of unidirectional AC–DC boost converters with an input inductor L and its parasitic resistor R is shown in Figure 10.4. Kirchhoff's voltage law with the source voltage v_g, the switch voltage vd, and the input line current yields to

$$v_g = R * is + L\frac{dis}{dt} + vd \tag{10.1}$$

v_g is the instantaneous value of the source voltage expressed as v_g sin(ωt). The switch voltage is always a major factor in determining the waveform of the input current. In other words, when producing a sinusoidal input current, the switch voltage has to emulate the source voltage identically, with the exact phase difference due to input impedance.

The average switch voltage over a switching cycle in continuous conduction mode

$$vd = (1 - \mathrm{d})vdc \tag{10.2}$$

The average on-time duty ratio of the switches is "d" and *v*dc is the DC output voltage. When the source voltage is in the negative half period, the sign of the input current and switch voltage will be opposite. Therefore, the source voltage can be considered as a rectified voltage, which can be expressed by the absolute sign. This indicates that the bridgeless PFC converter is identical to the general boost converter,

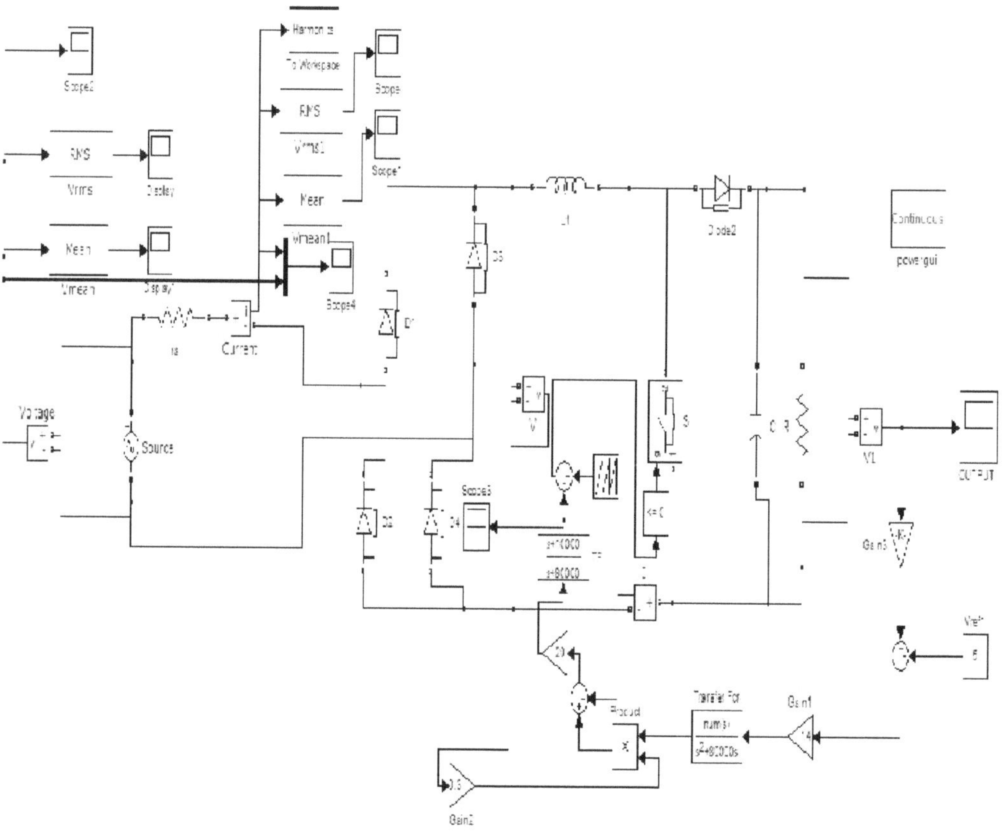

FIGURE 10.3 PFC boost topology.

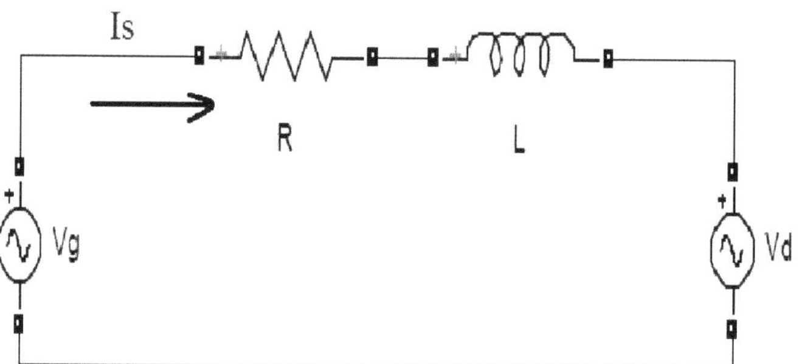

FIGURE 10.4 System equivalent.

using the single switch in rectified DC link, with regard to its operational principle. The general block diagram of PFC boost control is shown in Figure 10.5.

The duty ratio equation can be obtained as

$$\partial = \frac{1}{V_{\mathrm{dc}}}\left(R * is + L\frac{dis}{dt}\right) + \left(1 - \frac{|vg|}{V_{\mathrm{dc}}}\right) \tag{10.3}$$

Theoretically, the duty ratio should be generated for the ideal switch voltage as accurately as possible through adequate converter compensators to yield sinusoidal input current. In order to classify the duty ratio d of the system, the feedback duty ratio and the feed forward duty ratio can be considered

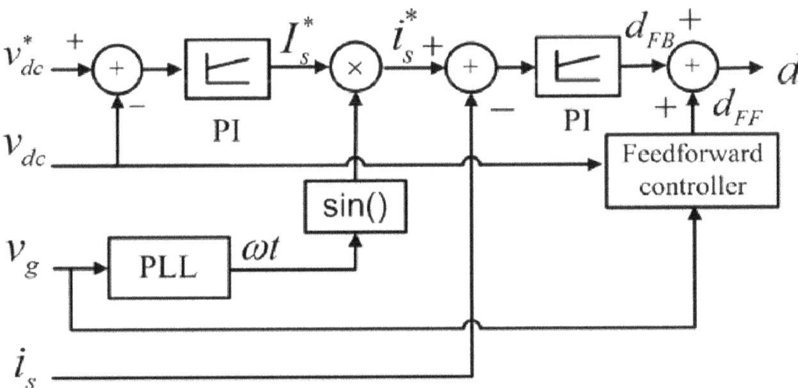

FIGURE 10.5 PFC boost control.

separately. dFB produces the exact phase difference between the source voltage and the average switch voltage. dFF produces the inverse of the source voltage waveform as the average switch voltage. Hence, the input current tracking is improved and the frequency range for which input admittance acts purely as a resistance can be extended to higher frequencies due to feed-forward control.

10.5 Reactive Power Compensation Control Mode

Unlike nonlinear loads, the current waveform of a linear load is sinusoidal at the frequency of the power system, but the power factor can be significantly exacerbated when the load is capacitive or inductive. The current waveform of a typical inductive load in a single-phase induction motor is shown in Figure 10.6. The current flow, consisting of the converter current with RPC and the load current *ir* consuming reactive power is also shown in Figure 10.7.

$$is = is\alpha - jis\beta \tag{10.4}$$

$$ir = ir\alpha - jir\beta \tag{10.5}$$

$$ig = ir\alpha + is\alpha - j(ir\beta - is\beta) \tag{10.6}$$

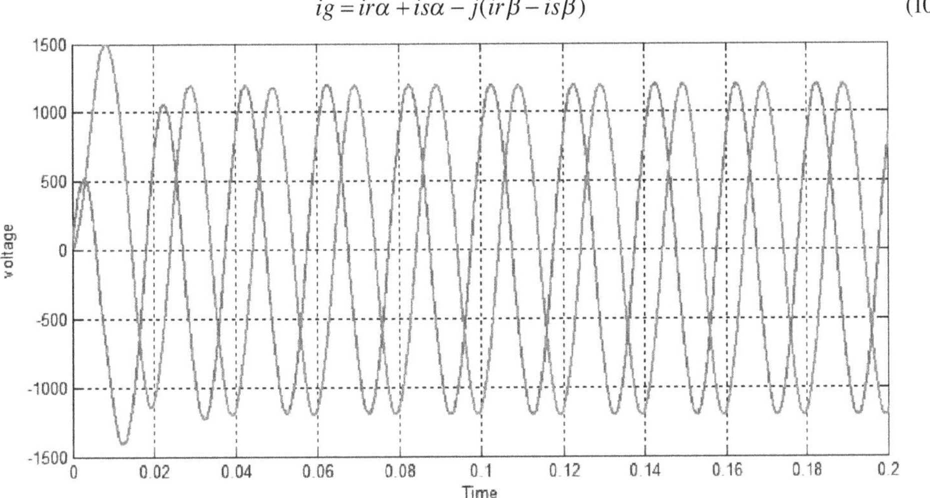

FIGURE 10.6 Reactive power consumed by non linear load. (*Note*: See color eBook for improved color differentiation in figures and labeling.)

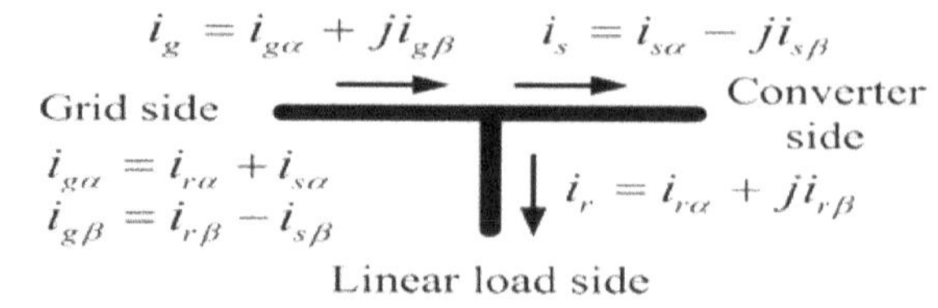

$i_g = i_{g\alpha} + ji_{g\beta}$ $i_s = i_{s\alpha} - ji_{s\beta}$

Grid side Converter side

$i_{g\alpha} = i_{r\alpha} + i_{s\alpha}$

$i_{g\beta} = i_{r\beta} - i_{s\beta}$

$i_r = i_{r\alpha} + ji_{r\beta}$

Linear load side

FIGURE 10.7 Current flow.

As a result, the grid power factor at the PCC can be improved by injecting reactive power from the converter as shown. However, it should be considered that the input current of the unidirectional converter becomes distorted due to the natural commutation of diodes; thus, the amount of reactive power generated by an individual converter should be restricted. A reactive current flow phasor is shown in Figure 10.8. Since the current waveform of the converter in RPC mode is not sinusoidal, the required phase angle of the current cannot be calculated by a simple reactive power equation. Thus, the phase angle reference to the input converter current needs to be generated by employing a proportional integral compensator as shown in Figure 10.9.

$$\varphi = Kpc(Q* - Q) + \int Kic(Q* - Q)dt \tag{10.7}$$

$$i*s = I*s\sin(\omega t + \varphi) \tag{10.8}$$

where *K*pc and *K*ic are proportional gain and integral gain of the reactive power compensator, respectively, and φ is the desired phase to be adjusted from the original current reference. It should be noted that the current magnitude reference *I*s* will be adjusted through the DC-bus voltage controller to feed active power to the DC load. The reactive power will be adjusted by changing the phase angle φ. Thus, initially *I*s* is determined by the DC-link voltage controller and actual active power will change as a result of generating reactive power with respect to the DC link command. However, since *I*s* will be

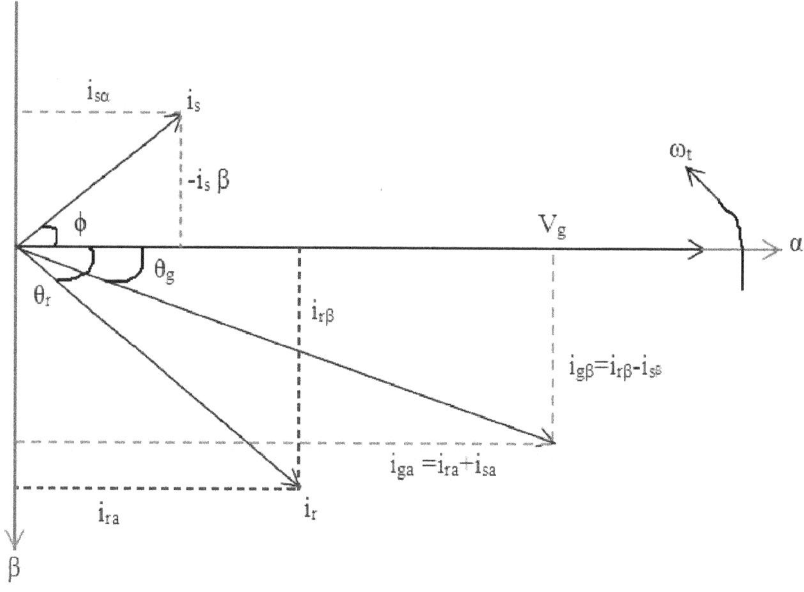

FIGURE 10.8 Reactive current flow phasor.

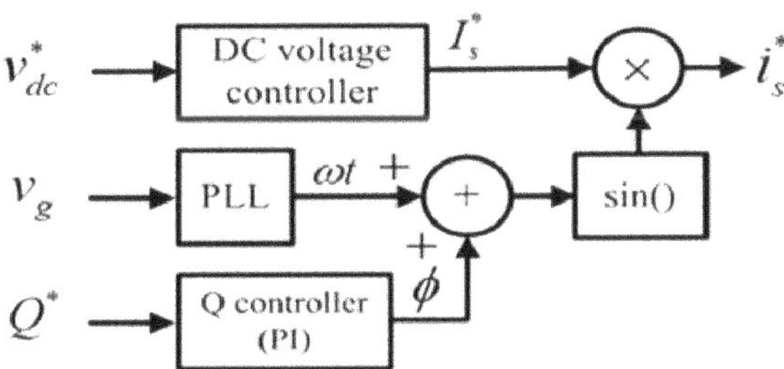

FIGURE 10.9 Reactive current flow control algorithm.

updated by the DC link voltage compensator, as the phase angle φ changes, the DC link voltage will be maintained.

10.6 Harmonic Current Compensation Control Mode

The current waveform of a typical nonlinear load in a single-phase diode rectifier is shown in Figure 10.10. Generally, the distorted load current i_{non} can be written in terms of its fundamental i_{fn} and harmonic $i$hn components as

$$i_{non} = I1\sin(\omega 1t + \theta 1) + \sum\nolimits_{n=2,3}^{\infty} I1\sin(\omega 1t + \theta n) \tag{10.9}$$

where $\omega 1$ is the line angular frequency and θn is the phase difference between the source voltage and input current. Assume that the input current from the unidirectional AC–DC boost converter operating in PFC mode is a purely sinusoidal waveform. The grid current ig includes $i$hn from a nonlinear load as shown in Figure 10.11a.

These harmonics are undesirable and should be removed. If the unidirectional AC–DC boost converter can generate the harmonic current capable of canceling the harmonics of the nonlinear load, the grid current will be composed of only fundamental components of the converter current and load current as shown in Figure 10.11b. Therefore, the new current reference for the current controller of the converter can be obtained as shown. Figure 10.12 depicts the harmonic current control mode.

$$is^* = Is^*\sin(\omega t) - i_{\text{hn}} \tag{10.10}$$

where Is^* is the magnitude reference provided by the DC-bus voltage controller.

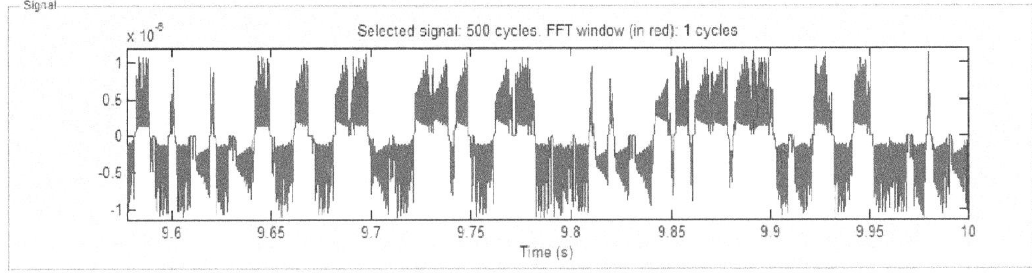

FIGURE 10.10 Example of harmonic current of a non-linear load.

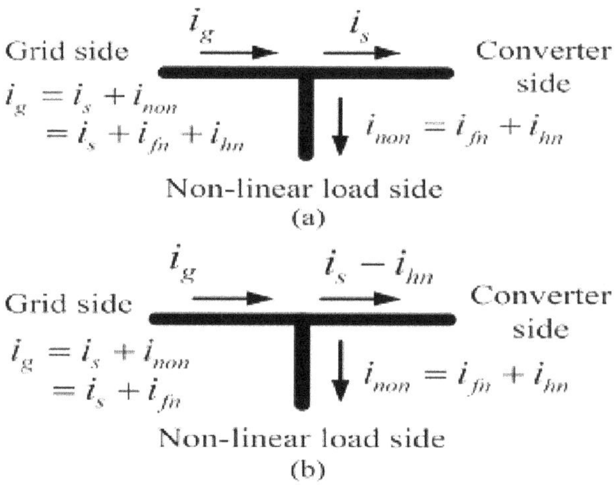

FIGURE 10.11 Harmonic current consumption of a non-linear load (a) with HFC and (b) without HFC.

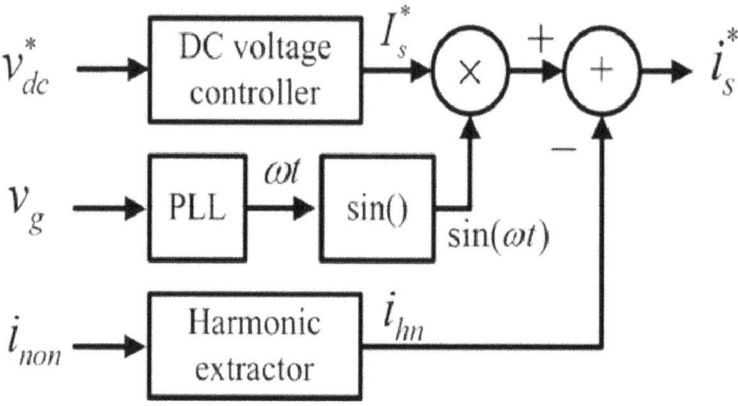

FIGURE 10.12 Harmonic current control mode.

10.7 HCC and RPC Combined Control Strategy

The control strategy of the unidirectional AC–DC converter including a feed-forward controller, HCC, and RPC is shown. Two control blocks for HCC and RPC have been added to the conventional control algorithm. Thus, the final current reference for a versatile control strategy is expressed as

$$is^* = Is * \sin(\omega t + \phi) - i_{hn} \tag{10.11}$$

Figure 10.13 shows the control strategy. The functionalities of HCC and RPC in unidirectional AC–DC boost converters are available only when these converters supply active power to its DC load. Thus, the current reference able to be used for HCC and RPC is highly dependent on its power rating and its existing loads. Since multiple unidirectional converters may be connected to the power system in residential applications, their RPC capabilities can be maximized by incorporating these aggregated converters.

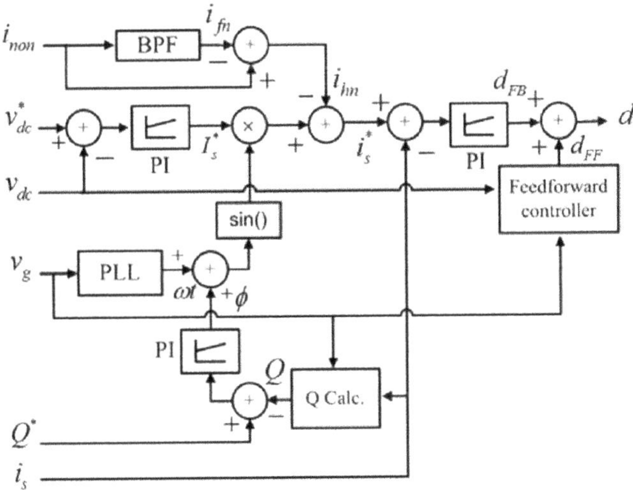

FIGURE 10.13 Control strategy.

Since numerous unidirectional converters are connected with AC power systems, existing unidirectional AC–DC boost converters can possess the ability to improve substantially the stability of AC power systems by maximizing the functionalities of aggregated unidirectional AC–DC boost converters. The control method of the unidirectional AC–DC converter has been presented to enhance the grid power quality through HCC and RPC. The effectiveness of the control method was validated through simulation and experimental results, showing improved power factor and total harmonic distortion of the grid. At the same time, it should be noted that due to the inherent limitations of the unidirectional AC–DC converter, the grid current can be distorted unintentionally when operating in RPC mode. Hence, the amount of reactive power injected from an individual converter to the grid should be restricted.

10.8 Simulations

Figure 10.14 shows the simulation result without PFC. Without closed-loop control, the power circuit was connected to linear and non-linear load and then THD was found in source current to be 55.99%. Figure 10.15 shows the output with PFC. A closed-loop strategy of harmonic current compensation and reactive power compensation together achieves a reduction in harmonics of about 20%. Figure 10.16 shows the Simulink block with PFC control. This is the control algorithm of conventional AC to DC boost converter topology. The duty cycle is being generated at d. Figure 10.17 shows the HCC control in Simulink. The control algorithm was modified with hysteresis controller along with feed-forward controller and duty cycle d was generated for HCC control. Figure 10.18 shows the HCC and RPC control in Simulink. This is the control strategy of duty cycle for the proposed system. The THD was found to be 13%. Figure 10.19 shows the output result with HCC and RPC control in Simulink with THD of 13.73%. The control method of the unidirectional AC–DC converter has been simulated to enhance the grid power quality through HCC and RPC. The THD was reduced by 20% in PFC and 13.73% by both HCC and RPC.

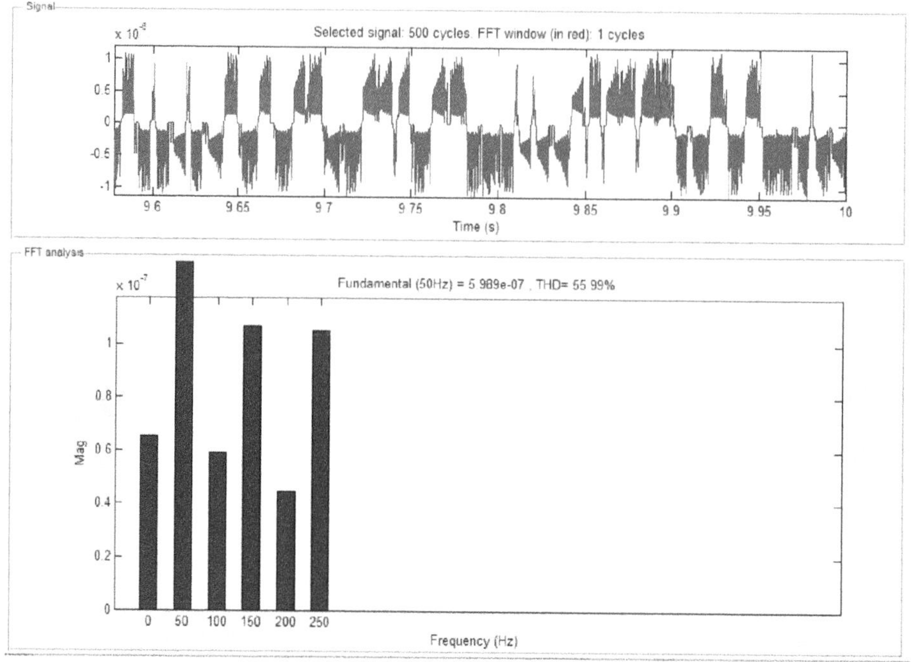

FIGURE 10.14 Without PFC.

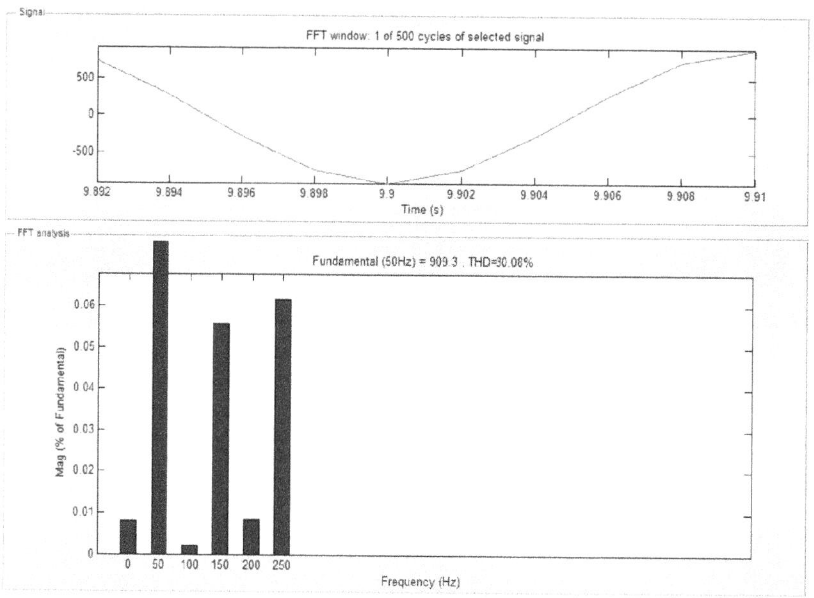

FIGURE 10.15 With PFC.

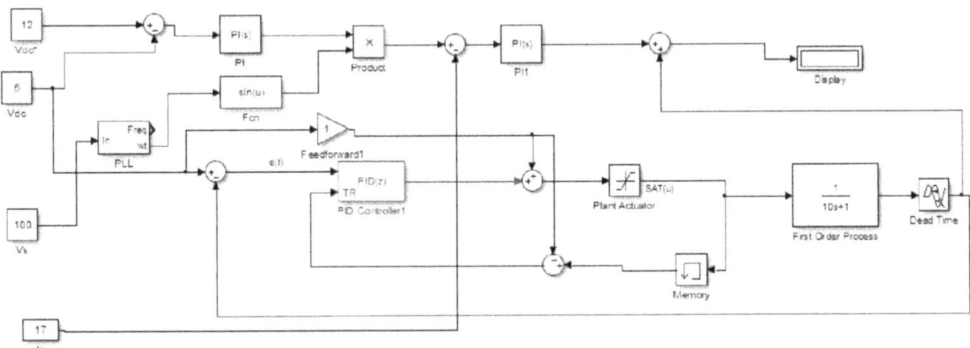

FIGURE 10.16 Simulink block with PFC control.

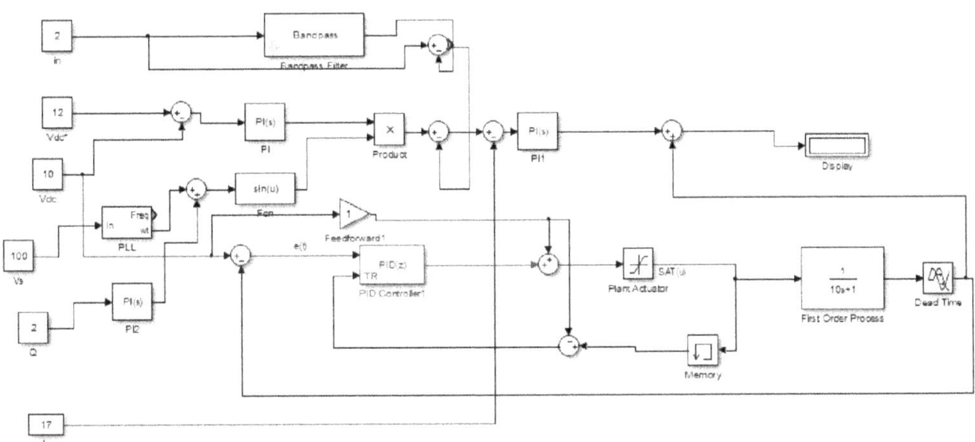

FIGURE 10.17 HCC control in Simulink.

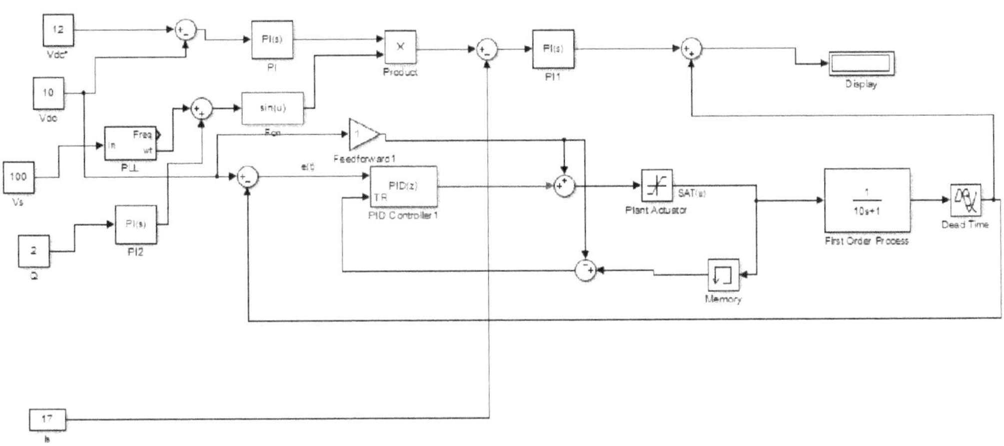

FIGURE 10.18 With HCC and RPC control in Simulink.

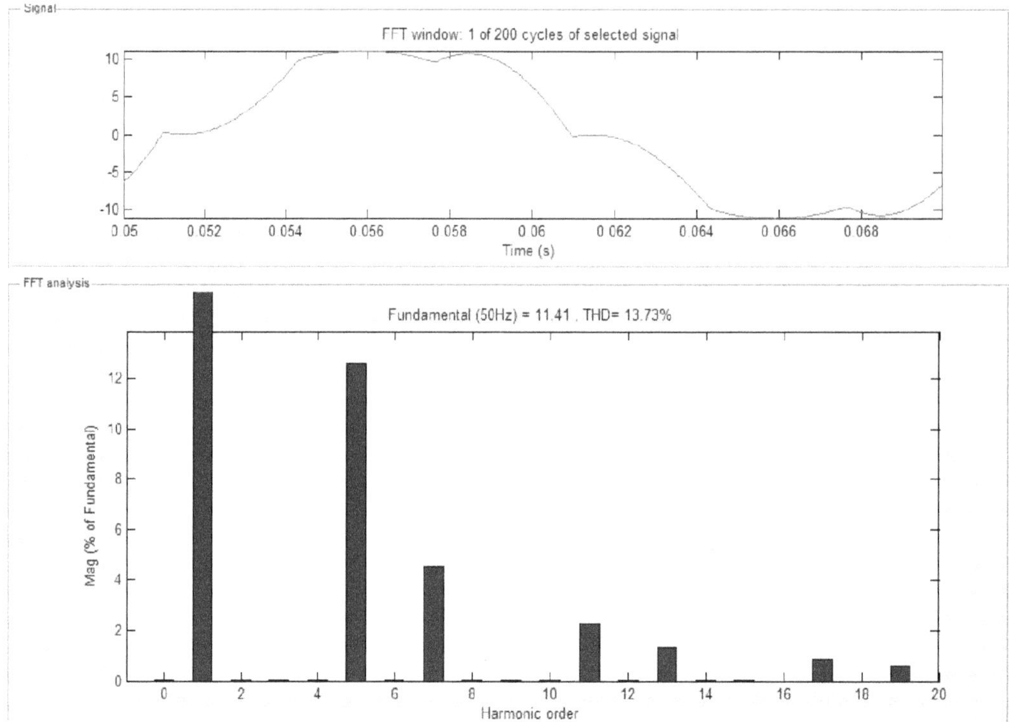

FIGURE 10.19 Output result with HCC and RPC control in Simulink.

11

Harmonic and Flicker Assessment of an Industrial System with Bulk Nonlinear Loads

11.1 Single Line Diagram of the System

The objective is to study and simulate the industrial system for harmonic and flicker assessment. The system consists of an AC arc furnace, DC and AC motor-drive loads, and the static var compensator. The increasing use of power electronic and arcing equipment, such as adjustable-speed motor drives (ASDs) and electric arc furnaces (EAFs), in the industrial environment has led to a growing concern for power-quality (PQ) disturbance and causing considerable impacts on the system and its equipment. Literature survey shows that PQ studies associated with ASDs and EAFs mostly focus on harmonic and flicker assessments and the modeling evaluation. Figure 11.1 shows the single-line diagram of the system.

11.2 System Modeling

In this model, an ideal three-phase AC voltage source model is used to feed electric power to the system, and several linear transformer models are connected to manage the voltage level at each transmission bus.

11.3 EAF Load Model

For a system with EAF loads, it is difficult to describe the arcing behavior of the EAF since the arcing is a non-stationary stochastic process that provides a highly nonlinear v–i characteristic of the EAF. Here, a harmonic voltage source model of EAF is adopted that is considered up to the ninth harmonic, behind lumped impedance (i.e., *Rc* and *Lc*), which represents the cable impedance from the EAF transformer secondary side to the electrode, and the effect of frequency dependence is ignored. However, it is found that the measured arc voltage v measured (t) in the real plant can only be obtained at the secondary side of the furnace transformer due to the difficulty of measuring the arc voltage at the furnace body. Figure 11.2 shows the model of EAF.

11.4 SVC Model

For compensating the reactive-power consumption of the EAF, the SVC, composed of delta-connected thyristor-controlled reactors (TCR) in parallel with fixed-capacitor banks, which often set up as *LC* filters to absorb harmonic currents produced by the TCR, is installed in the studied system. Once the EAF operates, the SVC acts like a shunt-connected variable reactance, which either generates or absorbs reactive power to regulate the voltage at the connection point to the inner network of the plant.

Figure 11.3 shows the single-phase equivalent of the SVC model with parallel passive *LC* filters. In the simulation, a fixed firing angle is adopted to control the TCR thyristors and to supply a constant reactive power for compensation.

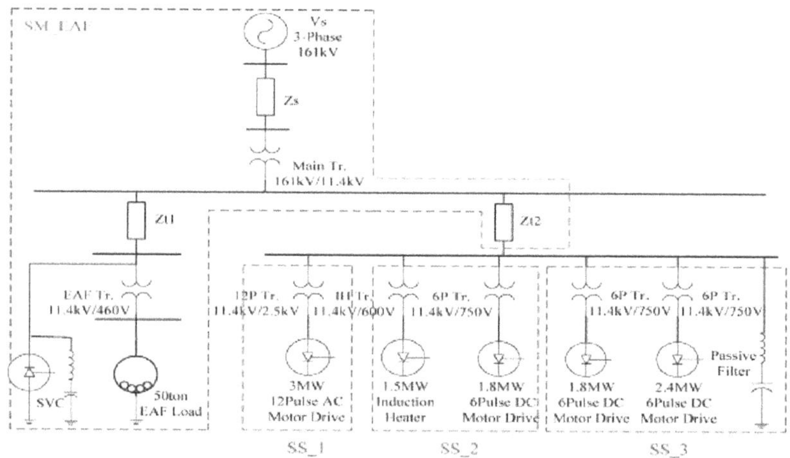

FIGURE 11.1 Single-line diagram of the system.

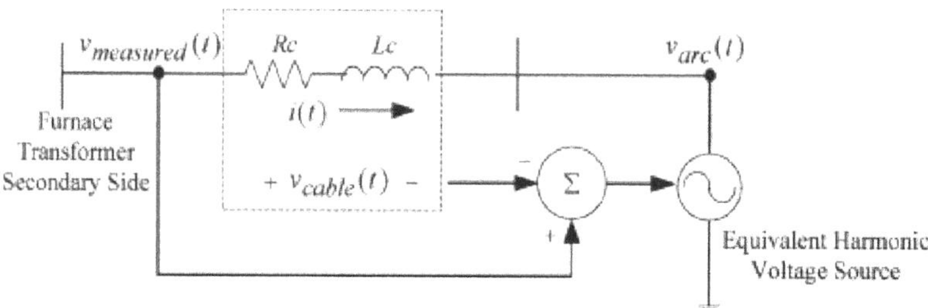

FIGURE 11.2 Model of EAF.

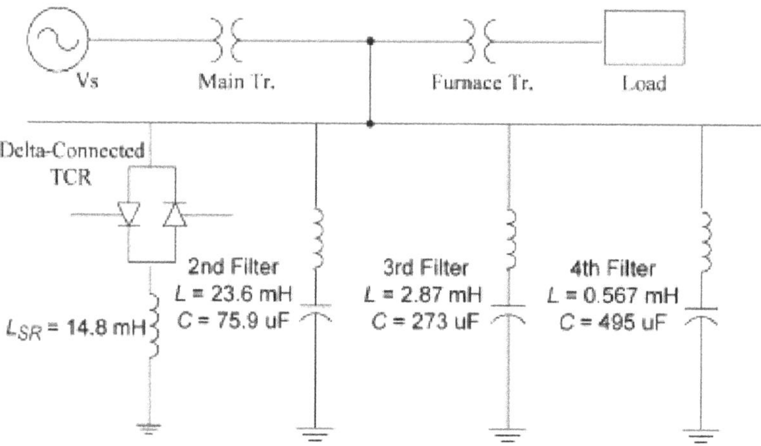

FIGURE 11.3 Equivalent single-phase representation of the SVC.

11.5 Thyristor Bridge Rectifier (6-Pulse)

Figure 11.4 shows the circuit of the 6-pulse AC/DC thyristor bridge rectifier for DC motor drives, where the motor load is modeled as equivalent impedance in series with a back EMF. The DC-link capacitor is also connected to form a low-pass filter for reducing the ripple.

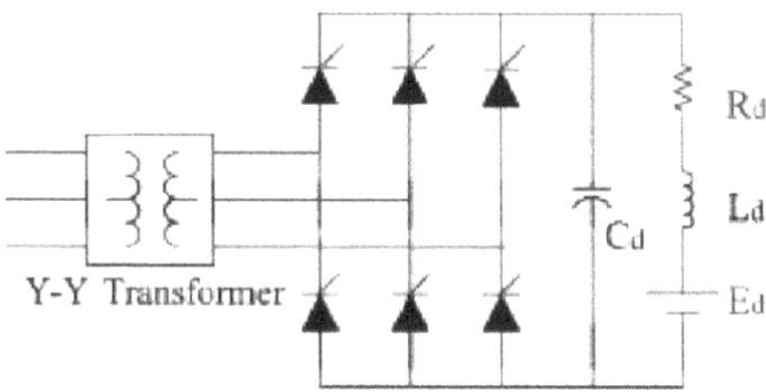

FIGURE 11.4 6-pulse AC/DC thyristor bridge rectifier.

11.6 Simulink Model

To design a thyristor-based MATLAB/Simulink model, choose the following blocks from the Library Simulink Browser:

AC voltage source

Simulink Library Browser>>Simscape>>SimPowerSystems>>Electrical Sources>>AC voltage source

Voltage and current measurement

SimulinkLibraryBrowser>>Simscape>>SimPowerSystems>>Measurements

Voltage measurements, current measurements

R, L, C

Simulink Library Browser>>Simscape>>SimPowerSystems>>Elements>>RLC branch

Transformer

Simulink Library Browser>>Simscape>>SimPowerSystems>Elements>>Three-phase transformer (two windings)

Line

Simulink Library Browser>>Simscape>>SimPowerSystems>>Elements>>Three-phase Pi-section line

Three-phase voltage measurement

Simulink Library Browser>>Simscape>>SimPowerSystems>>Measurements>>Three-phase voltage source measurement

Thyristor bridge

Simulink Library Browser>>Simscape>>SimPowerSystems>>Power electronics>>Universal bridge

Thyristor

SimulinkLibraryBrowser>>Simscape>>SimPowerSystems>>Power electronics

Thyristor

Simulink Library Browser>>Simulink>>User-defined functions>>MATLAB function

Synchronized 6-pulse generator

Simulink Library Browser>>Simscape>>SimPowerSystems>>Extra Library>>Control blocks>>Synchronized 6-pulse generator

Constant

Simulink Library Browser>>Simulink>>Commonly used blocks>>Constant

Simulink Library Browser>>Simulink>>Sinks>>Scope

Figure 11.5 shows the Simulink model of the system without filter and Figure 11.6 shows the Simulink model of the system with filter. Figure 11.7 depicts the Simulink model of EAF.

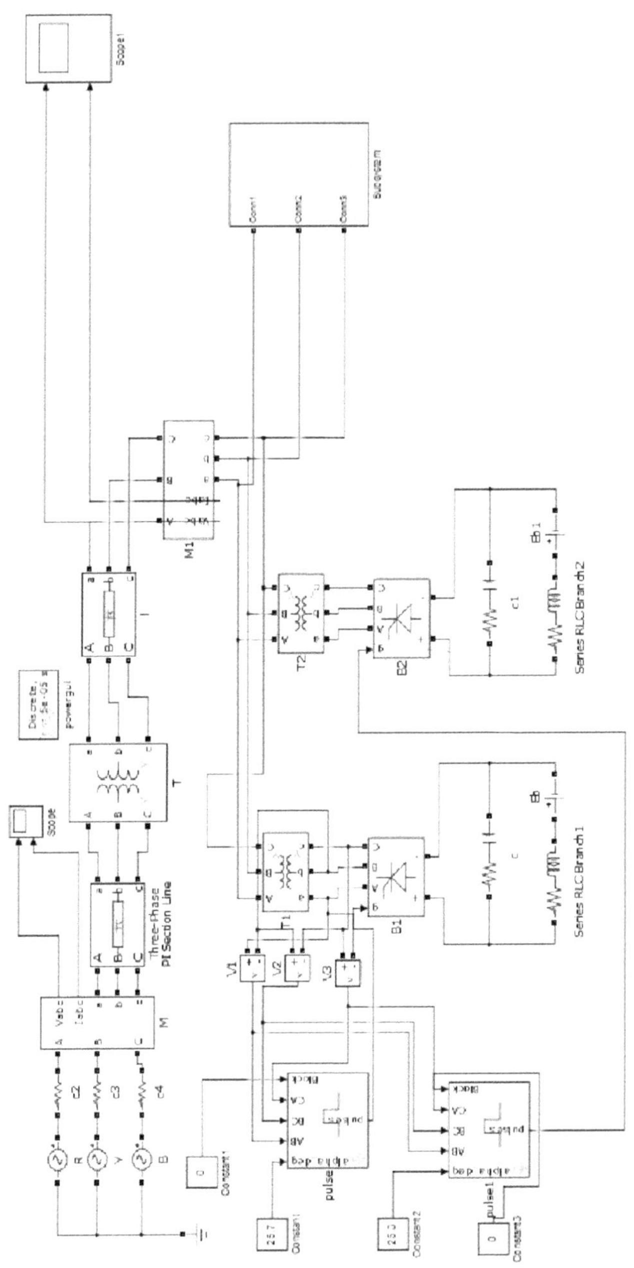

FIGURE 11.5 Simulink model of system without filter.

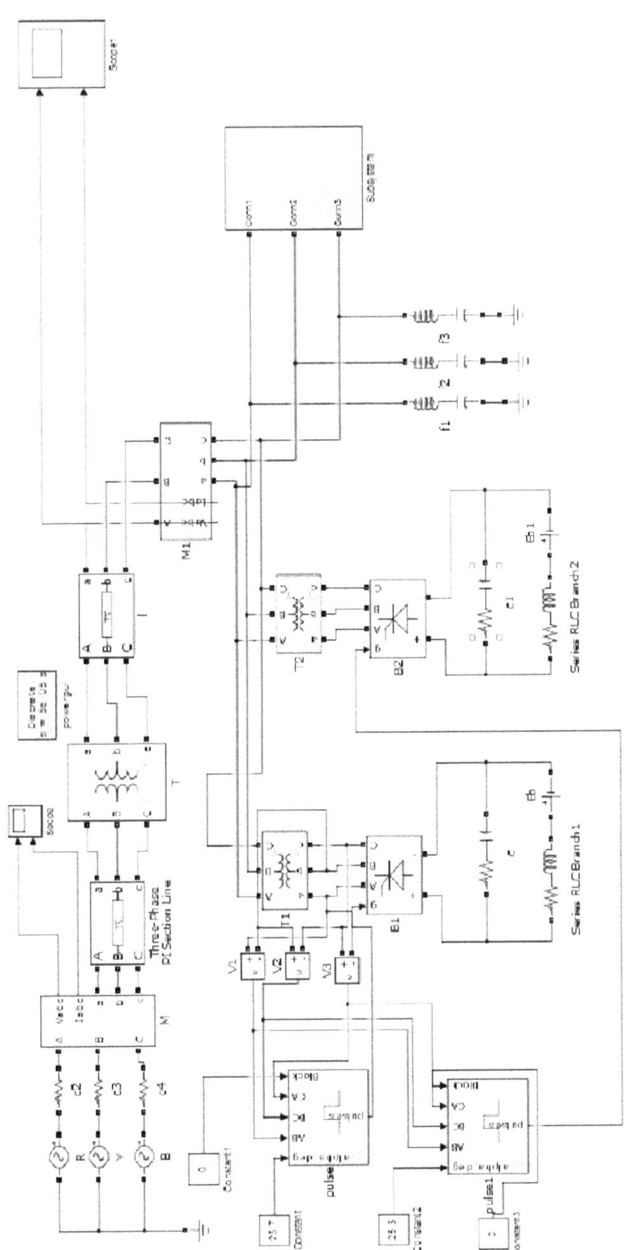

FIGURE 11.6 Simulink model of system with filter.

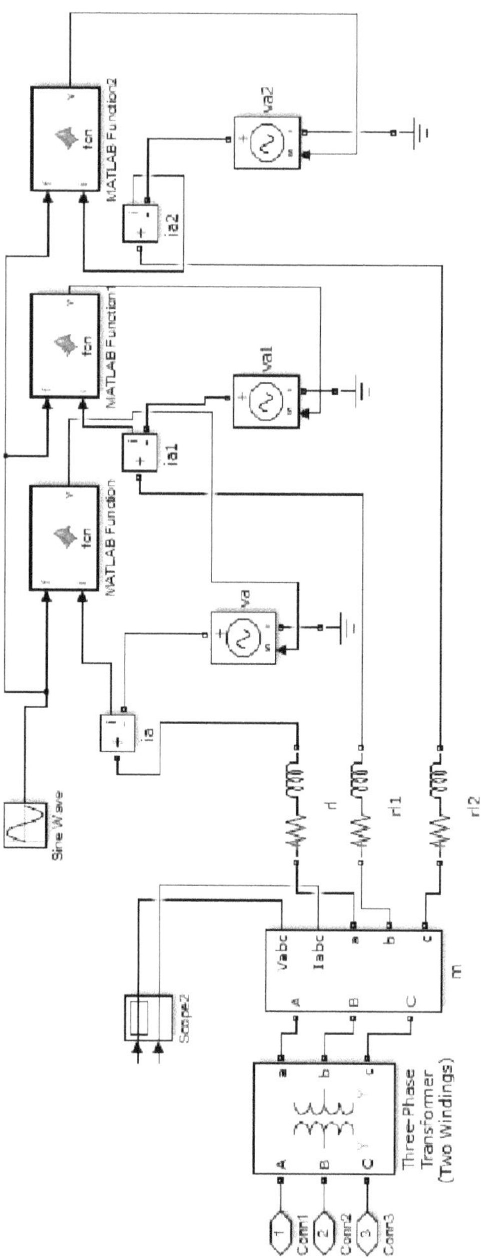

FIGURE 11.7 Simulink model EAF (electric arc furnace).

11.7 Waveforms

Without filter: Figure 11.8 shows the input voltage and current waveforms and Figure 11.9 shows the output voltage and current waveforms. **With filter:** Figure 11.10 depicts the input voltage and current waveform and Figure 11.11 shows the output voltage and current waveform. Thus, the industrial system with the bulk of non-linear loads, modeled after the EAF and simulated in the same way using MATLAB Simulink with and without filters and analyzed the waveforms of the system.

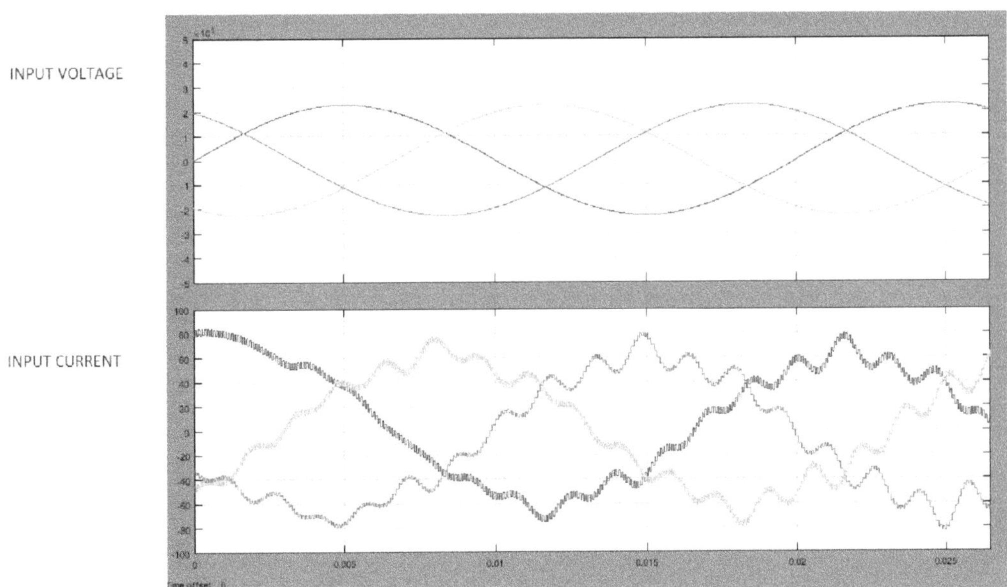

FIGURE 11.8 Input voltage and current waveforms (without filter).

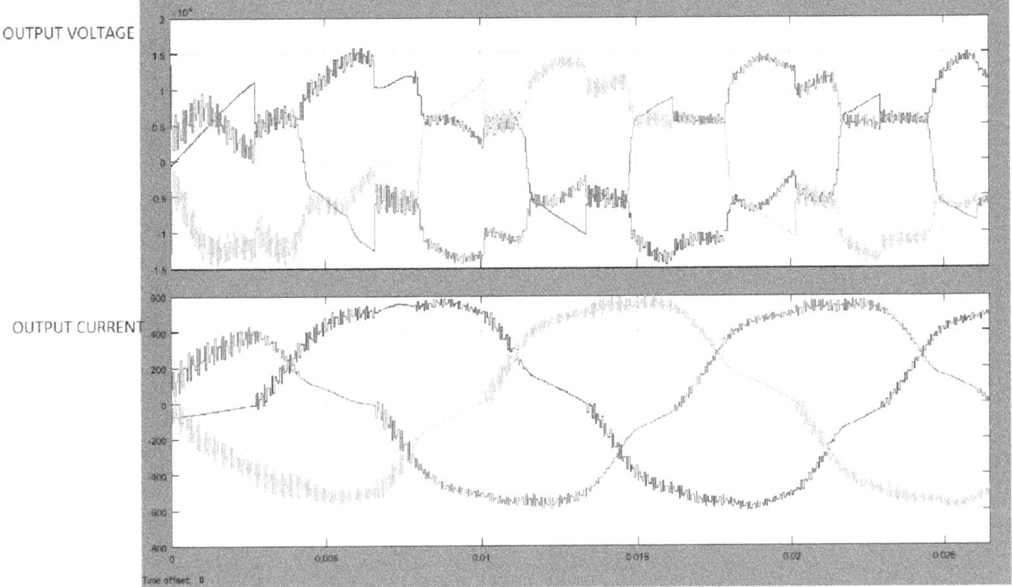

FIGURE 11.9 Output voltage and current waveforms (without filter).

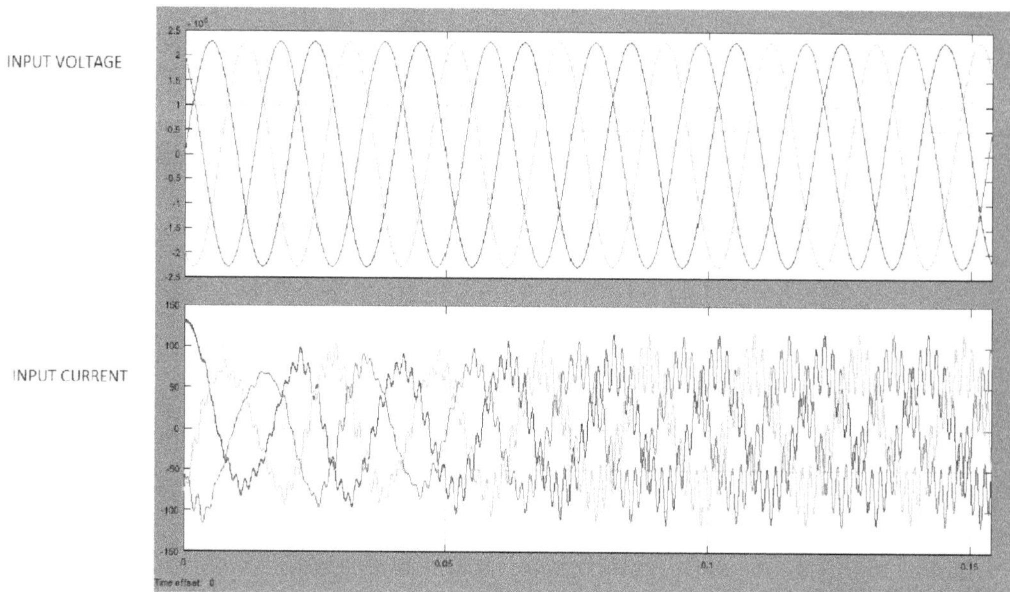

FIGURE 11.10 Input voltage and current waveforms (with filter).

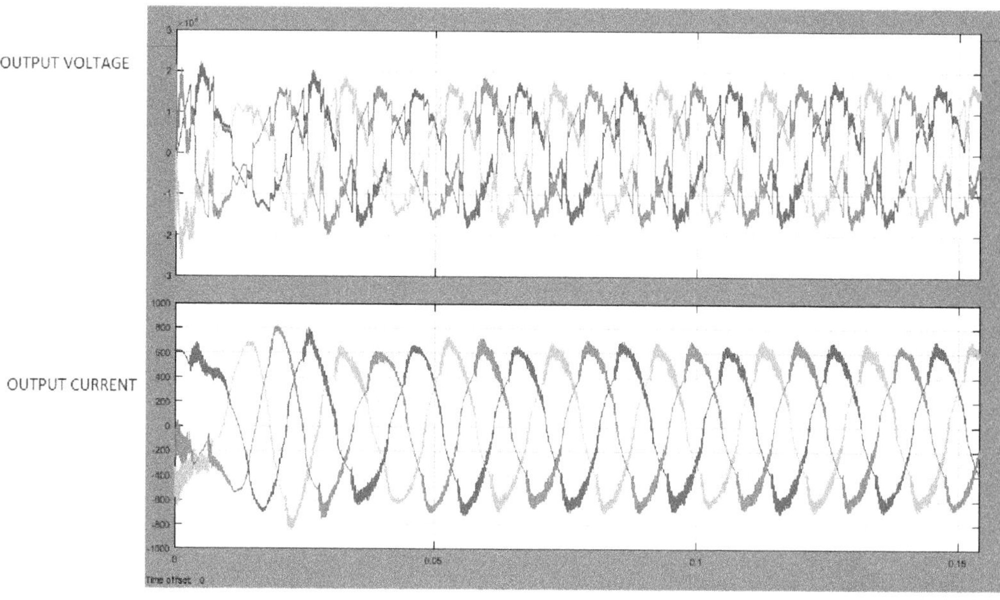

FIGURE 11.11 Output voltage and current waveforms (with filter).

12

LCL Filter Design for Grid-Interconnected Systems

12.1 Introduction

The growing demand for power and limited availability of conventional sources are the two key issues prompting researchers to think of alternative methods of generating power. That's why other non-conventional sources have become popular nowadays. Simultaneously, rising cost and complexity in existing electricity distribution systems and the inability of current systems to serve remote areas reliably has led to a search for alternate distribution methods. One viable solution is use of renewable energy sources directly at the point of load, which is termed as distributed generation (DG).

Most renewable sources of energy, like wind, solar, fuel cell, etc. are interfaced to the existing power supply by a power converter. This eliminates the transmission and distribution losses and improves reliability of the power supply. But use of power converters will also introduce undesirable harmonics that can affect nearby loads at the point of common coupling to the grid. Hence all such converters have a filter to eliminate these harmonics.

This chapter, with the three-phase PV grid-connected inverters topology, first analyzes the inductance, the ration of two inductances, selecting the filter capacitor, and resonance resistance. Based on these theories, an LCL filter is designed. The simulation result proved that the LCL filter achieved the best performance, and indicated the impacts on the stability and filtering property from the parallel resistor or the series resistor.

The present work is on design of such filters for high-power pulse-width modulated voltage-source converters for grid-connected converter applications. The conventional method to interface these converters to grid is through a simple first-order low-pass filter, which is bulky, inefficient, and cannot meet regulatory requirements such as IEEE 512-1992 and IEEE 1547-2008. The design of efficient, compact, higher-order filters to attenuate the switching harmonics at the point of interconnection to the grid to meet the requirement of DG standards of interconnection is studied. Also, different switching schemes for single-phase unipolar full-bridge inverters are studied and compared to get the switching scheme that gives less switching loss. The LCL filter is designed accordingly and optimal inductance and capacitance values are obtained. All the related models are simulated using the MATLAB software and graphs are studied.

12.2 Block Diagram

In Figure 12.1, L1 is the inverter side inductor, L2 is the grid-side inverter, and Cf is a capacitor with a series damping resistor Rf. Traditionally, a simple first-order L filter is used to connect the converter to the grid and reduce injected harmonic currents. However, implementation of third-order LCL filters has been recently proposed by some researchers. Despite many potential advantages, some important parameters must be carefully taken into consideration in design of an LCL filter. We are concerned with the investigation of LCL filters and their effects on the overall performance of a grid-connected VSC when the flow of power is from the DC side to the AC utility grid. Analytical expressions and plots are provided for better understanding of the filter behavior.

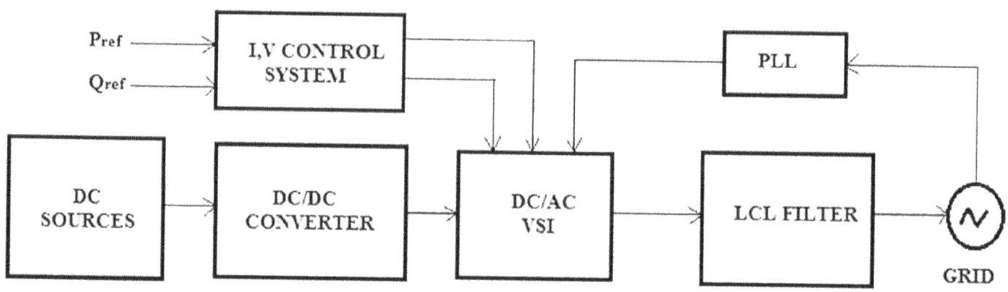

FIGURE 12.1 General schematic for grid-interconnected DC power source.

12.3 LCL Filter

The LCL filter is a third-order filter having attenuation of 60 db/decade for frequencies above full resonant frequency; hence, a lower switching frequency for the converter switches could be utilized. Decoupling between the filter and the grid-connected inverter having grid-side impedance is better for this situation, and lower current ripple over the grid inductor might be attained. The LCL filter will be vulnerable to oscillations too and it will magnify frequencies around its cut-off frequency. Therefore, the filter is added with damping to reduce the effect of resonance. Therefore, the LCL filter fits to our application. In the interim, the aggregate inductance of the received LCL filter is much less than with the L filter. Commonly, the expense is lessened. Besides, enhanced dynamic execution, harmonic attenuation, and decreased volume might be accomplished with the utilization of an LCL filter. The conduction and switching losses that are caused by the filter are calculated and optimized considering the level of reduction of harmonics.

12.4 LCL Filter Design

The following parameters are needed for the filter design: V_{L-L} line-line RMS voltage (inverter output), V_{ph}, phase voltage (inverter output), P_n: rated active power, V_{dc} DC link voltage, f_g: grid frequency, f_{sw}: switching frequency, and f_{res}: resonance frequency. The base impedance (Z_b) and base capacitance (C_b) are defined as

$$Z_b = \frac{E_n^2}{P_n} \tag{12.1}$$

$$C_b = \frac{1}{\omega_g Z_b} \tag{12.2}$$

The LCL filter design algorithm is given in Figure 12.2.

The maximum ripple current at the output of the inverter is

$$\Delta I_{L\max} = \frac{2V_{DC}}{3L_1}(1-m)mT_{sw} \tag{12.3}$$

where m is the inverter modulation factor (for a SPWM inverter)

It was observed that the maximum peak-peak ripple occurred at $m = 0.5$ then

$$\Delta I_{L\max} = \frac{V_{DC}}{6f_{sw}L_1} \tag{12.4}$$

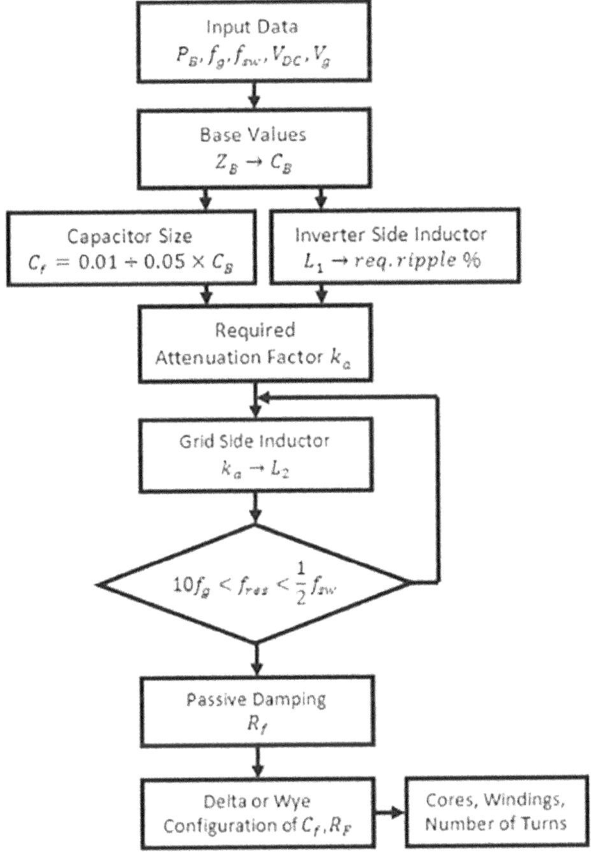

FIGURE 12.2 LCL filter design algorithm.

12.5 Filter Design Specifications

The following parameters are considered for the design of an LCL filter as given in Table 12.1. Inductor values considered: $L_1 = 2.33$ mH, $L_2 = 0.045$ mH.

TABLE 12.1

Parameters Considered

f_g	Grid frequency	60 Hz
f_{sw}	PWM carrier frequency	15 kHz
P_n	Nominal Power	5 kW
V_g	Phase grid voltage	120 V
V_{DC}	DC link voltage	400 V
L_1	Inverter-side inductor	2.33 mH
L_2	Grid-side inductor	0.045 mH
C_f	Capacitor filter Y/Δ	15 μF/5 μF
R_f	Damping resistor Y/Δ	0.55Ω/1.65Ω

12.6 Simulation

Figure 12.3 shows the complete system. A universal bridge is shown in Figure 12.4. The Universal Bridge block implements a universal three-phase power converter that consists of up to six power switches connected in a bridge configuration. The types of power switch and converter configuration are selectable from the dialog box. The Universal Bridge block allows simulation of converters using either naturally commutated (or line-commutated) power electronic devices (diodes or thyristors) or forced-commutated devices (GTO, IGBT, and MOSFET).

The Universal Bridge block is the basic block for building two-level voltage-sourced converters (VSC). The device numbering is different if the power electronic devices are naturally commutated or forced-commutated. For a naturally commutated three-phase converter (diode and thyristor), numbering follows the natural order of commutation as given in Figure 12.5. Voltage measurement block is given in Figure 12.6.

The Three-Phase V-I Measurement block is used to measure instantaneous three-phase voltages and currents in a circuit. When connected in series with three-phase elements, it returns the three phase-to-ground or phase-to-phase peak voltages and currents. The block can output the voltages and currents in per unit (pu) values or in volts and amperes.

If you choose to measure phase-to-ground voltages in per unit, the block converts the measured voltages based on peak value of nominal phase-to-ground voltage:

$$V_{\text{abc}}(\text{pu}) = \frac{V_{\text{phase to ground}}(V)}{V_{\text{base}}(V)} \tag{12.5}$$

If you choose to measure phase-to-phase voltages in per unit, the block converts the measured voltages based on peak value of nominal phase-to-phase voltage:

$$V_{\text{abc}}(\text{pu}) = \frac{V_{\text{phase to phase}}(V)}{V_{\text{base}}(V)} \tag{12.6}$$

The steady-state voltage and current phasors measured by the Three-Phase V-I Measurement block can be obtained from the Powergui block by selecting Steady-State Voltages and Currents. The phasor magnitudes displayed in the Powergui stay in peak or RMS values even if the output signals are converted to pu.

The Hysteresis control block is shown in Figure 12.7. The Relay block allows its output to switch between two specified values. When the relay is on, it remains on until the input drops below the value of the Switch Off point parameter. When the relay is off, it remains off until the input exceeds the value of the Switch On point parameter. The block accepts one input and generates one output. The Switch On point value must be greater than or equal to the Switch Off point. Specifying a Switch On point value greater than the Switch Off point models hysteresis, whereas specifying equal values models a switch with a threshold at that value.

The Data Type Conversion block shown in Figure 12.8 converts an input signal of any Simulink® data type to the data type you specify for the Output Data Type parameter. The input can be any real- or complex-valued signal. If the input is real, the output is real. If the input is complex, the output is complex.

To cast a signal of an enumerated type to a signal of any numeric type, the underlying integers of all enumerated values input to the Data Type Conversion block should be within the range of the numeric type; otherwise, an error occurs during simulation.

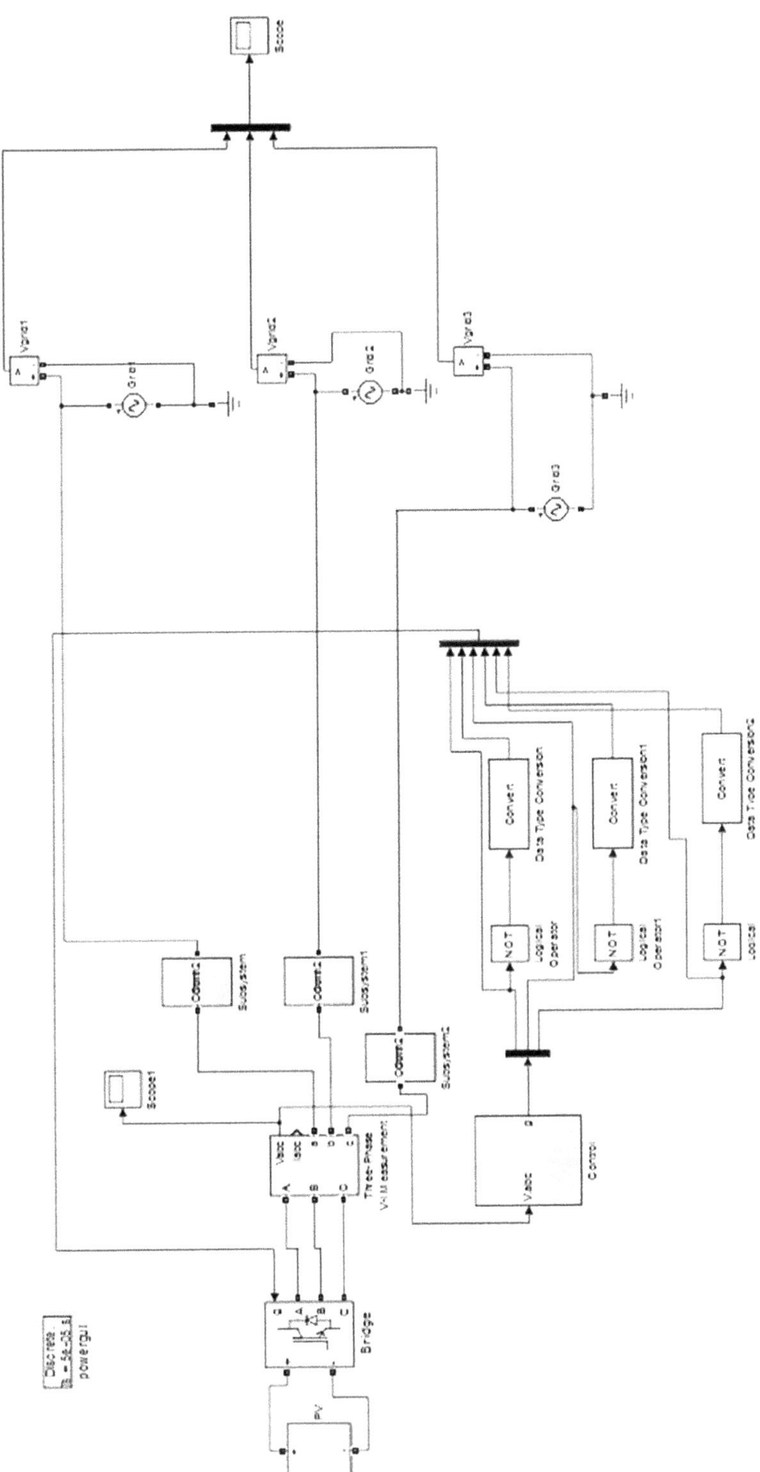

FIGURE 12.3 Complete system.

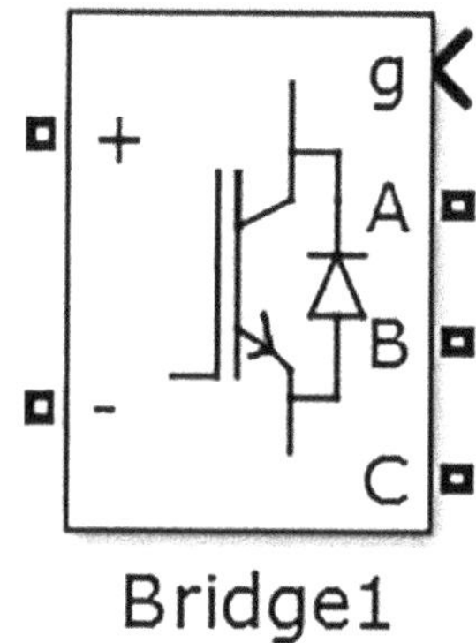

FIGURE 12.4 Universal bridge.

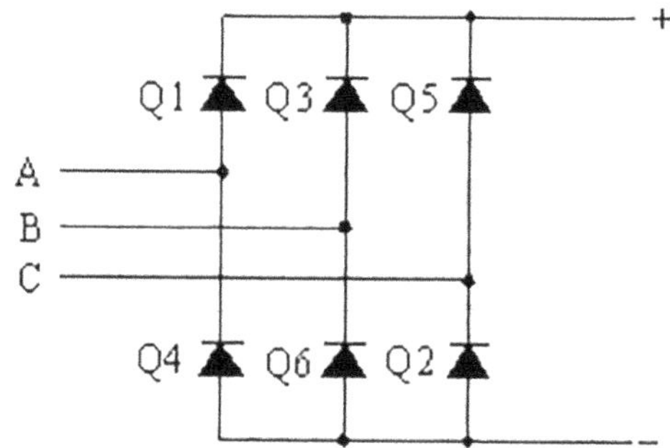

FIGURE 12.5 Full bridge.

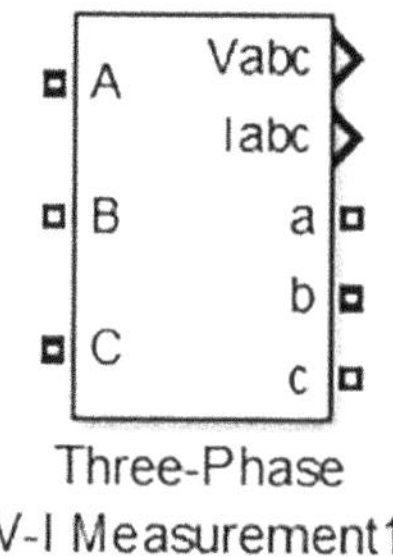

FIGURE 12.6 Voltage Measurement block.

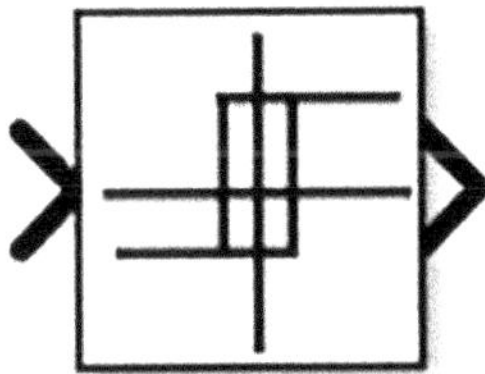

FIGURE 12.7 Hysteresis control block.

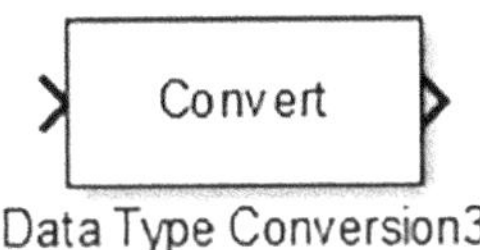

FIGURE 12.8 Data type conversion block.

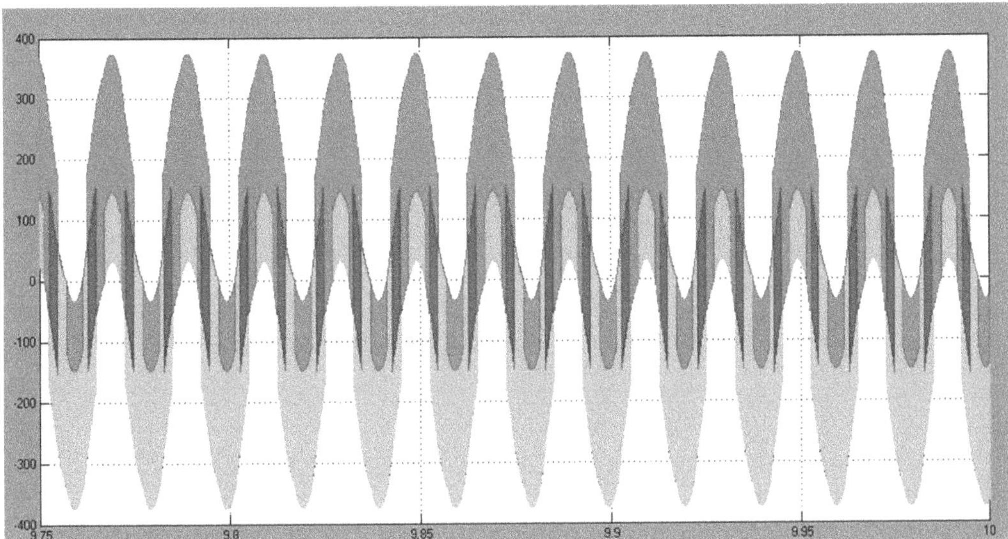

FIGURE 12.9 Inverter output voltage with harmonics.

Inverter output voltage with harmonics is shown in Figure 12.9 and filter output voltage in Figure 12.10. The output voltage harmonics are reduced using LCL filter, by calculating appropriate values for the inductor and capacitor values. This chapter postulate a systematic LCL filter design methodology for grid-interconnected inverter systems. The LCL filter reduces the switching frequency ripple and helps in coupling with a current-like performance to the utility grid.

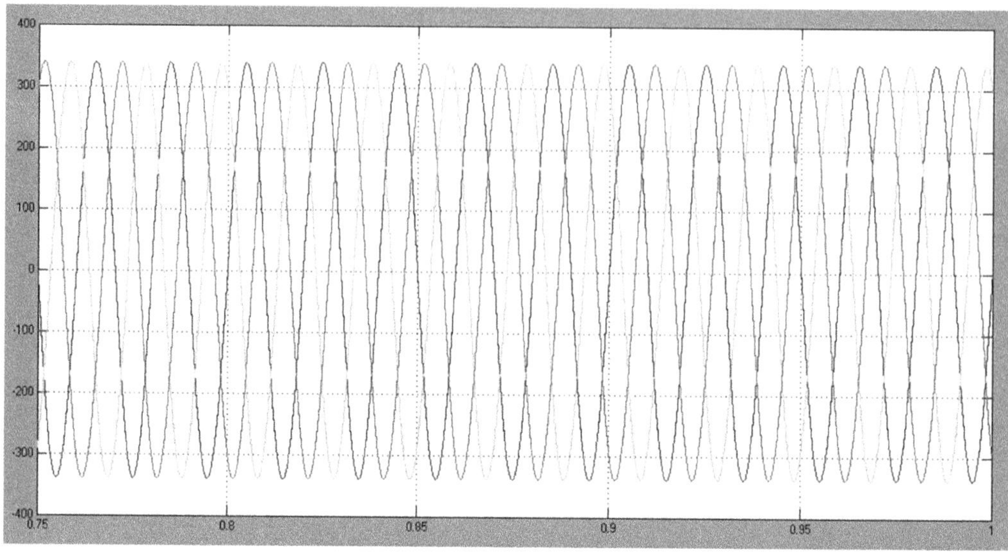

FIGURE 12.10 Filter output voltage.

13

Harmonics Mitigation in Load Commutated Inverter Fed Synchronous Motor Drives

13.1 Introduction

Power quality issues at AC mains are a major problem for load-commutated inverter (LCI)-fed synchronous motors. When operated as adjustable variable drives in high-power applications such as blowers, fans, pumps, and mill drives, commutatorless motor drives have several features like high efficiency, economic operation, and flexibility of operation. This work proposes several topologies for reduction of total harmonics for the mitigation of PQ problems in LCI-fed synchronous motor drives using multipulse AC–DC converters.

13.2 Synchronous Motor Drives

Synchronous motor drives are used in many industrial applications like induction motor drives and their application is growing. They are generally more expensive than induction motor drives, but the advantage is that the efficiency is higher, which tends to lower the lifecycle cost. The development of semiconductor variable frequency sources such as inverters and cycloconverters plays a vital role in variable-speed applications such as high power and high-speed compressors, blowers, induced and forced draft fans, main line traction, servo drives, etc.

Recently, many modern variable-speed drives have used a synchronous motor as well as an induction motor for precise and smooth control of speed with long-term stability and good transient performance. The silicon-controlled rectifier (SCR)-based inverter is now used to control the speed of these drives. A reliable control of the SCR-based power converter can be obtained by a synchronous firing control scheme by which its SCRs operate over a wide range of AC supply frequencies. From the very beginning, the conventional DC motors have been used in many industrial applications as variable speed drives.

However, for reliable operation of the system, it is not advisable to use a DC motor drive due to several drawbacks: the mechanical commutator needs regular maintenance, the additional weight of the commutator reduces the power/weight ratio, friction and sparking cause wear in the brush and commutator, the commutator construction increases the cost of the DC motor drive, and it is not suitable to operate in a dusty or explosive atmosphere.

13.3 Load-Commutated Inverters

The LCI is one of the earliest inverters developed for adjustable-speed drives. It mainly consists of a controlled rectifier, which feeds an adjustable DC current via a DC inductor to an LCI. Since the thyristor does not have self-extinguishing capability, it can be commutated by the load voltage with a leading PF. The inverter output current is a quasi-square wave. However, the motor voltage waveform is close to sinusoidal superimposed with voltage spikes caused by thyristor commutations. Hence, the motor current contains low-order harmonics, such as the 5th, 7th, 11th, 13th, etc. The torque pulsations and additional power losses are caused by harmonic currents in the motor and associated system.

An LCI uses the load voltage with a leading power factor (PF) for natural commutation of thyristors. The ideal load for the LCI is a synchronous motor operating at a leading PF. Therefore, the load commutation makes the drive system simple and reliable. However, the LCI-fed overexcited SM has problems at low speeds and at starting due to low back electromotive force (EMF) across stator terminals. One of the simplest methods is pulsed starting, in which LCI thyristors are commutated by interrupting the DC link current.

The LCI-fed SM drive features low cost and high efficiency due to the use of low-cost thyristors. The LCI is suitable for large drives with a power rating in megawatts, where the capital cost and running efficiency are of great importance. However, the input PF of the drive changes with its operating parameters and also the rectifier input current is highly distorted; therefore, an LCI-SM drive should be equipped with harmonic filters or any other compensation device to reduce line current total harmonic distortion (THD).

An LCI-fed synchronous motor with its excitation winding connected in series to the input of the inverter can be applicable as a variable-speed drive in place of conventional DC motor drives over a wide range of speed economically. The power circuit and firing control circuit configuration of the inverter are very simple in structure. The PQ concerns are more prominent in LCI-fed SM drives because of their high power ratings. The passive wave-shaping techniques are normally used, which are based on magnetics in three-phase AC–DC converters; one such system is known as a multipulse or multiphase converter. There are many configurations of multipulse AC–DC converters (MPCs), in 12 or many more pulses.

13.4 Converter Configuration

The current source converters with multipulse concept for an LCI-fed SM drive adopt load commutation with three-phase SM instead of an AC source, since the conventional drive system has a 6-pulse converter feeding to a three-phase SM. To avoid the complexity of triggering the control circuit, the operation is limited to a 12-pulse converter. Two topologies for 12-pulse LCI operation can be adopted. One uses phase-shifting transformers that combine the two 6-pulse LCI outputs to get a three-phase supply (12-pulse converter) for a conventional three-phase SM. Phase-shifting transformers are very expensive but have an advantage of being used with a conventional three-phase SM. Other topology uses an asymmetric six-phase SM in which two sets of three-phase windings displaced at an angle of 30° are employed. These topologies of SM can be fed from any rectifier set discussed earlier to meet the PQ standards, provided the cost, control complexity, and efficiency are within acceptable limits.

13.5 Requirements for the 18- and 24-Pulse Converters

The transformer design for an isolated 18-pulse AC–DC converter configuration requires 20°, 0°, and −20° phase shift, which is achieved by the use of two zigzags and a star winding in the secondary side of the transformer. The turns ratio of primary star winding to secondary star winding is 3:1, whereas, the turns ratio for $Y/Z-1$ and $Y/Z-2$ transformers to provide 20° and −20° phase shift, respectively, is given by $(N3)/(N3+N2)=0.227$ and $(N1)/(N2+N3)=1.959$, where $N1$ is the number of turns per phase in star winding, and $N2$, $N3$ are the number of turns of the zigzag winding. To achieve the 20° phase shift required for an 18-pulse converter with a non-isolated topology, the autotransformer may have various connections for, e.g., polygon, delta-polygon, zigzag, and T-connected transformers. The delta-polygon-connected autotransformer is used in this investigation. Figures 13.1 and 13.2 show the pulse converter-fed synchronous motor drive.

A 24-pulse AC–DC converter requires −15°, 0°, 15°, and 30° phase shift among four sets of voltages. For isolated transformer design, it is achieved by the use of two zigzag, one star, and one delta configurations in the secondary windings. The turns ratio for $Y/Z-1$ and $Y/Z-2$ transformers to provide 15° and −15° phase shift, respectively, is similar to a 12-pulse converter phase-shift transformer. However, for non-isolated converter to generate 15° and 30° phase shift, a hexagon-connected autotransformer is selected among various other connections, e.g., star, delta, hexagon, and T-connected configuration.

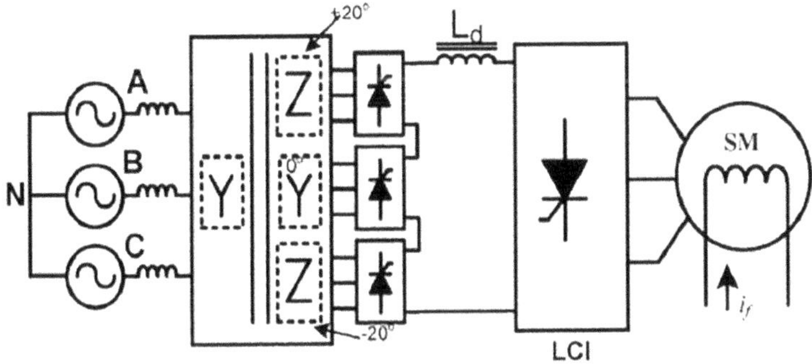

FIGURE 13.1 Pulse converter-fed synchronous motor drive (3 front end converter configuration).

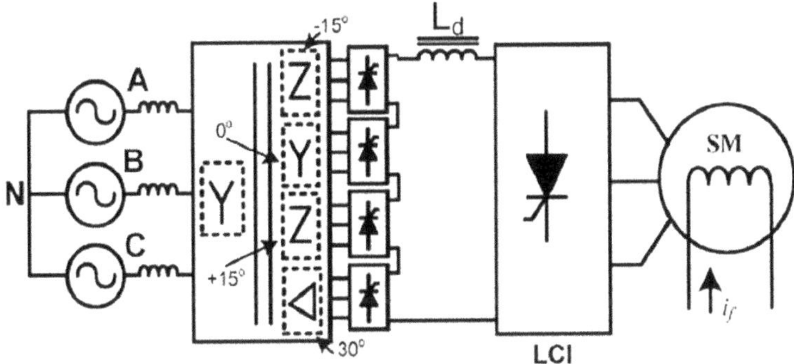

FIGURE 13.2 Pulse converter-fed synchronous motor drive (4 front end converter configuration).

13.6 Operation

13.6.1 6-Pulse Converters

Similar to a single-phase system, six thyristors are used in a 6-pulse full-wave controlled converter, which switches at 60° intervals sequentially from T1 to T6. When T1 and T2 are conducting v_{AN} and v_{BN} voltage with respect to star point of transformer appears at the load, i.e., $V_{AN} = V_{XN}$ and $V_{BN} = V_{YN}$. The load voltage $v_o = v_{XY} = v_{AN} - v_{BN} = v_{AB}$, which is the line voltage V_L (where $V_L = \sqrt{3}V$, $V_{Lm} = \sqrt{2}(V_L)$ and V is the rms value of the phase voltage). When $\omega t = \alpha + 60°$, T2 is triggered. At this condition, v_C is more negative than vB; therefore, due to conduction of T2, v_Y (negative bus) becomes equal to v_C. At the anode of T6 more negative voltage appears to make it reverse biased. Then load current transfers from T6 to T2 when T6 commutates. Again, when T3 is triggered it supplies a positive (higher) voltage ($v_B > v_A$) at the cathode of T1 to turn it off and the load current transfers from T1 to T3. For an inductive RL load six voltage pulses and the instantaneous output voltage (v_o) becomes negative. However, Vo is always positive except for R-L-E(–) loads where the operation of the converter happens in the fourth quadrant of the v_o-i_o plane. Figures 13.3 and 13.4 show the 6-pulse converter circuit without filter and 6-pulse converter circuit using passive filter.

13.6.2 12-Pulse Converters

In a 12-pulse AC–DC converter a 30° phase shift required can be provided by isolated star-delta transformer having 0° and 30° phase shift or isolated star–zigzag transformer combinations with phase shift of −15° and +15° with respect to supply voltage. The isolated star–zigzag transformer combination is balanced compared to the star–delta transformer combination. The turns ratios for

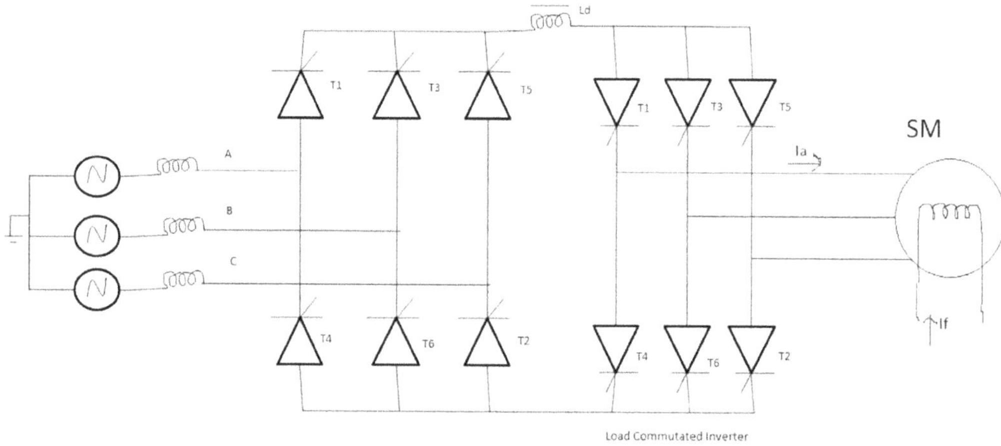

FIGURE 13.3 6-pulse converter circuit without filter.

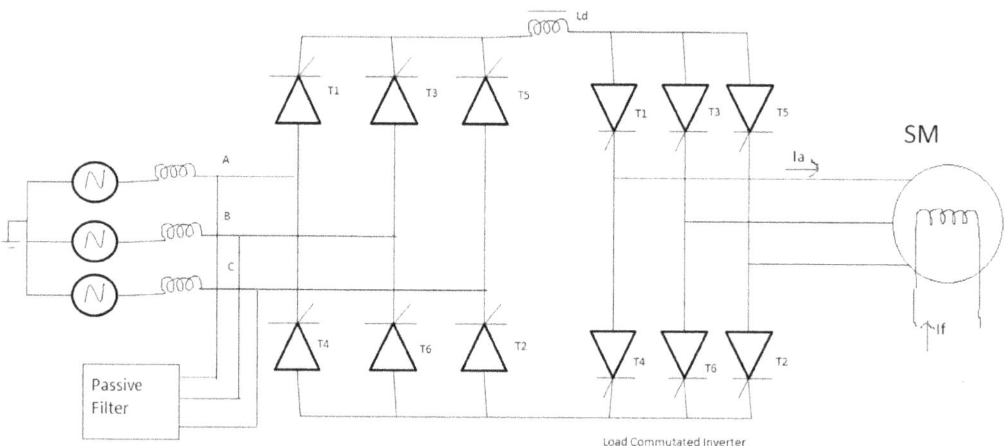

FIGURE 13.4 6-pulse converter circuit using passive filter.

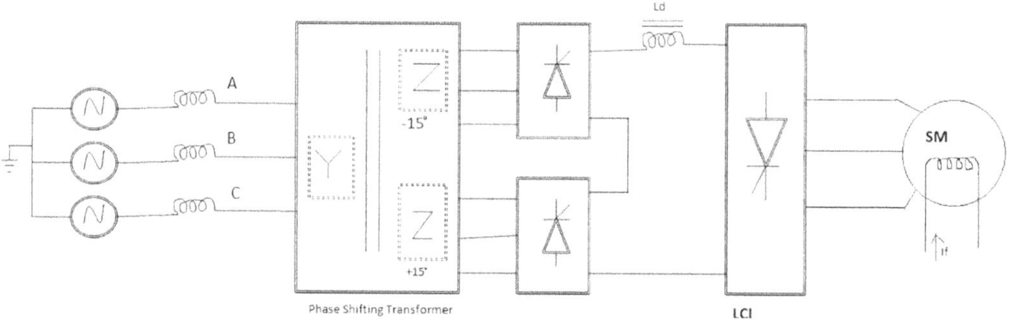

FIGURE 13.5 Block diagram of 12-pulse converter.

Y/Z–1 and *Y/Z*–2 transformers to provide +15° and −15° phase shift, respectively, are given by $N_3/(N_2 + N_3) = 0.366$ and $N_1/(N_2 + N_3) = 1.414$, where N_1 is number of turns per phase in star winding, and N_2, N_3 are number of turns of the zigzag winding. For non-isolated topologies, the autotransformer may have various connections for ±15° phase shift, e.g., polygon, delta-polygon, zigzag, and T-connected transformers. The circuit diagram of 12-pulse LCI-fed synchronous motor drive is as shown in Figure 13.5.

13.7 Design of Filters

At the input side of 6-pulse and 12-pulse converter-fed LCI-SM drives, a combined second-order damped passive filter tuned to 11th-order harmonics and a high-pass shunt passive filter is used. The passive filters have been designed for 11th-order harmonics and higher-order harmonics separately and connected in parallel in case of 12-pulse converters. For higher-pulse converters, passive filters tuned to other frequencies shall be used, e.g., 17th and high-pass filter for an 18-pulse converter and 23rd and high-pass filter in the case of a 24-pulse converter.

The impedance of the filter at any harmonic h is given by

$$Z_h = \left[\frac{R(hX_L)^2}{R^2 + (hX_L)^2} + j\left(\frac{R^2 hX_L}{R^2 + (hX_L)^2} \right) - \frac{X_c}{h} \right]$$
$$X_c = \left\{ \frac{h^2}{h^2 - 1} \right\} \left(\frac{V^2}{Q} \right) \tag{13.1}$$

where Q is the quality factor of the passive filter. The value of Q has been taken as 1 in this paper; however, for high-pass filters Q varies between 0.5 and 5.

The reactive power requirement decides the capacitance in the circuit, whereas the inductance is decided by the frequency to which the passive filter is tuned. The resistance for high-pass filters can be calculated by

$$R = \sqrt{\frac{L}{C}}. \tag{13.2}$$

13.8 Specifications

Synchronous machine

Nominal power: 85 kVA
Nominal voltage: 400 V
Nominal frequency: 50 Hz
No-load field current: 10 A
Stator armature resistance: 0.055 ohm
Stator leakage inductance: 0.3595 mH
d-axis mutual inductance: 12.82 mH
q-axis mutual inductance: 5.692 mH
Field resistance referred to stator: 0.03634 ohm
Field leakage inductance: 1.302 mH

Damper-winding parameters

d-axis resistance $R'_{kd} = 0.5075\ \Omega$
Leakage inductance $L'_{lkd} = 4.335$ mH
q-axis resistance $R'_{kq1} = 0.1167\ \Omega$
Leakage inductance $L'_{lkq1} = 0.5957$ mH
Inertia J = 4.22 kg-m^2
Friction factor 0.07 Nm·s
Pole pairs = 2
Rotor type: salient pole, wound field
Source impedance: 0.03 pu
Transformer leakage impedance: 0.03 pu
DC link inductor: 15 mH

High-Pass Filter

R = 0.4083 Ω
L = 0.1 mH
C = 600 μF

13.9 Simulation and Results

13.9.1 MATLAB Blocks

To design the MATLAB/Simulink model, choose the following blocks from the Library Simulink Browser:

AC voltage Source

Path: Simulink Library Browser>>Simscape>>SimPowerSystems>>Electrical Sources>>AC voltage source

Function: Ideal sinusoidal AC voltage source

Universal Bridge

Path: SimulinkLibraryBrowser>>Simscape>>SimPowerSystems>>PowerElectronics>> Universal Bridge

Function: This block implements a bridge of selected power electronic devices

Voltage and current measurement

Path: Simulink Library Browser>>Simscape>>SimPowerSystems>>Measurements>>Voltage measurements, current measurements

Function: Ideal voltage/current measurement

R, L, C

Path: Simulink Library Browser>>Simscape>>SimPowerSystems>>Elements>>RLC branch

Function: Implements a series branch of R, L, C elements

Synchronized 6-pulse generator

Path: SimulinkLibraryBrowser>>Simscape>>SimPowerSystems>>Extra library>>Control blocks>>>Synchronized 6-pulse generator

Function: Use this block to fire the 6 thyristors of a 6-pulse converter. The output is a vector of 6 pulses (0-1) individually synchronized on the 6 commutation voltages. Pulses are generated alpha degrees after the increasing zero-crossings of the commutation voltages.

Powergui

Path: Simulink Library Browser>>Simscape>>SimPowerSystems>>Powergui

Function: The Powergui block is necessary for simulation of any Simulink model containing SimPowerSystems blocks. It is used to store the equivalent Simulink circuit that represents the state-space equations of the model.

Scope

Path: Simulink Library Browser>>Sinks>>Scope

Function: Displays input signals with respect to simulation time

13.9.2 6-Pulse Converter without Filter Circuit

Figure 13.6 shows the Simulink model of the 6-pulse LCI-fed synchronous motor drive circuit without using filters at the input side. The configuration of the synchronous machine is a salient pole type with speed as mechanical input and no preset model. Universal bridges with thyristors are used as AC–DC converters and DC-AC inverters coupled by a series inductor in between.

13.9.3 6-Pulse Converter with Passive Filter Circuit

Figure 13.7 shows the Simulink model of a 6-pulse LCI-fed synchronous motor drive with the same configuration as a drive without a filter. It has a high-pass filter of RL at the input side, which reduces harmonics effectively. This is illustrated in FFT analysis and is figured below.

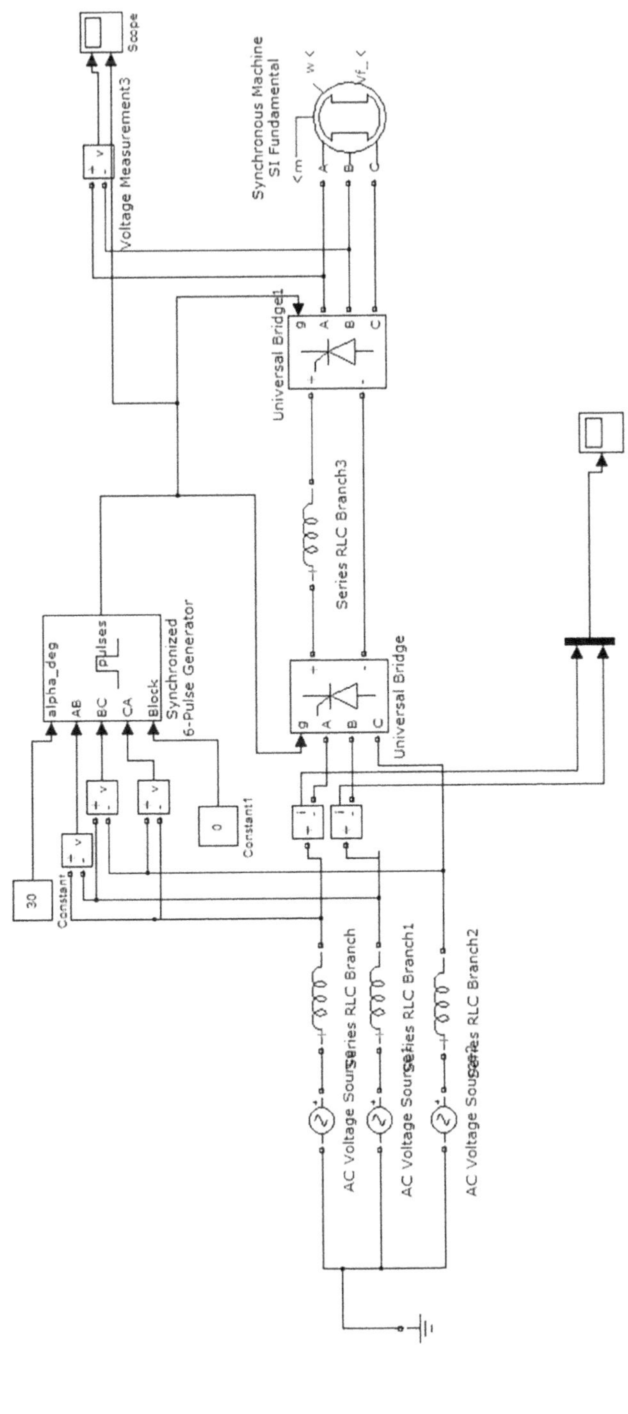

FIGURE 13.6 Simulink model of 6-pulse converter without filter.

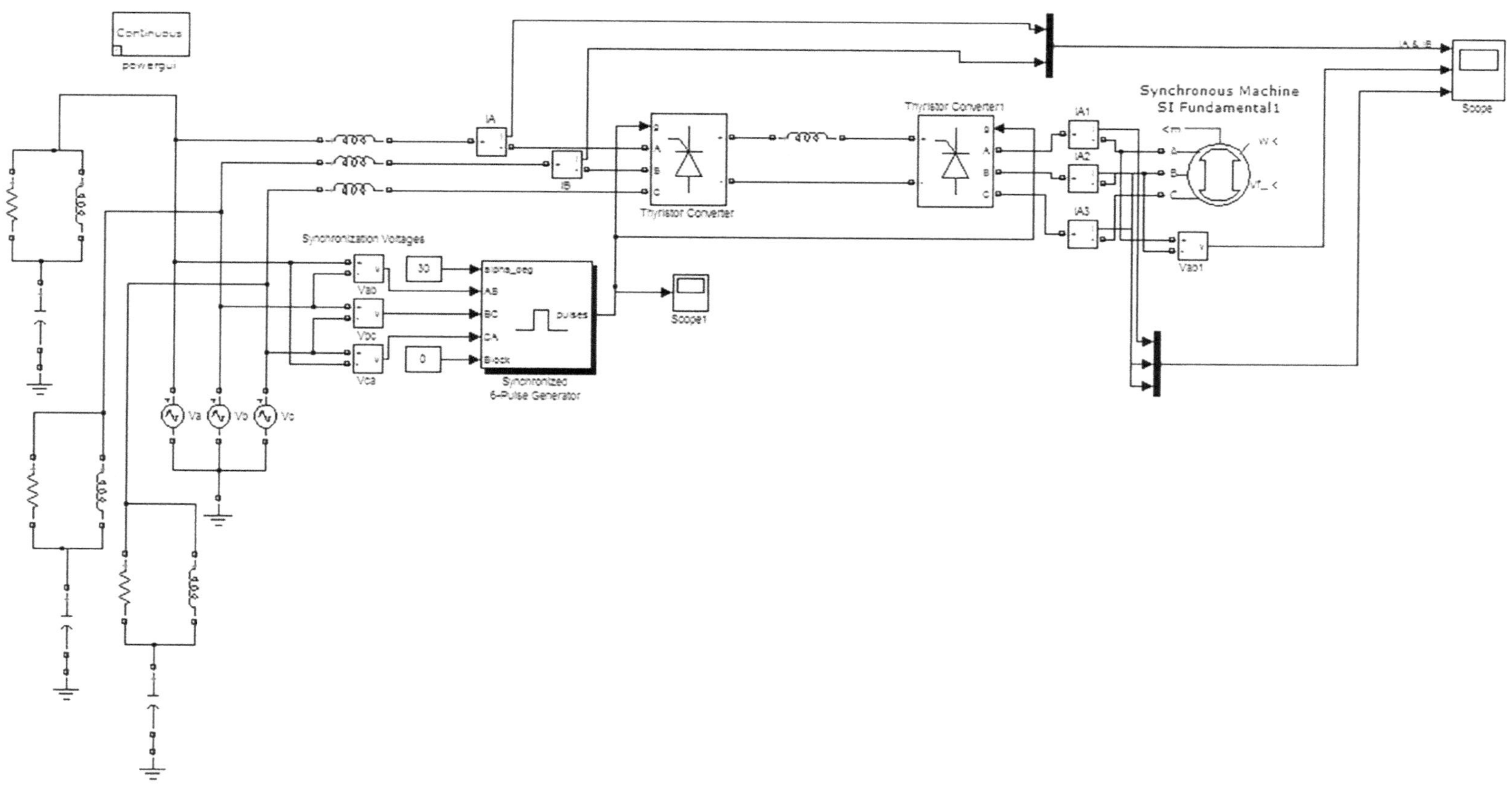

FIGURE 13.7 Simulink model of 6-pulse converter with passive filter.

13.9.4 12-Pulse Converter without Filter

Figure 13.8 shows the Simulink model of a 12-pulse LCI-fed synchronous motor drive without filter. The configuration of the synchronous machine is a salient pole type with speed as mechanical input and no preset model. The AC–DC converter is supplied through a transformer that has a single input and two outputs. The secondary side is taken as one star and one delta separately given to two thyristor converters that are connected in series. The universal bridge thyristor inverter is inputted by a series converter and the machine is run from its output. A synchronized 12-pulse generator is used for thyristor converters and a synchronized 6-pulse generator for thyristor inverters.

13.9.5 12-Pulse Converter with Passive Filter

Figure 13.9 displays the simulated model of a 12-pulse LCI-fed synchronous motor drive using a passive filter at the input side. The model uses two 6-pulse converters to form a 12-pulse converter. A zigzag phase-shifting transformer is used rather than a conventional transformer and the secondary winding is star grounded. The same RL filters are employed in each phase of supply at the input side and synchronized 12-pulse and 6-pulse generators are used for thyristor bridges in converter and inverter operation. Figures 13.10 and 13.11 show the Simulink models of an 18-pulse converter-fed and a 24-pulse converter-fed synchronous motor drive.

13.9.6 FFT Analysis

A common use of FFTs is to find the frequency components of a signal buried in a noisy time domain signal. In general, you use spectrum objects to perform spectral analysis, but if you want to create spectral analysis plots manually you start with an FFT. The FFT automatically does this if you provide an input argument specifying the length of the FFT. MATLAB does not scale the output of the FFT by the length of the input, so you scale it so that it is not a function of the length of x. Then, square the magnitude and since you dropped half of the FFT, multiply by 2 to retain the same amount of energy, but do not multiply the DC or Nyquist components by 2. You then obtain an evenly spaced frequency vector with NumUniquePts points. The FFT block in the Discrete Measurements library of **powerlib_extras** now implements the FFT analysis for a vector signal of any number of components and is no longer limited to the analysis of three components. The number of input ports has been reduced to one vector input. If you used this block in your old models, you must use a Mux block to combine the three input signals into one vector signal.

13.10 Graphical Results

13.10.1 6-Pulse Converter without Filter

Figure 13.12 shows the harmonic spectrum of a 6-pulse LCI-fed synchronous motor drive. The harmonic spectra are displayed for source-side current. From Figure 13.12, it can be seen that the THD of a 6-pulse converter without using filter is found to be 20.75%.

13.10.2 6-Pulse Converter Using Passive Filter

Figure 13.13 displays the harmonic spectrum of a 6-pulse LCI-fed synchronous motor drive. The harmonic spectra are displayed for the source-side current after including the passive filter at the source side. From Figures 13.12 and 13.13, it can be seen that the THD is reduced from 20.75% to 9.94% after using filters at the input side.

13.10.3 12-Pulse Converter without Filter

Figure 13.14 displays the harmonic spectrum of a 12-pulse LCI-fed synchronous motor drive. The spectra are displayed for source-side currents without using filters. The THD is found to be 17.79%.

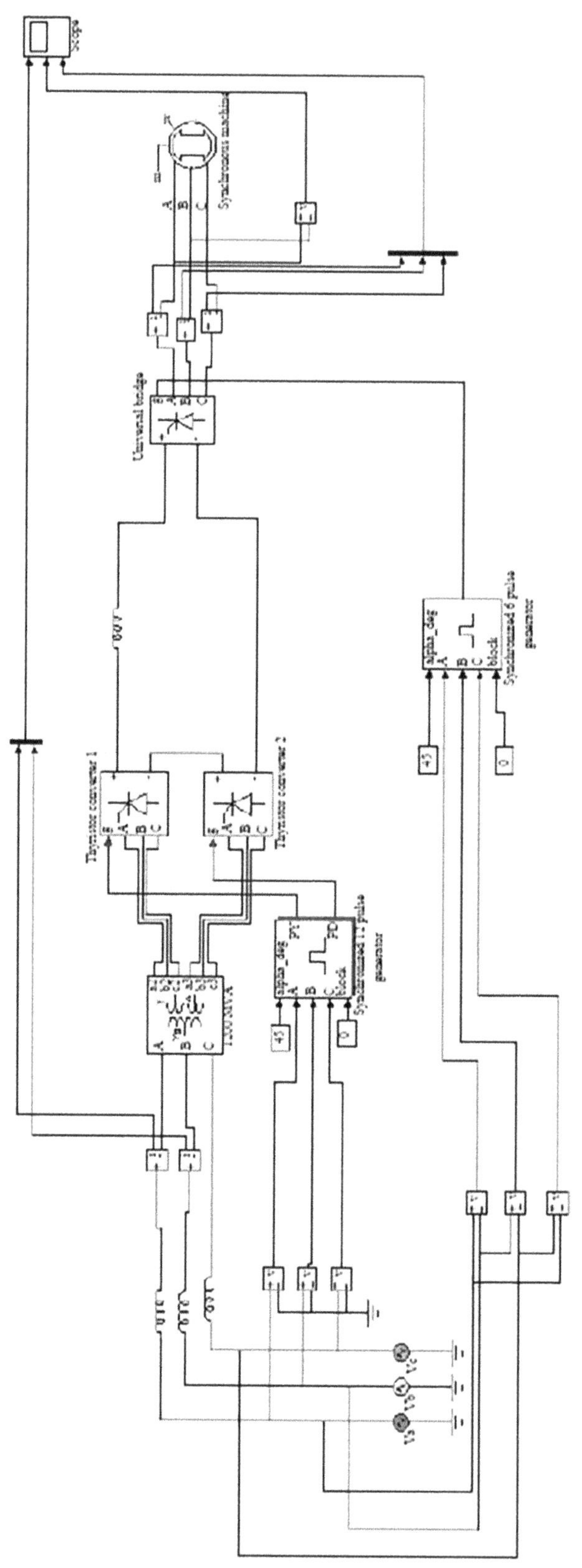

FIGURE 13.8 Simulink model of 12-pulse converter without filter.

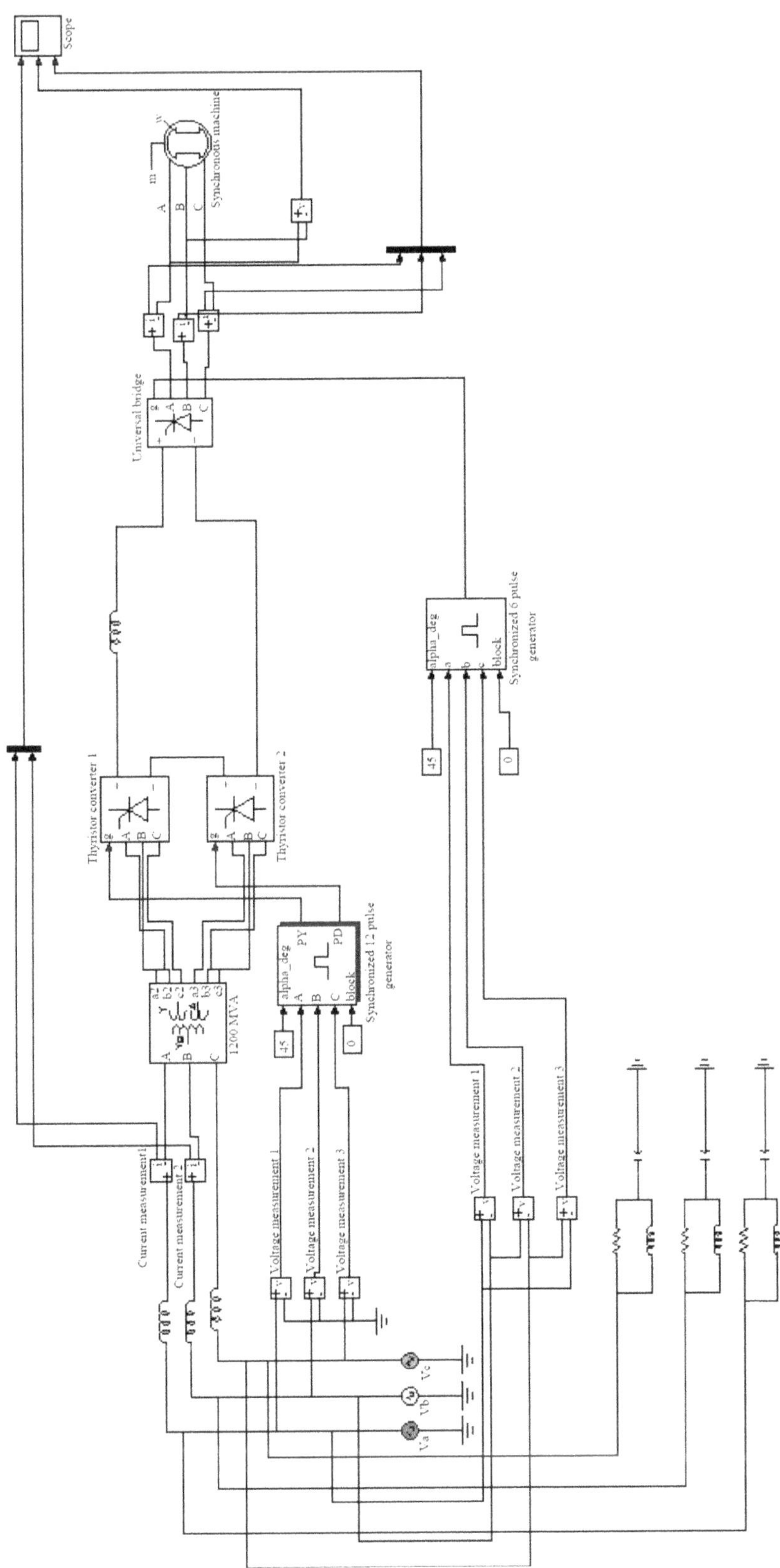

FIGURE 13.9 Simulink model of 12-pulse converter using filter.

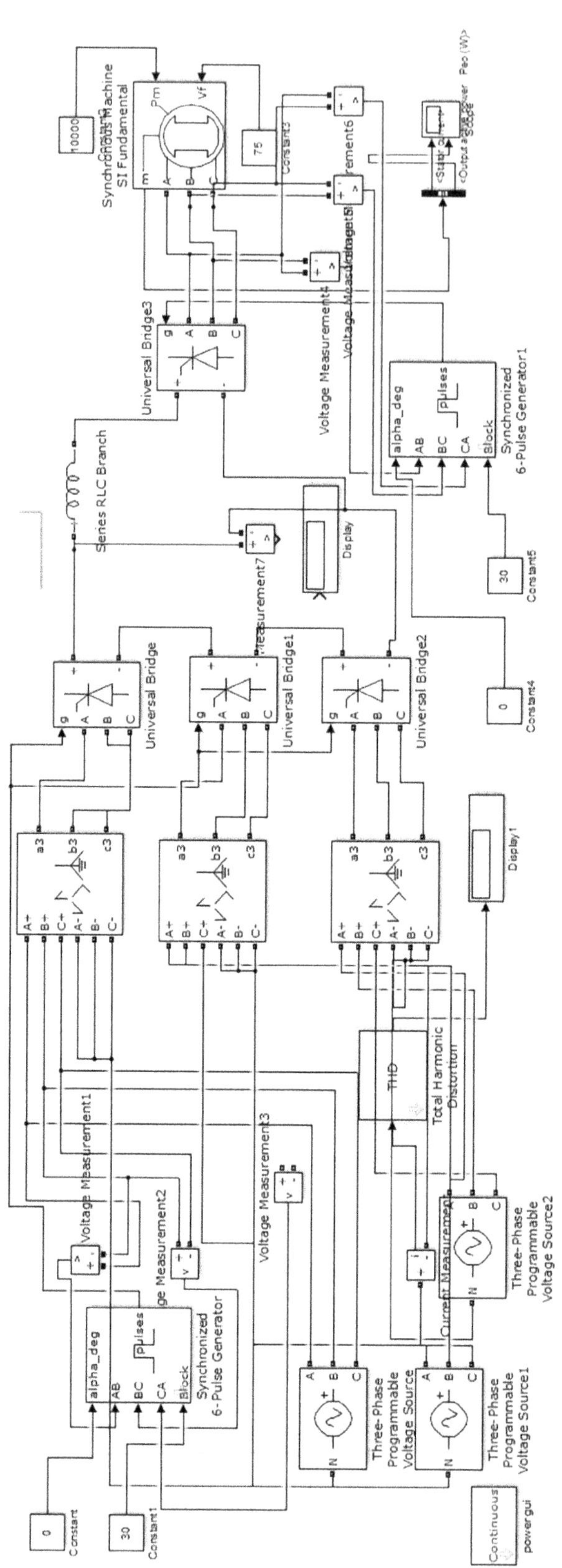

FIGURE 13.10 Simulink model of 18-pulse converter.

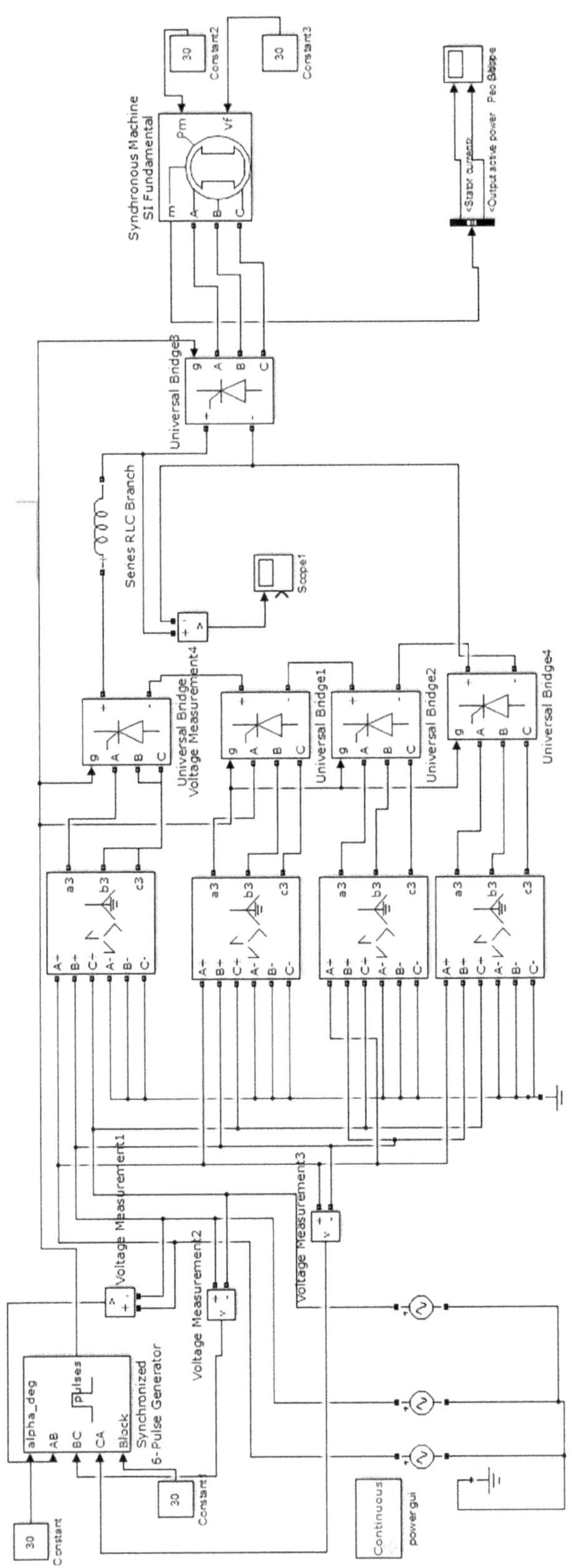

FIGURE 13.11 Simulink model of 24-pulse converter-fed synchronous motor drive.

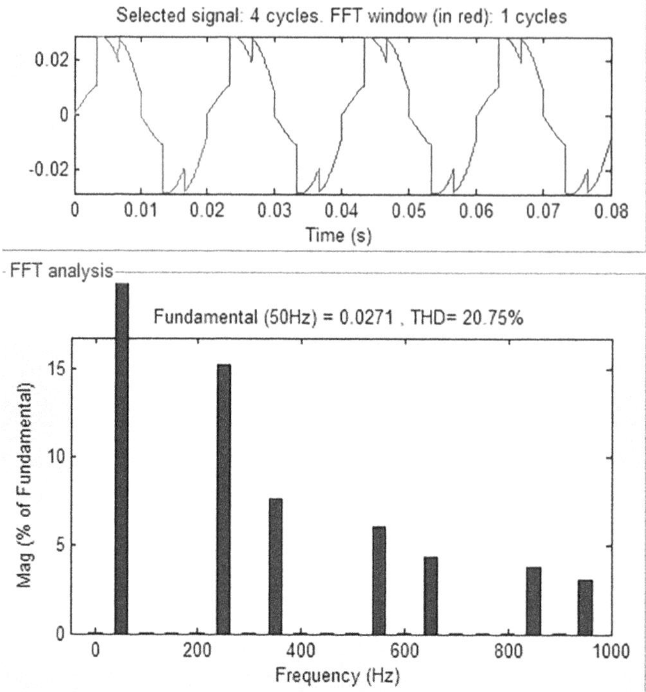

FIGURE 13.12 THD spectrum of 6-pulse converter without filter. (*Note*: See color eBook for improved color differentiation in figures and labeling.)

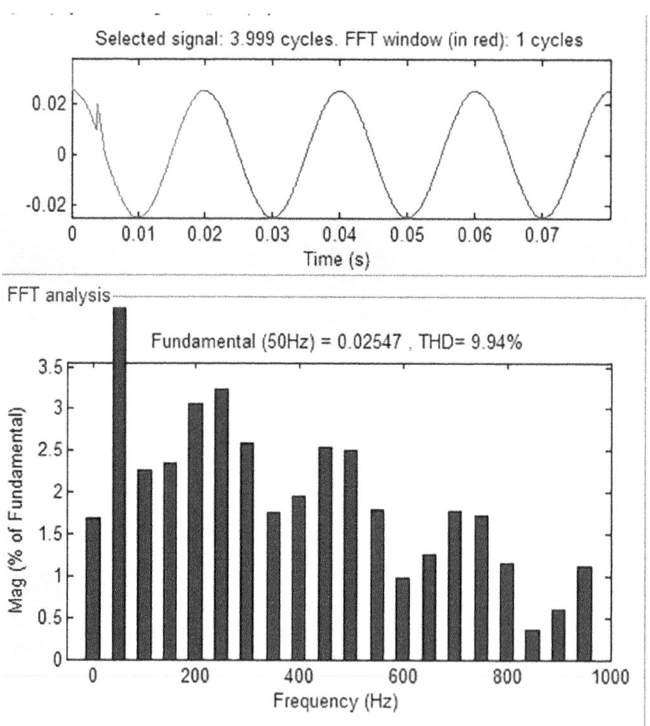

FIGURE 13.13 THD spectrum of 6-pulse converter using filter. (*Note*: See color eBook for improved color differentiation in figures and labeling.)

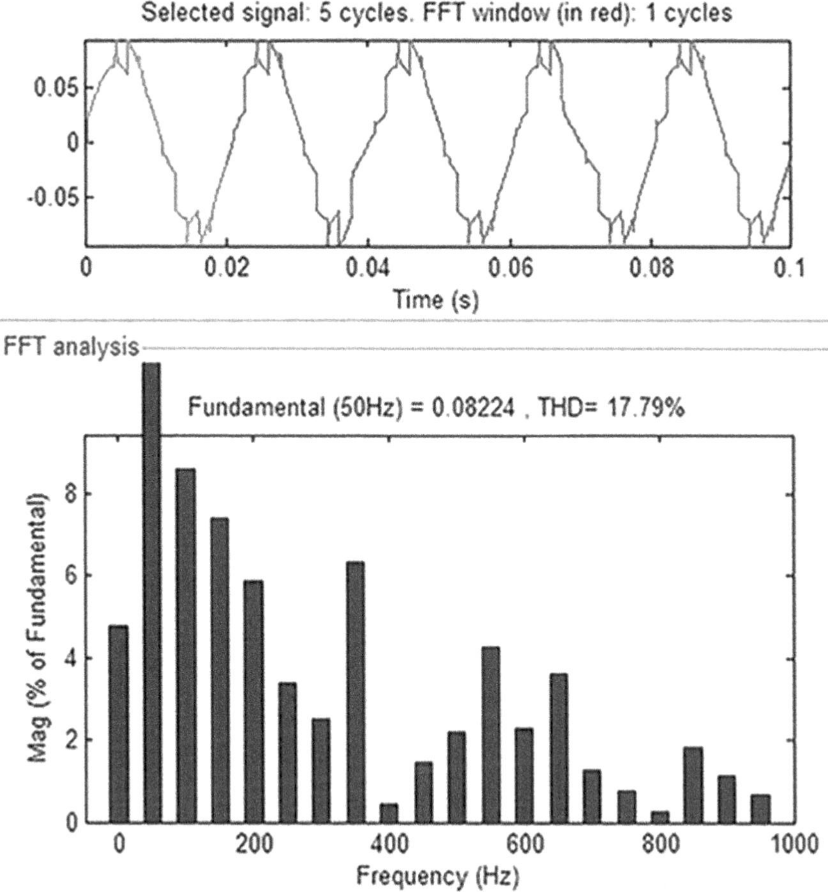

FIGURE 13.14 THD spectrum of 12-pulse converter. (*Note*: See color eBook for improved color differentiation in figures and labeling.)

13.10.4 12-Pulse Converter Using Filter

Figure 13.15 displays the harmonic spectrum of a 12-pulse LCI-fed synchronous motor drive. The harmonic spectrum is displayed for the source-side currents after using passive filters at the source side. From Figures 13.14 and 13.15, it can be seen that the THD is reduced from 17.79% to 5.01%.

13.10.5 18-Pulse Converter

From Figure 13.16, the current THD is reduced significantly; this can be observed above. Major lower-order harmonics are 3 and 5. From the above topology it can be observed that THD is reduced when compared to an 18-pulse converter for a 24-pulse converter as depicted in Figure 13.17.

A combination of passive filters and 6- and 12-pulse converters has been proposed for the LCI-SM drive, which shows performance improvement with reduced THD and magnitude of AC mains current. With shunt, the passive filter 12-pulse converter has added advantages of simple control and consistently improved PF in the wide operating speed range of the drive. To achieve improved power quality in an LCI-SM drive, the 12-pulse converter topology has been simulated at rated load with passive tuned filters and it is observed that this topology shows consistent improved power quality in a wide range of operation.

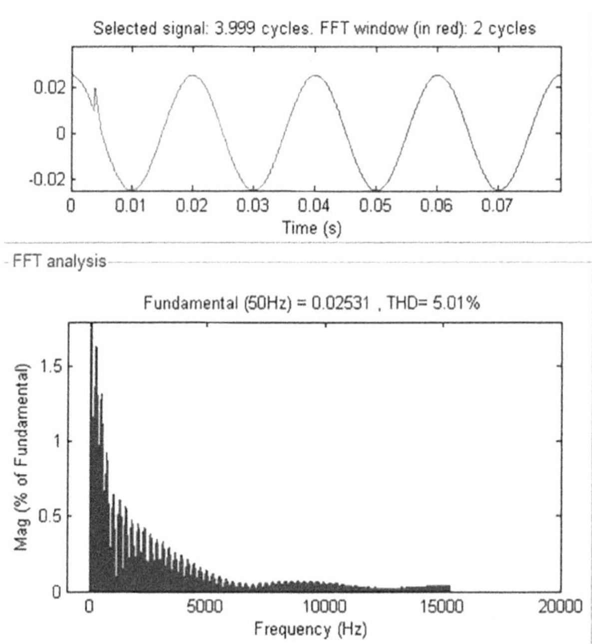

FIGURE 13.15 THD spectrum for 12-pulse converter using filter. (*Note*: See color eBook for improved color differentiation in figures and labeling.)

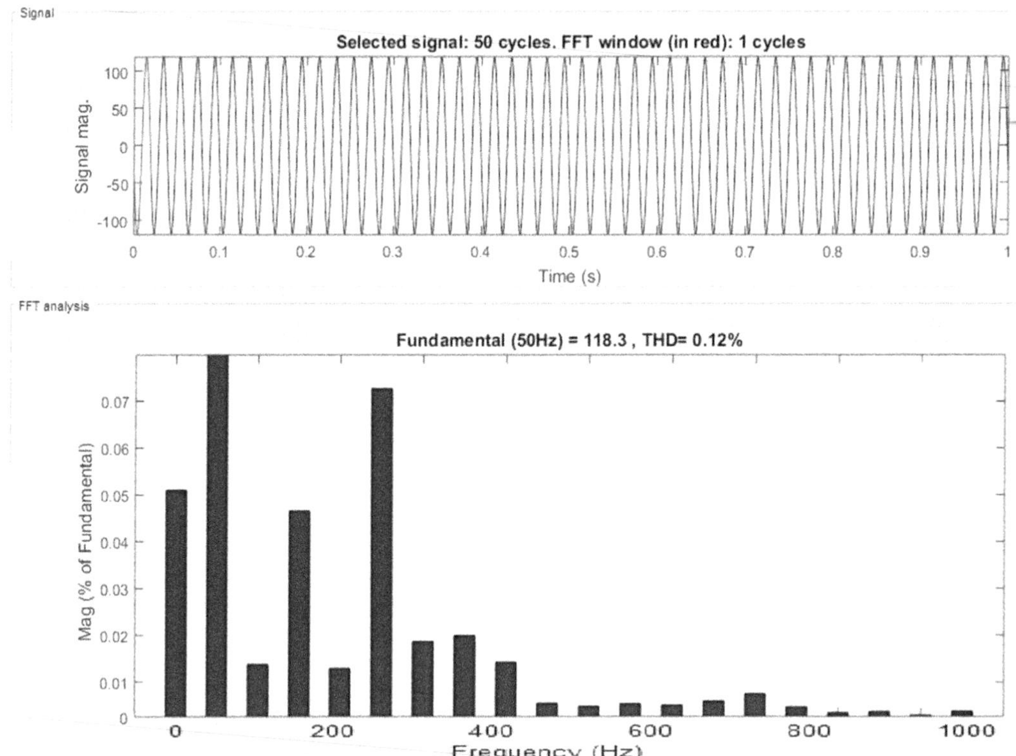

FIGURE 13.16 Current THD analysis of the converter-fed drive. (*Note*: See color eBook for improved color differentiation in figures and labeling.)

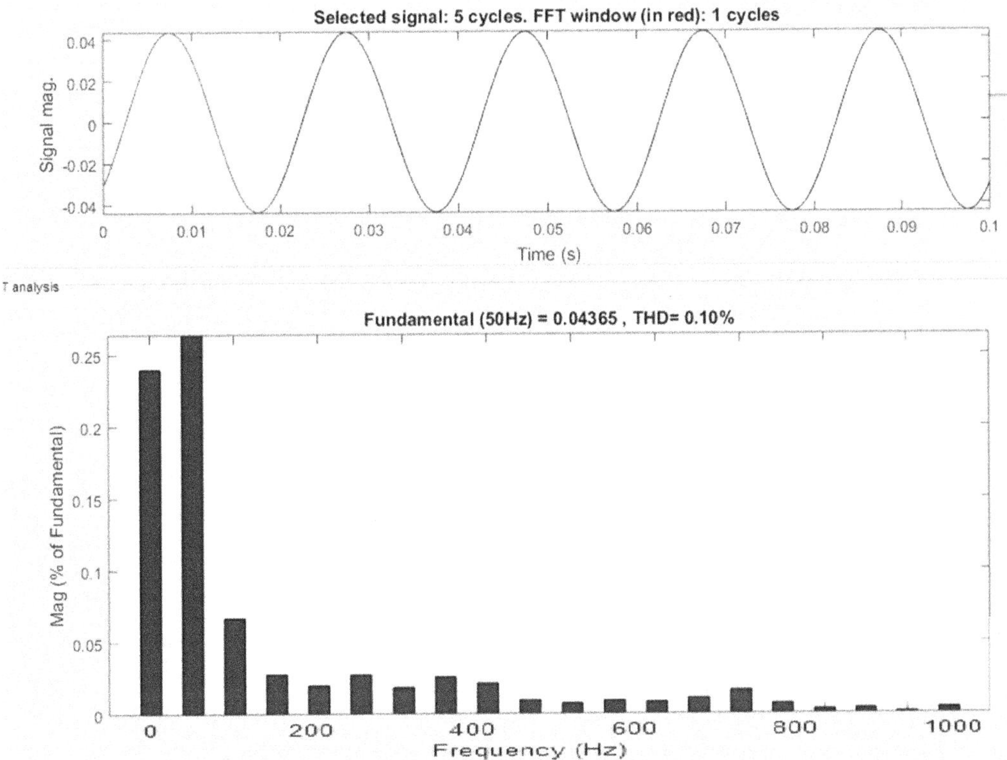

FIGURE 13.17 Current THD analysis of the converter proposed. (*Note*: See color eBook for improved color differentiation in figures and labeling.)

14

Power-Quality Improvements in Vector-Controlled Induction Motor Drives

14.1 Scalar Control

The name scalar control indicates variation of the control variable magnitudes only and disregards the coupling effect in the machine. For example, the voltage of an induction machine can be controlled to control the flux, and frequency or slip can be controlled to control the torque. However, flux and torque are components of frequency and voltage, respectively.

In scalar control both the magnitude and phase alignment of vector quantities are controlled. Scalar-controlled drives give somewhat poor performance, but they are easy to implement. Scalar-controlled drives have been widely used in the industrial sector. However, their importance has diminished recently because of the better performance of vector-controlled drives, which are demanded in many applications.

In many industrial applications, quality improvement requires variable speed and constant speed control. Several techniques have been developed to control alternating current (AC) power. The operation of induction motors in the constant volts per hertz (V/f) mode has been known for many decades, and its principle is well understood.

Scalar control means that variables are controlled only in magnitude and the feedback and command signals are proportional to DC quantities. A scalar control method can only drive the stator frequency using a voltage or a current as a command. Among the scalar methods known to control an induction motor, one assumes that by varying the stator voltages in proportion with frequency, the torque is kept constant.

$$\frac{V}{V_0} = \frac{f}{f_0} \tag{14.1}$$

The advantage of this control technique is its simplicity; it is easy and fast to program and requires only few calculations. The drawbacks are the very poor reaction time for load changes and low efficiency. A speed controller that takes into consideration torque changes and avoids undesirable trips can barely be achieved with a V/f open-loop control.

This method needs some characteristic plots to describe part of the control. One method to obtain these plots is to pre-calculate and store them in the memory, which requires additional silicon costs. The other way is to use a powerful processor and calculate in real time all the characteristics.

14.2 Vector Control

The vector control is referring not only to the magnitude but also to the phase of these variables. Matrix and vectors are used to represent the control quantities. This method takes into consideration not only successive steady-states but real mathematical equations that describe the motor itself; the control results obtained have better dynamics for torque variations in a wider speed range.

The space phasor theory is a method to handle the equations. Though the induction motor has a very simple structure, its mathematical model is complex due to the coupling factor between a large number of variables and the non-linearities. The field-oriented control (FOC) offers a solution to circumvent the need to solve high-order equations and achieve an efficient control.

This approach needs more calculations than a standard V/f control scheme. This can be solved by the use of a calculation unit included in a digital signal processor (DSP) and has the following advantages:

- Full motor torque capability at low speed
- Better dynamic behavior
- Higher efficiency for each operation point in a wide speed range
- Decoupled control of torque and flux
- Short-term overload capability
- Four quadrant operation

14.3 Representation of the System

Figure 14.1 represents the block diagram of the system. It consists of AC voltage source, zigzag transformer diode bridge, inverter, and vector-controlled induction motor drive. The purpose of increasing the number of pulses per cycle at the output is to reduce the harmonics at the input.

Some applications like fans in heating, ventilating and air conditioning systems, blowers, pumps for waste water treatment plants, etc. use a 6-pulse diode bridge rectifier to rectify the input AC. This induces current harmonics into the system, which in turn causes the distortion in supply voltage. So, to mitigate the harmonics different techniques are used to increase the rectification pulses.

Various pulse-based rectification schemes have been used in practice for the purpose of line current harmonic reduction. With the increase in number of pulses, current harmonics can be reduced further.

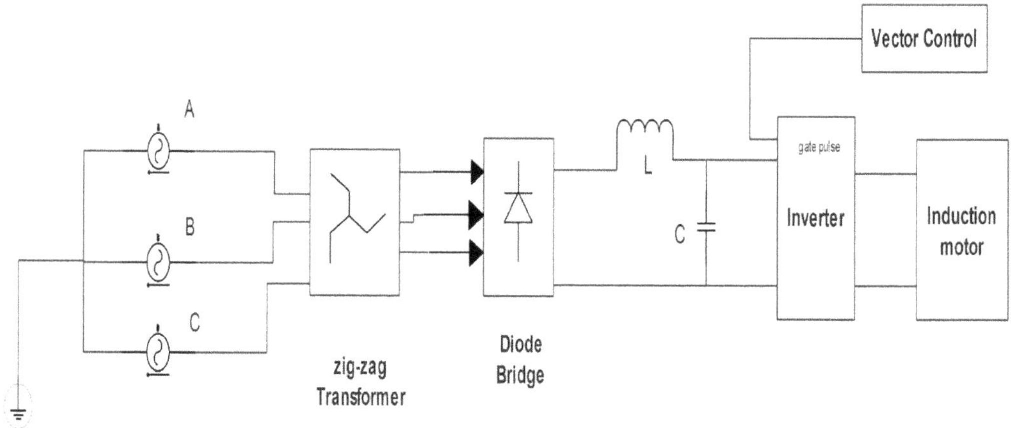

FIGURE 14.1 Block diagram of the system.

But this increases the cost and complexity. So, to increase the number of pulses without increase in cost and complexity, the DC ripple-rejection technique is used.

The system uses an autotransformer-based pulse AC-to-DC conversion with reduced magnetics rating and suitable for retrofit applications (where presently a 6-pulse diode bridge rectifier is used).

14.4 MATLAB Simulation

The triggering pulses to the gate terminal of the inverter are from a vector block. The input DC supply to the inverter is from the output of a 6-pulse AC–DC converter. Figure 14.2 shows the individual semiconductor device used. The load connected here is an induction motor as given in Figure 14.3. The diode bridge rectifier is used for the rectification process as given in Figure 14.4. The zigzag phase shifting transformer is used for isolation purposes as depicted in Figure 14.5.

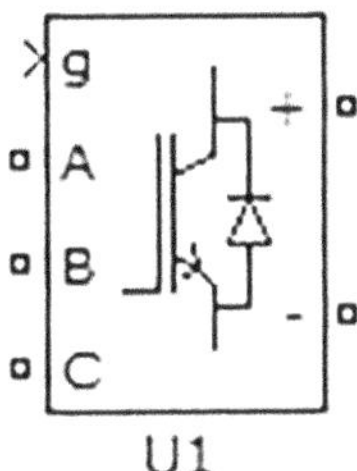

FIGURE 14.2 Individual semiconductor device used.

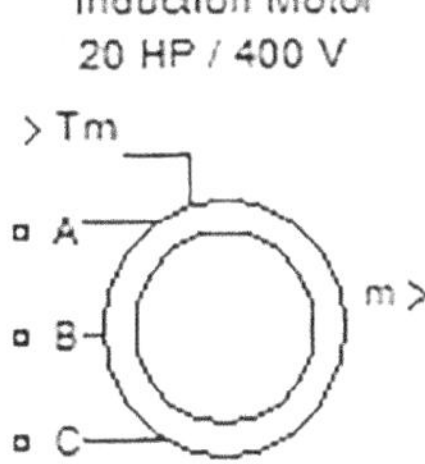

FIGURE 14.3 Induction motor.

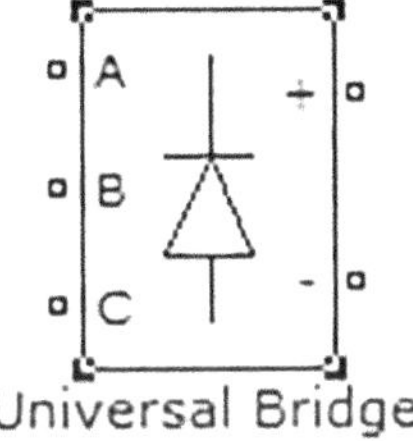

FIGURE 14.4 Diode bridge rectifier.

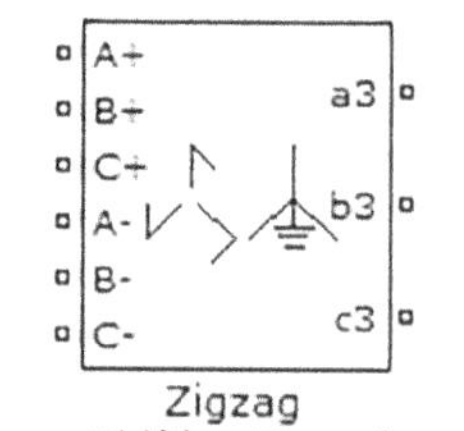

FIGURE 14.5 Zigzag phase-shifting transformer.

14.5 Simulink Model of Vector-Controlled Induction Motor Drive

Figure 14.6 shows the MATLAB Simulink model of a vector-controlled induction motor drive (VCIMD).

The modeling of VCIMD consists of

- Speed controller
- Vector controller
- Current controller
- Inverter
- Induction motor

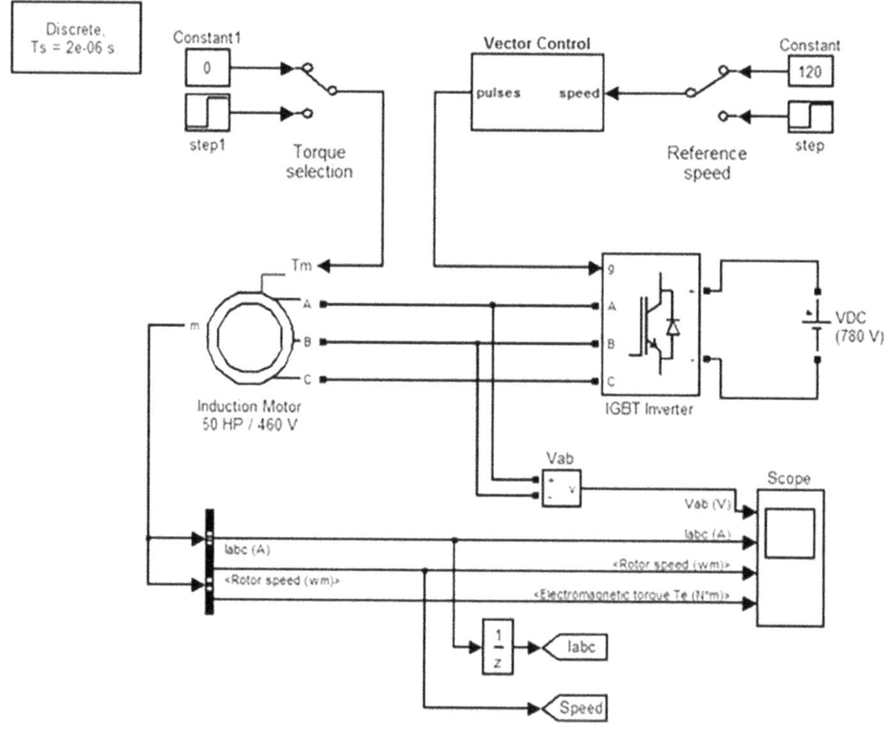

FIGURE 14.6 Simulink model of VCIMD.

Here, speed controlling is implemented in the speed-control block and it is enclosed in a sub-system named the speed-controller block. A PI controller is normally used as a closed-loop speed controller.

14.6 Subsystem Model of VCIMD

Figure 14.7 shows the modeling of a vector controller. The vector controller block comprises three stages wherein the flux component of the current vector (i_{ds}^{*}), the torque component of the current vector (i_{qs}^{*}), and the slip speed (ω_{ls}^{*}) are calculated as

$$i_{ds}^{*} = i_{mr}^{*} + \tau_r \frac{d_{mr}^{*}}{dt} \tag{14.2}$$

$$i_{ds}^{*} = \frac{T*}{k i_{mr}^{*}} \tag{14.3}$$

$$\omega_{ls}^{*} = \frac{i_{qs}^{*}}{\tau_r i_{mr}^{*}} \tag{14.4}$$

The conversion of the synchronous-speed rotating-frame vector-controlled currents to stationary-frame three-phase currents is made in DQ-to-ABC conversion sub-block. The error between the reference currents and the sensed winding currents is fed to the sub-system called the current controller.

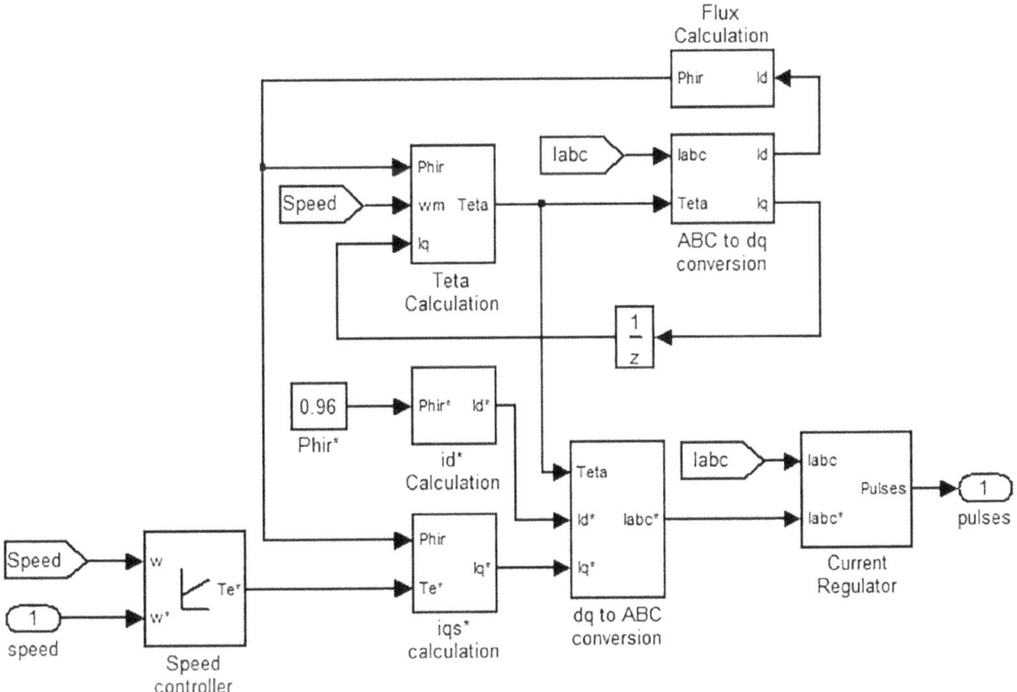

FIGURE 14.7 Subsystem model of VCIMD.

14.7 Simulation Results of VCIMD

Figure 14.8 shows the output voltage waveform of the vector-controlled induction motor drive. Figure 14.9 shows the three-phase output current waveform of the VCIMD, and the value of the current is nearly 150 A.

14.8 Speed Waveform

Figure 14.10 shows the speed waveform of the vector-controlled induction motor drive and the output is captured.

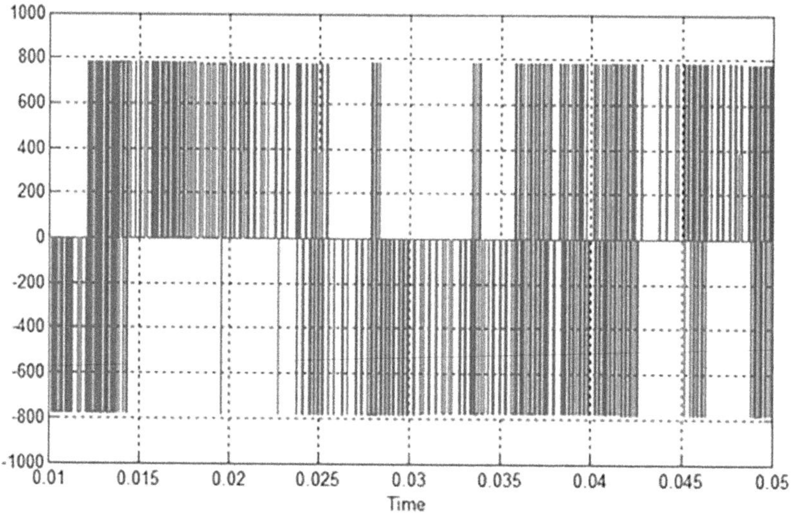

FIGURE 14.8 Output voltage waveform.

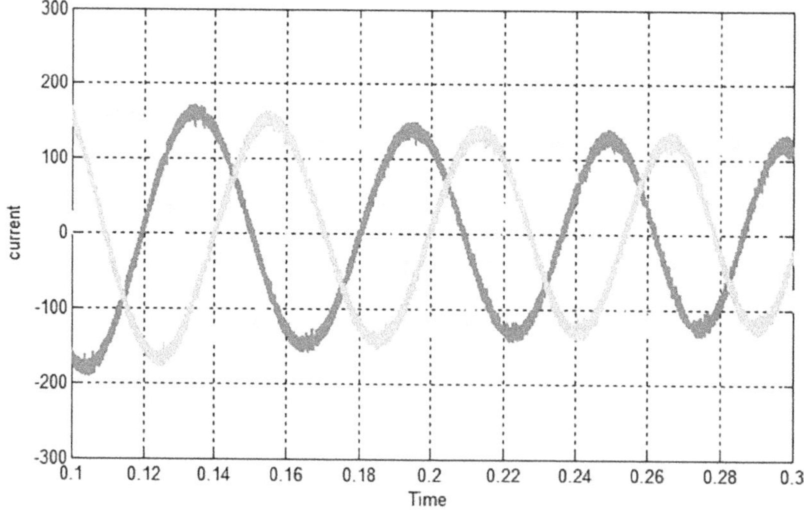

FIGURE 14.9 Output current waveform.

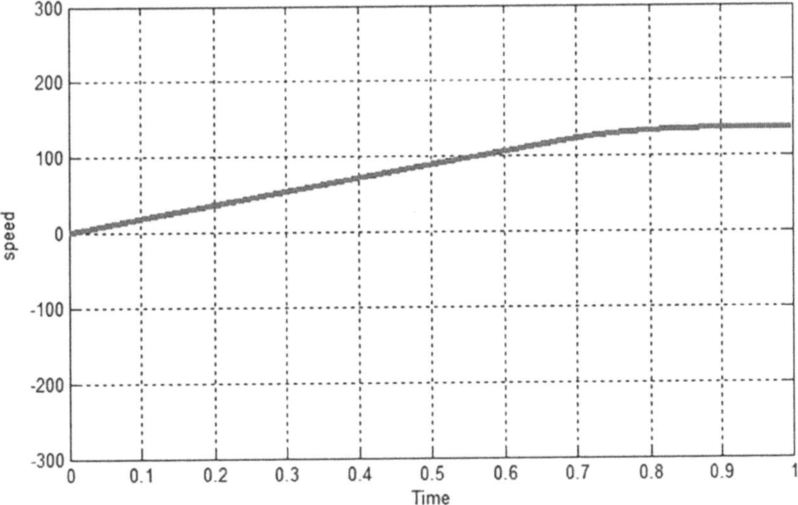

FIGURE 14.10 Speed waveform.

14.9 Torque Waveform

Figure 14.11 shows the torque waveform of the vector-controlled induction motor drive and the range of values at one cycle is determined.

14.10 6-Pulse Converter with VCIMD

Figure 14.12 represents the MATLAB Simulink model of 6-pulse AC–DC converter fed with a vector-controlled induction motor, and here VCIMD acts as a load. The output waveforms of the converter and the FFT analysis is done for the AC input mains current and the value of THD is displayed.

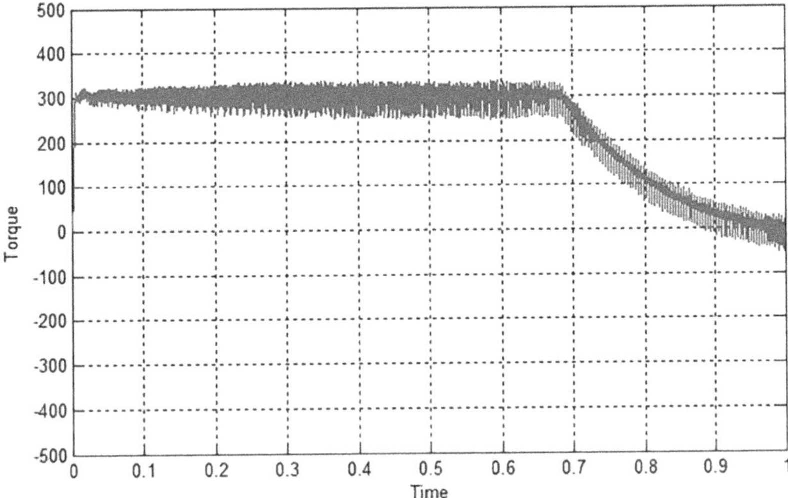

FIGURE 14.11 Torque waveform of VCIMD.

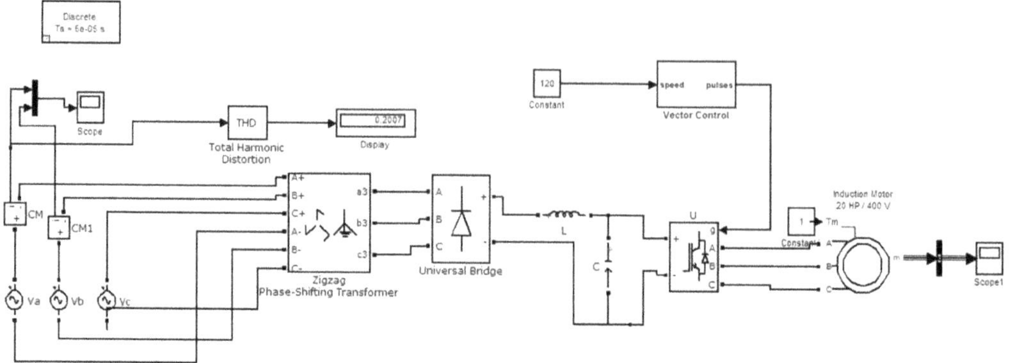

FIGURE 14.12 Simulink model of 6-pulse converter.

Figure 14.13 shows the input source current waveform of a 6-pulse converter. Figure 14.14 shows the THD analysis waveform of a 6-pulse converter, and the value of THD is 24.89%.

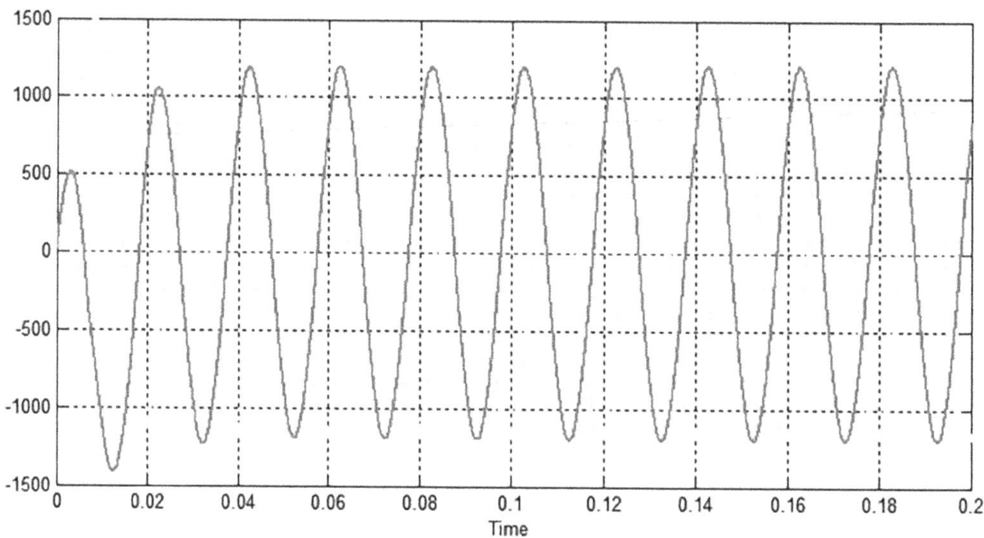

FIGURE 14.13 Input source current waveform.

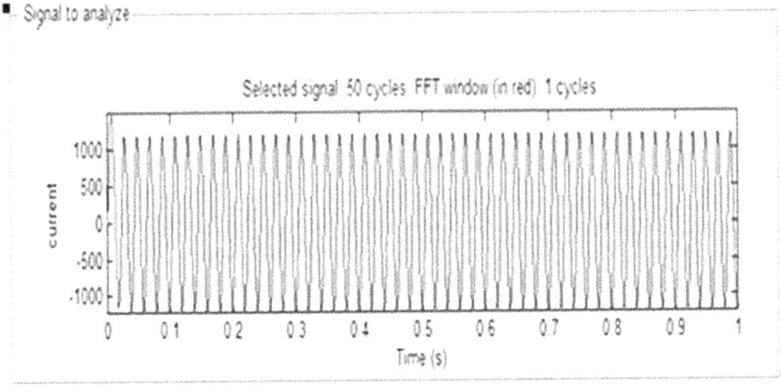

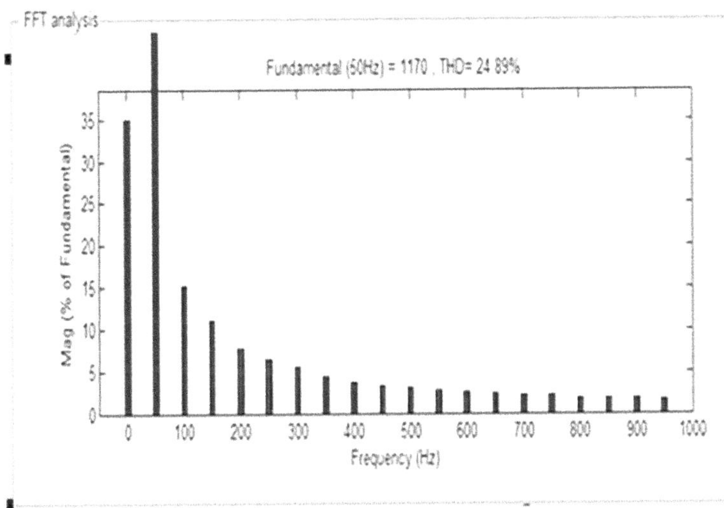

FIGURE 14.14 THD waveform for 6-pulse converter.

14.11 12-Pulse Converter

Figure 14.15 represents the MATLAB Simulink model of a 12-pulse AC–DC converter fed with a vector-controlled induction motor and here VCIMD acts as a load. The output waveforms of the converter and the FFT analysis is done for the AC input mains current and the value of THD is displayed. Figure 14.16 shows the THD analysis waveform of the 12-pulse converter is shown above and the value of THD is 14.40%

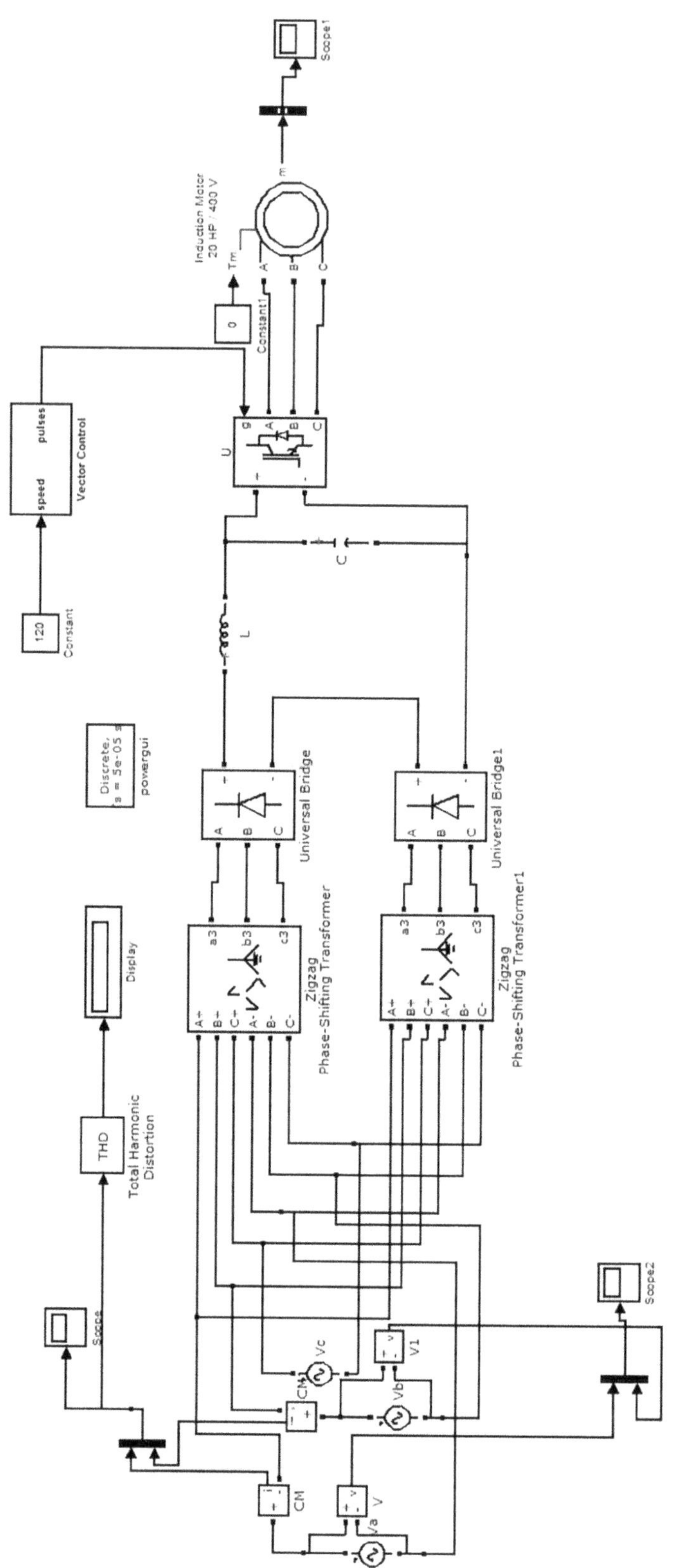

FIGURE 14.15 Simulink model of 12-pulse converter.

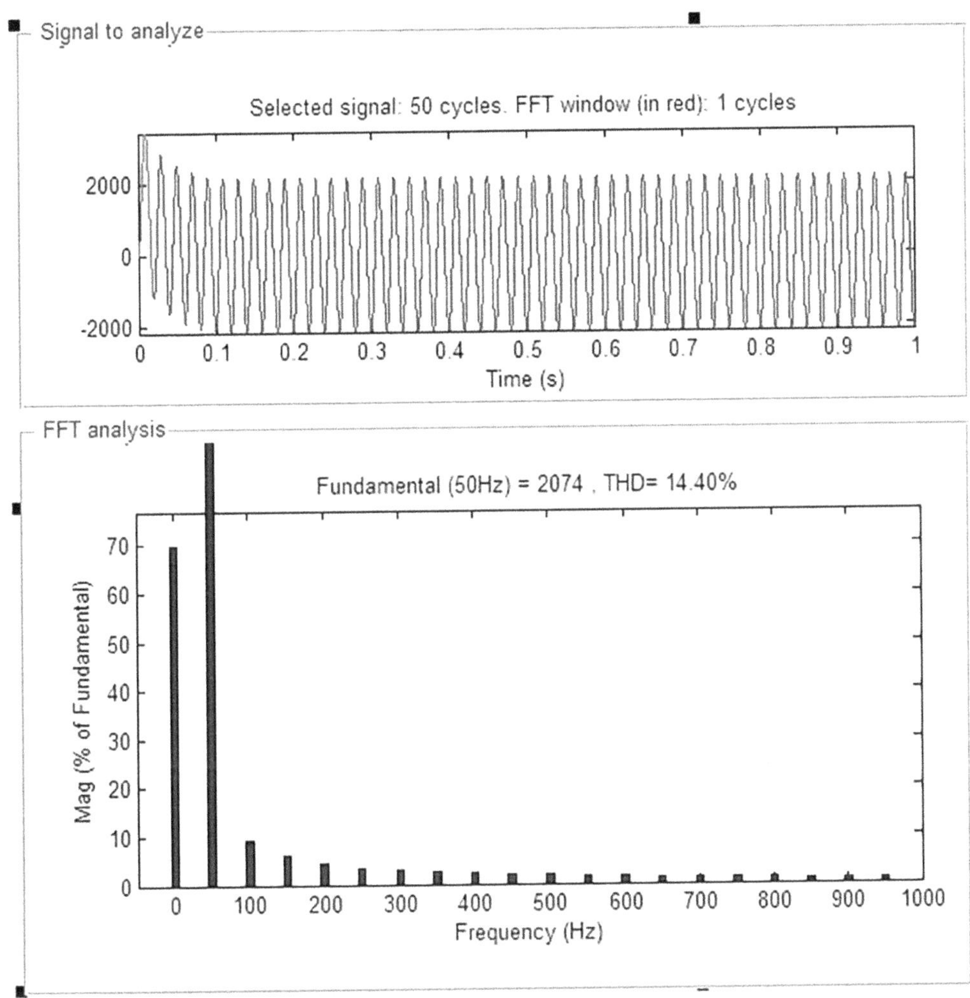

FIGURE 14.16 THD waveform for 12-pulse converter.

14.12 18-Pulse Converter

Figure 14.17 represents the MATLAB Simulink model of an 18-pulse AC–DC converter fed with a vector-controlled induction motor and here VCIMD acts as a load. The output waveforms of the converter and the FFT analysis is done for the AC input mains current and the value of THD is displayed. Figure 14.18 shows the THD analysis waveform of a 12-pulse converter and the value of THD is 12.74%.

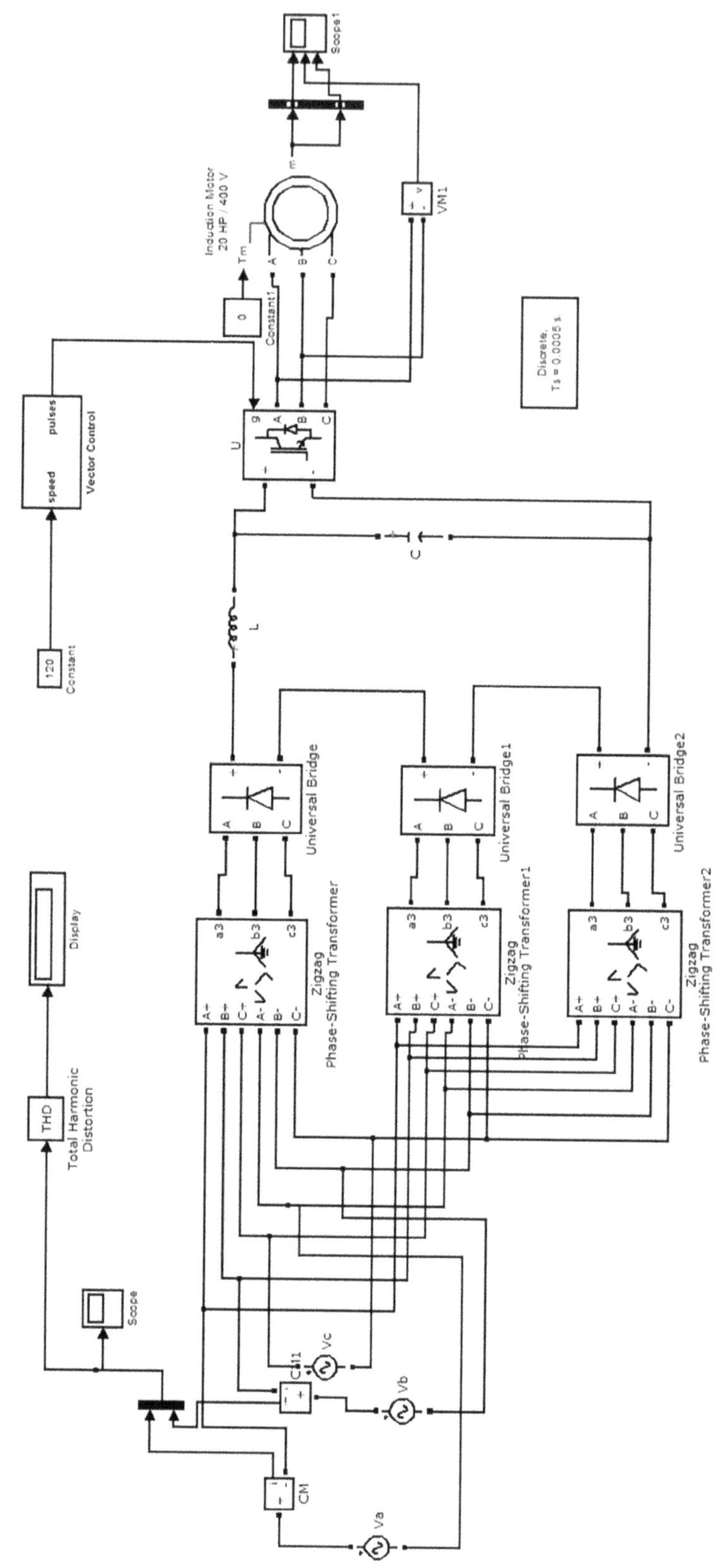

FIGURE 14.17 Simulink model of 18-pulse converter.

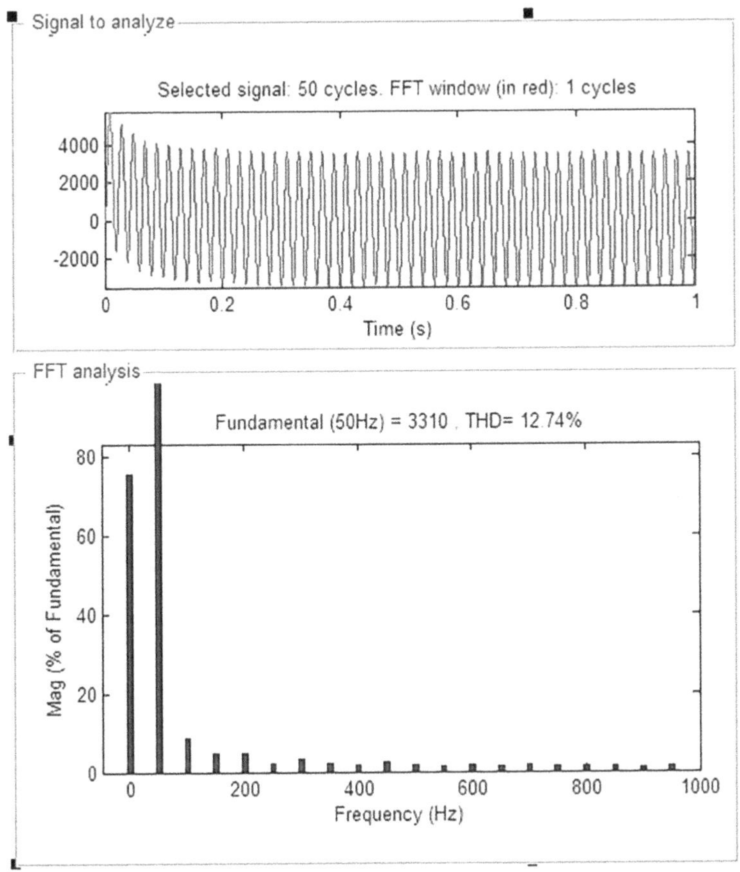

FIGURE 14.18 THD waveform for 18-pulse converter.

14.13 24-Pulse Converter

Figure 14.19 represents the MATLAB Simulink model of a 24-pulse AC–DC converter fed with a vector-controlled induction motor and here VCIMD acts as a load. The output waveforms of the converter and the FFT analysis is done for the AC input mains current and the value of THD is displayed. Figure 14.20 shows the THD analysis waveform of a 24-pulse converter and the value of THD is 9.38%.

14.14 Results and Conclusion

The performance of 6-pulse, 12-pulse, 18-pulse, and 24-pulse AC–DC converters for harmonic mitigation in vector-controlled induction motor drives are presented in this project. The four types of AC–DC converter modeled in SIMULINK are extensively analyzed and the performance characteristics are observed.

The THD analysis result of the 6-pulse converter is to be found as 24.39%, the analysis result of the 12-pulse converter is to be calculated as 14.40%, the analysis THD result of the 18-pulse converter is shown as 12.74%, and the result of the 24-pulse converter is reduced to 9.38%.

The various converters are shown to overcome some of the problems associated with lower pulse rectifier. By increasing the number of pulses, the harmonic values can be found to be improved. Thus, this technique of harmonic mitigation of various converters with vector-controlled induction motor drives is extensively analyzed and its performance is observed.

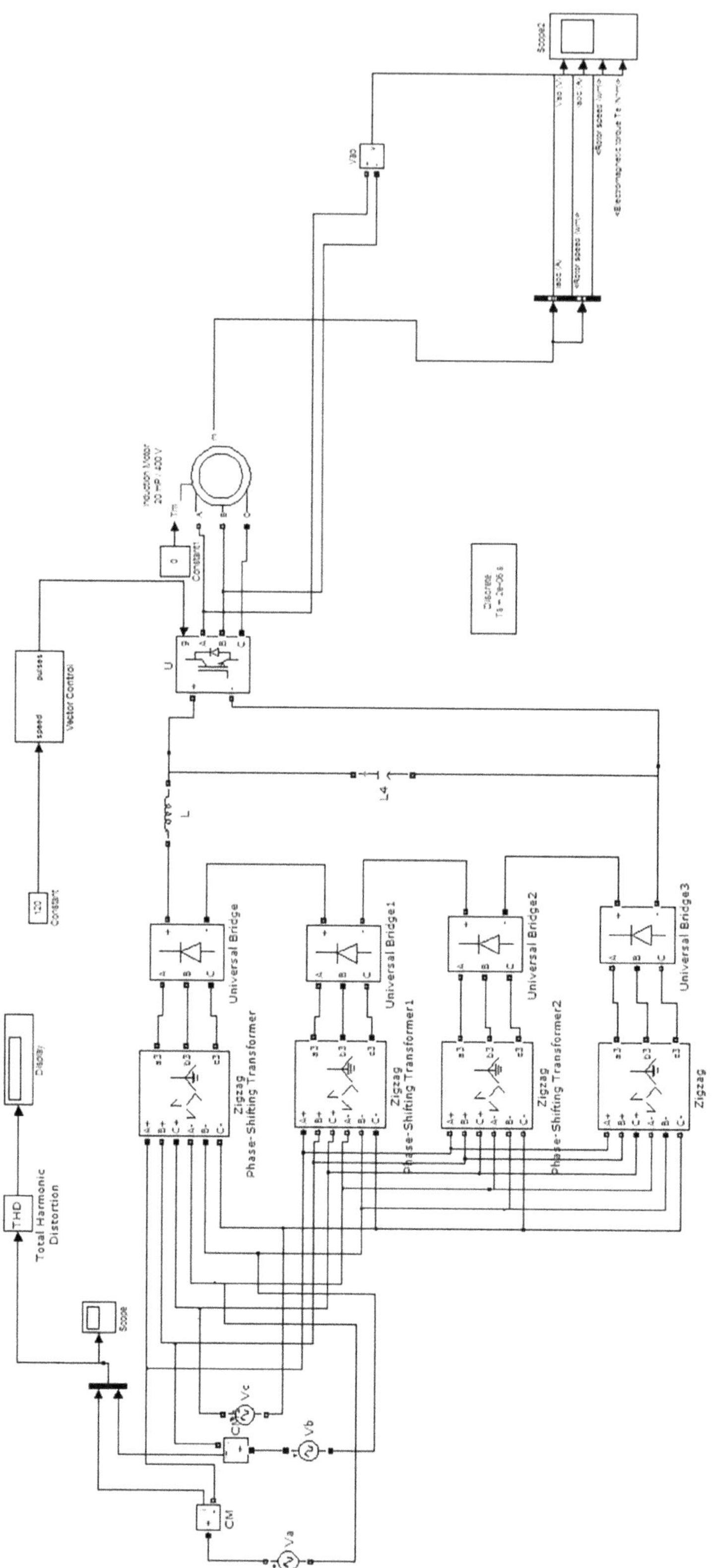

FIGURE 14.19 Simulink model of 24-pulse converter.

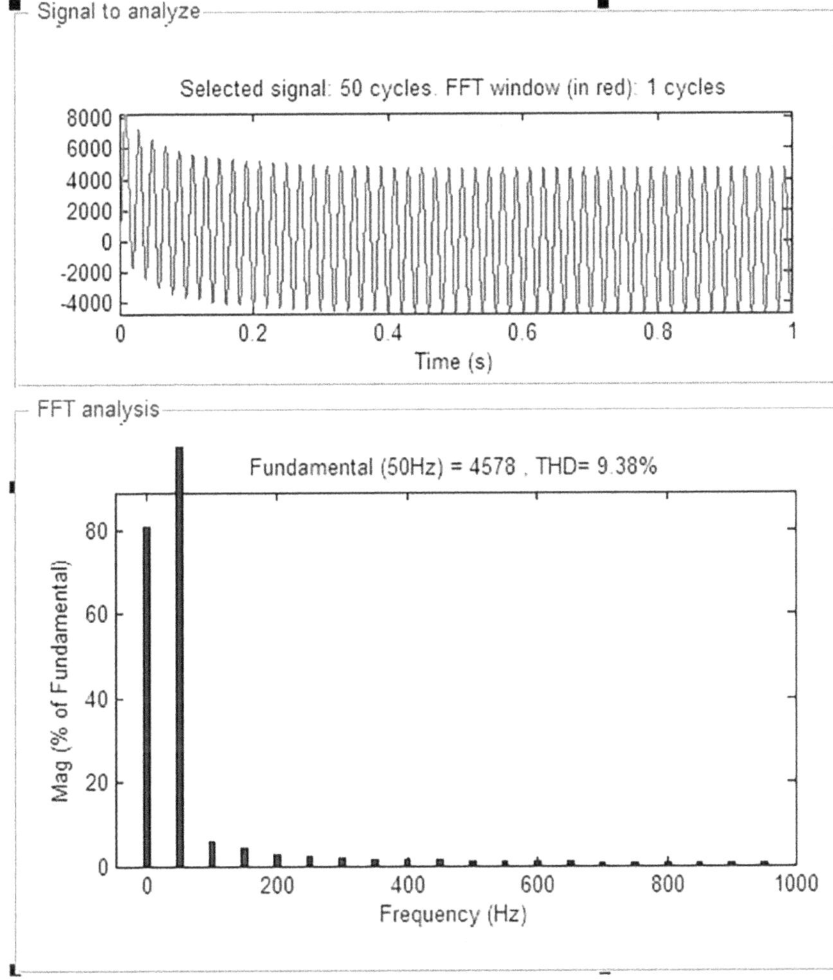

FIGURE 14.20 THD waveform for 24-pulse converter.

Index

Milton Keynes UK
Ingram Content Group UK Ltd.
UKHW051941071024
449327UK00026B/2118